PULSED LIGHT SOURCES

PULSED LIGHT SOURCES

I. S. Marshak

Springer Science+Business Media, LLC

Library of Congress Cataloging in Publication Data

Marshak, I. S.
  Pulsed light sources.

  Bibliography: p.
  Includes index.
  1. Electric discharge lighting. I. Title.
TK4371.M37   1984   621.32   84-12669
ISBN 978-1-4684-1649-7   ISBN 978-1-4684-1647-3 (eBook)
DOI 10.1007/978-1-4684-1647-3

© Springer Science+Business Media New York 1984
   Originally published by Consultants Bureau, New York in 1984
Softcover reprint of the hardcover 1st edition 1984

FOREWORD

The extensive development of electronics and automation equipment defines the scientific and technical significance of all types of high-density energy converters that are used in electronic and automatic equipment. A powerful pulsed discharge in gas is one type of conversion of electrical energy into extremely intense optical radiation. In order to characterize the possibilities inherent in such a discharge, it suffices to recall that high-power lasers were first developed by using flashlamps based on precisely this kind of discharge.

The correct use of existing types of flashlamps, work toward developing new types of flashlamps, and the solution of new problems by using such flashlamps require a knowledge of the physical processes that occur in them and the relationship between their technical characteristics and their design data and power-supply parameters. An acquaintance with the variety of existing flashlamps and the circuits used in the equipment employing them also is needed.

This book summarizes the findings of research on the physical and technical characteristics of pulsed discharges in gases and on the implementation of such discharges in pulsed light sources, which has been conducted by the authors and their co-workers over the last few decades. In view of the lack of sufficiently complete monographs on this subject, the authors decided not to confine themselves to a presentation of their own data, but to attempt wherever possible to summarize the findings of numerous papers by other investigators which have appeared in the scientific and technical literature in recent years. We thus intended to create a needed guide to systematic acquaintance with pulsed light sources for a constantly expanding group of diverse specialists who must now develop or use flashlamps in devices in various fields of science and engineering and who must study phenomena associated with short pulses of radiation at optical wavelengths.

The first part of the book is devoted to the physical processes that occur in pulsed light sources: the triggering of the pulsed discharge and the characteristics of its high-current stage. The second part is devoted to the technical characteristics of flashlamps, the principles of their design and inductrial production, and some problems associated with their use.

The first edition of this book was published in the early 1960s [0-1]. Since then, however, our understanding of the physical processes that occur in pulsed light sources and in the design and use of such sources has deepened considerably. This is due to the current intensive development of low-temperature plasma physics as a whole, and to the less intensive development of quantum electronics, in which pulsed light sources are used as one of the principal means of optical pumping. In particular, because of this it has become possible to calculate the characteristics of pulsed sources much more accurately than when the first edition was published. We can determine the time dependence of the radiation power in a specified spectral range (from given parameters of the external electric circuit and the properties of the gas-filled gap). We also can solve the inverse problem much more effectively and comprehensively, i.e., the problem of selecting the parameters of the circuit and the gas-filled gap according to the requirements of the specific technical application of the source. Therefore, the book had to be substantially updated by including current theoretical concepts and the latest experimental data on the physical characteristics of pulsed gas discharges, as well as information on the available selection of flashlamps* and their circuit diagrams.

---

* As in Reference 0-1, attention here is paid mainly to widely used pulsed light sources: flashlamps. It would be beyond the framework of the book for us to cover nonlamp-type pulsed sources specifically designed for laboratory use: discharges in the vapors of exploding metal conductors, high-power discharges with channel constriction due to the self-magnetic field, creepage over the surface of dielectrics, and others which are covered in the corresponding reviews (see, e.g., A. F. Aleksandrov and A. A. Rukhadze: "High-current discharge-type light sources," Uspekhi fizicheskikh nauk, 1974, No. 2, vol. 112, pp. 193-230; K. Fol'rat: "Spark light sources and high-frequency spark cinematography," in Fizika bystroprotekayushchikh protsessov (The Physics of High-Speed Processes), vol. 1, Moscow, Mir Publishers, 1971, pp. 96-199; S. I. Andreev, M. P. Vanyukov, and E. V. Daniel': "Surface discharge as a source of intense light flashes," Zhurnal prikladnoi spektroskopii, 1966, No. 6, vol. 5, p. 712; B. L. Borovich, P. G. Grigor'ev, V. S. Zuev, V. B. Rozanov, A. V. Startsev, and A. P. Shirokikh: "Experimental and theoretical research on the dynamics of high-power emitting electrical discharges in gases," Trudy FIAN im. P. N. Lebedev, 1974, vol. 76, p. 1; and I. V. Dvornikov, Yu. N. Kolpakov, V. A. Lakutin, and I. V. Podmoshenskii: Zhurnal prikladnoi spektroskopii, 1974, No. 2, vol. 21, pp. 227-234.

The revision affected practically all chapters, with minor changes in the general structure of the book.

In particular, new information on static and pulsed breakdown has been added in Chapter 1, which is devoted to the triggering of discharge. Instead of a separate (chapter-by-chapter) semiempirical treatment of the electrical, hydrodynamic, and optical characteristics of the plasma channel, the channel is considered as a whole in subsequent chapters on the basis of the concepts developed in recent years. Chapter 2 is devoted to processes in the expanding channel of a pulsed discharge, and Chapter 3 to the characteristics of the quasi-stationary discharge that is typical principally of tubular flashlamps. Recent data on processes near electrodes have been added in Chapter 4. Several chapters in the second part of the book describe various technical characteristics of tubular and spherical lamps: their radiative, load, and operating characteristics. These chapters, like those devoted to the industrial production and circuit diagrams of flashlamps, have been significantly supplemented with the latest data. The diversity of flashlamp applications has grown so much in recent years that describing them is beyond the framework of the book.

Most of the bibliographic references in [0-1], which pertained mainly to the history of the field, have been omitted from this edition in order to save space. The bibliography includes chiefly publications that have appeared in print since the publication of [0-1]. Earlier works are duplicated only in the most important cases, e.g., if information essential to understanding the text was drawn from them (say, the formulas derived in these papers), or if they contain a bibliography which expands on the information given here and thus is essential to today's reader.

In addition to I. S. Marshak, A. S. Doinikov (in Chapters 5 and 10), V. P. Zhil'tsov (in Chapter 7), V. P. Kirsanov (in Chapters 2 and 6), and L. I. Shchukin (in Chapters 5, 7, and 8) helped produce the new edition of the book. Chapter 3 was rewritten by R. E. Rovinskii, and Chapter 9 by V. P. Zhil'tsov and M. G. Feigenbaum.

CONTENTS

PART 2

GAS-DISCHARGE FLASHLAMPS

INTRODUCTION

Extremely high-intensity radiation is known to be required in
the solution of many scientific and technical problems. The instantaneous level is more important in these problems than the average
value over a prolonged period. Accordingly, a trend has appeared
in technology directed toward an increase in intensity at the expense
of continuity of the radiation, i.e., toward changing over to pulsed
bursts of radiation. In addition to increasing the intensity, the
use of pulses also opens up the possibility of pulse coding. This
is significant in many problems involving data transmission and
identification of pulses against a constant radiation background.

The trend toward conversion to pulsed radiation in the radio-
frequency band has become evident in fields of radio engineering
that developed at a relatively late date: radar, radio navigation,
etc. The same trend in the optical band has led to the development
of pulsed optical-radiation sources by the electric-lamp industry.
"Optical radiation" includes ultraviolet, visible, and infrared rays.
For brevity, such radiation sources are simply called pulsed light
sources.

The relation among the attained values of power, luminance, and
luminous flux of modern continuous-wave and pulsed light sources is
illustrated in Table 0-1.

The data in Table 0-1 confirm the advantage of pulsed light
sources in those instances of energy transfer and data transmission
where either a radiation detector and data-acquisition equipment
having a sufficiently short lag are used (e.g., a vacuum photocell
or a photoelectric multiplier with an appropriate circuit diagram)
or the transfer process itself must persist for an inherently short

1

Table 0-1. Comparison of the Most Powerful and Brightest Pulsed and Continuous Electric Light Sources

Type of source	Type of lamp	Peak power, kW	Peak luminance, $Mcd \cdot m^{-2}$	Peak luminous flux, klm
Continuous	Incandescent (projector) lamps	20	30	600
Continuous	Water-cooled tubular xenon lamps with vortex stabilization of discharge [0-2 and 0-2a]	500	1000	22,000
Continuous	Superhigh-pressure spherical xenon lamps [0-2a to 0-5]	30	6000	1300
Continuous	Open high-intensity arcs	100	1400	4500
Pulsed	Tubular quartz xenon lamps	200,000	10,000	10,000,000
Pulsed	Spherical xenon lamps	10,000	100,000	200,000

time (e.g., photography of a moving object or stroboscopic observation). It is this advantage which explains the trend mentioned
above toward the use of pulsed radiation as a method of improving
the parameters of optical equipment with light sources in the direction of increasing speeds, improved accuracy, expanded ranges, and
automation of processes in the latest equipment.

Short-duration light flashes also can be produced by using a
continuous source fitted with an optical shutter or by operating in
a discontinuous mode (e.g., incandescent lamps operating in an over-
heated condition for a short period of time, or briefly overloaded
xenon or mercury arc lamps [0-5a to 0-8a]). Here approximately an
order-of-magnitude increase in the power of the source over its rating
is tolerable. Pulsed light sources may be based on the use of a
chemical reaction (one-shot lamps [0-9 and 0-10], such as electronic
flashes with a metal foil that burns in an atmosphere of oxygen or
fluorine, or so-called magnesium photoflashes or photoflash bombs in
which a metallic powder burns instantaneously as a result of the
liberation of oxygen from an oxygen-rich salt mixed with the powder,
and lamps filled with a noble gas which produce a flash as a result
of the shock wave produced by an explosive [0-11]). They may be based
on the brief excitation of a luminescent solid (e.g., by using an
electron beam [0-12]) or on a short-duration electric discharge in
a gas or in the vapor of a metal: a condensed electric spark.

The specific features of a condensed spark discharge (high temperature and luminance, its ready controllability, the possibility
of frequent repetition of flashes, and the comparative simplicity of
the auxiliary devices) have made the last type of pulsed light sources
the most widely used ones.

A spark discharge was first used as a pulsed light source in the
mid-19th century, when Fox Talbot [0-13] made high-speed photographs
by using an electric spark for illumination.

In our time, pulsed (spark) discharges in gases have been studied
and put to use in an extremely large number of applications. Work
is being done on pulsed discharges in high-voltage engineering in
the areas of lightning protection and insulation. It is used as the
principal method of igniting fuel mixtures: in engines, for example.
Pulsed discharges are used extensively in spectroscopy to excite the
spectra of ionized atoms. It plays a significant part in many switching devices used in electrical engineering, radio engineering, and
electronics (discharges, trigatrons, etc.). Nuclear physics specialists also have become interested in the pulsed discharge because the
instantaneous formation of a gas plasma which accompanies the discharge is now the highest-temperature physical process that can be
implemented in a small volume (in contrast to explosive processes,
which use short-duration chemical or nuclear chain reactions that
encompass large volumes). They view it as a possible way to carry

out a controlled thermonuclear reaction or at least to study the
properties that govern such a reaction.

The pulsed electric discharge in gases is used most extensively
in a field that was among the first to find a practical application
for it: the production of pulsed light sources.

Up until the 1930s either an open spark discharge in air [0-14]
or a pulsed discharge in mercury-vapor-filled tubes [0-15] was used
as an electrical pulsed light source that did not find its way outside the laboratory. An advance in the development of such sources
was the use of pulsed discharges in tubes filled with noble gases
and discharges in vapors of a metal wire exploded by a current. A
start was made in the second half of the 1930s and the first half of
the 1940s in the work of Laporte, Edgerton, Wulfson,[†] Marshak, and
their co-workers [0-16 to 0-19]. The first series-produced pulsed
electric-lamp devices were introduced in engineering as a result of
that work. Extensive research has since been conducted in many
countries, resulting in design and manufacturing developments which
led to the rise of the industrial field of pulsed electrical light
sources in the 1950s.

Today hundreds of such sources are produced in the most advanced
countries. Among these we may list above all: (1) various tubular
flashlamps with various peak flash energies (from a few joules for
intracavity medical photography and portable electronic flashes, to
hundreds of thousands of joules for aerial flash spotting and the
optical pumping of lasers ([0-20 to 0-23]) and with various shapes
of the glow volume (straight ones to produce a flat fan-shaped light
beam, e.g., in lasers and cloud chambers [0-23 to 0-27]; spiral or
U-shaped ones to produce a conical beam; circular ones to produce
shadowless light beams; and so forth [0-23 and 0-28]);(2) stroboscopic
flashlamps (strobotrons) with a flash rate up to a few kilohertz for
stroboscopes and illuminators used in high-speed filming (from low-
power neon strobotron-thyratrons for strobotachometers to xenon lamps
with an average power of tens of kilowatts [0-23 and 0-29 to 0-32]);
(3) lamps with light-signal devices (with a frequency of 1-3 Hz, a
power of 10-500 W, and an operating life of up to a few million
flashes [0-23 and 0-33 to 0-35]); (4) lamps with an especially short
flash for photochemistry and various types of electronic equipment
[0-36 to 0-38]; (5) a series of strobotrons for computer and other
automatic devices (with a few watts of power and a frequency in the
hundreds of hertz [0-39 and 0-40]); and so forth. At the same time,
a wide variety of optical equipment using flashlamps is being produced. Here we may list various lasers, consumer electronic flashes,
illuminators for high-speed, medical, biological, and other special

---

[†] Translator's note: names marked with a dagger have been back-
transliterated from Russian and may not reflect correct spelling.

types of photography, various strobes, and others. Flashlamp devices
are used in automation and remote control (devices with light control
and data-transmission channels: remote-control optical contact devices, the "angle-number" sensors of computers, light-protection apparatus, transformer control on high-voltage DC transmission lines,
thickness gages, etc.). They are used in laser radar and optical
communications (cloud height indicators and other range finders,
optical telephony, etc.). More and more light-signal devices using
flashlamps are being produced (light tracers, beacons, navigation
lights for modern high-speed long-range aircraft, and other lighting
equipment used in transportation). A number of devices exist or are
being developed in which flashlamps are used to produce time marks,
for photographic recording, microfilming, time-lapse photography,
printing, photolithography, photometry, etc. [0-23 and 0-41 to 0-45].
Several types of movie projectors with flashlamps have appeared, and
these lamps have begun to be used for television transmission of
motion pictures and for illumination in television studios using a
scanning-beam transmission system. Work is being done on the use of
flashlamps in many fields, such as photochemistry (photolysis, photo-
synthesis, metalworking by etching a surface precovered with a photo-
sensitive varnish which is hardened by light at points that are not
to be etched), etc. There can be no doubt that the further development of science and technology also will open up other areas for the
use of these lamps.

The total production volume of pulsed light sources is characterized by the fact that, for example, Japan alone produced several
dozen kinds of zenon flashlamps for a total of 3.32 million lamps in
1973 [0-46 and 0-47]. At the same time, flashlamp production has
begun to acquire some stability and order in the last decade, after
its previous turbulent development. Increasingly often, newly developed lamps and devices use standardized components, and the development work is reduced to moderate modification of parameters without
implementing any fundamentally new physical or technical improvements.
Attention is being paid to the possible standardization of flashlamps
in terms of product variety, manufacturing processes, mating to other
equipment, documentation, and other areas. Under these conditions
an overall survey of the scientific principles of the field and of
the design and manufacturing methods used in it seems entirely feasible and timely. This book is devoted to such a survey.

PART 1

PHYSICAL PROCESSES IN PULSED DISCHARGES

CHAPTER 1

DEVELOPMENT OF DISCHARGE:
ELECTRICAL BREAKDOWN OF GAS AT NEAR-ATMOSPHERIC PRESSURES

1-1. Classification of Types of Breakdown

It is natural to begin an examination of the physical processes
in a pulsed (spark) gas discharge with the phenomenon of formation
of a highly conductive plasma channel in a gas-filled gap which
previously had good electrical-insulation properties. Achieving
high-intensity glow discharges in pulsed light sources requires a
high gas-particle density, so near-atmospheric gas pressure is used.

The term "gas breakdown at near-atmospheric pressures" encompasses a broad class of transient electrical-discharge processes in
gases and still does not identify a group of physical phenomena that
obey uniform laws.

The data amassed to date make possible a further classification
of the different forms of such breakdowns into several types which
are governed by different mechanisms characterizing each type. Among
these types we may list the following, which have the greatest scientific or practical importance:

A) Breakdown when the applied voltage across spark gaps of modest
   dimensions slightly exceeds the static breakdown voltage (breakdown voltages of up to a few hundred kilovolts), with slight nonuniformity of the field and weak external ionization ("short static
   breakdown").
B) Static breakdown of extremely long spark gaps (of the lightning
   type) and of gaps with a nonuniform electric field (also with not-
   too-strong external ionization).
C) Pulsed breakdown (breakdown resulting from a voltage pulse applied
   briefly to a gas-filled gap) when the pulse voltage greatly exceeds

the constant voltage required for static breakdown of the gap.
D) High-frequency breakdown (at a voltage oscillation frequency comparable to the collision frequency of the gas particles).
E) Breakdown during intense ionization of a gap by an auxiliary low-
   power high-voltage source which produces a glowing channel in the
   gas.

   Historically, the use of pulsed electric light sources began
with short spark discharges in air, in which type A and C breakdown
occur. Most of the physical mechanisms characterizing the formation
of a plasma channel are evident in such breakdowns. Therefore, a
brief summary of the pattern of such breakdowns, which has not been
determined quite well, will be presented first in this chapter.

   The question of type B breakdown is more special and is treated
in the pertinent specialized reviews [1-1 and 1-2]. A general treatment of the problem is difficult because it is characterized by many
factors (electrode shape and polarity, etc.). In this chapter the
subject is touched on only in connection with the treatment of the
boundary regions between breakdowns of types A, B, and C.

   The question of type D breakdown is entirely isolated and is
not considered here, since an extensive independent literature is
devoted to it [1-2 and 1-3].

   Finally, the question of type E breakdown, which is of great
practical importance for modern flashlamps, has not been paid sufficient attention to date. The current concept of the physical
mechanism of such breakdowns also will be examined in this chapter.

1-2. Breakdowns of Types A and C

   At high pressures (above 2500 Pa or 20 mm Hg) the transition
from a dependent to a self-sustained gas discharge always is accompanied by a rapid physical process that may be characterized by
the concept of "electrical breakdown." The turbulent nature of this
process, whose detailed mechanism may vary significantly under different conditions, is explained by two general features of electrical
phenomena in gases:

1. The exponential dependences of the characteristic parameters of
    the majority of elementary ionizing processes (for small absolute
    values of these parameters which are characteristic of short mean
    free paths) on particle energy or on the ambient conditions that
    determine it (electric field strength, temperature). Examples
    of such exponential relations are the dependence of the coef-
    ficient of impact ionization of atoms by electrons, α, on electric
    field strength; the analogous excitation function of the atoms,
    which affects step ionization; photoeffects on the cathode and

in the gas volume; and so forth. At high gas pressure and temperature there is a tendency toward thermodynamic equilibration
in the plasma due to the multitude of various collisions inherent
in these conditions. The corresponding temperature dependence
of the degree of ionization (a dependence which characterizes the
degree of all ionization processes in aggregate) for an equilibrium plasma ("Saha's equation") also is exponential.

2. The existence in the gas of repeated cyclical processes with a
very short cycle time. By analogy with chemical and nuclear
reactions, these also may be called chain reactions. Impact
ionization by electrons is one example of a chain process. The
number of free electrons involved in the process doubles with
each new cycle during such ionization, forming an "electron
avalanche." Another example can be seen in the development of
thermal ionization of a just-formed spark channel, leading to a
rise in current density and to increases in the dissipated power
and the temperature (here the thermodynamic equilibration time,
i.e., about 10^{-9} s, may be taken as the cycle time). This in
turn increases the degree of ionization of the gas.

These two features, together with the diversity of the elementary
processes of energy exchange between particles, are responsible for
the special problems that arise in experimental and theoretical
studies of gas breakdown at high pressures.

The complexity of the experimental study of breakdown is due
above all to the following factors:

(1) The short duration of the breakdown process. High-speed oscillographs and photorecorders are required to register its temporal
characteristics.
(2) The change of discharge parameters (e.g., current strength and
radiant intensity) by many orders of magnitude during breakdown.
Hence the instruments used to measure these parameters also must
have a very large dynamic range.
(3) The highly critical nature of the condition for onset of breakdown. Instruments for monitoring the parameters that determine
this condition must have high absolute precision.

Knowing how far laboratory equipment in these areas has advanced
in recent years, we can easily understand why only quite recently
(actually in the last two or three decades) has experimental equipment made it possible to penetrate sufficiently deeply into the gas
breakdown phenomenon at high pressures.

It is difficult to give a theoretical explanation of the breakdown phenomenon because the two general features of electrical
phenomena in gases mentioned above determine the turbulent nature
of breakdown and the critical nature of the conditions for its onset.
Thus, the actual values of the parameters that determine breakdown

can at first glance correspond to a whole series of specific mechanisms leading to increased electrical conductivity. This circumstance
creates favorable ground for the construction of speculative theoretical schemes of the breakdown mechanism that are insufficiently
substantiated. Such hypothetical schemes provided plausible explanations for the transition of a gas-filled gap from the nonconducting
to the conducting state and gave a satisfactory quantitative estimate
of the dependence of the breakdown voltage on the length of the gap
and on the gas density (an estimate which is little-dependent on the
assumed mechanism). Subsequently, however, these schemes had to be
discarded or subjected to fundamental revision as new data were accumulated.

In a simplified way the authors of these schemes tried to reduce
the extremely complicated phenomenon of breakdown (over virtually
its entire duration) to some single isolated mechanism such as interaction between α and β ionization (impact ionization by positive
ions), between α and thermal ionization, or, finally, between α
ionization and the field of the space charge of the head, and so
forth. The schemes may be divided into two groups within which only
the dominant elementary process changed, but the space-time pattern
of breakdown development was preserved:

(a) Schemes with Townsend "growth of ionization," in which the
     products of primary electron avalanches (products of α ioniz-
     ation, mainly near the anode) give rise in one way or another
     to secondary electrons near the cathode as a result of β or η
     ionization (ionization by photons) in the gas volume, and also
     as a result of γ, δ, or ε processes on the cathode (stripping
     of electrons from the cathode by positive ions, photons, or
     excited atoms respectively). These electrons are initiators
     of secondary electron avalanches, which in turn produce a third
     generation of electrons at the cathode, and so forth. If there
     is active balance in one such growth cycle (the formation on
     average of more than one secondary electron per primary ava-
     lanche), the process leads to unlimited current rise without
     constriction of the discharge in a narrow channel.
(b) So-called "streamer" schemes, in which an individual electron
     avalanche (produced by a single electron that appears in the
     cathode region) builds up in its path a large space charge in
     the head, with a diameter determined by electron diffusion
     (about 1 mm in air at atmospheric pressure). This space charge
     is so large that local photoionization or thermal ionization
     causes local distortion of the field, leading to the formation
     of a narrow column (equal in diameter to the head itself) of
     highly ionized plasma (a "streamer") which rapidly penetrates
     to the anode and cathode. Streamer penetration leads to a fast
     current rise in the discharge circuit and culminates in the
     formation of a constricted conducting channel between the gap
     electrodes.

In the period when these hypotheses were originally being considered, it seemed that the schemes of one group completely excluded
applicability of schemes of the other. As concepts of type A and C
breakdown were further developed, however, the simplified approach
to the phenomenon gradually was overcome, and the correct concept,
combining schemes from both groups, was worked out. According to
this concept, mechanisms characteristic of different theoretical
schemes play a decisive part in different stages and under different
conditions of breakdown. The story of how the details of one of the
most intricate, varied, and shortest phenomena in nature - gas breakdown at high pressure - were uncovered provides an example of col-
laboration among scientists of different countries and different
generations who used the most varied methods of research and the most
highly refined equipment.

The first steps in this field were taken by Pedersen, Burav,
Rogovskii, Flegler, and Tamm and by Bims, Torok, and co-workers.*
Immediately after the first cathode-ray oscillographs appeared,
these researchers measured the extremely short times required for
breakdown formation during overvoltages (about 10^{-8} s): times that
do not even correspond to a single growth cycle (about 10^{-7} s). It
was shown in the theoretical work of Rogovskii, Loeb, and von Hipple
and Frank that the growth of ionization, beginning with Townsend
"growth" (because of the interaction of α and β processes,** α and γ
processes, or, finally, α and δ processes), may be accelerated significantly in a later stage by the positive space charge formed during
the growth (the charge is in the entire plane of the gap, in contrast
to the concentrated space charge considered in "streamer" schemes).
A number of subsequent papers by Varney, White, Loeb, and Pozin and
by Rogovskii and co-workers which were devoted to the influence of a
plane space charge on the growth of ionization played a significant
part in the development of this new concept. These papers experimentally established and theoretically explained the proportionality
between the decrease in the breakdown voltage of a plane gas-filled
gap and the square root of the photocurrent from the cathode that is
generated in the gap by external illumination.

Constricted channels were studied during breakdown in the work
of White and Dannington, who used a Kerr cell; Loeb and co-workers,
who observed corona discharge; Flegler and Raether, who observed the
tracks of delayed pulsed discharges in a Wilson cloud chamber; and
in the accompanying theoretical papers by Raether, Loeb, and Meek.

---

* A detailed bibliography of these and the other works mentioned is
   given in [0-1].
** The study of the process of impact ionization by positive ions
   and of the energy distribution of the ions at breakdown electric-
   field strengths has demonstrated that this mechanism is ineffective
   at high pressures.

As a result, yet another vital link was found in the chain of processes that occur at different stages: the pattern of development
of the fine channel of highly ionized plasma called a streamer. It
was this work which served as the basis for the fact that for a long
period of time many authors totally rejected, without sufficient
grounds, the physical reality of the growth mechanism for breakdown
at high pressures (at least for gaps longer than about 1 cm).

The comparison made by Marshak [1-4] of the experimental data
available by the start of the 1940s (data that confirmed either ionization growth schemes or streamer schemes) and the acquisition by
Marshak of new experimental data on the dependence of the breakdown
voltage on the cathode material and of data on breakdown formation
times of about 10^{-5} s at small overvoltages (times which exceed the
growth cycle severalfold) made possible the development of a new hypothesis of type A breakdown and of its transition to type C breakdown. According to this hypothesis,* type A breakdown begins with
Townsend growth (with possible simultaneous involvement of one or
more secondary mechanisms whose relative efficiency depends on the
specific conditions). This leads to the formation of a plane positive
space charge. This charge increases the electric field strength E
in the cathode region, improving still further the active balance of
the growth. (Because of the positive value of the second derivative
of α with respect to E for breakdowns at near-atmospheric pressures,
$\int \alpha(E)dx$ increases with increasing nonuniformity of E.) At length,
the number of electrons in the head of an individual electron avalanche increases to such an extent that the conditions are created
for regeneration of one avalanche as a contracted streamer channel
which closes the gap between electrodes. The increase in the overvoltage across the gap during pulsed breakdown when the static breakdown voltage is slightly exceeded leads to a faster transition of
growth to streamer. At some overvoltage (the longer the gap, the
smaller the overvoltage) the very first avalanche is transformed into
a streamer (type C breakdown). For day air and a spacing of 1 cm,
the overvoltage required for breakdown with growth to become purely
streamer-type breakdown is about 10%. The spacing at which a negligible overvoltage should lead directly to streamer breakdown (the
lower limit of long breakdowns of type B) is estimated as 5-10 cm
for air at atmospheric pressure.

This hypothesis has been confirmed in a large number of experimental projects in the course of which the following factors were
investigated**:

---

* This hypothesis was outlined in incomplete form in papers by
  von Hippel and Frank [1-5] and by Raether [1-6].
** A deficient bibliography is given in [0-1]. Only the most recent
  papers are cited below (see Foreword).

(a) Breakdown formation time: work by Genger,[†] Fisher, Bederson,[†] Kachikas,[†] and Lessin[†]; Morgan, Eiked,[†] Bruce, and Tedford[†]; Retner,[†] Fel'dt,[†] Zost,[†] Dehne, Köhrmann and Lenne [1-7]; Däcke [1-8]; and Efendiev and others [1-9 to 1-14].

(b) The current in a steady-state self-sustained discharge close to breakdown: work by Llewellyn-Jones, Parker, Crompton, Haydon, and Dutton and others [1-15 and 1-16], Wilkes, Hopwood, and Peacock, and Fisher, De-Bitetto,[†] and Roze.[†]

(c) The influence of cathode material on breakdown voltage: work by Tramp,[†] Klaud,[†] Mann,[†] and Khanson[†]; Shreder,[†] Dutton, Haydon, and Llewellyn-Jones; De-Bitetto, and Fisher; and the authors listed above [1-13 and 1-14].

(d) The variation of current and voltage in the early stages of breakdown: work by Bandel',[†] Kojima and Kato; Retner, Shmidt,[†] Fogel',[†] Frommgol'd,[†] Klyukov, Shlumbom,[†] Mil'ke,[†] Dehne, Körmann, Lenne [1-7], Pfaue [1-17], Tholl [1-18], Richter [1-19], Muller [1-20], and Efendiev [1-21].

(e) Glow in the early stages of breakdown: work by Fisher, Kachikas, Lessin, and De-Bitetto, Retner, Legler,[†] Dibbern,[†] Shreder,[†] Tholl [1-18, 1-22, and 1-23], Driver [1-24], Raethjen [1-25], Teich [1-26], Hoffmann [1-27], Wagner [1-28 to 1-32], Sachs and Chippendale, Loeb and Hudson, Sroka [1-33], and Suleebka and Rau [1-34].

(f) Direct observation of the development of electron avalanches and the discharge channel by means of a Wilson cloud chamber, high-speed photography, and image converters: the work mentioned above by Raether and other researchers [1-8, 1-22, 1-28, and 1-32] and by Allen and Phillips, Doran, Mostl and Timm, Cavenor, Kekez, Barrow and Craggs, Chalmers and Duffy, and Reinghaus [1-35 to 1-42].

In recent years the corresponding concept of type A breakdown and of its transition to type C breakdown has been developed in detail in the theoretical papers and reviews by Loeb, Fisher, Bederson and Kachikas, Raether, Körmann, Legler and Klyukov, Meek and Craggs, Marshak, Llewellyn-Jones, Sachs, Davidson, Davis and Week, Mioshi, Dawson and Vinn,[†] Davis and Evans, Ward, Kline and Siambis, Oppenheimer, Michel and Nasser, Lozanskii, Kirilenko, and Matyushinykh [1-1, 1-2, and 1-43 to 1-54].

These papers summarize the experimental data which confirm the current concept of types A and C breakdown presented above, data which enable us to formulate the concept in a highly precise form. It is demonstrated that theoretical schemes which explain the development of static breakdown (the transition to self-sustained discharge) in terms of the growth of ionization are fully consistent with all experimental data for air and most other gases at electrode spacings of up to a few centimeters (at atmospheric pressure). On the other hand, purely streamer-type theoretical schemes of the initial development of breakdown when the static breakdown voltage is slightly exceeded, field nonuniformity is slight, spacings are not too long, and external ionization is weak are at odds with many experimental data and must be rejected. (Exceptions are the vapors of organic substances and, when the cathode is specially selected, oxygen, for which the secondary ionization coefficient δ is particularly small). Such static breakdown is observed in air for electrode spacings exceeding 2 cm under conditions of extremely intense instantaneous ionization of the gas by α particles, which produce about 10^4 electrons in a few nanoseconds [1-17].

The mechanisms of secondary ionization that lead to extended reproduction of electron avalanches during buildup (up to a significant current rise that leads to field distortion by the space charge) depend on the specific conditions of breakdown. The most important part among these mechanisms is played by the photoelectric effect on the cathode due to the radiation of atoms excited as the avalanches develop (the δ process). This is confirmed by:

(a) the corresponding values of the breakdown formation time at small overvoltages;
(b) the results of oscillograph traces of avalanche chains and current rise in the early stages of breakdown;
(c) the increase in the intensity of the diffuse glow as the streamer spreads, and data on the gas-ionizing radiation of the discharge [1-26, 1-33, 1-55, and 1-56].

In most cases this process of secondary ionization is not the only one (especially at extremely small overvoltages, at which the breakdown formation time runs to hundreds of microseconds). It is supplemented by the process of electron stripping when the cathode is bombarded by positive ions (the γ process), and perhaps by other processes [1-57 and 1-58].

Under the conditions indicated above, the mechanism of ionization growth acts only during the first stage of breakdown. This mechanism helps create in the gas-filled gap a sizable plane space charge which leads to constriction of the field in the cathode region. At the same time a diffuse discharge, which by nature is close to an anomalous glow discharge, becomes established for a short period of time. Then the second stage of breakdown, in which the streamer mechanism is dominant, begins in the region of elevated electric field strength next to the cathode. The concentrated ("point") space charge formed in the head of one of the electron avalanches that is developed in the most statistically favorable way is the predominant factor in this mechanism. This charge creates such a large local distortion of the field that phase advance of the elevated ionization front (an advance that is an order of magnitude faster than that of the avalanche) along the gap in the direction of the anode and cathode becomes possible. The fine gas-channel remaining behind this front is filled with highly ionized neutral plasma. This plasma ensures trans-

mission of an extremely high electric current after the channel closes
the gap between the cathode and anode and creates electrode discharge
regions which by nature are close to the corresponding regions of an
arc.

The specific nature of the phase advance of the enhanced ionization front in the field of the concentrated space charge of the
head has not been fully explained. This advance may be brought about
by: (a) thermal ionization of the gas (during which various elementary processes of atomic ionization are at work); (b) photoionization
of atoms located at some distance in front of the head, as a result
of which enhanced α ionization creates a daughter electron avalanche
along a short path, the charge in whose head is equal to the charge
in the head of the parent avalanche; (c) the accelerated motion of
electrons after they are stripped from atoms, until they acquire a
kinetic energy corresponding to the electron temperature; (d) impact
ionization of excited atoms located in front of the avalanche head.

The increase in overvoltage during pulsed breakdown creates the
conditions for the start of the second stage of breakdown: the development of a streamer when the field is less concentrated at the
cathode and hence the plane space charge formed as a result of the
ionization growth is weaker. This leads to a shortening of the
breakdown formation time, accelerated current rise in the early stage
of discharge, and advance of the region of streamer formation toward
the anode. At some overvoltage ΔU which decreases with increasing
gap length l (Figure 1-1), streamer formation becomes possible without
field concentration on the cathode region due to the plane space
charge. In this case breakdown may begin directly from the second
stage, i.e., the streamer-development stage, as assumed in theoretical
streamer schemes. The transition to a breakdown beginning directly
with the second stage is characterized by a number of features, among
which we may list the following: (a) the breakdown formation time
becomes approximately equal to the time of transit across the gap for
a single electron avalanche; (b) in the graph of the overvoltage dependence of formation time there is a discontinuity associated with
the transition to a new mechanism of formation; (c) because of the
absence of predistortion of the field by the plane space charge, the
streamer formation region advances toward the anode itself; (d) oscillograms of the potential difference between the spark-gap electrodes no longer exhibit a step corresponding to the diffuse ("glow")
discharge phase.

Recently, yet another mechanism has been found whose action
makes it possible, in some cases, to identify the manifestation of
features (a) and (b) with the transition from breakdown with Townsend
growth to streamer breakdown. [References 1-8, 1-24, 1-36, and 1-44].
We have in mind the mechanism of electron attachment to electrically
negative atoms such as oxygen. At the comparatively low electric
field strengths corresponding to static breakdown of plane gaps more

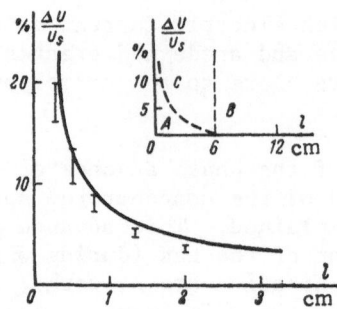

Fig. 1-1. Limit of overvoltage $\Delta U/U_s$ for spark gaps of different
lengths l above which breakdown develops in accordance
with the streamer mechanism, and below which in accordance
with the "growth" mechanism (for air). The curve is from
[1-4]. I) Range of scatter of experimental points from
[1-59]. At top is a schematic depiction of the boundaries
between breakdown of types A, B, and C.

than 6 cm long (air, atmospheric pressure), such attachment leads to
the escape of negative charges from the avalanche to the anode, delayed by 2 orders of magnitude. Accordingly, a high concentration
of the space charge of positive ions that is capable of causing development of the streamer toward the cathode is attained not during
the transit time of electrons across the gap, but in a time 2 orders
of magnitude longer, even though what occurs is essentially a single-
avalanche streamer breakdown without growth (the "modified streamer
mechanism"). As the overvoltage increases, electron attachment at-
tenuates substantially. Some overvoltage leads to an incremental
decrease in the breakdown formation time to values corresponding to
the electron transit time. This corresponds to a transition from
modified breakdown to conventional streamer breakdown.

The voltage at which breakdown development begins directly with
a streamer is the breakdown voltage from the standpoint of purely
streamer-type theoretical schemes. It must be in accord with all
the regularities implied by these schemes, including the extremely
large spread (about 1 kV) of the measured values of the breakdown
voltages. This spread is characteristic of the streamer mechanism.
Accordingly, the boundary of overvoltages beyond which the growth
stage of ionization disappears should not be sharp, but should occupy
a statistical range having a half-width of about 0.5 kV. This corresponds precisely to the range of overvoltages in which oscillograms
of the potential difference between spark-gap electrodes sometimes
have and sometimes do not have a glow-discharge step [1-59]. The
uncertainty of the location of the region of streamer formation at
large overvoltages is a manifestation of the statistical nature of
breakdown beginning directly with a streamer.

For small overvoltages at which discharge begins with growth of
ionization, pulsed breakdown in the uniform field of not-too-large
gaps is qualitatively analogous to the static breakdown of the same
gaps. Pulsed breakdowns of such gaps at a large overvoltage, in
which the discharge begins directly with streamer formation, differ
qualitatively from type A breakdowns and constitute the separate type
C.

The boundary between breakdowns of types A and C (Figure 1-1)
thus is a rather broad zone within which breakdown begins with the
growth of ionization in some cases, and directly with streamer
formation in others.

As the spacing increases, the overvoltage boundary which separ-
ates type A and C breakdown becomes lower. For dry air at atmospheric
pressure, overvoltages corresponding to the transition from type A
breakdown to type C breakdown become practically equal to the corresponding statistical spread at a gap length of approximately 6 cm.
Starting at this spacing, even static breakdown may begin directly
with streamer formation. Static breakdown begins in the same way in
highly nonuniform electric fields. Static breakdowns of long gaps
(over about 6 cm for dry air) in a uniform field and breakdowns in
highly nonuniform fields may be classified under a single type, type
B, in terms of the nature of their mechanism. The boundary between
type A and B breakdown also is quite blurry. Its width is determined
by the statistical spread in the development of electron avalanches.

Increased air humidity or special treatment of the cathode lowers
the second ionization coefficient. This may reduce the spacing at
which the region of type B breakdown begins, and also may decrease
the relative value of the overvoltage (while simultaneously increasing
the absolute value of the breakdown voltage) at which the region of
type C breakdowns begins.

1-3. Type E Breakdown

The processes described above characterize the formation of the
high-current gas discharge in pulsed light sources based on the use
of a two-electrode gas-filled gap similar to the spark gaps used for
illumination in the early stage of development of high-speed photography. The gap is connected to a storage capacitor (directly or
through a switch). However, three-electrode (in some cases, multielectrode) regulated flashlamps have become much more widely used.
In these lamps a high-current discharge between the primary current-
carrying electrodes is formed by means of a fine secondary plasma
channel produced by a high-voltage pulse on the control electrode
(sometimes on one of the primary electrodes which is separated from
the storage capacitor by an inductance) from a low-power auxiliary
source (usually a pulse transformer).

A secondary plasma channel is formed when a flashlamp is switched on in circuits similar to the ones shown in Figure 1-2. The closing of switch K produces a high-voltage pulse on the control electrode which breaks through the gap between the primary electrodes via one of the breakdowns described above (types A-D). If the lamp is a long glass tube of relatively small diameter, part of the secondary channel usually passes over the inner surface of the tube. The secondary breakdown as a whole may then be characterized by the equivalent scheme presented in Figure 1-3, which is outwardly reminiscent of the scheme in Figure 1-2b. Because a section of surface breakdown based on the mechanism of interaction with static charges on the walls is included [1-60 to 1-64], the secondary pulse voltage required to cover the large distance between primary electrodes is reduced substantially (see, e.g., the relations in Figure 1-4).

In view of the low power and short duration of the high-voltage pulse, however, auxiliary breakdown does not in itself place the flashlamp gap in a state of high conductivity, since it does not produce significant current when the supply voltage is low. If by the concept of "breakdown" we mean the transition of a gas-filled gap from a state of extremely low conductivity to one of extremely high conductivity, then this concept also should encompass the formation of the primary discharge in the flashlamp. The gradual increase in the voltage on the source, during which there are repeated secondary breakdowns from time to time, ultimately leads to a situation in which one of the secondary breakdowns becomes a primary breakdown with a rapid current rise in the circuit being powered by the source. It is this primary breakdown which is type E breakdown. The minimum voltage U_t on the source at which primary breakdown can occur (the "trigger voltage" of a gas-discharge tube) is analogous to the breakdown voltage during the static breakdowns discussed in the previous section. This voltage may be considerably lower than the ordinary static breakdown voltage of the same gas-filled gap in the absence of an auxiliary high-voltage pulse (the "self-breakdown voltage" U_{self} of the lamp).

The concept established for the condition of onset of type E breakdown [1-65] makes allowance for certain features of such breakdown, which are known from the development and operation of flashlamps based on it: (a) specificity of the trigger voltage; (b) a significantly lower value of U_t for devices filled with noble gases than for devices filled with molecular gases; (c) a sharp increase in U_t when small impurities of a molecular gas are added to the noble gas; (d) a decrease in U_t as the gas pressure and electrode spacing decrease and the power of the triggering pulse increases; (e) a decrease in the difference between the trigger voltage U_t and the extinction voltage U_e as the inside diameter of the discharge tube decreases. (As the tube's diameter approaches the diameter of the auxiliary discharge channel determined by the streamer mechanism, this difference tends toward zero.)

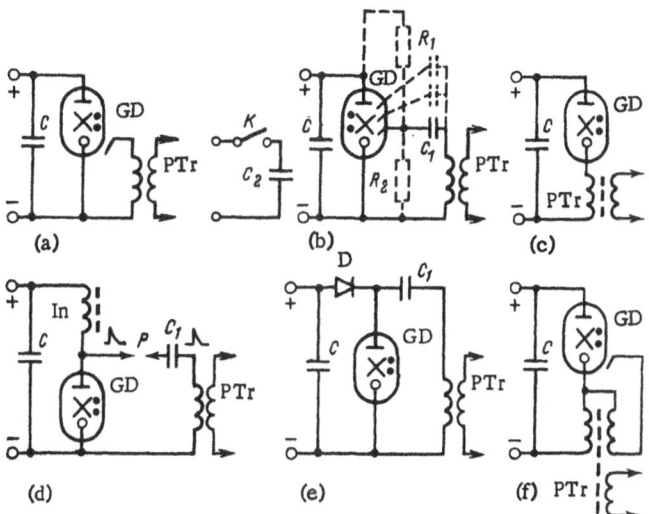

Fig. 1-2. Examples of switching circuits in devices using type E
breakdown. a, f) schemes with external trigger electrode;
b) scheme with one or more internal trigger electrodes;
c-e) schemes without an auxiliary electrode; GD) gas-
discharge device; C) high-current power source (capacitor,
grid); PTr) pulse transformer; C_2) capacitor powering the
pulse transformer (charged from C or from a separate
source); K) switch (synchronous contact, thyratron, auxiliary discharger, etc.); In) inductance (valve); C_1 ($C_1 \gg C$)
and P) coupling capacitor and spark gap; R_1, R_2) resistors.

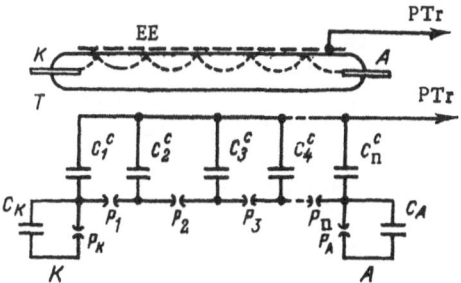

Fig. 1-3. Equivalent circuit of secondary breakdown in a tubular
lamp [1-60]. PTr) pulse transformer; T) tube; K) cathode;
A) anode; EE) external trigger electrode; P_K, P_A) gas-
filled gaps between primary electrodes and glass;
P_1, P_2, ...) gaps on glass surface; C_K, $C_A \simeq 10$ pF) capacitances between primary electrodes and glass; C_1^c, C_2^c,
..., $C_n^c \simeq 100$ pF) equivalent capacitances of surface gaps.

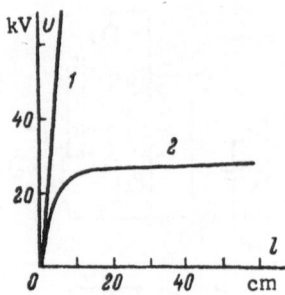

Fig. 1-4. Dependence of breakdown voltages on length of gas-filled
          gap (1) and surface gap (2) [1-60].

A description of the extinction of a discharge powered by a
capacitor can be outlined by using the concept of an energy balance
in the pulsed discharge formed. The discharge is confined by the
walls of the discharge tube. As the electric field strength E decreases during the discharge, the voltage drop across the capacitor
causes the fraction of losses at the walls to increase in the overall
energy balance, approaching 100%. At low electric field strengths
the power supplied to the discharge channel decreases rapidly because
of the characteristic sharp increase in plasma resistivity. At some
value of E it no longer compensates for energy losses at the walls.
As a result, the current is cut off quickly (cooling of the channel
causes its resistance to drop, which in turn leads to a further decrease in the power in the channel and to still greater cooling, and
so forth). Because the fraction of energy on the walls increases
with decreasing channel diameter, the discharge extinction voltage
is higher in narrow tubes than in wide ones.

The convergence of U_t and U_e for especially thin tubes allows
us to hypothesize that type E breakdown is a process inverse to the
extinction of a pulsed discharge. The plasma channel formed between
the primary electrodes of the gas-filled gap during secondary breakdown either may be deionized rapidly if the power of the discharge
induced by the primary power supply is insufficient to make up for
losses (negative power balance), or may begin to expand if the power
exceeds losses (positive power balance). Expansion of the channel
leads to an even greater excess of the power supplied (which is proportional to the square of the diameter) over the power lost (which
is proportional to the first power of the diameter). Thus, the
transition from negative to positive power balance is critical and
leads to the rapid current rise characteristic of all types of breakdown at high pressures. The criterion for occurrence of a positive
power balance may be identified with the criterion for the onset of
type E breakdown.

This concept is consistent with the peculiarities noted above
for this kind of breakdown. Indeed, the specificity of the value

of U_t stems from the specificity of the power balance in a specified plasma channel of the secondary breakdown and from the critical nature of the criterion for the onset of breakdown. The small cross section for the scattering of electrons by atoms of noble gases (the Ramsauer effect) explains why the electric field strength required to ensure a positive power balance is lower in noble gases than in molecular gases, and hence explains the correspondingly smaller values of U_t in noble gases. An increase in the secondary pulsed power causes an increase in the conductivity and diameter of the original plasma channel. Consequently, a positive power balance may become established in the channel at a lower electric field strength.

The value of U_t is determined by the equation for the input and dissipated power for the secondary discharge channel. We shall ignore the sum of the electrode potential drops U_{el} and the longitudinal nonuniformity of the secondary channel, due to the existence of channel sections that slide along the surface of the tube and of sections surround by gas. Of course, this is not fully justified [1-60]. The equation is then:

$$\pi r^2 \frac{U_t}{l} J = 2\pi r P, \qquad (1-1)$$

where r and l are the radius and length of the secondary discharge channel, J the current density in a channel acted on by U_t, and P the power dissipated across 1 cm^2 of channel surface area.

Taking into account the proportionality of J to the degree of ionization x, the numbers of atoms per cubic centimeter n (including ionized atoms), the electron mean free path λ_e, and the electric field strength U_t/l, and assuming that $\lambda_e = 1/n [xq_i + (1-x)q_0]$, where q_i and q_0 are the cross sections for scattering of electrons by an ion and by an atom, respectively, we obtain

$$U_t^2 = \frac{A l^2 P \sqrt{T}}{rx} [xq_i + (1-x) q_0], \qquad (1-2)$$

where A is a numerical factor and T is the channel temperature.

The simplified concept of the mechanism of energy losses by the original discharge channel makes it possible to obtain the dependences of U_t on the gas pressure p_0 and channel length l, and on the fraction θ of the content of molecular-gas impurity with cross section q_0' for scattering of electrons by atoms [1-65]:

$$U_t = \text{const } l \sqrt{p_0}, \qquad (1-3)$$

$$(U_t'/U_t)^2 \approx 1 + \frac{q_0' \dfrac{1-x}{x}}{q_i + q_0 \dfrac{1-x}{x}} \theta = 1 + K\theta, \qquad (1-4)$$

where U_t and $U_t^!$ are the trigger potentials in pure and contaminated gases, and K is a coefficient determined by the cross sections of neutral atoms of the inert gas and impurity, and by q_i and x, which depend on the temperature of the original channel, i.e., on the power of the triggering pulse.

Of particular interest is the case in which the degree of ionization x is so small that we may ignore xq_i in comparison with q_0. In this case,

$$K = q_0'/q_0. \qquad (1-5)$$

Equations (1-3) to (1-5) were checked in [1-66 and 1-67], in which the trigger voltages of tubular flashlamps with an external trigger electrode were studied (see the diagram in Figure 1-2a). The lamps were filled with xenon, argon, and neon to which were added different small amounts of nitrogen, hydrogen, and oxygen impurities. Shown in Figure 1-5 is an example of the dependence of $(U_t^!/U_t)^2$ on θ. For small θ the experimental points fit well on a straight line, according to relation (1-4). Figure 1-6 shows the dependences of K on the primary potential U_I of the triggering pulse transformer for six combinations of noble and molecular gases. The scale along the ordinate is 6 times smaller on the graphs for hydrogen (dashed lines) than on the graphs for nitrogen (solid lines). The graphs for the different noble gases were drawn to the same scale. The satisfactory similarity of the dashed lines to the solid lines attests to the experimental confirmation of equation (1-5).

The dependences of the trigger voltage U_t and extinction voltage U_e of a discharge in tubular (capillary) flashlamps on the gas pressure p_0, capillary length l, inside diameter d_i, and triggering pulse power are shown in Figures 1-7 to 1-10 [1-67]. These graphs are in agreement with equation (1-3). The discharge extinction voltage also turns out to increase linearly with l and $\sqrt{p_0}$. The slope of the lines $U_t = f(l)$ and $U_e = f(l)$ increases with increasing gas pressure and decreasing inside diameter of the tube. The slope of the lines $U_t = f(l)$ obviously increases severalfold with increasing triggering pulse power until the corresponding straight line begins to run almost through the origin (a further increase in triggering pulse power causes practically no decrease in the trigger potential).

Increasing the triggering pulse power to a value corresponding to $U_I = 800$ V appreciably lowers U_t, but a further increase in U_I has an extremely small effect on the trigger potential (in Figure 1-7 there is a common graph for $U_I = 800$ and 1200 V).

The capillary diameter has a slight effect on the trigger potential (in Figure 1-7 a common graph represents the relationship $U_t = f(p_0)$ for all diameters) and a much stronger effect on the

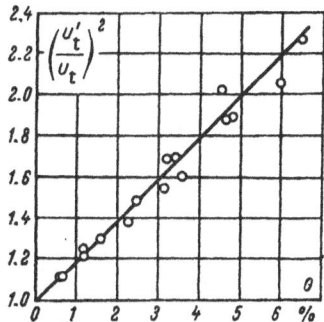

Fig. 1-5. Dependence of the square of the trigger voltage on
nitrogen content in xenon.

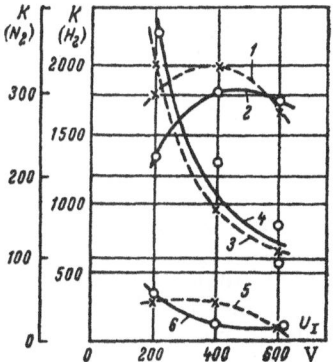

Fig. 1-6. Dependence of angular coefficient K on the voltage U_I.
1) Neon + hydrogen; 3) argon + hydrogen; 5) xenon +
hydrogen; 2, 4, 6) the same noble gases with nitrogen.

discharge extinction voltage. As the diameter decreases below approximately 0.2 mm, U_t and U_e tend toward a common limit.*

Apart from the regularities listed above, it has been noted
that the values of U_t and U_e have a rather large spread from lamp
to lamp (10-15%) and a small spread for a given lamp from measurement to measurement (1-2%). The value of U_t is nearly independent
of the shape of the external trigger electrode and of its position
relative to the discharge tube. (When the trigger electrode is
moved away from the outer surface of the tube to a distance of up
to 3 mm, U_t is practically unchanged. Moving it to a distance of

---

* The significant extinction voltage for narrow capillaries makes
  the correction for the unexpended energy $CU_e^2/2$ significant with
  respect to the energy $CU_0^2/2$ stored in the storage capacitor before
  the flash.

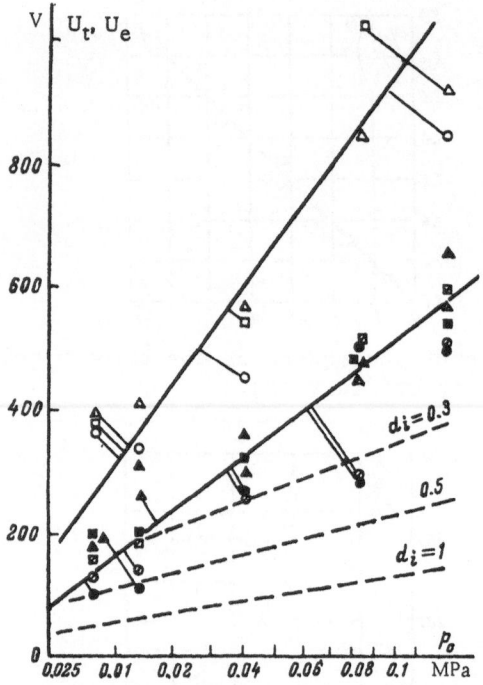

Fig. 1-7. Dependence of the trigger voltage U_t (solid lines) and
the extinction voltage U_e (dashed lines) on xenon pressure
p_0 (plotted to the $\frac{1}{2}$ power along the abscissa) in lamps
with a capillary length l = 70 mm and a capillary inside
diameter d_i = 0.3 mm (triangles), 0.5 mm (squares), and
1 mm (circles). The experimental points obtained for a
potential U_I = 1200 V on the primary coil of the triggering
pulse transformer are drawn solid in the figure. Points
obtained for U_I = 800 V have a slash drawn through them.
The open symbols are points obtained for U_I = 400 V.
(For the first two values of U_I there is a single plot of
U_t.)

8-10 mm causes an increase of approximately 30% in U_t.) It is noted
in [1-68] that as an annular trigger electrode is moved along the
tube, U_t reaches a minimum when the trigger electrode is located near
the anode.

Of practical interest are comparative experimental data [1-67]
on the values of the trigger voltage for type E breakdown and the
uncontrolled breakdown voltage of gas-filled gaps of the same lamps
with supply of a triggering pulse. In view of the large spread of
these values, the concept of two lower boundaries of the spread was
introduced:

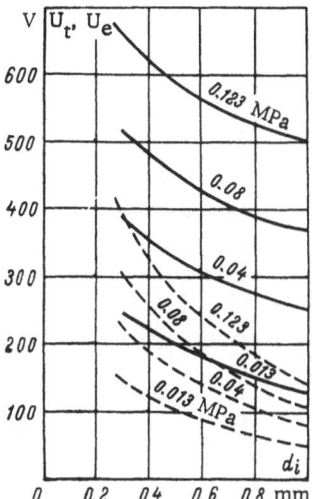

Fig. 1-8. Dependence of U_t (solid lines) and U_e (dashed lines) on inside diameter of discharge tube (l = 70 mm) at different xenon pressures in MPa (indicated in the figure). U_I = 1200 V.

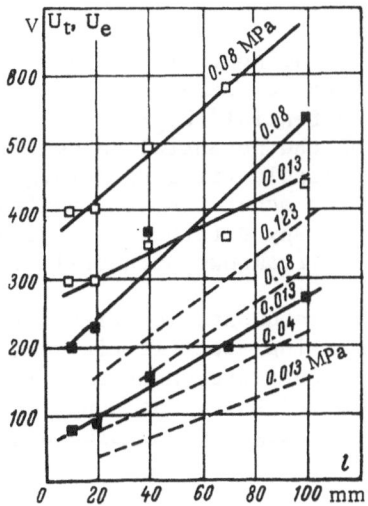

Fig. 1-9. Dependence of U_t (solid lines) and U_e (dashed lines) on length of capillaries with d_i = 0.5 mm at different xenon pressures. The solid symbols are experimental points obtained for U_I = 1200 V. The open symbols are for U_I = 400 V.

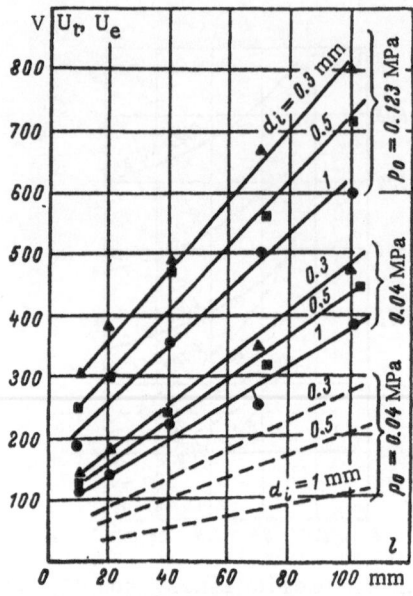

Fig. 1-10. Dependences of U_t (solid lines) and U_e (dashed lines) on
the length of capillaries with d_i = 0.3, 0.5, and 1 mm.
U_I = 1200 V. The xenon pressure is 0.04 and 0.123 MPa
(300 and 920 mm Hg).

(a) The lower boundary of the potential values at which uncontrolled
triggering of the discharge occurs in some cases. This is called
the "self-breakdown potential" U_{self}.
(b) The lower boundary of the potential values. When this potential
is reached slowly, reliable triggering (e.g., in 90% of the
cases) of the discharge occurs in all samples of lamps of one
type in the absence of any external electrical impulses. This
is called the "static breakdown potential" U_{st}.

In measurements of these quantities [1-67] it was noted that
the presence of rapidly varying electric fields (even weak fields,
e.g., from the sparking of contacts) near the "lamp-capacitor" circuit sharply reduces U_{self} and U_{st}. The dependences of U_{self} and
U_{st} on gas pressure and discharge tube inside diameter and length
are shown in Figures 1-11 to 1-13. As can be seen from these figures,
the measured values of U_{self} and U_{st} have a large spread which increases particularly in those cases where the trigger electrode is
removed from the lamp. In view of the large spread, the graphs in
these figures can be plotted only approximately, with an outline of
the behavior of the possible minimum values of U_{self} and the maximum
values of U_{st}. Available experimental data show that the self-breakdown potential U_{self} (when a trigger electrode is present) exceeds
by a factor of 5-15 the trigger potential for the least powerful of

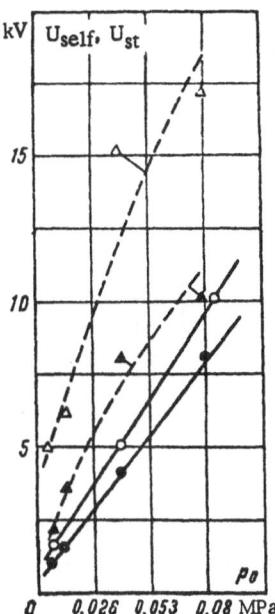

Fig. 1-11. Dependence of U_{self} (solid lines) and U_{st} (dashed lines)
on xenon pressure. The solid points are with a trigger
electrode and the open points without one. The inside
diameter of the discharge tube is $d_i = 0.5$ mm, and its
length is $l = 20$ mm.

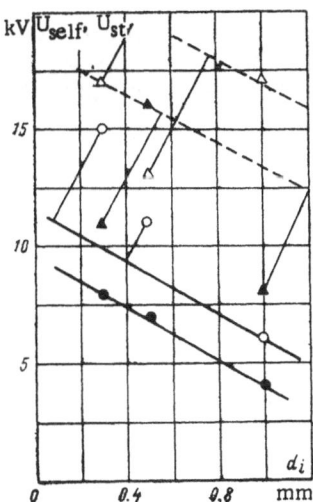

Fig. 1-12. Dependences of U_{self} and U_{st} on inside diameter d_i of
discharge tube. Symbols are the same as in Figure 1-11.
$l = 70$ mm, xenon pressure $p_0 = 0.08$ MPa.

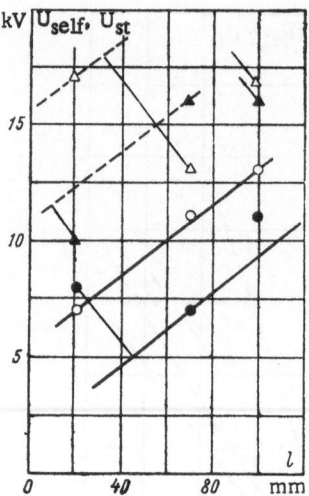

Fig. 1-13. Dependence of U_{self} and U_{st} on capillary length. Symbols
are the same as in Figure 1-11. d_i = 0.5 mm, p_0 = 0.08
MPa.

the triggering pulses used (the potential on the primary winding of
the pulse transformer is U_I = 400 V). Without a trigger electrode
U_{self} increases by an additional factor of 1.5-2. The self-breakdown potential increases approximately linearly with discharge tube
length and gas pressure and decreases with increasing tube inside
diameter. The static breakdown potential exceeds the self-breakdown
potential by a factor of about 2 and varies with the design data of
the lamps according to approximately the same laws. According to
the data of [1-69], a molecular gas impurity (3% nitrogen or oxygen)
increases U_{self} by a factor of 1.5.

CHAPTER 2

HIGH-CURRENT PULSED DISCHARGE WITH AN EXPANDING CHANNEL

2-1. Primary Current Rise Before Channel Expansion

It was noted in Chapter 1 that a rapid current rise occurs in the gas-filled gap after breakdown (triggering). This chapter examines the rise process in its entirety.

During the initial stage of discharge before streamer formation, the distribution of all physical quantities is uniform and there is practically no spatial motion of atoms or ions (mainly electrons and photons move), the exceptions being few and not very significant. The process usually is independent of the parameters of the circuit feeding the discharge (except for the supply voltage). The phenomena in the regions around the electrodes are of the same nature as in a stationary self-sustained discharge, combining individual elementary interactions that submit to separate study.

By virtue of these circumstances, with reasonable assumptions it is possible in principle to make a mathematical calculation of the time dependence of all physical parameters in the theory of discharge development [1-48 to 1-54]. As the discharge develops further, the phenomena become considerably more complicated and begin to be characterized by a variety of rapidly time-varying, interrelated features, namely: the radial distribution of the physical quantities in the constricted discharge channel (temperature, gas density, the degree of ionization of the gas, etc.), the energy flow to the surrounding space (radiation, gas-dynamic processes), the current density and strength, the longitudinal electric field strength, the voltage drops near the electrodes, and so forth.

As we know, a mathematical calculation of all characteristic

quantities is possible only for particular cases of discharge, and
not in general form, even for a stationary arc. It seems all the
more complicated to consider in general form all physical quantities
for the transient high-current stage of a pulsed (spark) discharge,
which is characterized by a much larger number of more intricately
interconnected, rapidly changing parameters.

For this reason the study of this stage first moved in the
direction of experimental and phenomenological investigation of the
time dependence of individual quantities. The most important targets
of this investigation were:

(a) the electrical parameters of the discharge: the voltage across
    the gas-filled gap, the current strength and density in the dis-
    charge, and the resistance (or conductance) of the discharge
    channel;
(b) the expansion of the discharge channel and concomitant gas-dynamic
    processes;
(c) the emission characteristics of the discharge: luminous inten-
    sity, luminance, and spectral composition;
(d) processes around the electrodes.

Basic data on the electrical parameters of discharge were obtained in the work of Rogovskii and co-workers, Kyun,[†] Rose, Dehne,
Körmann and Lenne, and Effendiev (sic) [1-7 and 1-9*], who determined
the rate of the voltage "dip" across the gas-filled gap; in the work
of Laporte, Murphy and Edgerton, Abramson and Marshak, and Wulfson,
who detected the presence of finite resistance of a pulsed discharge
with a channel bounded by the walls of the discharge tube, and determined the presence of a limiting current density in the bounded and
unbounded discharge channels; and in a number of very recent papers
which studied in greater detail the pattern of the time dependence
of all electrical characteristics (including discharge power and resistance) for different parameters of the gas-filled gap and power
circuit.

The expansion of the discharge channel has been studied in the
work of Mandel'shtam, Abramson, Gegechkori, Drabkina, Braginskii,
and others. They established the analogy between the expansion and
an explosive-type hydrodynamic process during which a brief liberation
of a tremendous amount of energy occurs in the narrow plasma channel,
the volume of heated gas increases, and a shock wave propagates.

The emission characteristics of the discharge were determined
in the work of Laporte, Edgerton, Bogdanov, Wulfson, Marshak, and a

----

* A bibliography of the older papers is given in [0-1], but here we
  cite only the most recent publications and give the names of the
  authors of basic works.

number of others who investigated chiefly the radiation from a discharge bounded by a discharge tube, and in the work of Wulfson,
Charna,[†] Libin, Vanyukov, Mak, Andreev, and others who investigated
mainly the radiation of an unlimited discharge and who demonstrated
the existence of a limit on its luminance. Many spectral and time
characteristics of limited and unlimited discharge also have been
obtained by Mandel'shtam, Glazer, Fischer, Früngel, Craggs, Mik,[†]
and a number of others.

Pulsed discharge phenomena on electrodes have been studied by
Frum,[†] Somervell,[†] Blevin,[†] Raiskii, Zimin, Mandel'shtam, and others.

As we showed in Chapter 1, the rise in current density at the
start of breakdown occurs through a number of increasingly rapid
"chain" processes that are set in motion one after another: impact
ionization by electrons (α ionization), the interaction between α
and δ ionization (secondary processes at the cathode), the effect
of the plane space charge on α and δ ionization, the interaction
among α ionization, photoionization in the gas volume, and the concentrated space charge of the streamer head, and others. A longitudinal electric field strength, equal to the ratio of the potential
difference across the gap (less the voltage drops between the heads
and the corresponding electrodes: after the channel spreads over
the entire length of the gas-filled gap, these drops become the potential drops near the electrodes) to channel length, becomes established in the ionized gas column (the plasma channel) formed behind
the head (or between the heads if the streamer develops toward the
anode and cathode simultaneously). If the power supply feeding the
discharge has enough power for the potential difference across the
gap greatly to exceed the sum of the potential drops near the electrodes, even when the discharge current is strong, then yet another
highly intense chain process which is characteristic of high current
density and which lasts just a few tens of nanoseconds (during which
the channel is practically unable to expand) occurs in the channel
as a result of the significant longitudinal voltage gradient. It is
already quite difficult to identify the different kinds of elementary
interactions of atoms, ions, electrons, and photons in this stage
because of the significant degree of excitation and ionization of
the gas in the channel and the strong simultaneous influence of many
particles on each other. Therefore, it makes sense to speak of the
overall resultant thermal ionization of the gas (or of some analogue
thereof in the case of thermal equilibrium which does not manage to
become established because of the rapid change in characteristics).*

---

* Obviously, the very length of this chain process is determined by
  the time required for some leveling-out of the energy distribution
  with respect to the degrees of freedom of the plasma particles as
  a result of the most varied elementary interactions.

The chain process that occurs in the channel as a result of the longitudinal strength of the electric field E consists in the mutual acceleration of the thermal ionization of the gas (which grows as a result of heating of the gas due to the energy dissipated in the channel) and the power developed by the discharge (this power increases as a result of the rise in the current density j as the degree of ionization x increases). This process is capable of leading to a primary rise of current density by many orders of magnitude in a few tens of nanoseconds. This process can only be halted or slowed by the following factors:

(1) Cessation of the increase in the electric power dissipated in the discharge as a result of the fact that a further increase leads to a decrease in E such that growth of j becomes impossible. When this occurs, the decrease in E may be related to:

   (a) the characteristics of the external discharge circuit (its resistance, the inductance or low capacitance of the storage capacitor);
   (b) the increase in the sum of the near-electrode voltage drops u_{ac}.

(2) A pronounced slowing of the rise in current density due to an increase in the degree of ionization x (the overall ion cross section) such that electrons begin to be scattered mainly by ions rather than by atoms, and their mean free path becomes inversely proportional to x.
(3) An increase in channel temperature to a level at which the increase in electric power that accompanies the heating of the channel begins to lag behind the increase in energy losses to the surrounding space that accompanies that same heating (losses due to radiation, hydrodynamic expansion, etc.).

Experimental data on the time dependence of the electrical characteristics (the potential u across the gas-filled gap and the current strength i in the discharge, as well as various functions of these quantities: power, current steepness, etc.) under various conditions have made it possible to refine the picture of the process and to determine the true reasons for its termination.

Solving this problem by oscillograph recording of the electrical characteristics of a freely expanding discharge was complicated by the fact that the current strength continues to increase rapidly because of channel expansion, even after the rise in current density comes to a halt for one of the reasons indicated. For example, in the case of air breakdown at atmospheric pressure, the original discharge channel has a diameter equal to that of the streamer (about 0.1 mm). If the fast process of primary growth of the current density j in the channel lasts about 50 ns and if the channel expands with the velocity of the shock waves (about 10^5 cm/s, corresponding to a

rate of increase in diameter of 2×10^5 cm/s), then the channel diameter d increases by approximately 0.1 mm during the 50 ns following
cessation of the rise in j. Hence, even at constant j the current
strength continues to increase at a rate on the same order of magnitude as during the primary process of the rise in j at constant diameter. Therefore, the most graphic results were provided by oscillograph recording of the electrical characteristics in experiments
using discharges that were artificially limited with respect to the
channel diameter.

If the discharge circuit has significant resistance (tens of
ohms or more), then the rapid process of primary current rise causes
an extremely fast (tens of nanoseconds) voltage drop ("dip") in the
discharge to the usual arc voltage at a current strength determined
by the values of the voltage across the storage capacitor U_0 and the
ballast resistance R_b (limiting of j for reason 1a). The discharge
toward the end of the dip is essentially the same as an ordinary arc,
although it is characterized by an elevated current density (several
thousand A/cm^2) and gas pressure. The discharge channel expands in
the subsequent period of time, and the current and gas densities become consistent with the usual values for an arc discharge. As this
occurs, the electrical resistance of the discharge decreases, but
this has practically no effect on the current strength since the resistance is much lower than R_b immediately after the dip. The process
as a whole corresponds to curve 1 in Figure 2-1.

The picture we see when the circuit has low resistance and
significant reactance is close to this. If R_b is negligible and the
capacitance of the storage capacitor is sufficiently high, then the
potential drop across the spark gap is approximately L di/dt (where
L is the circuit's inductance) as a result of the fast rise in the
current strength i. If L is large, the potential across the gap
"dips" to 200-300 V during the primary rapid increase in current density. Its subsequent variation and the variation of the current
strength then quickly become consistent with the regime of a high-
current AC arc powered by an LC circuit. Here too the halt in the
growth of current density is related mainly to reason 1a. There is
also a short transitional period of channel expansion between the
time when the dip ends and the conventional arc discharge becomes
established.

Entirely different processes take place when gas breakdown occurs
at negligibly small R_b and L and high capacitance C of the storage
capacitor. If L is small enough that the value of L di/dt is significantly lower than U_0 at the maximum di/dt associated with the rapid
primary growth of current density and the subsequent expansion of the
channel, then the voltage curve for the gap at the start of the high-
current stage has the shape illustrated graphically in Figure 2-1 by
segment abc of curves 2 and 3. If the discharge channel is unbounded,
then the voltage after the "dip" to level U_1 varies according to

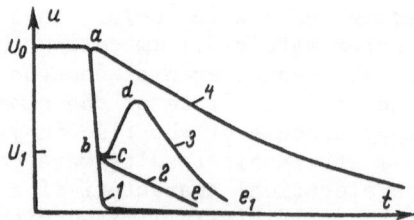

Fig. 2-1. Voltage curve for a pulsed discharge powered by a discharge
           circuit with high external resistance (curve 1) and with
           low external resistance and inductance and high capaci-
           tance (curve 2 is for an unbounded channel, and curve 3
           for a channel of limited diameter), and for a discharge
           with an artifically increased electrode spacing (curve 4).

segment bce of curve 2, which reflects the gradual discharge of the
capacitor as the current strength continues to increase for a significant period of time (with a slightly varying di/dt) because of
channel expansion. If the channel diameter is artificially restricted
at time c, then the current strength at this time also is "saturated"
at a current density which has previously reached some "saturation"
value. When this occurs, di/dt decreases to zero and the potential
across the gap rises to $U_0 - \frac{1}{C} \int i \, dt - R_b i$ (point d on curve 3; the
voltage rise may not be instantaneous since channel expansion may
come to a halt gradually, and moreover the increase in the potential
across the gap causes an additional increase in current density).
The voltage traverses segment de_1 as the subsequent quasi-stationary
discharge of the capacitor progresses. Thus, oscillograms of the
voltage during a discharge with an unbounded channel, having the
form of abce in the graph, attest to the possibility that the rise
in current density may not come to a halt solely as a result of insufficient power from the power supply (reason 1a), but additionally
for one of the other reasons listed above. However, oscillograms of
the type $abcde_1$ for a discharge with a channel of limited diameter
prove that the current density has a limit due to "self-saturation"
of the primary chain process described above. If such a voltage
curve were observed when the discharge region around one of the electrodes had a limited diameter (due to the small area of the open
portion of the electrode), this would indicate the existence of a
limiting current density in this region and the necessity of increasing the corresponding near-electrode potential drop in order to obtain high densities (reason 1b). The experimentally established
lack of influence of electrode-area limitation on the oscillograms,
as well as the appearance of oscillograms of type $abcde_1$ when the
discharge column is confined by the walls of the surrounding tube,
have proved that reasons 2 and 3 are the main factors in the "saturation" of current density following its rapid primary rise.

Figure 2-2 shows sample oscillograms made during open breakdown in air. The curve of the voltage u in this figure matches curve 2 in Figure 2-1, which pertains to a discharge in which saturation current density is reached during the potential "dip" with a significant voltage U_1 remaining in the discharge, and the current strength continues to increase after the "dip" with a nearly constant di/dt due to the ensuing expansion of the channel. The increase in the difference $U_0 - u$ after the maximum di/dt apparently is related to the capacitor discharge, which is already significant ($\frac{1}{C_0}\int_0^t i\ dt$ becomes comparable to L di/dt). According to experimental data, the potential U_1 that remains across the gap at the end of the "dip" increases with decreasing inductance of the discharge circuit, increasing U_0, and increasing electrode spacing (this dependence of U_1 on the parameters is in agreement with concepts of the limitation of current density and the mechanism of channel expansion, as will be evident from what follows). The limitation of the discharge column diameter in the air by the walls of the surrounding capillary leads to oscillograms of an entirely different kind, samples of which are shown in Figure 2-3. The voltage curve in this figure is in agreement with curve 3 in Figure 2-1, which relates to a discharge in which the limiting current density is reached after the "dip", in u, and the limiting current strength also is reached after the internal cross section of the capillary is filled by the channel. This is followed by a slowly varying ("quasi-stationary") discharge. The oscillograms of i and di/dt also have a corresponding form. As the capillary diameter decreases, the saddle bc (Figure 2-1) becomes narrower and the hump d becomes higher.

The discharge in tubular flashlamps filled with noble gases is analogous to a short spark discharge in air with a channel that is bounded by walls. In these lamps the narrow discharge channel formed by the auxiliary high-voltage pulse, with a current density that very quickly reaches "saturation," usually manages to fill the entire internal cross section of the discharge tube uniformly (with the same current density) within approximately 10^{-5} s. If the capacitance of the storage capacitor is sufficiently high, the channel expansion time is small compared with the remaining duration of the discharge, during which the characteristics of the channel change extremely slowly and the discharge may be considered quasi-stationary. Such a discharge has the following essential distinguishing features:

(1) It occurs in noble gases, which are characterized by the Ramsauer effect: a small cross section for scattering by atoms for electrons having velocities corresponding to a temperature of around 10,000 K. Because of this, the total cross section of ions begins to exceed that of atoms even when the degree of ionization is comparatively low ($x \approx 10^{-4}$). The scattering of electrons by ions rather than by atoms begins to predominate

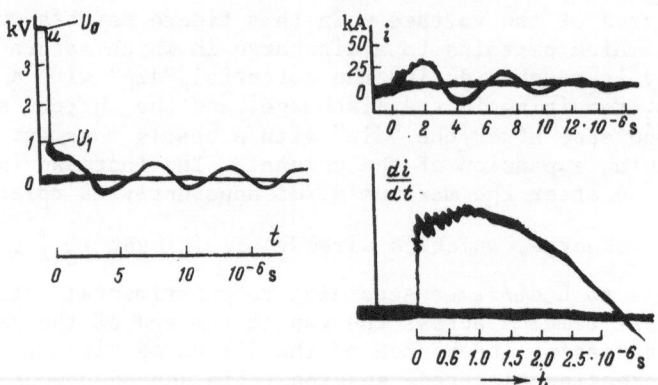

Fig. 2-2.   Sample oscillograms of u, i, and di/dt for a discharge
with an unbounded channel (air, p_0 = 0.1 MPa, l = 2.5 mm,
C = 6 μF, U_0 = 4.5 kV, L = 0.16 μH).

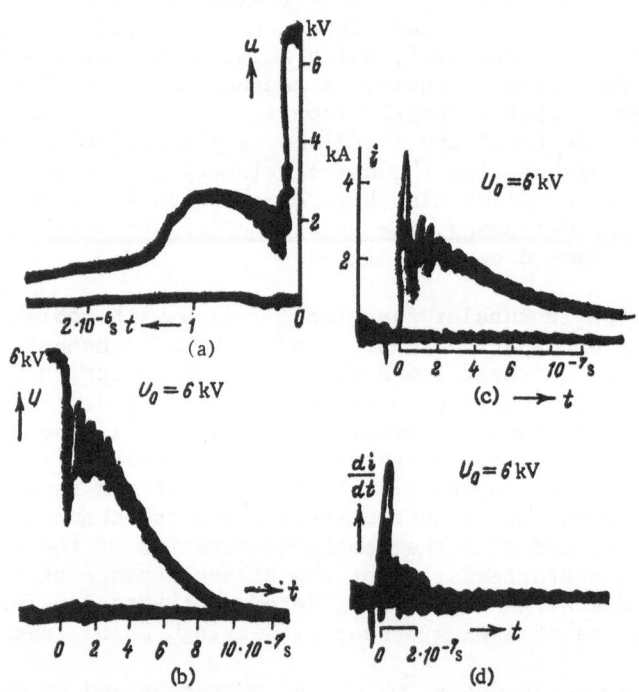

Fig. 2-3.   Sample oscillograms of u, i, and di/dt for a discharge
in a capillary (air, p_0 = 0.1 MPa, l = 4 mm).   a) C ≈
6 μF, U_0 = 6.8 kV, L = 0.16 μH, capillary diameter d_i =
0.6 mm; b-d) C = 0.25 μF, U_0 = 6 kV, L = 0.08 μH, d_i ≈
0.2 mm.

as x increases further. This should lead to termination of the
direct dependence of current density on the degree of ionization
when x $\approx 10^{-4}$ (reason 2 for current saturation following the
rapid primary rise in current density).

(2) The length of the discharge channel has been increased artificially by 1-2 orders of magnitude (compared with the spark discharge in molecular gases at the same supply voltages) in tubular
flashlamps through the use of triggering by means of an auxiliary
high-voltage pulse, which is especially effective in noble gases.
Thus, the initial longitudinal electric field strength is significantly reduced in these lamps (the voltage curve in the
discharge at small L is represented by curve 4 in Figure 2-1,
with a scarcely noticeable spur of the "dip" near point a).
The maximum current density and strength, which are limited by
reason 2, have lower values in this case. Thus, these reduced
saturation values can be achieved at not-so-low inductance and
resistance of the discharge circuit.

(3) The comparatively low electric field voltages (sic) and current
strengths correspond to the small amount of electric power dissipated per unit length of the discharge channel. This electric
power is so small that the channel can act on the walls of the
discharge tube for a considerable period of time without causing
the tube to fail. By increasing the capacitance, one can obtain
a quasi-stationary discharge time in tubular lamps which is
several orders of magnitude longer than that of a spark discharge
in an air-filled capillary.

The second of the features indicated above also characterizes
another kind of pulsed discharge in gases with an artificially elongated channel: discharge in the vapors of a fine metal wire exploded
by a current. Research on the dynamics of such a discharge has showed
that it is similar to the discharge in tubular flashlamps. For example, Toepler images made with an electrooptical shutter reveal a
fine, glowing current column which appears approximately 2 μs after
the explosion of a copper wire. By that time the highly charged
vapor of the metal fills a cylinder about 6 mm in diameter. Like
the walls of the tube, this cylinder is bordered by a dense shock
front. Approximately 3 μs later, the gas discharge column fills the
entire cylindrical cavity inside the shock front, whose diameter increases to about 8 mm. The current density in the discharge changes
slowly after quickly reaching some value, as is the case with tubular
flashlamps.

Figure 2-4 shows summary graphs which give the dependences of
the saturation current density and the corresponding conductivity
σ and resistivity ρ on the electric field strength in the pulsed
discharge column. As we see from the figure, the points for discharges in totally different media satisfactorily fit on common
graphs which can be represented by analytic expressions (in units V,
A, Ω, and siemens) over a very broad range of voltage and current
density:

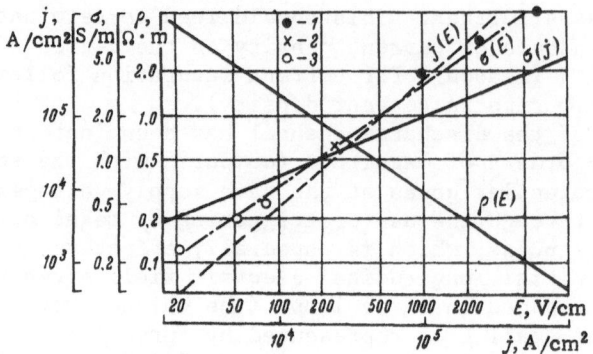

Fig. 2-4.   Dependences of current density j, plasma resistivity ρ,
            and plasma conductivity σ on electric field strength E,
            and the dependence of ρ on j. 1) σ(E) for a discharge in
            air bounded by a capillary; 2) for discharge in vapors of
            a current-exploded wire; 3) for discharge in tubular xenon
            flashlamps [2-1 to 2-5]. The lines represent empirical
            relations (2-1) and (2-2) (see Chapter 3 for more details).

$$\sigma = 3.17\,E^{2/3}, \quad \rho = 0.317\,E^{-2/3}, \quad j = 3.17\,E^{5/3}. \tag{2-1}$$

These correspond to a relation that is convenient for practical application:

$$\sigma = 2j^{0.4}. \tag{2-2}$$

To judge by the existence of common graphs, the fact that the
saturation of current density after the rapid primary rise is seemingly independent of the nature of the gas atoms enables us to select
reason 2 from the possible causes listed above. Reason 2 explains
the saturation in terms of the changeover from the scattering of
electrons by atoms (a changeover which is essentially different for
different atoms) to the scattering of electrons by ions, whose electric field is the same for all gases.

It should be noted that the electrical characteristics of the
pulsed discharge do not always correspond to relations (2-1) and
(2-2) by the end of the rapid primary rise in current density. Indeed, substantially higher plasma resistivities than would follow
from (2-1) were obtained when flashlamps were fed by limited power
supplies (e.g., by a capacitor with considerable series-connected
ballast resistance or by an inductive storage device) or by a low-
voltage AC grid, with triggering near the peak voltage (this is even
possible for a 127-V grid because of the features of noble gases
that were noted in Section 1-3) and extinction as the voltage passes
through zero, or with repeated triggerings in each half-period.
Here $\rho \sim E^{-3/2}$ is observed instead of $\rho \sim E^{-2/3}$, with a simultaneous
anomalously elevated sensitivity of ρ to small impurities of molecular

gases [2-6]. A calculation in [2-7] shows that even the heat removal
through the walls leads to significant constriction of the channel
at a power density of less than 2000 W/cm^2, ignoring other kinds of
energy dissipation. All these circumstances indicate that the cessation of the primary increase in current density that occurs when
the discharge is powered by limited power supplies is related to
reason 3: the lag of the increase in electric power behind the increase in concomitant energy losses. They also indicate that the
degree of plasma ionization in these cases is still insufficient for
scattering of electrons predominantly by ions rather than by atoms.
This determines the saturation of current density during primary
growth of a sufficiently powerful pulsed discharge.

2-2. The Theory of Discharge Expansion

     Channel expansion does not play the same part and on the whole
proceeds differently for the two types of pulsed discharge: discharge
bounded by walls, and discharge not so bounded. However, the onset
of expansion only differs quantitatively and may be considered jointly
for both.

     The concept of discharge channel expansion based on the concepts
of gas-dynamic explosion that were developed by Zel'dovich, Sedov,
and Raizer [2-8 and 2-9] has been expanded by Drabkina, Braginskii,
and others [2-10 to 2-12]. The purely gas-kinetic theory of Drabkina
is based on the simplified assumption that the channel expands as a
result of the significant amount of rapidly liberated energy in the
fine channel (about 0.1 mm in diameter) whose formation was described
in Chapter 1 and Section 2-1. The theory does not take into account
the parameters of the discharge circuit or certain parameters of the
gas-discharge plasma (e.g., plasma conductivity and radiation). Here
the assumption is made that the boundaries of the heated gas column*
in which thermodynamic equilibrium is established act on the surrounding gas like a cylindrical "piston" and move with supersonic velocity,
causing a shock wave to be formed ahead of the piston. The theory
ignores the diffusion of electrons and ions, heat conduction, and
convection as being incapable of explaining the supersonic speeds of
channel expansion, which requires a pressure jump on the order of
tens of atmospheres.

     This very simple theory outlines the correct qualitative and
quantitative picture of channel expansion. However, it does not include the other parameters that determine the radiation of the discharge. In particular, the assumptions made in the theory do not

---

* By the term "channel" we mean the entire region of disturbed gas,
  and by the term discharge "column" we mean the conducting region
  proper of highly ionized gas heated to high temperatures.

take into consideration the following facts:

(a) The possible distortion of the temperature and gas-density dis-
     tributions inside the column due to the nonuniform conductivity
     of different layers of the highly ionized gas.
(b) The analogous influence of the skin effect and magnetic pressure.
     Reference [2-11] shows that this influence is small in the range
     of parameters that are characteristic of discharges in flash-
     lamps, though the skin effect and magnetic pressure should have
     a significant effect at much lower initial gas densities and ex-
     tremely high electric field strengths and discharge energies
     and lengths for substantially larger column radii. This actually
     is observed in long lightning-type discharges and pulsed dis-
     charges used in purely laboratory-type facilities to produce
     superhigh temperatures or superpower light pulses [2-14 to
     2-18]. The particularly pronounced instability of the column
     during short discharges in hydrogen [2-19] also may be explained
     by the significant role of magnetic forces.
(c) The existence of types of energy transfer other than gas-dynamic
     expansion (formal allowance for the power carried off by radi-
     ation by equating the column with a black body of the same size
     [2-20] is unproductive). It is not hard to see that the types
     of energy transfer not included in the theory should help de-
     crease the large gradients it predicts for the temperature T
     and gas density δ inside the plasma column [2-12].

In order to fill these gaps in the theory, [2-11 and 2-12] made
approximate allowance for the conductivity of the discharge column
and introduced a radiation correction on the assumption of hydrogen-
like atoms, at the expense of complicating the self-similarity principle. The dependences obtained for the channel radius and the temperature on δ_0 and the discharge current, given an optically transparent plasma and assuming uniform distribution of the parameters T,
δ, and p (the pressure) in the channel, are in agreement with experiment to approximately the same extent as the results of a calculation
based on the gas-kinetic theory. Reference [2-21] considered the
possibility that the channel expands solely as a result of radial
diffusion of electrons and radiation, while [2-11] showed that heat
transfer can occur through heat conduction at comparatively low temperatures (low supply voltages or high inductance L) and through
radiation at high temperatures. The energy transfer in the hot part
of the channel is now ascribed to three mechanisms, depending on discharge conditions with allowances for the findings of [2-12 to 2-16,
2-22, and 2-23]: electron thermal conduction and radiative conduction (the heat conduction coefficients depend on T) and "radiative
transfer" in an optically thin plasma. The gas temperature may
change significantly in the last case at the mean free path of the
radiation quanta, in contrast to the first two mechanisms.

Papers dealing with the theory of the expansion of high-power

discharges in air that are initiated by the electrical explosion of
thin wires in which the discharge column is optically opaque hold a
special position [2-14 and 2-15]. In these papers allowance was
made for all types of energy transfer through an even greater compli-
cation of the self-similarity principle. The model treated in [2-14]
applies to discharges which occupy an intermediate position, in terms
of conditions of energy transfer, between an optically transparent
electric spark, in which expansion can be explained mainly within
the framework of gas dynamics, and a thick layer of air heated
quickly to a high temperature (around 300,000 K) at which the transfer of thermal energy to the peripheral layers proceeds at a speed
significantly higher than the velocity of gas particles, as determined by gas dynamics. Radiation absorption in the peripheral layers
becomes a significant factor when the temperature is comparatively
low (tens of thousands of degrees Kelvin) and the thickness of the
heated air column is sizable, so that both mechanisms of expansion
of the boundary of the plasma column are at work. This intermediate
case is called a "thermal wave of the second kind."

Finally, yet another model of the expansion of the discharge
channel (the model considered in [2-12]) is based on the fact that
the boundary of the highly ionized column expands during electrical
discharge like the detonation and deflagration front of fuel gases
[2-9]. It is believed that the motion of the boundary (or "thermal
wave") occurs in two stages, as it were: heating of the gas surrounding the column to a temperature of $10-15 \times 10^3$ K by the transfer
mechanisms indicated above (at this temperature there is an incremental increase in the conductivity σ to a value comparable to the
σ inside the column), followed by further heating of this layer by
the current to a temperature equal to that of the column. The analytic expressions obtained have been confirmed experimentally for
long discharges (tens of centimeters) in air at relatively low energies. This model might serve as a basis for the construction of a
general theory for the different types of discharges (including discharges in noble gases).

All theories that have been refined over the gas-kinetic theory
can be tested experimentally quite well by using discharges in air.
Such a comparison is complicated for the noble gases (and for mixtures of such gases with molecular gases) by the fact that some of
the physical constants are unknown. Thus, theory awaits further
refinement for the discharges that are used most extensively in
flashlamps.

It is advisable to interpret the expansion of discharges in
flashlamps within the framework of the gas-dynamic theory, which is
in satisfactory agreement with experiment, because of the lack of a
sufficiently complete theoretical mechanism applicable to such discharges.

The corresponding calculation uses the system of equations of continuity of motion and the adiabats for the case of cylindrical symmetry, as well as equations that characterize the boundary conditions: continuity of mass flow, momentum flux, and energy flux at the boundary of the disturbed region (the shock front):

$$\left.\begin{array}{l} \dfrac{\partial \delta}{\partial t} + v\,\dfrac{\partial \delta}{\partial r} + \delta\,\dfrac{\partial v}{\partial r} + \dfrac{\delta v}{r} = 0, \\[2mm] \dfrac{\partial v}{\partial t} + v\,\dfrac{\partial v}{\partial r} + \dfrac{1}{\delta}\,\dfrac{\partial p}{\partial r} = 0, \\[2mm] \left(\dfrac{\partial}{\partial t} + v\,\dfrac{\partial}{\partial r}\right)\dfrac{p}{\delta^{\gamma}} = 0, \end{array}\right\} \qquad (2\text{-}3)$$

$$\left.\begin{array}{l} \dfrac{\delta_0}{\delta_f} = \dfrac{\gamma-1}{\gamma+1}, \quad v_f = \dfrac{2}{\gamma+1}\,D, \\[2mm] p_f = \dfrac{2\delta_0}{\gamma+1}\,D^2, \end{array}\right\} \qquad (2\text{-}4)$$

where v is the gas velocity; r and t the cylindrical coordinate and the time; D = dR/dt the velocity of a shock front of radius R; γ the adiabatic constant; and subscripts "0" and "f" the initial value of a given parameter and its value in the shock front.

It is known from hydrodynamics that the solution of this system for the case of instantaneous energy release in a vanishingly thin column of an ideal gas has the property of self-similarity: the distribution of δ/δ_0, p/p_f, and v/v_f with respect to the dimensionless coordinate $\xi = r/R$ is stationary. The expression that relates R to the explosive energy has the form:

$$R = \left(\frac{\alpha W_0}{\delta_0}\right)^{0.25} t^{0.5}, \qquad (2\text{-}5)$$

where W_0 is the explosive energy per centimeter of column length, and α is a dimensionless constant that depends solely on γ and can be calculated from the energy integral.

We assume that the equation of state for a real gas in the temperature range of interest (10–30 ҡ 10^3 K) is satisfied approximately when expressed in the following form:

$$\varepsilon = A\delta^a T^b, \qquad (2\text{-}6)$$

where ε is the specific energy, and A, a, and b are numerical coefficients that can be calculated by the trial-and-error method so that equation (2-6) is obtained from the expression (for a diatomic gas) $\varepsilon = 3R_0 T(1 + x) + 2x\varepsilon_{ion} + \varepsilon_{dis}$ (where ε_{dis} and ε_{ion} are the dissociation and ionization energies of 1 mol of gas, R_0 the gas constant, and x the degree of ionization, as calculated from Saha's formula*). It can be reduced to the form

$$p = (\gamma - 1)\,\varepsilon\delta. \tag{2-7}$$

This expression formally matches the corresponding expression for an ideal gas, although here γ is simply an arbitrary quantity related to the exponents in equation (2-6) by the expression:

$$\gamma = 1 - \frac{a}{b-1} \tag{2-8}$$

(the numerical values of a, b, α, and γ are given in Table 2-1). Given such formal correspondence of the equations of state for real and ideal gases, the adiabatic equation for a real gas has the conventional form $p\delta^{-\gamma}$ = const. Hence the solution of system of equations (2-3) and (2-4) also possesses the property of self-similarity. Figure 2-5 gives the corresponding pressure and density distributions for air, for which γ = 1.22. In the region ε = r/R << 1 these distributions are expressed approximately by the following first terms of the rapidly converging expansions of the corresponding functions into series:

$$\left. \begin{aligned} \frac{\delta}{\delta_f} &\approx \left(\frac{\gamma}{2}\right)^{\frac{2}{2-\gamma}} \xi^{\frac{2}{\gamma-1}} \left[1 + \frac{2}{\gamma+1}\,\xi^{\frac{2\gamma}{\gamma-1}}\right], \\[2ex] \frac{p}{p_f} &\approx \frac{\gamma+1}{2\gamma}\left(\frac{\gamma}{2}\right)^{\frac{\gamma}{2-\gamma}} \left[1 + \frac{1}{\gamma+1}\,\xi^{\frac{2\gamma}{\gamma-1}}\right]. \end{aligned} \right\} \tag{2-9}$$

We can change over from instantaneous energy release in an infinitesimally narrow column to gradual energy release by transforming (2-5). To do this we use the proportionality between p_f and the average pressure, which in turn is proportional to $w(t)/R^2$ (where $w(t)$ is the energy released by time t). We find $w(t)/R^2$ = const $(dR/dt)^2$ by using the last equation of system (2-3). After extracting the root and integrating, we get the revised expression for R in place of (2-5);

$$R = K\left[\int_0^t w^{0.5}\,(t)\,dt\right]^{0.5} = \left(\frac{\alpha}{\delta_0}\right)^{0.25}\left[\int_0^t w^{0.5}\,(t)\,dt\right]^{0.5} \tag{2-10}$$

(we select the constant K so that (2-10) becomes (2-5) when there is instantaneous release).

_____

* There is some doubt as to the applicability of Saha's formula because of the possible absence of thermodynamic equilibrium.

Table 2-1. Numerical Values of the Coefficients in Formulas (2-4) to (2-15)

Gas	p_0, MPa	δ_0, kg/m³	α	A	α	b	γ	K	L	M	N
Air	0.1	1.29	0.55	$6.9 \cdot 10^4$	−0.122	1.55	1.22	4.55	0.19	0.125	0.376
Air	0.026	0.339	0.55	$6.9 \cdot 10^4$	−0.122	1.55	1.22	6.35	0.292	0.125	0.376
Air	0.3	3.87	0.55	$6.9 \cdot 10^4$	−0.122	1.55	1.22	3.46	0.133	0.125	0.376
Argon	0.1	1.78	0.332	1.35	−0.125	2.75	1.075	3.7	1.1	0.043	0.46
Hydrogen	0.1	0.0899	5.4	$4.33 \cdot 10^6$	−0.093	1.37	1.25	8.25	0.44	0.138	0.363

Note: $\alpha \approx 0.3$

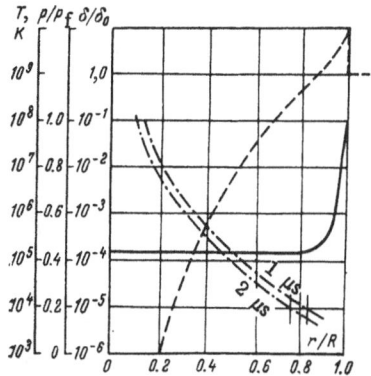

Fig. 2-5.   Calculated distributions of pressure (solid line), density
(dashed line), and temperature (dot-dash lines) in the
channel at t = 1 and 2 µs. Air, 0.1 MPa. Some 5 J of
energy is released in the column in 1 µs. The vertical
marks are the conditional boundaries of the column at the
corresponding times.

From (2-10) and the last equation of (2-4) we find:

$$D = \frac{dR}{dt} = \frac{K}{2} w^{0.5}(t) \left[\int_0^t w^{0.5}(t)\, dt \right]^{-0.5}, \tag{2-11}$$

$$p_f = \frac{(\alpha\delta_0)^{0.5}}{2(\gamma+1)} w(t) \left[\int_0^t w^{0.5}(t)\, dt \right]^{-1}. \tag{2-12}$$

By converting (2-6) and (2-7) to $T^b = p\delta^{-(a+1)}/(\gamma-1)A$ and
inserting (2-9), (2-10), and (2-12), we can find the temperature
distribution inside the channel:

$$T^b = \frac{1}{4A\gamma} \left(\frac{\gamma}{2} \right)^{\frac{\gamma-2a-2}{2-\gamma}} (\gamma+1)^{-a+1} (\gamma-1)^a\, \alpha^{\frac{\gamma+a}{2\gamma-1}} \times$$

$$\times \delta_0^{\frac{a-\gamma-2a\gamma}{2(\gamma-1)}}\, r^{\frac{2(a+1)}{\gamma-1}}\, w(t) \left[\int_0^t w^{0.5}(t)\, dt \right]^{\frac{2+a-\gamma}{\gamma-1}}. \tag{2-13}$$

Figure 2-5 gives the corresponding graphs of the temperature
distributions for typical discharge conditions at a time t = 1 µs
(immediately after the cessation of energy release) and 1 µs later.

We shall assume that the boundary of the highly ionized conducting column is characterized by the fact that the temperature inside it exceeds some minimum temperature T_{bound} which may be, for
example, 10^4 K (the specific value selected for T_{bound} is insignificant because of the very steep drop of T with r). Inserting this
value into (2-13), we can find the radius r_c of the conducting

column that corresponds to the given time. The derivative of r_c
with respect to time is equal to the rate of expansion of the column.
The following expressions are obtained from (2-13):

$$r_c(t) = L w^M(t) \left[\int\limits_0^t w^{0.5}(t)\, dt \right]^N , \qquad (2\text{-}14)$$

$$\frac{dr_c(t)}{dt} = L w^{M-1}(t) \left[\int\limits_0^t w^{0.5}(t)\, dt \right]^N \left\{ M\, \frac{dw(t)}{dt} + \right.$$

$$\left. + N w^{0.5}(t) \left[\int\limits_0^t w^{0.5}(t)\, dt \right]^{-1} \right\} , \qquad (2\text{-}15)$$

where M and N are coefficients expressed in terms of γ and a (they
are independent of the gas density), and L is expressed in terms of
γ, a, b, A, α, δ_0, and T_{bound}. The values of all the coefficients
are given for some gases in Table 2-1, which was taken from [2-10].

Finally, if we take into account that the condition $p_f \gg p_0$ is
no longer fulfilled in the later phase of channel expansion, then we
should use the following equations instead of the first and last
equations of (2-4):

$$\left. \begin{aligned} \frac{\delta_f}{\delta_0} &= \frac{(\gamma+1)\,p_f - (\gamma-1)\,p_0}{(\gamma-1)\,p_f + (\gamma+1)\,p_0} , \\[2mm] D^2 &= \frac{(\gamma+1)\,p_f - (\gamma-1)\,p_0}{2\delta_0} . \end{aligned} \right\} \qquad (2\text{-}16)$$

Equations (2-16) take into consideration the fact that the external pressure of the undisturbed phase may not be ignored in the
final phase. It follows from them that a shoulder should form
gradually on graphs of the drop in density and pressure from the
shock front toward the axis of the discharge. A large density gradient with sign opposite to the negative sign of $d\delta/dr$ at the leading
edge will be observed behind the shoulder.* In other words, the
gas region with elevated density should be separated by a jump in δ
from the inner conducting "hot" region (the column). The radius of
this jump zone, which is called the channel "shell," is larger than
that of the column as determined by expression (2-14).

Thus, the picture of the development of a powerful pulsed discharge, based on the simplified calculation given above, is as follows in the stage of the freely expanding channel:

---

* It was erroneously pointed out in [2-10] that the density should
  be approximately constant between the shock front and the zone of
  large positive $d\delta/dr$. Since T increases continuously from the
  front toward the discharge's axis, the pressure drop in this
  direction (Figure 2-5) should qualitatively correspond to the
  same decrease in gas density [2-24].

(a) The most rapid propagation is that of the outer zone of the channel: the shock front which forms the density and pressure jumps (from high values to values corresponding to an undisturbed medium). The boundary of a highly heated, highly ionized plasma (the conducting column proper) nearly coincides with the shock front at the very start of the discharge, but then is pinched away from it, expanding at a smaller velocity. A zone of an inverse density jump (called the "shell") that expands with an intermediate velocity should appear between the boundary of the column and the shock front as the pressure in the front decreases.

(b) The gas density inside the conducting column should be several orders of magnitude lower than δ_0, decreasing continuously toward the channel axis. Consequently, a zone with anomalous conductivity may be formed on the column's axis.

(c) The expansion of each of the zones occurs with gradually decreasing velocities whose initial value (which is common to all zones) increases slightly with increasing α (which characterizes the type of gas), decreasing initial gas density, increasing instantaneous power, and increasing total energy dissipated in the channel.

By using the formulas obtained and the dependence (2-1) of the conductivity on the electric field strength, we can make an approximate estimate of the initial rate of channel expansion and a rough calculation of the influence of some parameters of the gas-filled gap and power circuit on the initial rate. The voltage U_1 in the discharge immediately after the end of the rapid primary rise in current density also is calculated at the same time.

If we assume the instantaneous release of all energy W_0 (to occur) by the end of the effective time interval t_e (on the order of the time of the primary rise in current density), equation (2-11) can be used to obtain:

$$D = 158 \left(\frac{\alpha W_0}{\delta_0} \right)^{0.25} t_e^{-0.5}. \tag{2-17}$$

where D in in cm/s, δ_0 in kg/m^3, t in seconds, and W_0 in J/cm.

Assuming that the voltage U_1 in the discharge at time t_e after the end of the increase in current density is equal to the difference between the voltage U_0 across the capacitor and the voltage drop across the inductance $L(di/dt)_e$, and assuming

$$i = 3.17 \, \pi \, (U_1/l)^{5/3} \, r_c^2,$$
$$\frac{di}{dt} = 20 \left(\frac{U_1}{l} \right)^{5/3} r_c D,$$

according to (2-1), we find:

$$\frac{U_1}{l} = \frac{U_0}{l} - 20r_c \frac{L}{l} D\left(\frac{U_1}{l}\right)^{5/3} .$$ (2-18)

The following approximate expression may be inserted in formula (2-17):

$$W_0 = \frac{U_1}{l} i_e t_e \approx 10r_c^2 t_e \left(\frac{U_1}{l}\right)^{8/3}$$

(We ignore the radiation, whose effect is negligible when the channel diameter is small at the start of channel expansion [2-20].) From this we find a second equation, which relates U_1 to D:

$$D = 280 \left(\frac{\alpha}{\delta_0 t_e}\right)^{0.25} r_c^{1/2} \left(\frac{U_1}{l}\right)^{2/3} .$$ (2-19)

We can use this system of equations to find D and U_1/l for various given values of U_0/l and L/l by inserting the numerical values of the constants from Table 2-1 into (2-18) and (2-19), and assuming the actual values of the effective time and initial column radius to be $t_e = 5 \times 10^{-8}$ s and $r_c = 10^{-2}$ cm, respectively. The corresponding graphs are presented in Figure 2-6.

Figure 2-6 can be used to derive empirical formulas that are more convenient for practical estimates than system of equations (2-18) and (2-19). These can be used to summarize the experimental data* presented in the next section:

$$D_0 = 325 \left(\frac{\alpha l}{\delta_0 L}\right)^{0.25} \left(\frac{U_0}{l}\right)^{0.32} ,$$ (2-20)

$$\frac{U_1}{l} = 0.032 \left(\frac{L}{l}\right)^{-0.42} \left(\frac{U_0}{l}\right)^{0.5} ,$$ (2-21)

where D is expressed in cm/s, L in henrys, U_0 and U_1 in volts, l in cm, and δ_0 in kg/m^3.

This calculation of the dynamic characteristics of the discharge can be checked and a practical estimate of the other characteristics (including radiation characteristics) can be made on the basis of experimental work.

---

* To judge by the data of [2-2], expression (2-21) is in satisfactory agreement with experiment in the stage of initial channel expansion if the constant 0.07 is inserted instead of 0.032.

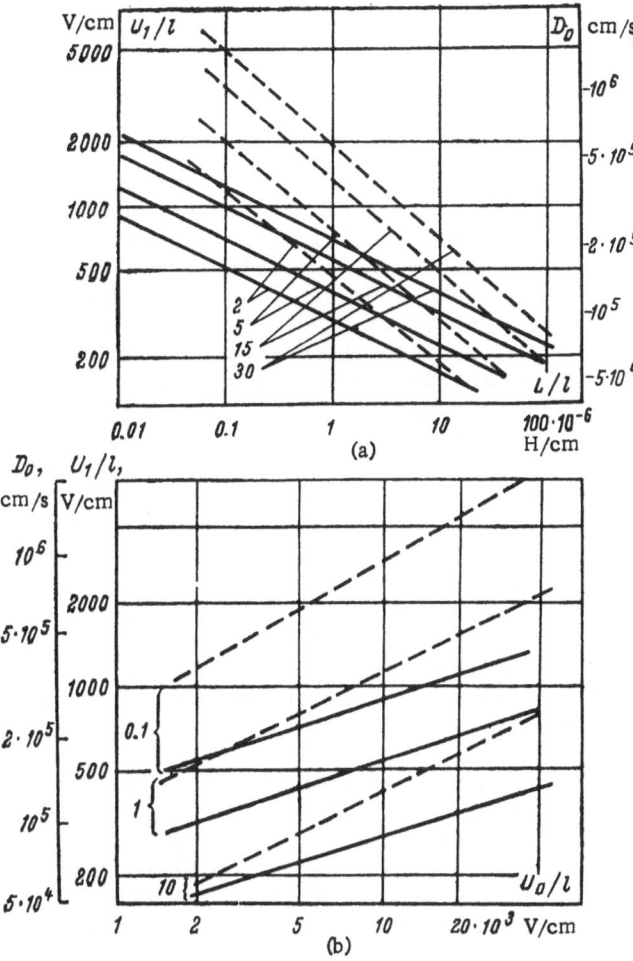

Fig. 2-6. Calculated dependences of the initial rate of channel
           expansion D_0 (solid lines) and the electric gradient
           U_1/l (dashed line) at the end of the rapid increase in
           the current density j on the inductance of the discharge
           circuit (a) and the voltage across the storage capacitor
           (b). Air, 0.1 MPa. The numbers indicate the correspond-
           ing values of U_0/l (kV/cm) in graph a, and the value of
           L/l (μH/cm) in graph b.

2-3. Channel Expansion in Tubular Lamps

        The initial and subsequent stages of the expansion of a dis-
charge confined by walls and the expansion of the channel of an unlimited discharge have been investigated by means of:

(a) photographic scanning (including the use of a spectrograph) of
    images of the channel cross section (as defined by a slit pos-
    itioned ahead of the channel);
(b) the same scans, with simultaneous intensification by the Toepler
    schlieren method, using a second spark in the space surrounding
    the channel;
(c) a scan on an image-converter screen of the image of the narrow
    slit positioned ahead of the channel and perpendicular to its
    axis, with simultaneous intensification and with spectrally
    selective filters, which make it possible to observe the radiation
    of the channel in various spectral regions, mounted between the
    slit and the image converter;
(d) rapid-framing photography of the discharge channel using an opto-
    electronic shutter, an electrooptical shutter, a motion-picture
    camera, or a scanning camera;
(e) high-speed interferometric investigation of the gas density in
    the channel and the surrounding region by placing the discharge
    under study in one of the arms of the interferometer and briefly
    illuminating it with a synchronized auxiliary discharge with a
    glow of very short duration;
(f) observation of the distribution of the spots formed on the elec-
    trodes by pulsed discharges of different durations;
(g) determination of the time dependence of the particle concentration
    and temperature from the change in the refractive index of a
    section of channel placed in one of the arms of a three-mirror
    laser interferometer.

    Channel expansion in tubular lamps has been investigated experi-
mentally by Klupo,[†] Leconte, and Edgerton and in a number of later
works for low capacitance of the storage capacitor (the duration of
the discharge slightly exceeds its expansion time). It was found in
the first of these studies that the hottest gap region, which usually
appears near the wall (at the location of the external trigger electrode) rapidly expands until it fills practically the entire cross
section of the tube. The initial rate of expansion was estimated as
6×10^4 cm/s for low circuit inductance, a lamp outside diameter of
3.5 mm, an initial xenon pressure $p_0 = 0.04$ MPa (300 mm Hg), and an
initial electric field strength $E_0 \approx 200$ V/cm. This rate increased
with increasing E_0 and decreased p_0. To judge by the original data,
it obeyed the following law (where the velocity is in cm/s, and E_0/p_0
is in V/(cm x MPa)):

$$\frac{dr}{dt} \approx 12 \frac{E_0}{p_0} . \qquad (2\text{-}22)$$

    The velocity increases somewhat as the atomic number of the gas
decreases (by approximately 20% as we go from xenon to krypton and
from krypton to argon). The channel front nearly reaches the opposite
wall of the tube approximately 5 μs after triggering of the discharge,
pressing against the wall the region of small cross-sectional area

(about 10% the cross section of the tube) of cold (nonglowing) gas compressed to high pressure. This region then alternately expands with a period of approximately 2.5 μs in such a way that the channel cross section decreases (by about 20%) and once again decreases until the fluctuating boundary between the hot and cold regions is gradually erased by the intermixing of the gas.

At low pulse energies the discharge does not at all fill the cavity of the tube, and the glow column follows the bends of the external (spiral) trigger electrode, remaining pressed against the wall [2-25 and 2-26]. An analysis of the photographs of a discharge that are presented in [2-26] shows that under these conditions the maximum attainable column diameter d_{c-max} is independent of the diameter of the discharge tube and is determined by the pulse energy per unit length of the discharge. For an initial xenon pressure p_0 = 0.04 MPa (300 mm Hg), this relation is expressed as

$$d_{c-max} \sim \left(\frac{CU_0^2}{2l}\right)^{0.4}. \qquad (2-23)$$

The nature of the development of the channel remains similar to the one described in its general features, even for discharges under the conditions that are usually used in tubular flashlamps (with large storage capacitors and substantially lower strengths). The study of such discharges also has detected the appreciable influence of the method used to trigger the lamp on the development of the discharge [2-27 and 2-28]: when the external trigger electrode is located along the forming tube, a straight channel appears right at the wall along this electrode. The cold gas region compresses the hot plasma much more strongly and mixes with it several times more slowly after the shock front is reflected from the opposite wall than in the case of the spiral trigger electrode that is usually used. For example, plasma completely fills the tube in about 140 μs when a spiral trigger electrode is used in a lamp 19 mm in diameter and 320 mm long (the xenon pressure is 0.04 MPa, capacitance 2100 μF, supply voltage 2.4 kV, and the inductance tens of microhenrys), whereas about 30% of the cross section of the tube still remains unfilled after 250 μs when a straight trigger electrode is used. An analogous process of compression and subsequent expansion of the cold gas also is observed in the direction of the tube's axis. Longitudinal compression and expansion waves propagate with approximately the same velocities as transverse waves (several hundred meters per second [2-29 and 2-30], causing short-duration longitudinal inhomogeneities of the channel luminance (an increase in luminance begins several microseconds after peak compression of the plasma). The duration of the longitudinal plasma oscillations depends on the tube's length and substantially exceeds the duration of transverse oscillations.

The picture of channel expansion - channel velocity - velocity, which depends on many factors (such as method of triggering, power-

supply parameters, the size of the lamp, and how and with what the
lamp is filled, and the frequency of fluctuations of the compression
waves, luminance and temperature distributions, and others [2-27,
2-28, and 2-31 to 2-33]), is in accord with the concepts of gas dynamics. For this reason it is possible, for example, to estimate
the plasma temperature from the propagation velocity of sound waves
in the plasma. For instance, a value of 940 m/s, corresponding to
a plasma temperature of 8400 K, was obtained in [2-29] for the speed
of sound in the plasma at a predischarge electric field strength of
120 V/cm and considerable inductance of the discharge circuit (82 µH).
The radial expansion velocity under such "mild" conditions is close
to the speed of sound in xenon at room temperature (177 m/s). By
contrast, under "severe" conditions the plasma acquires a fine structure (filamentary structure) in some cases. For example, "two-phase"
combustion of the arc was observed in [2-33] because of instability
at an atomic density of about 10^{18} cm^{-3} and a field strength of 150
V/cm in a cavity with a rectangular cross section, and the average
current density increased with E as a result of the increase in the
number of filaments of elevated luminance whose temperature exceeded
the background temperature by approximately 15%. As we can see from
Figure 2-7, the shock wave reflected from the opposite wall cuts the
glow column into two semicylinders in a straight tube with an external
trigger electrode located along the generatrix. The motion of different channel fronts is illustrated by the schematic diagram in
Figure 2-8 for this case with comparatively mild conditions. It can
be seen from the schematic diagram that the average expansion velocity of the column is lower by a factor of approximately 1.5 than
the shock velocity. (Both velocities remain more or less constant
during the first passage of the shock wave if the discharge energy
is a comparatively high 15 J/cm). The conducting column initially
expands with a velocity that depends solely on the electrical energy
introduced, $w(t) = \int_0^t iu_l dt$, which is independent of the tube's inside
diameter d_i, according to [2-26], in which oscillograph records were
made of the time dependence of the current strength i and the voltage
on the lamp u_l (xenon, 0.04 MPa), and the time dependence of the diameter d_c of the conducting column was determined from a plasma conductivity formula similar to (2-2). After $d_{cr} = 0.7d_i^{1.15}$ is reached,
the column expands much more slowly (Figure 2-9). The time dependence
of $d_c = 2r_c$ is described in these two periods by the respective
formulas:

$$d_c(t) = 1.5\,(w(t)/l)^{0.6}\,, \tag{2-24}$$

$$d_c(t) = 0.77d_l\,(w(t)/l)^{0.077}, \tag{2-25}$$

The critical concentration by volume of the energy released in the
channel (J/cm^3) at which channel expansion begins to be restrained
by the influence of the walls is

$$4W_{cr}\,/\pi d_i^2\,l = 0.36d_l^{-0.09}. \tag{2-26}$$

Fig. 2-7. Framing photography of a glow column cut off by a shock
           wave reflected from the tube wall [2-28]. Xenon, 0.04
           MPa, $U_0 = 3\,kV$, $C = 100\,\mu F$, $L \approx 20\,\mu H$, $l = 340$ mm, $d_i =$
           19 mm; the time interval between frames is 32 μs.

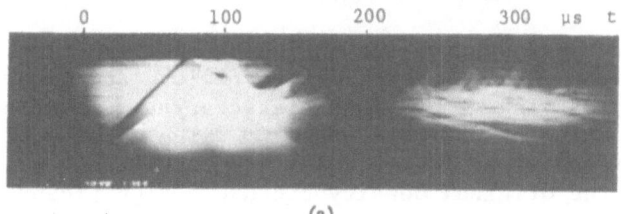

(a)

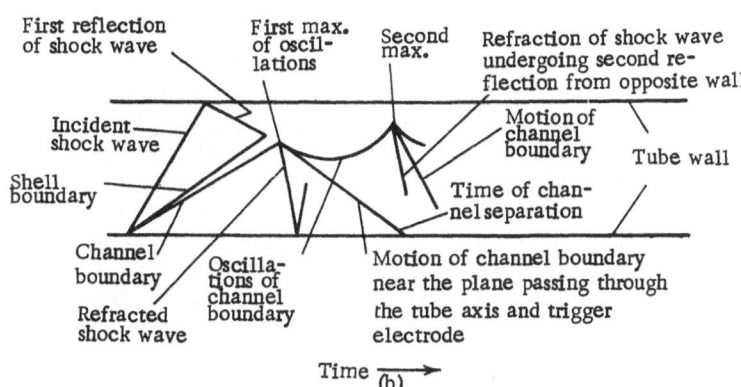

Fig. 2-8. Dynamics of discharge development (for shock waves and
           glow column) [2-28]. a) Photoscan; b) schematic repre-
           sentation of the process.

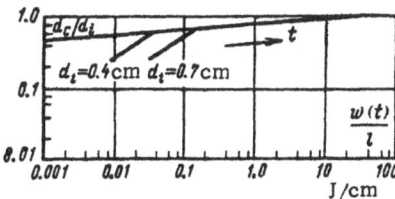

Fig. 2-9. Dependence of the glow column diameter $d_c(t)$ (for two
           tube inside diameters d_i) on the energy $(w(t)/l)$ intro-
           duced into the discharge by time t [2-26].

   Figure 2-10 summarizes the experimental data whose conditions
of acquisition are given in Table 2-2, with respect to the dependence
of the free expansion velocity of the channel (the shock wave and
column) on its length, in comparison with a calculated graph corresponding to expression (2-20). As we see from the figure, the
values of D_0 exceed the corresponding $(dr_c/dt)_0$ by approximately 40%.
Also plotted here is a graph of the averaged experimental data with
respect to the column expansion velocities. The graph corresponds
to a formula that differs slightly from expression (2-20):

$$\left(\frac{dr_c}{dt}\right)_0 = 390\left(\frac{\alpha l}{\delta_0 L}\right)^{0.25}\frac{U_0^{0.32}}{l^{0.5}}. \qquad (2-27)$$

   Taking into consideration the influence of triggering conditions,
optical distortions due to the small tube radius, and other inac-
curacies, we may say that there is satisfactory agreement between
the gas-dynamic calculation and experiment. The density of heavy
particles (atoms and ions) in the column is significantly reduced in
comparison with the original density because a significant fraction
of the gas is pressed against the walls during expansion. For the
low energy concentrations and short durations for which d_c remains
much smaller than d_i until the end of the discharge, the entire discharge cycle occurs at low particle density. To some extent this
determines the low radiative efficiency of such discharges. However,
a significant fraction of the gas escapes to cold regions beyond the
electrodes. The particle density by the end of the pulse n_∞ always
remains lower than the initial density n_0 (usually by a factor of
2-3 for lamps with $l < 10$ cm), even for high energy concentrations
that greatly exceed the critical concentrations (as determined from
(2-26)), and also for long pulse lengths (about 10^{-3} s) [2-5, 2-34,
and 2-35]. The gas escape velocity v_{es} into the cold regions of the
lamp beyond the electrodes is about 5×10^4 cm/s, corresponding to
the speed of sound at the gas temperature, averaged over the volume
of the channel. From this we can estimate the gas-dynamic equilibration time τ_{eq}, which should essentially depend on the length l
of the discharge gap [2-36]:

$$\tau_{eq} = \frac{l \cdot (1 - n_\infty/n_0)}{2v_{eq}}, \qquad (2-28)$$

where n_∞ is the density of heavy particles as $t \to \infty$.

   The ratio of the transelectrode volume V_e to the working volume
V_l lies in the range 0.04-1 for the usual designs of tubular flashlamps. The following approximate expression is valid for this
range [2-35]:

$$\frac{n_\infty}{n_0} = 0.65\frac{1 + V_e/V_l}{1 + 4V_e/V_l} \qquad (2-29)$$

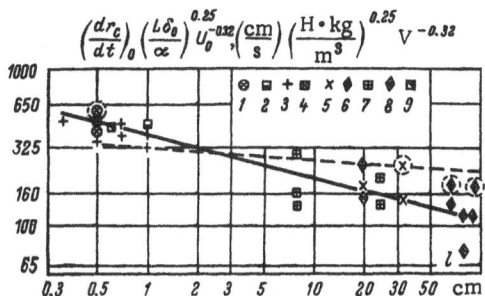

Fig. 2-10. Dependence of the channel expansion velocity on the type
and density of the gas, the supply circuit inductance,
and U_0 and l (composite data). For the meanings of the
symbols see Table 2-2 (a dashed line encircles the values
of the initial shock velocity, while the column front
velocity is not encircled). The dashed line is the graph
calculated from (2-20), and the solid line is the averaged
experimental graph corresponding to (2-27).

(the "arcing" of the discharge column at high pressures - e.g., 0.08
MPa for $d_i = 0.7$ cm - that is mentioned in [2-37] may have an analogous effect on the increase in V_e).

It follows from these estimates that the gas-dynamic equilibration time may reach 10^{-3} s for large l (over 100 cm) and that the
maximum pressure that can be attained in the second half of the flash,
according to the data of [2-37], should be determined by the relation
between τ_{eq} and the pulse length τ.

[Reference 2-36] derived a formula for determining the heavy-
particle density n_τ for a discharge time τ based on the approximate
estimate $v_{eq} \approx 5 \times 10^4$ cm/s from (2-28):

$$\frac{n_\tau}{n_0} = 1 - \left(1 - \frac{n_\infty}{n_0}\right) \exp\left[-\frac{3 \cdot 10^{-5} \, l(1 - n_\infty/n_0)}{\tau}\right]. \qquad (2-30)$$

This formula is in agreement with data on the tube-breaking pressures
reached in studies of the load characteristics of flashlamps
(Chapter 6).

2-4. Free Expansion of the Discharge

Most experimental data confirm, as a first approximation, the
correctness of the picture of the movement of the channel boundaries
that follows from the gas-dynamic theory, as well as the possibility
of using approximate formulas (2-20) and (2-27). These formulas

Table 2-2. Conditions of Measurement of the Initial Column Expansion Velocity in Different Papers

Item in Ref. 2-10	Ref. cited	Method of measurement	Gas	p_0, MPa	δ_0, kg/m³	l, cm	L, µH	U_0, kV	c, µF	$\frac{CU_0^2}{2}$, J
1	2-20	Shadow photoscan	Air	0.1	1.3	0.5	2-64	15-20	0.003-0.25	0.4-28
2	2-40	Photoscan	Ar	0.1-0.48	1.78-8.5	1.0	0.6	2-7	1.1	2.2-13.6
3	2-41	Scanning camera	Xe	0.3	17.5	0.3-1.0	0.2-0.02	2.5-7	0.5	1.5-12.5
4	2-41	Same	Ar	0.3	6	0.5	0.2-0.02	4.5-6	0.5	5-9
5	2-27*, 2-28*	Photoscan	Xe	0.04	2.3	20-34	20	3	100	450
6	2-14 – 2-16	»	Air	0.1	1.3	20-100	1	20-50	30-300	10^3-10^5
7	2-41*	»	Xe	0.01-0.04	0.6-2.3	8-25	5-96	1-2	600-1000	800-2000
8	2-41*	»	Air	0.001	0.13	20	2	2-10	50-100	100-1250
9	2-41	Scanning camera	Xe+30% N_2	0.09	4.1	0.7	0.2-0.02	5-4.5	0.5	5

* Discharge in sealed quartz tube.

were derived on the basis of theory and relate the initial expansion
velocity to the discharge parameters. The observed difference between
the average velocities of the channel and column boundaries does not
exceed 50%. Reference [2-16], which is devoted to high-power discharges in air, relates this difference to the initial conditions
under which the discharge develops. It remains minimal (10%) under
experimental conditions (l = 40 cm, W_0/l = 1.2 x 10^3 J/cm, a pulse
length of about 70 µs) with a current rise steepness (which is determined approximately by the expression $I_p/\sqrt{LC}$, where I_p is the peak
current strength) of over 1.4 x 10^{10} A/s, reaching 60% for $I_p/\sqrt{LC}$
= 0.2 x 10^{10} A/s (for spherical flashlamps $I_p/\sqrt{LC}$ usually lies in
the range 10^{10}-10^{11} A/s, and for tubular lamps it does not exceed
10^8 A/s).

At the same time, various data on the plasma column parameters
do not confirm the existence of the gradients of density and temperature that follow from the gas-dynamic theory (see Figure 2-5). This
means that the characteristics of the column cannot be calculated
without allowing for processes which are not taken into account by
this theory.

The expanding boundaries of the shock front, the shell, and the
conducting column can be seen in Figure 2-11. Such photographs enabled Gegechkori, Mandel'shtam, Dolgov, Vanyukov, and others [2-20,
2-24, 2-38, and 2-39] to note the following qualitative peculiarities
of the expansion of the shock wave and the "shell" which agree with
the gas-dynamic theory and at the same time supplement its conclusions:

(a) the shell is not a shock front, since its velocity may be either
    higher or lower than that of sound;
(b) the shell is destroyed a few tens of microseconds after the
    initiation of discharge because by that time the gas pressure
    on axis, which is 0.5 the pressure in the front, becomes lower
    than the initial pressure, and a flow of gas toward the channel
    axis commences.
(c) secondary shock waves due to repeated energy releases in the
    column during each half-period of the current are observed in a
    damped oscillatory discharge. These waves seemingly originate
    from the shell, as they are invisible inside the shell because
    of the low gas density and the high shock velocity;
(d) inside the shell there are a comparatively cold zone with a low
    (degree of) ionization of the gas and an extremely hot, highly
    ionized zone (the discharge column) whose expansion slows sharply
    at the time of the first current peak. (Obviously, the shock
    wave continues to be supplied with the energy being liberated
    in the channel after the shock wave separates from the column
    because of small disturbances whose propagation velocity exceeds
    the shock velocity D). The less-heated zone has a sharp external
    boundary (which coincides with the shell). It emits mainly arc-

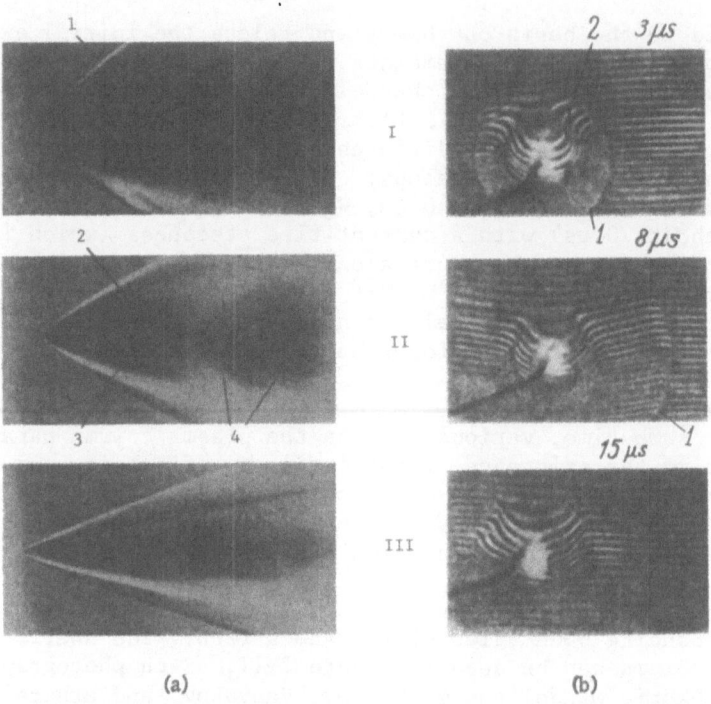

Fig. 2-11.   Sample photographic scans (a) and interferograms (b)
             obtained by the Toepler method [2-20 and 2-24].   1)
             Shock front, 2) shell, 3) discharge column, 4) secondary
             shock waves.   Air, 0.1 MPa, 0.25 µF.   a) U_0 = 15 kV,
             I) L = 2 µH; II) L = 12 µH; III) L = 64 µH; b) U_0 = 10
             kV, L = 2 µH, R_b = 6 Ω.   The pictures were taken along
             the channel 3, 8, and 15 µs after the start of discharge.

type infrared spectral lines as a purely surface-type emitter
(its luminance is the same over the entire cross section and
does not vary with current oscillations).   Its expansion continues
for several periods of current oscillations.   The boundary between
the less-hot and the high-temperature zones appears blurred
(this is explained by either the decrease in current density at
the edge of the column or the considerable depth of the glow
layer).   The hot zone emits mainly a visible and shortwave continuous spectrum that corresponds to a blackbody temperature of
$20-50 \times 10^3$ K, depending on the concentration by volume of the
electric power and on the type of gas.   The radiation intensity
of the hot zone increases appreciably at times of maximum current.
The expansion of the hot zone (the column) ends after one or two
current oscillations;
(e) extinction and an increase in the luminance of individual regions
    of the column are observed with local contractions in these

regions which are due, for example, to the focusing of reflected waves from the walls of a spherical discharge bulb, the super-position of a shock wave travelling from an auxiliary discharge, or the action of magnetic forces after a decrease in gas density.

Measurements of the shock velocity under various conditions make possible a quantitative comparison of the conclusions of gas-dynamic theory with experiment. Figure 2-12 shows graphs of the time dependence of the shock velocity in air at atmospheric pressure for different discharge conditions, and Figure 2-13 shows similar graphs for various gases and for air at different pressures. The dashed line in Figure 2-13 shows some of the corresponding graphs as calculated from the formulas of gas-dynamic theory by using data on the electric power. Figure 2-14 shows the composite data of various authors from measurements of the initial expansion velocity of the discharge column. This velocity practically matches the initial shock velocity. These data were obtained by different methods under the conditions characterized in Table 2-3. They have been combined into a single graph with allowance for relation (2-20), in which the inductance (referred to a segment 1 cm long) is taken as the inde-pendent variable. The dashed line in the same figure indicates the curve of relation (2-20), and the solid line shows the analogous relation with an altered coefficient (440 instead of 325). This re-lation may be considered a rough average of the experimental data.

Despite the large spread of the points with respect to the cal-culated graph, which is natural because of the different measuring methods and the approximate derivation of formula (2-20), the form of Figure 2-14, like the graphs of the increase with time in the column diameter for different discharges that are shown in Figure 2-15, allow us to assume that the dependences of D on various para-meters that are outlined by theory are satisfactorily confirmed both qualitatively and quantitatively in extensive experimental data (this looks even more convincing if we examine the distribution of the experimental points obtained by a given investigator for a single gas).

Figures 2-12 and 2-13 show in exactly the same way that:

(a) according to gas-dynamic theory, the decrease in the capacitance of the storage capacitor does not affect the initial expansion velocity and shows up only in a faster decrease in D with time (see Figure 2-12, curves 1 and 4);

(b) the general appearance of the experimental and calculated graphs is close enough (especially with allowance for radiation) for us to say that there is satisfactory agreement between the approxi-mate theoretical formulas and experiment.

In order to use the data presented in approximate calculations, it also is useful to note the following regularities, which stem

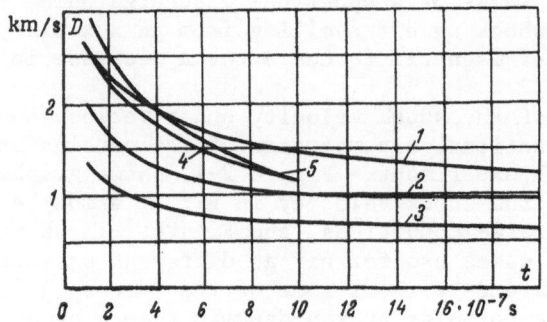

Fig. 2-12. Relation of time dependence of D to supply parameters.
Air, 0.1 MPa. 1) $L = 2$ μH; 2) $L = 12$ μH; 3) $L = 64$ μH
(for all three curves $C = 0.25$ μF, $U_0 = 15$ kV); 4) $U_0 =$
15 kV; 5) $U_0 = 20$ kV (for both curves $L = 2$ μH; $C = 0.01$
μF) [2-20].

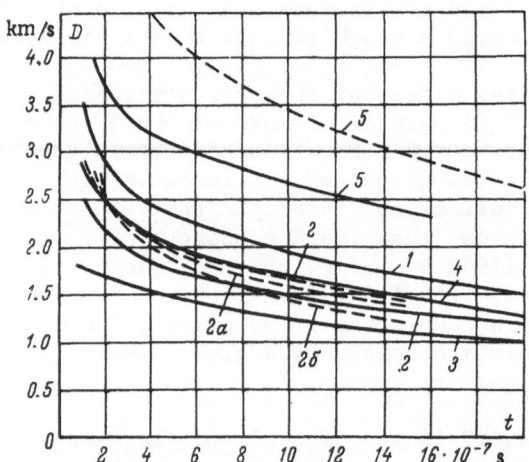

Fig. 2-13. Relation of time dependence of D to type and pressure
of gas. The solid lines represent experimental data,
and the dashed lines are calculated. $L = 2$ μH, $C = 0.25$
μF, $U_0 = 15$ kV. 1) $p_0 = 0.026$ MPa; 2) $p_0 = 0.1$ MPa;
3) $p_0 = 0.3$ MPa (these curves are all for air); 4) Ar;
5) H_2 (for both curves $p_0 = 0.1$ MPa; curves 2a and 2b
were calculated with allowance for losses to radiation
from a blackbody with dimensions equal to those of the
column at temperatures of 15×10^3 and 20×10^3 K [2-20].

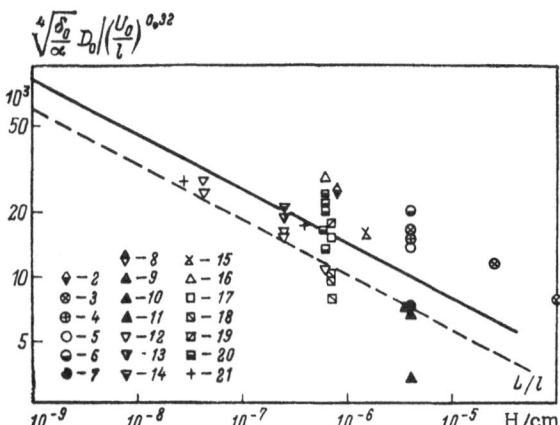

Fig. 2-14. Dependence of initial velocity of channel expansion D
on the type and density of gas, supply circuit inductance,
l, and U_0 (composite data). For the meanings of the
symbols see Table 2-3. The dashed line represents a
graph calculated from (2-20), and the solid line is an
averaged rough experimental plot.

from Figures 2-11a and 2-15. The shock velocity remains practically
constant (the diameter of the wave is proportional to t) when the
storage capacitors have comparatively high capacitances ($C \geq 1$ µF)
during the release of most of the electrical energy (up to half the
first half-period of the current), and the expansion velocity of the
highly ionized discharge column decreases slightly (for discharge
gaps about 1 cm long, r_c is proportional to $t^{0.7}$). After the energy
release ends (for practical purposes, when $t \simeq 2.6\sqrt{LC}$), damping of
the shock velocity occurs in accordance with gas-dynamic theory (the
diameter of the shock wave becomes proportional to $t^{0.5}$). The end
of column expansion (the attainment of r_{max}) occurs by the time $t =$
$(4-5)\sqrt{LC}$.

Another kind of quantitative test of the theoretical calculation
of the shock velocity D was carried out in [2-24] by determining the
change in the refractive index and hence in the density from the
shift of the interference bands of an interferometer in one of whose
arms there was a discharge gap. (Measurements were made in the less-
heated zone near the front, for which we may use the usual relation
between the refractive index and gas density, ignoring ionization
and dissociation.) The experimental values of δ_f (Figure 2-16) can
be converted to the corresponding values of p_f and D by using formulas (2-16). For example, the highest value of δ_f that could be
found for the accuracy of the method at $t < 1$ was $8\delta_0$ (the maximum
theoretical value for $p_f \gg p_0$ for air is $\delta_f = 10.1\delta_0$). The corresponding values of the pressure and shock velocity are $p_f = 3.8$

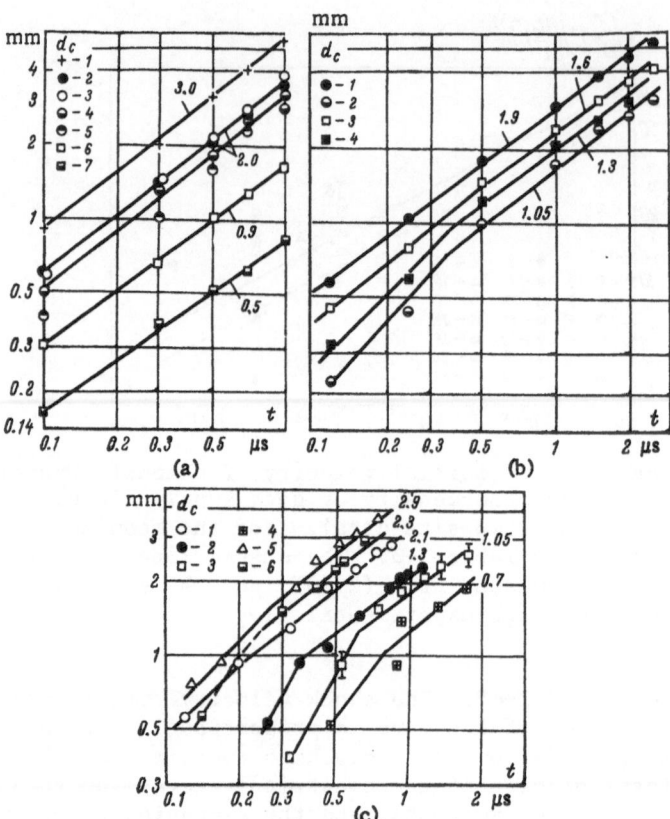

Fig. 2-15.   Expansion of the diameter of the glow column ($l \leq 1$ cm,
$w_0/l \approx 10$ J/cm).   The numbers on the graphs indicate
$(dr_c/dt)_0$, km/s.
a)   [Reference 2-20], 15 kV, 0.25 μF, $l = 5$ mm; 1) H_2,
0.1 MPa; 2) Ar, 0.1 MPa; 3) air, 0.025 MPa; 4) air, 0.1
MPa; 5) air, 0.3 MPa; 1-5) 2 μH; 6) air, 0.1 MPa, 12 μH;
7) air, 0.1 MPa, 64 μH.
b)   [Reference 2-40], 1.09 μF, 0.6 μH, $l = 10$ mm, argon;
1) 0.1 MPa, 5 kV; 2) 0.1 MPa, 2 kV; 3) 0.48 MPa, 7 kV;
4) 0.48 MPa, 5 kV.
c)   [Reference 2-41], 0.5 μF, $l = 5$ mm; 1) Ar, 0.32 MPa,
0.02 μH; 2) Ar, 0.3 MPa, 0.2 μH; 3, 4) Xe, 0.3 MPa, 0.2
μH; 5) Ar + 30% N_2, 0.1 MPa, 0.02 μH; 6) Xe + 30% N_2,
0.091 MPa, 0.02 μH; 1-3, 5, 6) 4.5 kV; 4) 2.5 kV.

MPa and D = 1.8 km/s.   The results of such a calculation of the values
of D based on the results of interferometric measurements of different
times are compared in Table 2-4 with direct measurements of D using
mirror scanning and with the values of D calculated from formula
(2-11) on the basis of the measured electric power of the discharge
(without allowance for radiation).

Table 2.3. Conditions of Measurements, by Various Workers, of the Channel Expansion Rate and of the Maximum Column Radius r_{max} (1–20, see [2.1], 21 — [2.41])

Labels on Figs. 2-14 and 2-18a	Measurement method	Gas	p, MPa	δ, kg/m³	l, mm	L, μH	U, kV	C, μF	$\frac{cv_0^2}{2}$, J
1	ICMPP*	Air	0.1	1.3	8.4	0.01	20	0.005	1
2	Photoscanning with SP†	Ar	0.44	7.8	2	0.15	6	0.57	10.2
3		Air	0.1	1.3	5	2—64	15—20	0.0035	0.4—0.28
4	Ditto	»	0.026	0.35	5	2	15	0.25	28
5	»	»	0.3	3.9	5	2	15	0.25	28
6	»	Ar	0.1	1.78	5	2	14	0.25	28
7	»	H₂	1.1—1.7	0.09	7—11.2	0.36	10—14	0.25	28
8	Photoscanning	Ar	0.1	19.5—30	11	3.5	10—35	0.1—0.2	5—20
9	»	Air	0.033	1.3	11	3.6	11	0.06	3—40
10	»	»	0.1	0.43		3.6	17	0.06	3.6
11	»	H₂	0.1	0.09		0.02—0.3	5—7	0.06	8.7
12	Raster camera	Xe+H₂	0.3	15	5	0.12	7	0.005—0.1	0.05—2.5
13	Ditto	Kr+H₂	0.3	10	5	0.12	7	0.1	2.5
14	»	Ar+H₂	0.3	4	2.3	0.37	14	0.1	2.5
15	ICS‡	Air	0.1	1.3	6	0.35	24	0.1	10
16	ICS‡	Xe	0.4	23	10	0.66	8	4	29
17	Photoscanning	Kr	0.2	8	10	0.65	4—8	0.1	128
18	»	Kr	0.4	16	10	0.65	8	4	0.8—3.2
19	»	Kr	0.6	24		0.6	2—7	0.1	128
20	»	Ar	0.1—0.48	1.78—8.5		0.2—0.02		1.1	2.2—13.6
21	Raster camera	Xe	0.3	17.5	5—7		4.5	0.5	5

*ICMPP — Image-converter motion-picture photography.
†SP — Shadow photography.
‡ICS — Image-coverter scanning.

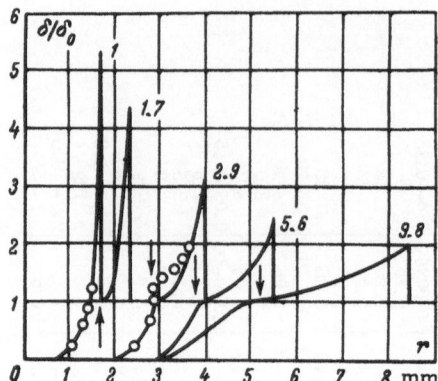

Fig. 2-16.   Graphs of the experimental radial distribution of air
            density (solid lines) and the calculated radial distri-
            bution (circles) at different times (in microseconds);
            10 kV, 0.25 μF, 2 μH, 0.1 MPa [2-24]. The arrows indi-
            cate the positions of the shell.

This table also indicates the satisfactory agreement between
the approximate theory and experiment.

Finally, data like those presented in Figure 2-16 for the radial
distribution of gas at different times are comparable to the corresponding theoretical data. The circles in Figure 2-16 indicate
the results of a theoretical calculation of the behavior of δ at
time t = 1 μs (using the first formula in (2-9)) and at t = 2.9 μs
(using the second) on the assumption that the usual equation of
state $p = R_0 \delta T$ is valid for the less-heated zone and that the temperature in this zone is constant. This check also satisfactorily confirms the methods, based on gas-dynamic theory, used to calculate
phenomena in the shock front.

It is considerably more difficult to compare theory with data
on the high-temperature zone, i.e., the discharge column. As indicated above, the column's boundary is poorly defined after it separ-
ates from the front. Hence the data of various investigators regarding the late stages of column expansion are in substantial con-
flict, depending strongly on the criterion selected for the boundary
(e.g., the photographic density).

The experimental data indicate (see Figure 2-15) that the time
dependence of the column boundary r_c is close to $t^{0.7}$ when the capacitance of the storage capacitor is relatively high ($C/U_0 \simeq 10^{-4}$
μF/V) and is near-linear for an even high capacitance ($C/U_0 \simeq 10^{-3}$
μF/V) (Figure 2-17).

Table 2-4. Values of D (km/s) at different times after
            the start of discharge in air (atmospheric
            pressure, U_0 = 15 kV, C = 0.25 μF, L = 2 μH)
            obtained in different ways

Type of measurement	t, μs				
	1	1.7	2.9	5.8	9.8
Direct measurements	1.5	1.29	0.9	0.57	0.5
Measurements of δ_f	1.75	1.28	0.88	0.57	0.49
Measurements of w(t)	1.65	1.4	–	–	–

It is most convenient to characterize the expansion of the
column in terms of the initial expansion velocity, which practically
matches the shock velocity, and the column radius r_{c-max} after the
maximum increase in the radius. The column radius r_c' at the time
$t = 2.6\sqrt{LC}$, corresponding to the cessation of the release of most of
the energy in the discharge, can serve as a characteristic quantity.
Composite data on the influence of the discharge conditions on r_{c-max},
which were taken from the work of many investigators (Table 2-3), are
presented in Figure 2-18 as graphs of the dependences of r_{c-max} and
r_c' on the energy of the power circuit. As we see from this figure,
the points on the graphs fit tightly enough that the general dependence of r_{c-max} on the parameters can be outlined, despite the differences in experimental methods and column-boundary criteria. The
energy W_0 has the most pronounced effect on r_{c-max}. The following
empirical formula may be adopted for air at atmospheric pressure:

$$r_{c-max} = 0.05 \left(\frac{CU_0^2}{2} \right)^{0.4}, \qquad (2-31)$$

where r is in cm, C in μF, and U_0 in μV. The energy dependence of
r_c' turns out to be similar to (2-31).

Reducing the gas pressure and the discharge circuit's inductance
by nearly an order of magnitude leads to an increase of just 15-30%
in r_{c-max}. The capacitance and voltage separately (at constant W_0)
have practically no effect on r_{c-max}. It is hard to say what influence the type of gas has because of differences in the nature of
the glow and transparency of the column in the late stage of channel
expansion. For example, given the same parameters, the boundary of
the discharge channel in argon is heavily blurred and the column
diameter turns out to be smaller than in xenon and krypton, which
initially expand much more slowly. (The significant deviation of
curves 10-12 from the other data in Figure 2-18 probably is explained
by a similar factor.) The discharge column is so unstable in hydrogen

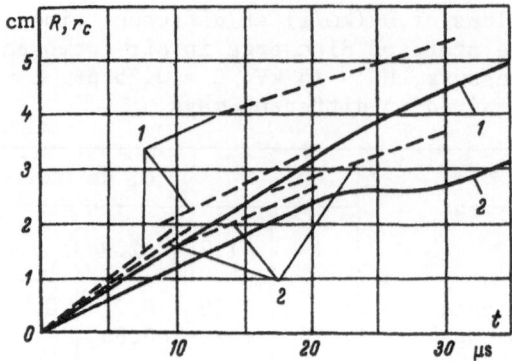

Fig. 2-17. The development of a shock front (1) and of the boundary
of the "hot" column (2) of a high-power discharge in air
at 0.1 MPa [2-14] induced by the explosion of a fine
wire: $l = 67$, 30 µF, 1 µH, 50 kV. The solid lines
indicate experimental values. The dashed line indicates
self-similarity theory.

that after just 1-1.5 µs we may no longer say that it has a diameter
at all, to judge by photographs obtained by means of an image converter [2-19] (this may be explained by the action of the magnetic
field).

Nonetheless, the position of the plots for different gases in
Figure 2-18 shows that the difference in the corresponding values of
r_{c-max} is not too large (it does not exceed 60%, all other conditions
being equal). Thus, formula (2-31) may be used to estimate the discharge conditions that are encountered in practice. If the factor
0.065 is inserted in this formula instead of 0.05, the deviations of
the actual values of r_{c-max} from the calculated ones generally will
not exceed 20-25%.

The available experimental data make it possible to form some
assessment of the internal characteristics of the column, in addition
to data on the column boundaries. Processing of interferograms
similar to those shown in Figure 2-11b enabled the authors of [2-24]
to make a crude estimate of the gas density in the inner part of the
column (by subtracting the amount of gas concentrated in the comparatively cold high-density zone from the total mass of the disturbed gas) and the corresponding electron concentration (from the
negative quantity $\mu - 1$, where μ is the measured refractive index).
As a result, they obtained approximately equal concentrations of
atoms and electrons (about 10^{17} cm^{-3}) that are approximately constant
over the entire cross section of the hot part of the column. Confirmation of this conclusion would mean that the extremely large
gradients of density and temperature in the column (which increase
steadily as the axis is approached) that follow from the purely gas-

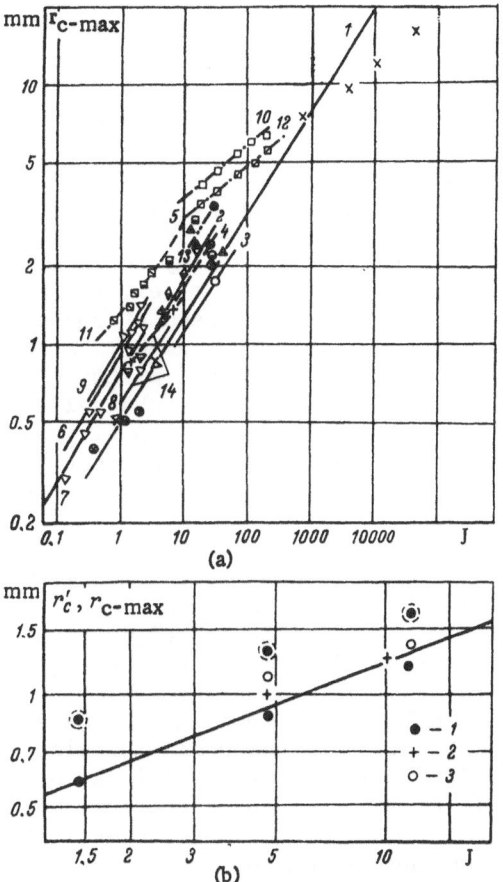

Fig. 2-18. Composite graphs of the dependence of the radius of the
glow column of a discharge on pulse energy.
a) for maximum radius (X is column 1 in Table 2-3; the
data are from a paper by J. W. Flowers: Phys. Rev., 1943,
vol. 64, p. 225; the other points have the same meanings
as in Figure 2-14): 1, 2, 3) air at 0.1, 0.03, and 0.3
MPa respectively; 4, 5) Ar at 0.1 and 0.3-1.1 MPa; 6, 7,
8) Xe with a H_2 admixture at 0.3 MPa, and L is 0.02,
0.12, and 0.3 µH respectively; 9) Kr with a H_2 impurity
at 0.3 MPa and 0.12 µH; 10, 11, 12) Kr at 0.2, 0.4, and
0.6 MPa, 0.6 µH; 13) H_2 at 0.1 MPa; 14) Xe at 0.3 MPa
and 0.2 µH (curve 1 also corresponds to formula (2-31));
b) r_c' at time $t = 2.6\sqrt{LC}$, 0.5 µF, 0.2 µH, l = 5 mm; 1)
Xe at 0.3 MPa; 2) Ar at 0.32 MPa; 3) Xe + 30% N_2 at
0.091 MPa.
Dashed circles surround points for the dependence of
r_{c-max} on W_0 (corresponding to points 1); the line represents the function of $W_0^{0.4}$.

dynamic theory (see Figure 2-5) actually are absent or are much
smaller than at the column boundary. Then it would be natural to
relate the leveling-out of the density and temperature to diffusion-
type processes, including radiative heat conduction.

The theoretical and experimental data presented in [2-12] also
argue in favor of the absence of large gradients of density and temperatures. Reference [2-12] presents the expanding plasma column
model mentioned above, based on the analogy to processes of fuel-gas
flash (Figure 2-19). According to this model, the expansion velocity
of the column is always less than the speed of sound in the surrounding shell, the temperature inside which is $10-15 \times 10^3$ K. Therefore,
the column boundary is a sort of permeable piston whose energy is
transferred to the shock front through the channel shell by sound
waves (their speed is about 1.5-2 km/s for an air spark, i.e., it is
comparable to the shock velocity). It can be seen from Figure 2-19
that the distribution of δ and T over the column cross section does
not have any large gradients.

However, it is difficult to tie these conclusions in with the
findings of Sommervell[†] and others who studied the space-time distribution of the spots on the anode of a pulsed discharge (in air, p_0
≈ 0.1 MPa, l is a few millimeters). In these experiments the anode
was covered with an insulating film on which electrons built up
during the discharge until local breakdowns of the dielectric occurred. The anode drop then "focused" arriving electrons in "holes,"
making them expand and the metal melt. The location and size of the
spots made it possible to estimate the electrical conductivity of
the adjacent portion of the column. These investigators found that
during the period of channel expansion "holes" are formed first near
the axis. These later cease to function, as if because of the decrease in plasma conductivity in this region, and new "holes" closer
to the periphery of the column are formed at the same time. The
radial velocity of the band of "acting holes" matches the calculated
values of D from gas-dynamic theory (formulas (2-11) and (2-17)) for
the first tenth of a microsecond. According to these investigators,
this confirms the notion that the instantaneous position of the
"holes" corresponds to the conducting zone of the column. Confirmation of the existence of a "hollow" zone near the column's axis
would indicate that the gas density and temperature distributions
inside the column have sizable gradients.

Thus, the question of the density and temperature distribution
inside the column of a freely expanding discharge, and the related
question of the role of physical processes inside the column for
different types of discharges that are not taken into account by
gas-dynamic theory, require further study.

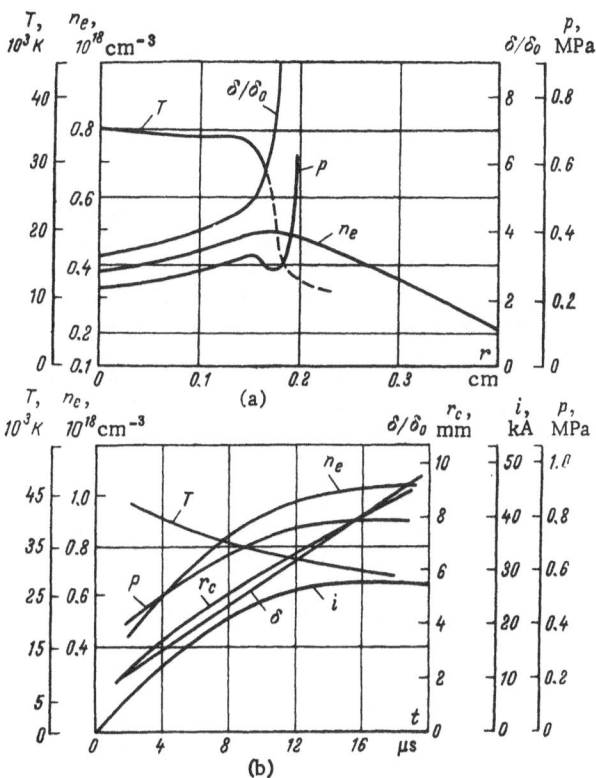

Fig. 2-19. Radial (a) and time (b) dependences of the density δ/δ_0, pressure p, temperature T, and electron concentration n_e in an ionized column having the length of a spark in air (0.1 MPa, $l = 50$ cm) [2-12].

2-5. The Luminance of a Freely Expanding Discharge

The main characteristics from the standpoint of using an expanding discharge as a pulsed light source are the luminance $L_v(t)$ (the spectral radiance is $Le_\lambda(t)$ and the column dimensions are its length l and radius $r_c(t) = \int_0^t \dfrac{dr_c(t)}{dt} \, dt$). If the relation of $L_v(t)$ and $r_c(t)$ to the discharge circuit characteristics (U_0, C, L, l, R_b) and to the gas pressure and type of gas were known, then we could calculate the peak radiant intensity I_{e-p} or the peak luminous intensity I_p, the radiated energy $Q_e = \Omega_{eq} I_{e-p} \tau$ ($\Omega_{eq} \approx 10$ sr is the equivalent solid angle, and τ is the effective duration of the flash at a level of about 35% of I_{e-p} or I_p) or the luminous energy $Q = \Omega_{eq} I_p \tau$, and the efficiency of the source $\eta = Q_e/W_0$ or its luminous efficiency $\eta_v = Q/W_0$ (see Chapter 5).

If the column emits as a blackbody, then a luminance graph can be obtained for the temperature range 20–70 x 10^3 K by using Planck's formula and the relative spectral luminous efficiency curve. This graph can be represented quite well by the convenient expression:

$$L_v \approx 5.9 \cdot 10^{-3} (T - 8300) \text{ Gcd/m}^2. \tag{2-32}$$

The spectral radiance is determined by the absorption coefficient κ_λ and the plasma temperature. This is equal to the luminance of a blackbody for a uniform, optically dense volume ($2r_c \kappa_\lambda \geq 2.3$), and is approximately equal to the luminance of a blackbody multiplied by $2r_c \kappa_\lambda$ for an optically transparent plasma ($2r_c \kappa_\lambda \leq 0.35$).

However, theoretical estimates of the radiation characteristics of the column are even more problematic than estimates of the related dynamic characteristics of an expanding discharge.

The luminance and blackbody temperature of a freely expanding discharge have been investigated by Vanyukov, Mak, Wulfson, Charna, Fischer, and other [1-68 and 2-42 to 2-48]. The main methods were:

(a) photoelectric and photographic measurements of the time dependence of the spectral luminance $L_{e\lambda}$;
(b) photoelectric measurement of the time dependence of the luminance L_v using a detector with a correcting absorber.

The published data make it possible to put together the following general picture of the dependence of the luminance on various parameters.

The luminance L_v, which is maximal in time and space at high electric field strengths (from several hundred V/cm and pressure of 13 kPa (100 mm Hg), increases even more when the electric power density increases (as the supply voltage U_0 increases and the inductance and electrode spacing decrease), but only to some limit. The same phenomenon of luminance saturation is observed as the pressure p_0 of the gas filling the lamp increases. The value of the saturation luminance L_{v-sat} obtained by varying some single parameter may undergo a further increase if the other parameters are varied toward increasing power density in the discharge column, and then may cease to be dependent on them. The maximum luminance attained, $L_{v-sat} = L_{v-abs}$, is characteristic of the given gas as the limiting luminance. The smaller the atomic mass of the gas, the larger L_{v-abs}, but for heavier gases L_{v-sat} and L_{v-abs} begin to be reached at circuit parameters that correspond to a lower power density. The blackbody temperatures calculated from the values of L_{v-abs} turn out to be close to the plasma temperature values determined from the spectral radiance in different regions of the spectrum. The saturation of $L_{e\lambda}$ occurs earlier in the longer-wave portion of the spectrum.

The picture described is illustrated in Figures 2-20 and 2-21 and in Tables 2-5 and 2-6. The following notations are used in the tables: L_{v-abs} is the limiting luminance, as estimated from equation (2-34) (see below); T_{L-abs} is the corresponding blackbody temperature; U_{iI} and U_{iII} are the first and second ionization potentials of the gas; and A is the atomic mass.

Figure 2-22 shows the space-time pattern of the luminance distribution under conditions which ensure luminance saturation in a spherical flashlamp similar in design to the ISSh100-3 lamp. The data presented in the figure were obtained by imaging a discharge at a magnification of approximately 15 with dimensions of 4 x 4.7 mm for the entrance slit of the luminance-measuring device (reducing the slit width to 2 mm has no effect on the measured luminance values).

It can be seen from the figure that the section of the column around the electrode is wider and has somewhat reduced luminance. (In the second half-period the broader portion and the less-bright portion of the column are adjacent to the right electrode, which becomes the cathode at this time. The channel seems somewhat constricted in the central portion to an eye averaging the picture.) Increasing the inductance from 0.09 to 0.2 µH somewhat decreases the steepness of the luminance rise front and requires an increase in voltage for luminance saturation. For pure gases, a decrease in the atomic mass leads to an increase in the limiting luminance and in the front steepness. However, using a xenon-nitrogen mixture with an average atomic mass of about 80 ($p_0 \approx 0.1$ MPa) makes it possible to obtain higher values of dL_v/dt and L_v than can be produced in pure argon ($p_0 = 0.3$ MPa). The graph of the transverse distribution of L_v approaches a Π-shape at a voltage that exceeds the voltage required for luminance saturation. This indicates the considerable optical thickness of the column. Figure 2-23 shows typical dependences of the peak column luminance on the parameters (lamps with disk seals were used to reduce the inductance to 0.02 µH [2-45]).

Figures 2-24 and 2-25 show the dependences on the atomic mass of L_{v-abs} and the supply voltage at which the peak luminance reaches 90% of the limiting luminance. (The effective values of p_0 and A for molecular gases were adopted, allowing for molecular dissociation. For mixtures, A was calculated as the average, again allowing for dissociation.)

An inspection of the experimental data enables us to draw the following conclusions:

(a) $U_{0.9}$, the quantity that indicates attainment of a state close to luminance saturation (90% of L_{v-abs}), can be related (Figure 2-25) to the main parameters by the expression

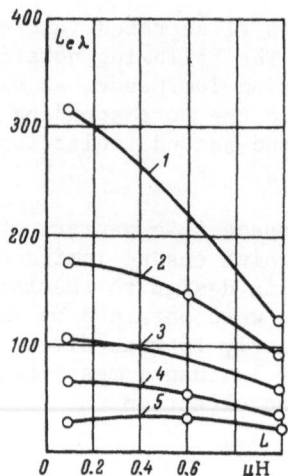

Fig. 2-20. Dependence of the peak spectral luminance $L_{e\lambda}$ (W/(cm^2·nm·sr)) of a discharge in argon on the inductance L. 0.4 MPa, 0.011 μF, 12 kV. 1) λ = 468 nm; 2) 554 nm; 3) 652 nm; 4) 723 nm; 5) 887 nm.

$$U_{0.9} = \frac{152}{A} \left(\frac{Ll}{p_0 C} \right)^{0.25}$$

(2-33)

where L is in henrys, C in farads, p_0 in megapascals, and l in millimeters.

(b) The dependence of the limiting luminance (Gcd/m^2) on the atomic mass is described satisfactorily for pure gases by the empirical formula

$$L_{v\text{-abs}} = 640 A^{-1/3}.$$

(2-34)

As yet no generally recognized explanation has been worked out for the luminance saturation of a freely expanding discharge. There have been a number of experiments attesting to the failure of the continuous background radiation to reach blackbody luminance (the absence of self-reversed lines of the ionized atoms), to the fact that the luminances observed from the end and perpendicular to the axis of a capillary discharge are equal, to the absence of temperature gradients in the column during interference measurements, and to the fact that the blackbody temperatures estimated from the spectral radiances in different spectral regions are equal (these regions reach saturation, as mentioned above, under conditions of increasing severity as the wavelength decreases). These experiments seemingly make it possible to characterize the state of the plasma by a single temperature value (at least for a transparent plasma). The luminance increases with increasing power density as a result of the increase

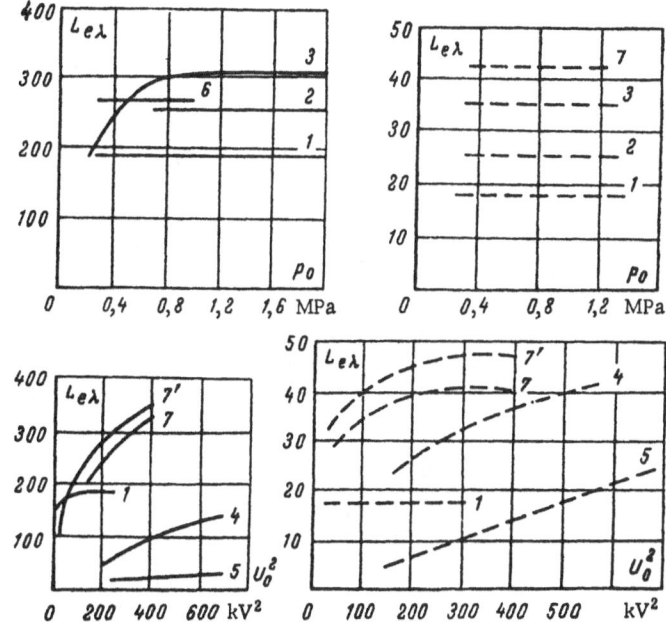

Fig. 2-21. Dependence of peak $L_{e\lambda}$ (W/(cm^2·nm·sr)) of discharges in
various gases on p_0 and U_0^2; C = 0.2 µF, L = 0.8 µH. The
solid lines are for λ = 500 nm, the dashed line for 900
nm. 1) Xe; 2) Kr; 3) Ar; 4) Ne; 5) He; 6) O_2; 7) N_2;
U_0 (as p_0 is varied) up to 26 kV, p_0 (as U_0 is varied)
is 0.2 MPa (7' is at 0.3 MPa).

in the "blackness" of the column at the critical temperature T_{cr},
which is practically constant for a given gas. This temperature is
reached at comparatively low power densities ($U_0 < 0.5U_{0.9}$). This
is in agreement with the weak dependence of the plasma temperature
on the current density j and also with the strong dependence of the
absorption coefficient κ on T [2-5, 2-23, and 2-49]. This explains
the wide variation of the radiated power and the blackbody temperature
when there is a slight change in plasma temperature. (It should be
noted that the question whether it is correct to use the concept of
"plasma temperature" with respect to a rapidly expanding column requires more precise definition in each specific case from the standpoint of whether thermodynamic equilibrium is present). In this
case, luminance saturation should be linked to the attainment of
blackbody luminance, a condition which is fulfilled when $2r_c\kappa_{av} \geq 2.3$.
The absence of a further rise in plasma temperature could be explained
hypothetically here in terms of an increase in the effective heat
capacity so significant that it would prove practically impossible
to exceed T_{cr} substantially.

However, in some papers, e.g., [2-47], saturation is explained
in terms of the blocking of radiation in the column by the lower-

Table 2-6. Relation between the atomic constants and
L_{v-abs} for various gases

Type of gas	A	U_{iI}, V	U_{iII}, V	L_v abs, Gcd/m^2	T_L abs 10^3 K
Xe	131	12.1	21.2	126	29.5
Kr	83.8	13.9	24.6	146	33
Ar	39.9	15.7	27.6	187	40
Ne	20.2	21.5	41	236	48.3
N$_2$	14	14.5	29.6	257	51.8
He	4	24.5	54.4	400	75.3

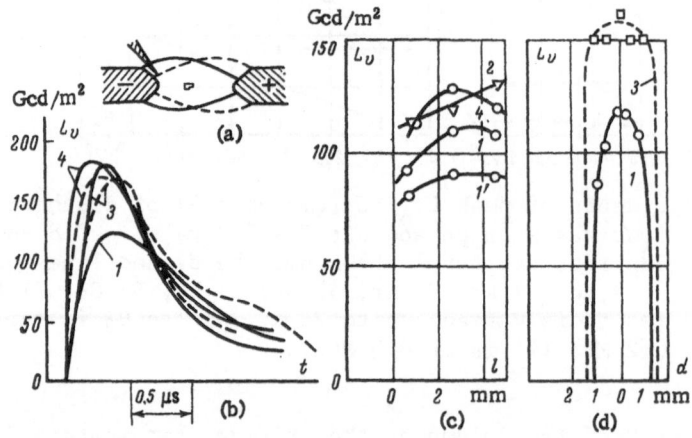

Fig. 2-22. Pattern of distribution of luminance L_v in a spherical
flashlamp with a tube diameter of 26 mm and l = 5 mm.
a) shape of column at time of first current maximum
(solid line) and second current maximum (dashed line)
(the figure in the middle is an image of the entrance
slit during the luminance measurement, drawn to the same
scale; the polarity of the electrodes at the first current maximum is indicated);
b) oscillograms of the luminance in the middle of the
column at U_0 corresponding to the attainment of L_{v-sat};
c, d) distributions of L_v along the longitudinal and
transverse axes;
1) Xe, 0.3 MPa, 2.5 kV, 0.5 μF, 0.2 μH (1' is for 0.15
MPa);
2) Kr (same parameters);
3) Ar (5-7 kV; other parameters the same; the dashed
line is for 0.09 μH);
4) mixture: Xe (66 kPa) + N$_2$ (26 kPa) (parameters the
same).

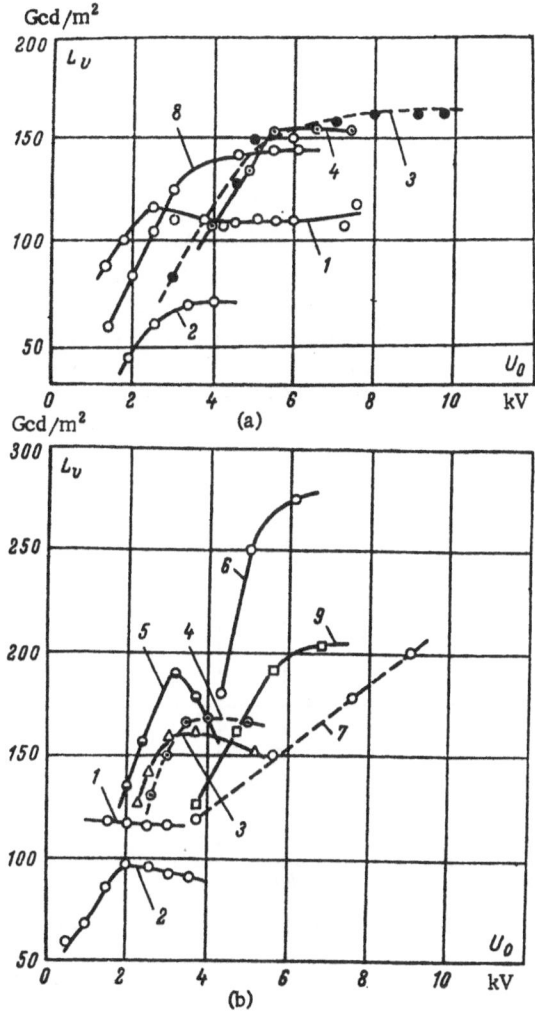

Fig. 2-23. Dependence of peak luminance on supply voltage; l = 5 mm.
a) 0.2 µH; b) 0.02 µH; 1) Xe, 0.3 MPa, 0.4-0.5 µF; 2) Xe,
0.066 MPa, 0.4-0.5 µF; 3) Ar, 0.4 MPa, 0.4-0.5 µF; 4) Xe
+ 30% N_2, 0.092 MPa, 0.4-0.5 µF; 5) the same, but at 2
µF; 6) N_2, 0.3-0.35 MPa, 2 µF; 7) the same, but at 0.1
µF; 8) Kr, 0.3 MPa, 0.5 µF; 9) air, 0.1 MPa, 2 µF [2-44].

temperature opaque outer layer. Temperature measurements in strong
shock waves (in gases) produced by an explosion lead to an analogous
hypothesis of radiation "blocking" [2-50]. This hypothesis also is
in agreement to some extent with the attainment of temperatures
higher than T_{L-abs} in discharges in air that are confined by capillary walls [2-51] for which a luminance of about 500 Gcd/m^2

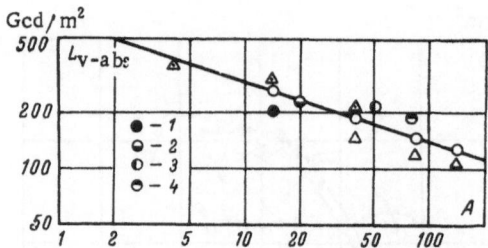

Fig. 2-24. Dependence of the limiting luminance L_{v-abs} on the atomic mass A. The circles are from [2-44], the triangles from [2-43], and the dotted triangles from [2-42]. 1) air, 0.1 MPa; 2) Ne (0.3 MPa) + N_2 (0.025 MPa); 3) Kr (0.066 MPa) + N_2 (0.025 MPa); 4) Xe (0.066 MPa) + N_2 (0.025 MPa); the other points are for pure gases; the line is a graph of relation (2-34).

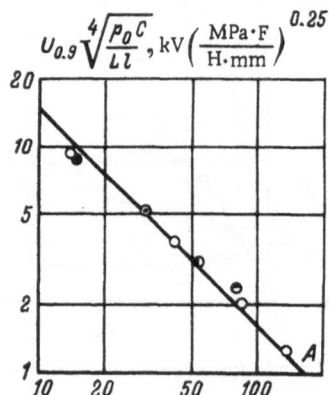

Fig. 2-25. Dependence of the voltage $U_{0.9}$ on the atomic mass. The dotted circles are for Ar (0.08 MPa) + N_2 (0.025 MPa); the other symbols are the same as in Figure 2-24. The line is a graph of relation (2-33).

($T_L \approx 90 \times 10^3$ K) apparently was obtained at a tube diameter of 0.4 mm (l = 10 mm, C = 0.011 µF, U_0 = 29 kV). The second hypothesis is supported by the facts noted in [2-46 and 2-52]. These facts attest to the storage of a considerable amount of energy in the column when $U_0 > U_{0.9}$. This energy cannot be attained at the column temperature T_{L-abs} observed from the outside, with allowance for the low particle concentration (about 10^{17} cm^{-3}). For example [2-52], the peak radiation power $P_p \approx 0.5$ MW is reached 0.5 µs after the peak of P_{el} for a discharge in xenon (5 mm, 0.3 MPa, 0.5 µF, 4.5 kV, 5 J, 0.2 µH) at a peak electric power $P_{el} \approx 6$ MW and a pulse length of about 0.8 µs. The pulse length of P_r is about 4 µs, indicating that over 1 J of energy is stored in the column (at a density of over 100 J/cm^3).

This corresponds to a temperature of about 120 x 10^3 K at a particle
concentration of about 10^{17} cm^{-3}, although the T_{L-abs} for xenon is
about 30 x 10^3 K. The pulse length of P_r is 0.05 μs at around a
0.01-μs length for the pulse of P_{el}, exactly as it is for the lighter
noble gases [2-46]. This corresponds to the storage of an extremely
large amount of energy in the column. The experimental data show
that an appreciable lag of the maximum of P_r with respect to the
maximum of P_{el} occurs when $U_0 > 0.5U_{0.9}$.

We may assume that the thickness of the layer that blocks the
radiation is close to the minimum so-called Rosseland path length
l_R (the effective range of radiative energy transfer), which is
reached at a plasma temperature close to U_{i-II} [2-8] (e.g., for air
at p_θ = 0.3 MPa and T = 43 x 10^3 K, $l_R \approx 0.1$ mm). The mechanism of
luminance saturation is illustrated in this case by the schematic
diagram of the temperature profiles of discharges with different
densities of input energies shown in Figure 2-26. The zones that
determine the apparent luminance of the column are hatched in this
figure. T_l designates the lowest temperature at which there is still
a substantial contribution to the radiation, and T_h the temperature
of the plasma boundary layer, radiation from which still reaches the
outside. The apparent luminance of the column is determined by the
temperature, which is close to T_h. As the input energy density and
the temperature on the column axis (designated 0) increase, T_h initially coincides with the on-axis temperature T_0, and then stops
increasing because of the rapid shortening of l_R (which intensifies
even further as the outward-radiating zone moves away from the axis
because of the increased gas density at the periphery). The apparent
luminance also ceases to increase as the increase in T_h comes to a
halt (L_v becomes equal to L_{v-abs}).

The temperature T_h of the deepest layer "visible" from the outside (the temperature at which the Rosseland path length has a minimum) should increase together with the second ionization potential
of the gas as we go from a heavier gas to a lighter one.

As we see from Table 2-6, the proportionality of the limiting
value of T_h (i.e., T_{L-abs}) to the potential U_{i-II} which follows from
calculated estimates, is confirmed by the relation:

$$T_{L-abs} \approx 1400 U_{iII} . \qquad (2-35)$$

Relation (2-35) may assign physical meaning to the relation
between the saturation luminance and the atomic mass, as expressed
by formula (2-34), by virtue of the approximate proportionality of
U_{i-II} to the value of $A^{-1/3}$ (this proportionality is characteristic
of atoms of noble gases). The anomalously high value of L_{v-abs},
which does not correspond to U_{i-II}, and the low value of $U_{0.9}$ for
nitrogen and nitrogen-xenon mixtures may be attributed to the effect
of shortwave reradiation at the periphery of the column as longwave

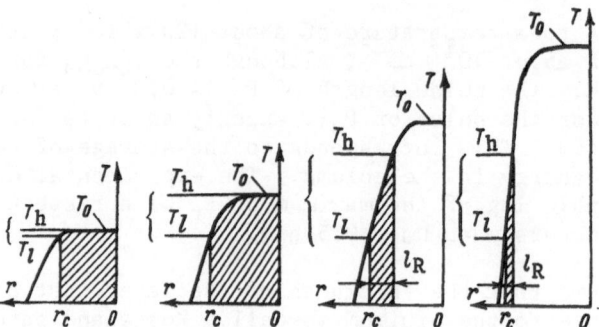

Fig. 2-26. Illustration of the mechanism of luminance saturation
            at various increasing (from left to right) power den-
            sities. The portion of the temperature profile of the
            column up to l_R thick, which makes a contribution to
            the radiation observed from the outside, is hatched.

quanta. This effect is due to the high absorption coefficient of
ultraviolet rays in nitrogen [2-12, 2-14, and 2-15]. It finds practical application in spherical enhanced-luminance flashlamps.

A fundamental theory of a freely expanding pulsed discharge
10^{-8}-10^{-4} s long should take into account the existence of three
main types of pulsed discharge which may be classified as follows:

(a) "normal" discharge with a transparent column and simultaneity of
    the electric power pulse (P_{el}) and the radiated power pulse (P_r).
    Such a discharge is characterized by the fact that $U_0 < 0.5U_{0.9}$,
    where $U_{0.9}$ is determined by expression (2-33). This type of
    discharge allows the production of the shortest light pulses.
(b) "Saturated" discharge, in which the plasma column is opaque for
    a time of about $2.6\sqrt{LC}$, and the maximum pulse of P_r is appreciably
    time-shifted with respect to the maximum of P_{el}. This discharge,
    which is characterized by the fact that $0.5U_{0.9} \leq U_0 \leq 1.5U_{0.9}$,
    has a luminance close to L_{v-abs} for a short period of time.
(c) "Supersaturated" discharge (which is most readily observed in
    heavy noble gases at high pressures with a short channel), in
    which the plasma column is opaque for a period of time longer
    than $2.6\sqrt{LC}$, and the P_r pulse differs markedly from the P_{el}
    pulse in shape, being much longer. This discharge, which is
    characterized by $U_0 \geq 1.5U_{0.9}$, has a luminance L_{v-abs} for a
    significant period of time.

The transition from the first to the second type of discharge
and from the second to the third should be related to an increase
in the initial current steepness, to the increase in the energy introduced per unit length of the discharge, to increasing pressure and
atomic mass of the gas, and to a decrease in the ionization potential
of the gas.

CHAPTER 3

PULSED DISCHARGE OF THE QUASI-STATIONARY TYPE

3-1. Local Thermodynamic Equilibrium in a High-Pressure Discharge

The stage of the formation and expansion of the discharge channel
enters the "quasi-stationary stage" if the current rise time in the
pulsed discharge appreciably exceeds both the relaxation times of the
basic plasma processes and the characteristic times of nonstationary
gas-dynamic phenomena (channel expansion, the motion of gas masses, etc.).
The geometry of the discharge channel goes practically unchanged throughout nearly all of the quasi-stationary stage, there is no noteworthy
motion of gas, and all of the electric field energy liberated in the
plasma goes to compensate for losses to the surrounding medium. Therefore at every instant the energy balance of a pulsed discharge is similar (as a first approximation) to the energy balance of a stationary
arc discharge under corresponding conditions, and the instantaneous
state of the pulsed plasma is similar to the state of a stationary plasma.

In order to study or understand correctly the phenomena which occur
during the quasi-stationary stage, it is important to determine whether
local thermodynamic equilibrium (LTE) is reached in the discharge channel. Here we may speak of instantaneous equilibrium states, since the
parameters characterizing the plasma change slowly on the whole.

Strictly speaking, the concept of complete thermodynamic equilibrium is applicable to a plasma in a state of energy isolation when the
plasma has a uniform temperature distribution. A Maxwell distribution
of particles of each species with respect to velocities and a Boltzmann
distribution with respect to potential energies become established
under these conditions. There also is a detailed balance, meaning that
the rates of all direct and inverse elementary processes are equal.

In its general features, the picture of the elementary processes
in a gas-discharge plasma appears as follows, according to current

81

concepts. An external electric field applied to an ionized gas interacts mainly with free electrons, imparting excess energy to them. Col-
liding with each other, the electrons effectively exchange energy and
some velocity distribution becomes established among them. Under equilibrium conditions this is a Maxwell distribution, which can be described by the following expression for particles of species s:

$$dn_s = 4\pi n_s \left(\frac{m_s}{2\pi k T_s}\right)^{3/2} \exp\left(-m_s v_s^2/2kT_s\right) v_s^2 \, dv, \tag{3-1}$$

where n_s, m_s, and v_s are the concentration, mass, and velocity of the particles, respectively; T_s is their temperature and k is the Boltzmann
constant. The kinetic energy of electrons, for example, may be characterized by the parameter T_e, which is called the electron temperature.

The interaction of electrons with heavy particles is efficient,
provided that they have the high collision frequency characteristic
of high pressures, so that in a single elastic collision an electron
transfers only a fraction $2m_e/m_a$ of its excess energy to a heavy
particle (where m_e and m_a are the mass of the electron and the atom,
respectively). Energy exchange between heavy particles in turn leads
to a Maxwell distribution with respect to velocities, according to
equation (3-1). The concepts of ion temperatures T_i, the temperature
of neutral atoms T_a, and so forth are introduced in this manner. All
these temperatures are equal under equilibrium conditions, and the
plasma is characterized by a single temperature.

Inelastic collisions of electrons with heavy particles lead to
the excitation and ionization of atoms. These processes always are
accompanied by inverse processes: collisional deexcitations and
recombinations of ions with electrons. Under conditions of thermodynamic equilibrium inelastic collisions lead to a Boltzmann distribution of particles over all possible energy states:

$$n_k = n_0 \frac{g_k}{g_0} \exp\left(-eU_k/kT\right), \tag{3-2}$$

where n_k and n_0 are the concentration of particles in the k-th and
the ground state respectively; U_k is the potential of the k-th state;
g_k and g_0 are the statistical weights of the states, which determine
the number of physically different states having the same energy;
and e is the charge of the electron.

Processes involving photons occur in a plasma in addition to
elementary collisional processes. Photons are emitted upon spontaneous transitions of excited particles to lower energy states
(bound-bound transitions), upon the recombination of electrons with
ions (free-bound transitions), during breaking of electrons in the
field of ions or atoms (free-free transitions), and during the slowing
of some heavy particles in the field of others. The interaction of
photons with matter leads to the inverse processes: photoexcitation,

photoionization, acceleration of electrons in the field of ions and atoms due to the absorption of photons, and so forth.

When there is detailed balance, the number of direct and inverse processes of each kind is strictly the same per unit volume of the system per unit time. The equal rates of the processes of collision ionization and recombination lead in particular to an equation which relates the equilibrium concentrations of ions n_i, electrons n_e, and neutral atoms n_a to the plasma temperature T and the ionization potential U_i. This equation is known as Saha's formula. For single ionization

$$\frac{n_e\, n_i}{n_a} = 4.9 \cdot 10^{15} \cdot \frac{G_i}{G_a}\, T^{3/2} \exp\left(-eU_i/kT\right) \qquad (3\text{-}3)$$

(the concentrations are in cm^{-3}, the temperature is in degrees Kelvin, and G_i and G_a are the statistical sums of the ground states of the ion and atom). The equality of the direct and inverse processes involving photons implies the applicability of Kirchhoff's law, which relates the emission coefficient ε_ν [J/($cm^3 \cdot sr$)] to the absorption coefficient κ_ν (cm^{-1}):

$$\frac{\varepsilon_\nu}{\kappa_\nu} = \frac{2h\nu^3}{c^2} \left(e^{\frac{h\nu}{kT}} - 1\right)^{-1}, \qquad (3\text{-}4)$$

where ν is the radiation frequency and h is Planck's constant.

The presence of thermodynamic equilibrium in the plasma eliminates the necessity of resorting to the kinetic theory of gases to obtain the macroscopic characteristics of the plasma from the elementary processes that occur in it. In this case the state of the plasma and its mechanical, electrical, and thermal properties in the absence of external fields are determined by the type of gas and by two thermodynamic parameters (usually the temperature and pressure). All concepts and methods of thermodynamics are fully applicable to an equilibrium plasma. A phenomenological treatment of an equilibrium plasma thus becomes possible without a detailed analysis of the elementary processes that transpire within it.

Here we will present a list of the basic thermodynamic functions used in investigations of a plasma in thermal equilibrium and in calculations of its parameters. All these functions can be expressed in terms of the statistical partition function

$$G = \Sigma g_\varkappa \exp\left(-eU_\varkappa/kT\right). \qquad (3\text{-}5)$$

The free energy of the plasma per particle is

$$F = -kT \ln G; \qquad (3\text{-}6)$$

the entropy is

$$S = -\frac{\partial F}{\partial T} = k \ln G + kT \frac{\partial \ln G}{\partial T};$$ (3-7)

the internal energy is

$$u = F + TS = kT^2 \frac{\partial \ln G}{\partial T},$$ (3-8)

and the enthalpy (heat content), the thermodynamic potential, and the chemical potential respectively are:

$$H = kTn\left(1 + \frac{\partial \ln G}{n \partial \ln T}\right), \quad \Phi = F + kTn,$$

$$\mu = \left(\frac{\partial \Phi}{\partial n}\right)_{p,T} = \left(\frac{\partial \Phi}{\partial n}\right)_{T,V},$$ (3-9)

where n is the total number of particles in the plasma, and V is the volume of the plasma.

We should note that, strictly speaking, relations (3-2), (3-3), and (3-5) are valid on the assumption of an ideal equilibrium gas. If the degree of ionization has an appreciable value, the ideality of the gas is violated because of Coulomb interactions of charged particles. One consequence of the violation of ideality is a decrease in the ionization energy of atoms in a plasma, compared with the corresponding value for an ionized atom. This decrease results from the expansion and coalescence of the upper energy levels bordering on the ionization limit. The decrease depends on the electron concentration and the plasma temperature. This decrease has been determined in a number of papers, and relations for calculating it have been suggested. The relations differ in structure and in numerical coefficients, depending on the approach taken by different authors. Therefore, the calculated values of the decrease in U_i differ appreciably. As yet the decrease in ionization potential has not been observed experimentally because of the insufficient accuracy of existing spectral methods. The best-substantiated relations are those proposed in [3-1] and [3-2]. In the latter, the following formula was derived for determining the decrease in ionization potential (ΔU_i is in volts):

$$e\Delta U_i = 1{,}21 \cdot 10^{-6} n_e^{1/3} + 2{.}5 \cdot 10^{-6} (n_e/T)^{1/3}.$$ (3-10)

The conditions for complete thermodynamic equilibrium are absent under real conditions because of the finite dimensions of the gas-discharge tube. A plasma always has fluxes of heat, radiation, and charges which disrupt thermal equilibrium and lead to nonuniform temperature, nonuniform particle density, and irreversibility of

some processes. Additionally, the state of the plasma of a pulsed
discharge is unsteady in time.

Nonetheless, on certain assumptions we may say that there is a
quasi-equilibrium state at a given point in the plasma at a given
time. By a "point" is meant a local region of plasma large enough
that macroscopic concepts are applicable to it, but not so large that
nonuniformities of the plasma parameters are noticeably evident within
its boundaries. If the change in the amount of energy at every point
in the plasma per unit time is small compared with the total internal
energy at this point, then there is a Maxwell distribution of particles
of each species over velocities at every point at a given instant, even
though the plasma parameters vary from point to point and from one
instant to the next. Equality of the most probable energies also is
ensured for all components of the plasma, and the occupation numbers
of all bound and free states that directly interact with radiation
correspond to the thermal-equilibrium values at a unique local temperature. This plasma state, which is extremely close to thermal
equilibrium, is called local thermodynamic equilibrium (LTE).

The existence of LTE usually is observed in a collisional
plasma in which the following conditions are fulfilled: (1) the
frequency of collisions of electrons with heavy particles is such that
the electrons are able to transfer to these particles all excess energy
acquired from the external electric field; (2) ionization and recombination processes occur within the plasma volume, and the number
of charged particles diffusing out of the volume is negligibly small;
and (3) deactivation of excited atoms occurs mainly as a result of
damping collisions, while most of the radiated energy does not travel
outside the plasma, being absorbed in it.

We select as the criteria for the existence of LTE the conditions under which a deviation from the equilibrium state due to
some factor may be considered a sufficiently small perturbation. The
desire to obtain reliable criteria for the existence of LTE is dic-
tated not only by the desirability of using available methods of
calculating the properties of a given type of discharge, but also
by the need to know the state of the plasma in order to select appropriate methods of measuring its parameters and to interpret the
results of optical studies. Therefore, the problems of local thermal
equilibrium and of selecting appropriate criteria are treated in
all major works devoted to the properties of electric discharges in
gases and to the methods used to investigate them [3-3 to 3-7].

An external electric field of strength E is one factor that
disrupts the state of equilibrium of a gas-discharge plasma. The
method used to transfer energy from the field to the plasma via
electrons unavoidably leads to a difference between the electron
temperature T_e and the temperatures of ions T_i and neutral atoms T_a.

Assuming $T_i \approx T_a$, we may use the following condition as a criterion for the permissible excess of the electron temperature over the temperature of heavy particles [3-3]:

$$\frac{T_e - T_a}{T_e} = \frac{m_a}{4m_e} \frac{(\lambda_e \, eE)^2}{\left(\frac{3}{2} \, kT_e\right)^2} \ll 1, \qquad (3\text{-}11)$$

where m_e and m_a are the masses of the electron and of a heavy particle of the gas, and λ_e is the free path of the electron. Condition (3-11) means that the energy acquired by an electron from the field along its free path ($\lambda_e eE$) should be significantly lower than the electron's energy of thermal motion $^3/_2kT_e$.

The nonuniform distribution of temperature in a gas-discharge plasma and its tailoff toward the edges of the arc column also should lead to disruption of equilibrium due to the appearance of heat fluxes. The criterion for the permissible deviation from equilibrium due to this factor is the condition

$$\lambda_s \, \text{grad} \, T_s / T_s \ll 1, \qquad (3\text{-}12)$$

This condition means that the temperature difference along the free path of a particle is small with respect to the temperature. The free paths of electrons λ_e and atoms λ_a in the plasma of a high-pressure discharge are very short. Therefore, for such a plasma condition (3-12) is known to be fulfilled for the ordinary temperature gradients in the discharge column. An appreciable deviation from the equilibrium state may be observed in the boundary regions of the discharge column only in cases where the configuration of the discharge is characterized by a sharp temperature drop at the edge and, additionally, the specific features of the external electromagnetic field lead to an appreciable separation of the electron temperature from the gas temperature in the boundary regions. According to experimental data, this pattern is observed, for example, in the outer regions of an inductive high-frequency discharge when there is a strong skin effect.

The particle density gradients that exist in the plasma and the flows of diffusing particles that result from them do not cause noticeable deviation from LTE if the change in density along the free path of a particle is much smaller than the density of the particles in question. Here the criterion for local thermodynamic equilibrium has the form of (3-12), with the temperature replaced by the density of particles of the given species. In a high-pressure plasma these criteria generally are satisfied with room to spare.

The most complicated problem is that of local thermodynamic equilibrium under conditions of intense radiation and appreciable interaction of the radiation with matter. A distinction is drawn

between the cases of an optically transparent plasma and an optically
dense one. In an optically transparent plasma processes of photon
emission are not balanced by the inverse processes. This violates
the principle of detailed balance. The consequences of this fact
understandably are inconsequential if the energy carried away from
the given plasma region by photons amounts to an insignificant fraction
of the energy involved in collisional processes in this region. The
criteria for existence of LTE in an optically transparent plasma
have been examined in general form in [3-4, 3-5, and 3-8]. These
criteria are shaped by the following three requirements: (1) the
behavior of the gas is determined by collisions; (2) the relative
changes in gas temperature and density are small at distances equal
to the free paths and for times equal to the relaxation times for
bound states and free electrons; and (3) there are no external sources
of perturbation which are capable of appreciably distorting the velocity distribution of electrons or the populations of bound states
with respect to values corresponding to LTE. For a plasma that
satisfies these conditions, the criterion for permissible deviation
from the equilibrium state reduces to the requirement that the electron density be no lower than a specified value. The corresponding
criterion for the existence of LTE (no more than a 10% deviation from
the Boltzmann distribution) reduces to the inequality [3-4 and 3-5]:

$$n_e \geqslant 1.6 \cdot 10^{12} \, T_e^{1/2} \left(eU_{p,q} \right)^3, \tag{3-13}$$

where $U_{p,q}$ is the difference between levels p and q, radiative transitions between which can cause a deviation from equilibrium.

Uniformity of the populations of energy levels is maintained
by radiation in an optically dense collisional plasma that has significant absorption of self-radiation [2-8 and 3-8]. This is true
regardless of the electron concentration, provided that temperature
changes at distances on the order of the photon path length are
sufficiently small. If the role of collisions in an optically dense
plasma is not decisive, then the question of the existence of LTE
and the applicability of equilibrium versions of the radiation transport equations becomes complicated, and each case must be analyzed
separately.

The two extreme cases (an optically transparent plasma and
an optically dense one) do not exhaust all the situations encountered
in practice. In studying low-temperature plasma we often must deal
with intermediate cases in which the plasma is optically dense with
respect to certain spectral regions, while it is optically transparent
to others. For example, a low-temperature gas-discharge plasma in
noble gases is transparent in the continuous ultraviolet and partially
transparent in the visible region, but has appreciable absorption in
the continuous and line spectra of the infrared band. Particularly
strong absorption by such plasma is observed at resonance lines whose
emission usually turns out to be "blocked" (the path length of the
corresponding quantum is much shorter than the characteristic dimension of the plasma volume).

The theoretical determination of the nature of the population of individual levels in relation to the conditions in the discharge is a rather complicated task. In general form, we may draw the conclusion that deviations from the Boltzmann distribution should be expected in the ground energy level in a high-pressure plasma, whereas an equilibrium state is readily reached in the upper levels. It should be emphasized that a violation of the equilibrium population of the ground level when there is LTE for all other parameters is not of basic importance for many practical problems, since it does not disrupt use of the relations that follow from the criterion for the equilibrium state (without setting more stringent criteria than are required by the specific circumstances).

These criteria for local thermodynamic equilibrium in a gas-discharge plasma should be considered approximate, since assumptions that restrict the generality of the results were made in their derivation. Therefore, it is useful to compare the results of theoretical estimates of the range of LTE with experimental research on local thermodynamic equilibrium in specific types of gas-discharge plasma.

Because of the special role that a high-pressure discharge in argon plays in practical applications, nearly all known research on the equilibrium state in noble gases has been done in argon arcs. References [3-9 and 3-10] adopted as the criterion for local thermodynamic equilibrium the quality of the different "temperatures": the electron temperature T_e, the gas temperature T_a, the distribution temperature T_d, the population temperature T_p, and the ionization temperature T_i. Each of these temperatures is a parameter which characterizes a particular equilibrium state. For example, the temperatures T_e and T_a are determined from relation (3-1) on the condition that a Maxwell distribution of velocities is established among electrons and atoms (molecules). This distribution differs for particles of each species in the values of the most probable kinetic energy. The distribution temperature T_d is formally found from relation (3-2) if the nonequilibrium plasma has a limited group of atomic energy levels whose population is quasi-Boltzmann (exponential) in character. Understandably, in the general case the parameter T_d cannot be extended to the entire set of energy levels in a nonequilibrium plasma. The population temperature T_p is determined for each individual level from relation (3-2), and does not always match T_d in a nonequilibrium plasma. Finally, the ionization temperature T_i for a nonequilibrium plasma is formally determined by relation (3-3) for known values of n_e and n_a (or of the pressure p). Upon the transition to the equilibrium state, all of the temperatures indicated become equal and the plasma is characterized by a single temperature in the thermodynamic sense of the term. A detailed analysis of the dependence of these temperatures on discharge conditions has showed that LTE is reached in argon on the condition [3-10]

$$n_e \geqslant 5 \cdot 10^{15} \, \mathrm{cm^{-3}} \text{ and } \quad T \approx 8500 \, \mathrm{K}.$$

The use of other criteria for the equilibrium state leads to noticeably different estimates of the range in which it is reached. For example, it was found in [3-11] that LTE can be reached in an argon arc plasma only at pressures p > 0.3 MPa and electron concentrations $n_e \geq 10^{18}$ cm^{-3}. This means that equilibrium actually is violated under most of the discharge conditions used, which to now have been assumed to be equilibrium conditions. An analysis of the criterion used by the authors of [3-11] shows that it is determined by the requirement that the equilibrium population of the ground state of Ar$_I$ be reached. An analogous result also was obtained in [3-12].

The data of [3-11 and 3-12] show that when equality is established for the overwhelming majority of the particular temperatures, the population of the ground state of the argon atom still may remain non-equilibrium. Conversely, the population of the upper excited states has a Boltzmann distribution as early as $n_e > 3 \times 10^{14}$ cm^{-3}, even when there is an appreciable separation of the electron temperature (7000–10,000 K) from the gas temperature (2320–5400 K) (this was demonstrated experimentally in [3-12]). New data presented in [3-14 and 3-15] on the nature of the population show that LTE obtains for all states of Ar$_I$, including the lower state, at pressures p > 0.05 MPa and electron concentrations $n_e > 5 \times 10^{16}$ cm^{-3}. When p < 0.05 MPa and $n_e \approx 3 \times 10^{16}$ cm^{-3}, the ground state of Ar$_I$ is over-populated by a factor of 2–7 compared with the equilibrium population.

The results of experimental research on the equilibrium state of an argon plasma show how important it is to determine correctly the required degree of stringency of the LTE criterion under conditions where complete thermodynamic equilibrium is unattainable. Slight violation of the equilibrium population of the ground state of the atom is not of basic importance in the solution of problems of plasma diagnostics and in determining whether relations (3-2) and (3-3) can be used to compute the plasma parameters. Therefore, we may assume with an accuracy that is adequate for practical purposes that LTE is reached in an argon discharge when $n_e \geq 10^{15}$ cm^{-3} and T $\approx$ 8000 K. LTE is reached at even lower values of the temperature in a xenon discharge because of the lower ionization potential than that of argon.

3-2. Diagnostics of the Quasi-Stationary Stage of a Pulsed Dis-
       Charge in Tubular Lamps

The duration of the discharge in tubular flashlamps usually is on the order of a millisecond. This period of time suffices for a quasi-stationary regime to become established, at least on the descending segment of the current dependence on time (the relaxation [times of the basic interaction processes in the plasma at the particle densities and temperatures characteristic of flashlamps are on the order of 10^{-9} s, while the characteristic times required for gas-dynamic processes to become established in small flashlamps are 10^{-5} s]. While a high-pressure pulsed discharge bounded by walls is a source of intense optical radiation of broad spectral composition, at the same time it is a convenient object for comprehensive study of the properties of a plasma with appreciable deviations from the ideal state.

Three factors are essential for characterizing the quasi-stationary stage of a pulsed discharge in tubular lamps. First, the attainment of instantaneous LTE makes it possible to utilize the entire repertory of theoretical means which pertain to the case of an equilibrium plasma. Second, the absence of appreciable displacements of the gas makes it possible to use experimental techniques which are excluded during previous stages of the discharge because of the presence of gas-dynamic processes. This ensures the possibility of determining the composition and parameters of the plasma. Third, at every instant the parameters are constant throughout the entire volume occupied by the plasma for relatively narrow tubes whose cavity is entirely filled with plasma during the quasi-stationary stage.

The fact that there is a uniform temperature distribution in the plasma in the quasi-stationary stage of a pulsed discharge in tubes was confirmed experimentally only quite recently. It was shown in [3-16 to 3-18] that the instantaneous values of the temperature and electron concentration are uniform along the discharge axis, except for the narrow regions around the electrodes. The size of these regions is negligibly small compared with the length of the discharge gap. This allows us to ignore the temperature nonuniformity present in these regions.

Difficulties associated with the inadequate spatial resolution of the methods employed arose in attempts to measure the radial distribution of temperature. This prevented measurements of the temperature in direct proximity to the curved walls of the tube, where the nature of the radial temperature profile should be especially evident. A tube with a run of rectangular cross section was used in [3-19]. This made temperature measurements possible in the wall region, and helped get some idea of the temperature profile in a rectangular tube. However, the presence of angles in the tube should have caused the plasma to be disturbed in boundary regions, i.e., at precisely those points where there is a significant change in temperature. Therefore, it would be invalid to apply the results obtained for a tube of rectangular cross section to a discharge having a round cross section [3-20]. A through radial slot in the one of the electrodes was used for this purpose. The results of the determination of the radial temperature profile that were obtained in the papers indicated are presented in Figure 3-1. The temperature profile in a round discharge tube shows practically no difference from the ideal Π-shaped profile.

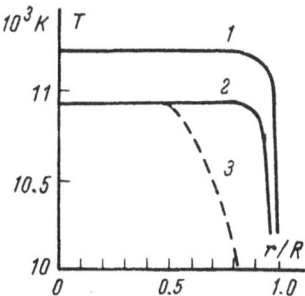

Fig. 3-1. Temperature profile in a cross section of a high-pressure
           pulsed discharge in xenon. 1, 2) Measurements in a round
           tube [3-20 and 3-21]; 3) measurements in a rectangular
           tube [3-19].

The deviation from the ideal in a rectangular tube can be seen well
at the scale selected for the figure, but the deviation is quite small
when the temperature profile is considered as a whole (rather than the
upper portion of the profile alone).

The Π-shaped temperature distribution in tubular lamps also is
confirmed by observations of the change in the contours of the two
self-reversed lines of xenon (Xe_I823.2 nm and Xe_I881.9 nm) under
various discharge conditions [3-22]. It can be seen from Figure 3-2
that the contour of the line Xe_I 823.2 nm has a gap in the center due
to self-reversal at low discharge energies. According to the results
of high-speed photography, the cross section of the tube is not completely filled with plasma at such energies. As the energy put into
the discharge increases, the gap smooths out and vanishes. The same
pattern was observed for the line Xe_I 881.9 nm. For both lines the
gap began to disappear under the same discharge conditions. This
means that the relatively cold boundary layers of plasma that exist
under low-power conditions disappear as soon as the entire cross
section of the tube is filled by the discharge.

The composition of a pulsed plasma and its temperature can be
determined by a number of methods. For a plasma in LTE, the temperature can be determined from the results of time-resolved absolute
and relative measurements of the radiation intensity of individual
lines and the continuous background. It also can be determined by
absolute measurements of the radiation intensity and absorption coefficient at the same wavelength. In the latter case, which often
is used in studies of flashlamp plasmas, the temperature is determined
from the following relation, assuming that the radiative and absorptive
properties are uniform in the direction of the measurements:

$$\frac{I_{\nu r}}{I_{\nu d}}\left(1 - e^{-\tau}\right) = \frac{e^{h\nu/kT} - 1}{e^{h\nu/kT}r - 1} \, ,$$

(3-14)

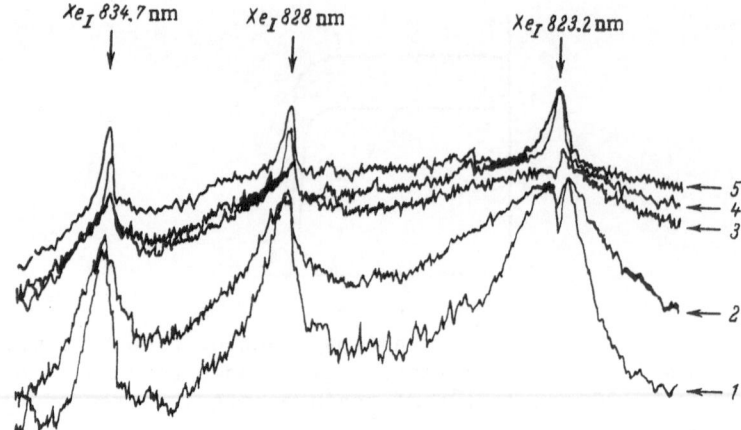

Fig. 3-2.   Microphotograms of the emission spectrum of an IFP 800
in the spectral range 820-835 nm. L = 80 µH.
1) U_0 = 500 V, w = 75 J;   2) 750 V, 170 J;   3) 1000 V,
300 J;   4) 1250 V, 470 J;   5) 1500 V, 680 J.

where $I_{\nu d}$ and $I_{\nu r}$ are the measured values of the luminous intensity of
the discharge and of the reference radiation source used, respectively;
τ is the optical depth of the discharge in the direction of the measurements; and T and T_r are the temperatures of the discharge under study
and the reference source.

The exponential dependence of the temperature on the measurable
quantities $I_{\nu d}$ and τ requires high-precision determination of plasma
absorption and emission, so that the accuracy of determination of the
temperature T is acceptable.

The electron concentration is estimated from the collisional
broadening of the contours of the lines of the main gas or impurities.
The concentrations of electrons and atoms also can be determined by
means of interferometers, including laser interferometers.

The main advantage of spectral methods of measuring the plasma
temperature and electron concentration which are based on the use of
plasma self-radiation is the fact that they do not introduce disturbances into the target of the measurements. However, when using them
one should keep in mind first that reabsorption at the wavelength of
the measurements must be negligibly small (otherwise a significant
error is introduced into the results of the measurements). Second,
during a determination of the discharge temperature the charged-
particle concentration is estimated either through independent
measurements (e.g., from the broadening of spectral lines) or by
computations using the values obtained for the temperature. In the
latter case we must additionally know the concentration of neutral

atoms n_a at the working gas-pressure p, which is related to the concentrations of plasma components and to the temperature by the equation of state

$$p = \sum_q n_s \, kT.$$

(3-15)

The charged-particle concentration then is determined by using Saha's formula (3-3). Direct measurements of n_a or p entail major experimental difficulties. Hence these parameters often are estimated indirectly (e.g., by using a calculation of the composition of a plasma in LTE). Such estimates cannot always be made with the required accuracy, especially for a pulsed discharge for which the time dependence of the plasma parameters must be determined.

Interferometric methods of determining the plasma parameters, which involve probing the plasma at two wavelengths, make possible a direct determination of the concentrations of charged and neutral particles. In this case the plasma temperature can be calculated with high accuracy from Saha's formula. A three-mirror laser interferometer (Figure 3-3) is especially convenient for studying a pulsed discharge in tubular lamps. This interferometer makes it possible easily to acquire time-resolved information on the concentrations of charged and neutral particles. Interferometric methods are based on the fact that any change in the state of a plasma is accompanied by a change in its refractive index. For electron, ion, and neutral gases the respective refractive indices μ_e, μ_i, and μ_a have these dependences on particle concentrations [3-23]:

$$\mu_e - 1 = -4.94 \cdot 10^{-14} \, \lambda^2 n_e,$$

(3-16)

$$\mu_{i(a)} - 1 = a \left(1 + \frac{b}{\lambda^2} \right) n_{i(a)},$$

(3-17)

where λ is the wavelength of the probing radiation (cm), and a and b are coefficients that depend on the type of gas (Table 3-1).

According to calculations [3-4 and 3-24], as a first approximation the polarizability of argon and xenon ions is equal to the polarizability of the neutron atoms (for argon the average polarizability of atoms is 1.65×10^{-24}, while the calculated polarizability of the ions is 1.36×10^{-24}; a similar correspondence also was obtained for xenon). For this reason we use one relation which takes into account the total contribution of the heavy-particle concentration $n_0 = n_a + n_i$ to refraction,* instead of two relations of the same

---

*Strictly speaking, we also should take into account the polarizability of the excited atoms, which may exceed the polarizability of atoms in the ground state by an order of magnitude or more. However, the relatively low concentration of excited atoms and the exponential decrease in the population with increasing energy of the level allow us to ignore this contribution.

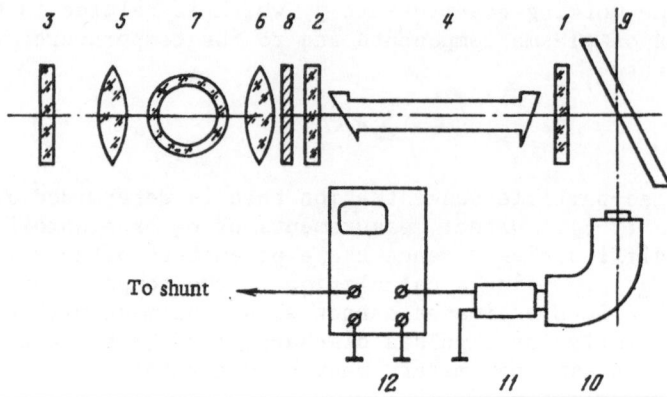

Fig. 3-3. Improved three-mirror interferometer [3-25]. 1, 2, 3)
          Mirrors of optical cavities; 4) He-Ne laser; 5, 6) long-
          focus lenses; 7) the flashlamp under study; 8) replaceable
          light filter; 9) tilting mirror; 10) monochromator;
          11) photoelectric multiplier; 12) oscillograph.

Table 3-1. Values of the Coefficients in Equation (3-17) for Various
           Gases [3-4]

Gas	$a \cdot 10^{24}$	$b \cdot 10^{11}$
He	1.3	2.3
Ne	2.5	2.4
Ar	10.0	5.6
Kr	16.0	. 7.0
Xe	25.0	10.0

type as (3-17). Hence the total refractive index μ for xenon plasma
and argon plasma (and also, we must assume, for a krypton plasma) is
determined by the expression

$$\mu - 1 = (\mu_e - 1) + (\mu_0 - 1) = -4.94 \cdot 10^{-14} \lambda^2 n_e + a\left(1 + \frac{b}{\lambda^2}\right) n_0. \qquad (3\text{-}18)$$

    If by selecting the wavelength of the probing radiation we can
ensure that the second term in the right-hand side of equation (3-18)
is much smaller than the first, then the electron concentration can be
calculated from the measured change in the refractive index of the
plasma by using relation (3-18).

    It is convenient to consider the outer cavity (Figure 3-3),
in which the plasma under study is placed, as a compound mirror

having a reflection coefficient r_{23}. Upon a monotonic change in the refractive index of the plasma, the value of r_{23} varies periodically, reaching a minimum when the lasing frequency matches one of the normal modes of the outer cavity. The number of cycles of modulations of f(t) which corresponds to a change of $\Delta\mu(t)$ in the value of μ is equal to

$$ f(t) = 2\Delta\mu(t)\, l/\lambda, \tag{3-19} $$

where l is the geometric length of the ray path in the plasma.

During two-wave probing of a plasma, the change in the electron and heavy-particle concentrations is determined by a system of equations that follows from relations (3-18) and (3-19):

$$ l\,\Delta n_e(t) = 10^{14}\frac{\lambda_2 f_2(t) - \lambda_1 f_1(t)}{8.96\left(\lambda_1^2 - \lambda_2^2\right)}, \tag{3-20} $$

$$ l\,\Delta n_0(t) = \frac{\lambda_1\lambda_2\left[\lambda_2 f_1(t) - \lambda_1 f_2(t)\right]}{4\pi\alpha_a\left(\lambda_2^2 - \lambda_1^2\right)}, \tag{3-21} $$

where $f_1(t)$ and $f_2(t)$ are the number of modulation cycles by time t at wavelengths λ_1 and λ_2, respectively; and $\alpha_a = a(1 + \frac{b}{\lambda^2})/2\pi$ is the polarizability of atoms in the ground state. If the contribution of the electron component to the refractive index is predominant, then it follows from relations (3-16) and (3-19) that

$$ l\,\Delta n_e(t) = -1.1\cdot 10^{13} f(t)/\lambda. \tag{3-22} $$

Figure 3-4 presents characteristic interferograms for the pulsed-discharge plasma in an IFP-800 lamp (the inside diameter is 7 mm, the electrode spacing is 80 mm, the "cold" xenon pressure is $p_0 = 0.053$ MPa (400 mm Hg), the capacitance of the supply capacitor is C = 600 μF, the voltage is $U_0 = 750$ V, and L = 80 μH). These interferograms were obtained by probing at two wavelengths (with a helium-neon laser). We can see quite well the "reversal points" which fix the time of disruption of the monotonic change* of the refractive index of the plasma due to the transition from the stage of discharge-current rise to the stage of current fall. At a wavelength of 0.63 μm the "reversal point" is shifted significantly with respect to the maximum current strength because of the difference in the sign of refraction for the electron gas and the heavy-particle gas.

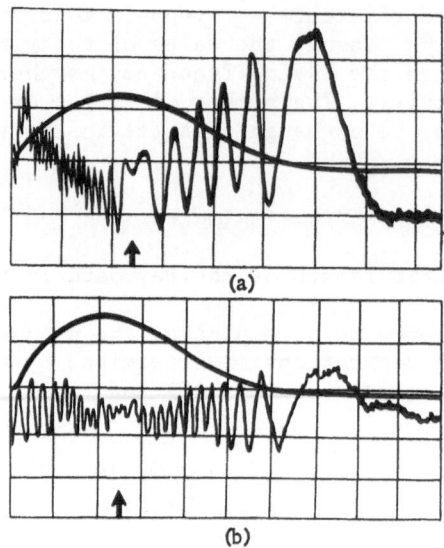

Fig. 3-4. Oscillograms of the modulation of the probing beam of a
three-mirror laser interferometer in a xenon plasma, and
of the current strength (the smooth curve) in a discharge
[3-26]. a) The laser wavelength is 0.63 μm; b) the laser
wavelength is 3.39 μm. The time scale is 200 μs/division.
The arrows point to the "reversal points."

The lower bound of the electron concentrations at which a laser
interferometer can be used is determined by the value of the modulation
cycle at the given probe wavelength. For example, the lower bound is
$(n_e l)_{min} = 1.75 \times 10^{17}$ cm^{-2} for the electron component at a wavelength
of 0.63 μm, but 3.3×10^{16} cm^{-2} at 3.39 μm. This limit can be lowered
by using schemes which provide reliable recording of fractions of the
modulation cycle. The upper bound for the use of an interferometer
is determined by the decrease in the contrast of the interference
pattern due to radiation absorption by the plasma. It can be seen
from Figure 3-4 that the amplitude of the modulation cycles decreases
appreciablyin the region of the maximum current when probing is done
at a wavelength of 3.39 μm. This is due to the absorption of the
probing radiation by the plasma. The same effect is evident in denser
plasmas at a wavelength of 0.63 μm, where the absorption coefficient
is lower. According to the estimates in [3-22] at 0.63 μm the contrast
of the interference pattern becomes unacceptable when
$(n_e l)_{max} \approx 5 \times 10^{18}$ cm^{-2}.

Figure 3-5 shows the pattern obtained from interferometric
measurements of the change in the electron and heavy-particle concentrations in xenon during a discharge-current pulse. For the lamp
in question, the heavy-particle concentration decreases by 50-80% by
the onset of the quasi-stationary stage compared with the initial

gas density due to the displacement of xenon atoms into the ballast volumes of the lamp. These volumes are formed by the gaps between the electrodes and the walls of the discharge tube. The heavy-particle concentration is practically unchanged throughout the entire period of current fall (only the number of ions and neutral atoms changes, while their sum remains constant), indicating the absence of appreciable motions of the gas in the tube. After the end of the discharge the original density is restored in a period of time on the order of 100 ms (according to calculations, precisely this amount of time is required for the discharge tube to cool).

Qualitatively analogous results were obtained in a study of the plasma composition in discharges in argon [2-49] and krypton [3-27]. The amount of gas expelled is practically independent of the type of gas, and is determined by the ratio of the ballast volume to the net volume and by the pulse length and intensity. Measurements of the electron concentration using a three-mirror interferometer have an error estimated at 7%, while the error for measurements of the heavy-particle concentration is 25%.

The results of measurements of the plasma composition in pulsed discharges were used to find the empirical dependence of the instantaneous values of the electron concentration n_e on the density of the discharge current j for xenon, krypton, and argon (Figure 3-6). We may note the following features in the relation obtained: (1) the effect of the type of gas and of the pressure is extremely weak; (2) as a first approximation, the electron concentration is proportional to the current density.

This fact has important consequences. Since

$$j = e n_e \bar{v}_e, \tag{3-23}$$

where $\bar{v}_e$ is the average drift velocity of electrons, the proportionality between n_e and j means that the electron drift velocity is constant over a broad range of discharge conditions. According to the data in Figure 3-6,

$$j \approx (2.6 \pm 0.4) \cdot 10^{-15} n_e. \tag{3-24}$$

For a specified current density (in A/cm^2), this makes it possible to determine the values of n_e with an error not exceeding 15%.

It follows from relations (3-23) and (3-24) that the electron drift velocity is $(1.5 \pm 0.3) \times 10^4$ cm/s. This is on average an order of magnitude lower than the values found in [3-28 and 3-29] for nonionized noble gases at the corresponding values of the parameter E/p. An appreciable dependence of the drift velocity on the type of gas also is observed in nonionized gases, in contrast to a plasma. This difference may be attributed to the nature of the predominant

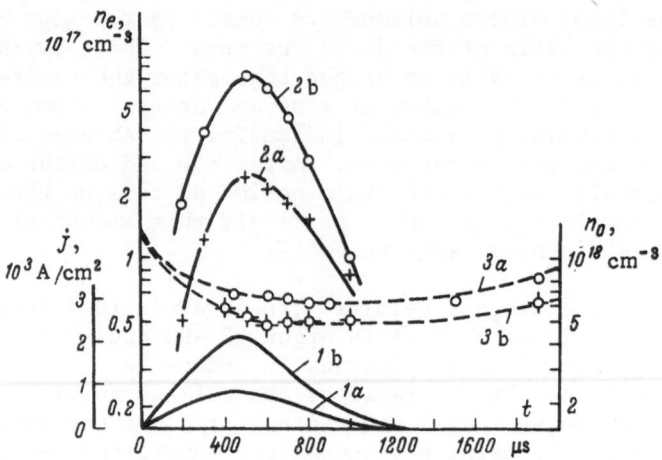

Fig. 3-5.   Time dependence of the concentration of components of a
             pulsed-discharge xenon-plasma [3-26]. The initial pressure
             is p_O = 0.053 MPa. 1) Current density j;   2) electron
             concentration n_e;   3) heavy-particle concentration n_O =
             n_a + n_i. a) U_O = 500 V;   b) U_O = 1000 V;   C = 600 μF,
             L = 80 μH.

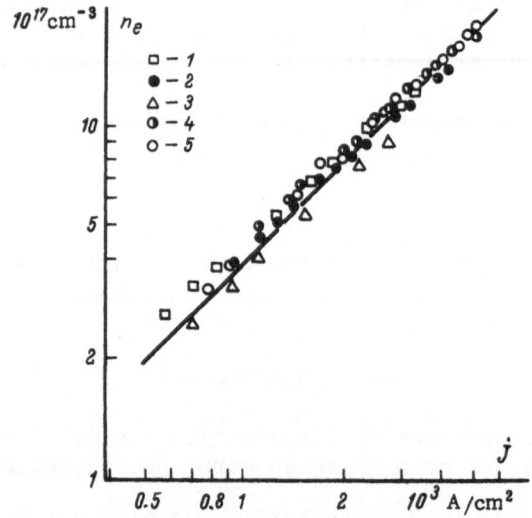

Fig. 3-6.   Dependence of electron concentration on current density
             [2-5]. The experimental points are: 1, 2, 3) discharges
             in xenon at p_O = 0.08, 0.053, and 0.013 MPa, respectively;
             4, 5) discharges in krypton and argon at p_O = 0.053 MPa;
             the solid line represents relation (3-24).

electron-particle interactions, as mentioned in Chapter 2: the electron drift velocity in a plasma with a noticeable degree of ionization
is determined by the interaction of electrons with ions, whose effective cross section is relatively large, being practically the same
for various noble gases. In a nonionized gas electrons interact with
neutral atoms, and the effective cross section of such interactions
is much smaller than that of Coulomb interactions, depending on the
type of gas.

     The values of the time dependences of the charged- and neutral-
particle concentrations make it possible to calculate the plasma
temperature from Saha's formula and the pressure from the equation
of state. The theoretical error in the calulated values of the
temperature does not exceed 4% for the errors indicated above in
the determination of plasma composition, and the pressure is determined with the same error as the neutral-particle concentration. The
results of temperature measurements using two independent methods are
compared in [3-30] for the discharge conditions mentioned. One method
is based on data from an interferometric determination of the plasma
composition, and the other is based on absolute measurements of
radiation intensity and absorption [relation (3-14)]. Both methods
gave matching values of the temperature throughout the entire range
of pulsed-discharge conditions studied (variation of current density
from 10^3 to 4.3×10^3 A/cm^2). This agreement indirectly confirms
the validity of Saha's formula in the quasi-stationary stage of a
pulsed discharge, i.e., the existence of at least ionization equilibrium in this stage.

     One important characteristic of a plasma in LTE is the gas
pressure. Direct measurements of the gas pressure, which entail
considerable technical methodological difficulties, can be made, for
example, by using pressure pickups mounted inside the gas-discharge
tube. The results of such measurements are presented in Figure 3-7,
together with calculated data from the results of interferometric
measurements. On the whole, the different experimental methods of
determining p give results which are in satisfactory agreement.

3-3. Elements of the Energy Balance, and the Electrical Properties of a Quasi-Stationary Plasma

     A knowledge of the elements of the energy balance of a pulsed
discharge is of considerable importance for understanding its basic
features. The incoming portion of the balance is the energy of the
external electric field that is released in the discharge (Joule
energy), while the outgoing portion consists of losses of all kinds.

     The value of the energy going into the discharge can be deter-
mined by measuring the time dependences of the current and voltage.
Since the discharge has ohmic resistance, the product of the instantaneous values of the current and voltage determines the instantaneous

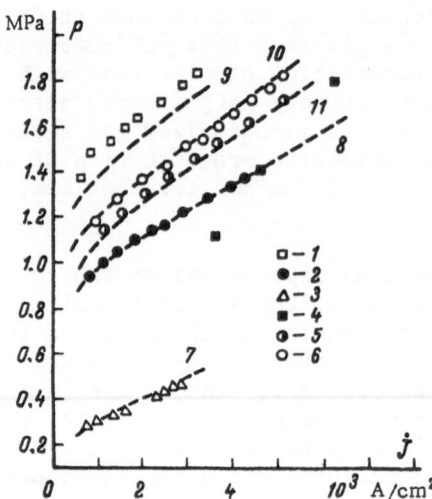

Fig. 3-7. Dependence of working pressure in an IFP-800 lamp on
current density. The experimental points are: 1, 2, 3)
Xe at p_0 = 0.08, 0.053, and 0.013 MPa, respectively [2-5];
4) Xe at p_0 = 0.053 MPa [3-31]; 5, 6) Kr and Ar at p_0 =
0.053 MPa [2-5]. The dashed lines represent values
calculated by the method given in Section 3-5; 7, 8, 9)
Xe at p_0 = 0.013, 0.053, and 0.08 MPa, respectively;
10, 11) Ar and Kr at p_0 = 0.053 MPa.

value of the power transferred to the lamp. Part of this power is
lost at the electrodes, but the electrode losses are comparatively
small (see Chapter 4).

The other losses ultimately reduce to two components: the
radiation escaping through the lamp walls, and losses on the walls
which cause wall heating. Measurements show that radiation accounts
for most of the energy consumption under optimal conditions (this
accounts for the value of such a pulsed discharge as a light source),
while heat losses in the walls of quartz lamps amount to just 15-25%
of the pulse energy (the losses in glass lamps are somewhat higher
because of the lower transparency of glass).

The integrated energy balance for the entire period of the discharge may differ from the power balance in the quasi-stationary
stage. This is due to the fact that during the formation of a discharge part of the energy goes to gas-dynamic processes and to a
change in the internal energy of the plasma. Calorimetric study of
the time dependence of wall losses (by cutting off the current at
different times during the discharge) and simultaneous measurements
of the instantaneous values of the current, voltage, and radiation

losses (using a radiation thermocouple) have made it possible to
estimate the instantaneous values of the energy balance throughout
the pulse. Figure 3-8 shows the time dependence of wall losses for
discharge energies less than 0.6 the limiting energy. At these energies
there is practically no vaporization of the walls (wall losses increase
quickly at higher energies, running to 40-50%; see Chapter 6). Over
60% of all the energy absorbed by the walls by the time of the maximum current, which comes approximately 450 µs after the start of a
pulse with a total length of 1.2 ms. The losses on the walls of a
quartz lamp as the discharge current density is varied from 1.5 x
10^3 to 5 x 10^3 A/cm^2 in a discharge tube with an inside diameter of
7 mm amount to 15 ± 3% of the instantaneous values of the electric
power. This is perhaps somewhat less than the same losses from
measurements of the time-integrated balance.

While the mechanism of radiative losses in a discharge is known
by and large, the mechanism of wall losses has not been studied very
thoroughly. Quite reliable estimates practically exclude heat conduction as the mechanism by which a significant portion of the enrgy
is transferred from the plasma to the walls. What remain are the
mechanisms of absorption of part of the discharge radiation by the
walls in the ulatraviolet, and ambipolar diffusion of particles
(ions and electrons), accompanied by an energy release upon recombination.

The quartz glass used to make the envelopes of tubular flashlamps,
which are the most powerful and efficient lamps, are opaque to radiation with a wavelength shorter than 200 nm. Depending on the purity
of the quartz, this limit may shift by ± 20 nm. The fraction of the
radiation from pulsed discharges in heavy noble gases that is absorbed

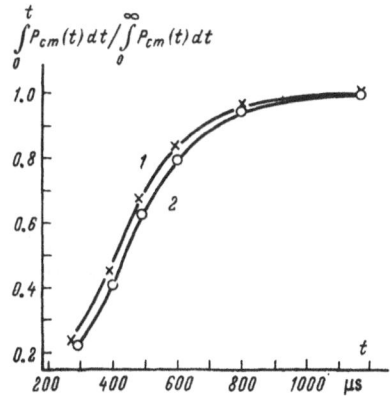

Fig. 3-8. Time dependence of losses in the walls of an IFP-800
lamp (no vaporization of walls) [3-32].
1) j_M = 2.2 x 10^3 A/cm^2;   2) j_M = 3.4 x 10^3 A/cm^2.

by quartz is estimated as 10-15% of the total radiant flux. A significant portion of this radiation lies in the spectral region adjacent to the resonance lines of these gases, near 120 nm. If the heating of the wall is accompanied by significant vaporization of the wall material, then the effect of "reversible opacity" is observed at the same time [3-33 and 3-33a]. This effect consists in the shifting of the transmission cutoff threshold into the longwave region, causing an additional increase in the radiant energy absorbed by the walls. There are no experimental data on the involvement of of particles in energy transfer to the walls, and estimates of the ion flows onto the walls are inapplicable to high pressures, since they are borrowed from calculations for a highly ionized low-density plasma in which there is a significant gap between the electron and ion temperatures. However, a comparison of the experimental data on energy losses on the walls with an estimate of the possible fraction of the discharge radiant energy absorbed by the walls enables us to assume that the absorption of the shortwave UV radiation of the discharge has the main effect under optimal conditions: pressures that are not too low (an initial pressure of over 10^4 Pa in the tube) and discharge-tube diameters that are not too small (at least 5 mm). The significant decrease in the luminous efficacy of a discharge as we move away from the optimal parameters (reduced pressure, a small diameter of the tube, and insufficient energy to fill the tube and establish a quasi-stationary discharge) may indicate an increase in other types of energy losses.

The most important physical parameter of a gas-discharge plasma, which determines its electrical properties, is the specific electrical conductivity. This is a macroscopic characteristic of charge transfer by the plasma which is related to two other electrical characteristics of a discharge (the current density j and the gradient of the electric field E) by the well-known equation

$$j = \sigma E. \tag{3-25}$$

Transport phenomena in a gas reflect the direction of kinetic processes. This direction results from external forces and is displayed against the background of random thermal motion of particles. The transport properties of a plasma in one of two extreme states have been studied in the most detail within the framework of the kinetic theory of gases: a slightly ionized plasma of comparatively low density, and a fully ionized plasma, which retains ideal properties. The first case is characterized by a rapid tailoff of interaction forces between particles with distance, and by the possibility of treating electron-atom interactions as pair collisions. The process of pair collisions is described by the Boltzmann kinetic equation. Enskog's method, based on the solution of the Boltzmann equation, was used in Ref. [3-34] to derive an expression for the electrical conductivity of a weakly ionized ideal plasma:

$$\sigma = 0.532 \frac{e^2}{(m_e kT)^{1/2}} \frac{x}{q_{ea}},$$

$$\tag{3-26}$$

where $x = n_e/(n_a + n_e) \approx n_e/n_a$ is the degree of ionization of the gas, q_{ea} is the electron-atom interaction cross section (determining q_{ea} is a separate task), and m_e the electron mass.

Coulomb interactions between charged particles have a decisive effect in a fully ionized gas. Such interactions exclude the hypothesis of short-range forces, and hence the hypothesis of pair collisions. The interactions become collective in character. Nonetheless, when certain conditions are met, Coulomb interactions may be thought of as equivalent electron-ion pair collisions which are described by the Boltzmann kinetic equation [3-35 to 3-38]. The conditions are: 1) predominance of multiple electron-ion collisions having a large impact parameter (distant collisions), each of which results in a small change in particle velocity; 2) limitation of the range of Coulomb interactions by the charge screening effect (a sphere having the Debye radius $r_D = (kT/4\pi n_e e^2)^{1/2}$ may be adopted as the limit of long-range action); and 3) the presence of a sufficiently large number of particles in the Debye-radius sphere ($n_D \approx n_e r_D^3 \gg 1$).

If these conditions are fulfilled, the frequency ν_{ei} of electron-ion collisions and the conductivity of a fully ionized ideal plasma are determined by the relations derived by Spitzer [3-39]:

$$\nu_{ei} = 1.8\, Z n_e \ln \Lambda / T^{3/2}, \qquad\qquad (3\text{-}27)$$

$$\sigma = 1.55 \cdot 10^{-4}\, T^{3/2}/Z^2 \ln \Lambda, \qquad\qquad (3\text{-}28)$$

where Z is the charge of an ion, and $\ln \Lambda$ is the "Coulomb logarithm" (this logarithm is slightly dependent on the type of gas, density, and temperature, having a value of about 3 under ordinary conditions; $\Lambda = 1.25 \times 10^4\, T^{3/2}/Z n_e^{1/2}$). Relations (3-27) and (3-28) also can be used for a partially ionized ideal plasma in which the total cross section of electron-ion interactions $n_e q_{ei}$ is much larger than the total cross section of electron-atom collisions $n_a q_{ea}$. One way of calculating the conductivity of a partially ionized plasma with allowance for both electron-ion and electron-atom collisions was proposed in [3-40] and developed in a number of subsequent papers. This method consists in constructing interpolation formulas which become relations (3-26) and (3-28) in the limiting cases of a weakly and a fully ionized plasma, respectively. This method makes it possible to obtain relatively simple relations which are convenient for computations, although there is no substantiation for their validity in the intermediate region between the two limiting cases.

Another method consists in the numerical solution of the Boltzmann equation for a partially ionized gas. This entails major computational difficulties, but does provide substantiated results for any values of the degree of ionization. This method has been used to calculate the transport coefficients for the noble gases [3-41 to 3-46]. The calculated dependence of the conductivity in argon on the

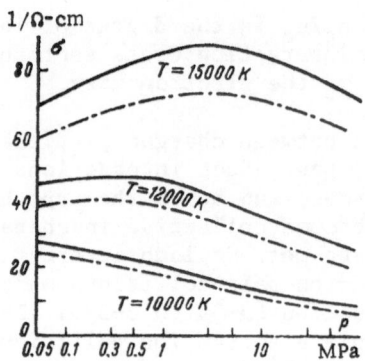

Fig. 3-9.   Calculated dependences of conductivity in argon on gas
            pressure (solid lines), as obtained by Devote, and the
            relations which follow from Spitzer's theory (dashed line).

gas pressure is shown in Figure 3-9, which is taken from [3-44], for
different values of the temperature. We may note that the results of
[3-44] are in qualitative agreement with a calculation using formula
(3-28), but quantitatively exceed it somewhat, throughout the entire
range of pressures (0.05-100 MPa) and temperatures (10,000-17,000 K).
This may mean that the contribution of electron-atom collisions is
insignificant, compared with that of electron-ion interactions, in a
refined calculation of the conductivity of a partially ionized gas for
these temperature and pressure ranges.

     These results no longer are valid when ideality of the plasma is
violated. A deviation from the ideal state occurs when the energy
of the Coulomb interactions between charged particles becomes commensurate with the energy of thermal motion of the particles. The
dimensionless factor

$$\gamma = Ze^2/kTd = 1.67 \cdot 10^{-3} n_e^{1/3}/T \qquad\qquad (3\text{-}29)$$

(where d is the mean distance between ions) is used as a measure of
the nonideal state of the plasma. This factor is equal to the ratio
of the potential energy of Coulomb interaction to the kinetic energy
of the random motion of the particles.

     The plasma parameters of pulsed discharges in tubular lamps that
were described in Section 3-2 make it possible to estimate the degree
of nonideality of the plasma and the applicability to such a plasma
of existing methods of calculating the transport coefficients. Depending on discharge conditions, the factor γ varies over the range
0.1-0.15. In every case the number of particles in a sphere having
the Debye radius is 1.5-3. Let us point out for comparison that
γ varies from 0.04 to 0.07 and n_D = 6-10 in atmospheric arcs at the
same temperatures. The plasma state in arcs thus does not differ
markedly from the ideal, and the condition $n_D \gg 1$ is fulfilled quite

well. By contrast, the deviation from the ideal is quite noticeable
in high-pressure pulsed discharges, and the number of particles in the
Debye sphere is commensurate with unity. A plasma having such parameters, for which the main conditions for applicability of existing
methods of calculating the transport coefficients are violated, is
called a slightly nonideal, non-Debye plasma.

The kinetic theory of a nonideal plasma, which substantiates the
methods used to calculate its transport properties, is at the very
start of its development. As a review of the problems and attained
level of theory notes [3-52], many problems of the theory have not
been solved completely, and many have not even been posed. The importance of experimental determination of the transport coefficients
in the plasma of a pulsed discharge understandable increases under
these conditions.

The peculiarities of a pulsed discharge in tubular lamps that
were noted in the previous section (namely, the attainment of LTE
in the quasi-staionary stage and constancy of parameters throughout
the entire volume occupied by the plasma) make it possible to determine the plasma conductivity for each instant formula (3-25). This
is done on the basis of the values of the current density j in the
discharge and the electric field strength E, which are taken from
experimental current-voltage characteristics in the quasi-stationary
stage of the type shown in Figure 3-10.

The error in the conductivity determined in this way consists
first of the error in the determination of j and E, and second of
the error introduced by the assumption of a uniform temperature
distribution in the discharge. According to the data of [3-20 and
3-21], the observed deviation of the temperature profile from the
ideal (see Figure 3-1) ensures that the value of the conductivity,
averaged over the diameter of the discharge, differs from the maximum
by no more than 3% (this deviation increases to 8% for the temperature
profile measured in a rectangular tube). As for the error in the
determination of current strength, tube diameter, and electric field
strength, they allow a determination of σ relation (3-25) with an
error not exceeding 15%.

It has been found as a result of a systematic study of the
dependence of the conductivity on discharge parameters that the
conductivity for a specific lamp is unequivocally dependent on the
current density and very slightly dependent on the tube radius, the
type of noble gas, and the initial pressure, decreasing somewhat as
p_0 and the atomic number increase.

Some results of measurements of the dependence of the conductivity on the current density are presented in Figures 3-11 and 3-12.
Approximation relations (2-2) and the relation $\sigma = 0.885j^{0.5}$, which
was given in [3-16], are presented in Figure 3-11. The approximation
formulas for the dependence of the conductivity on current density do

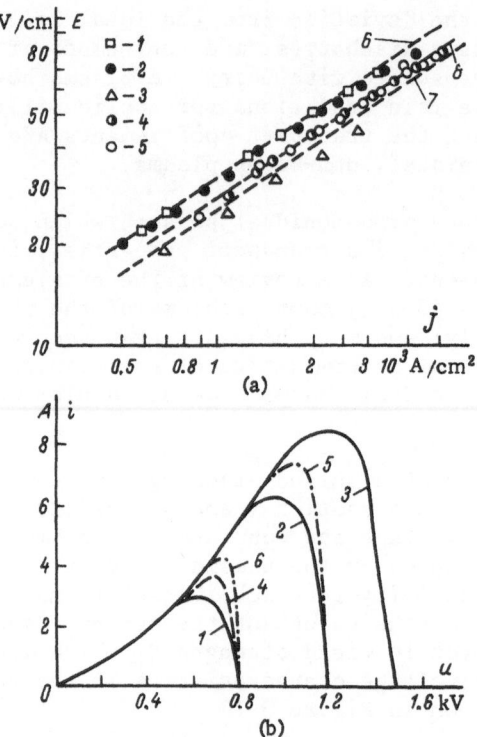

Fig. 3-10. Current-voltage characteristics of a pulsed discharge in
            the quasi-stationary stage.
            a) in an IFP-800 lamp [2-5]. The experimental points are
            1, 2, 3) Xe at p_0 = 0.08, 0.03, and 0.013 MPa, respectively;
            4, 5) Kr and Ar at p_0 = 0.053 MPa. The dashed lines were
            calculated by the method given in Section 3-5; 6) Xe, p_0 =
            0.08 MPa and 0.053 MPa; 7) Xe, p_0 = 0.013 MPa, and Ar,
            p_0 = 0.053 MPa; 8) Kr, p_0 = 0.053 MPa.
            b) current-voltage characteristics of a pulsed discharge
            [1-67 and 2-7]. Xe, p_0 = 0.013 MPa, l = 7 cm, d_i = 0.34 mm;
            1, 2, 3) C = 0.1 μF; 4, 5) 0.25 μF; 6) 0.5 μF.

not take into account the effect of the type of gas and the pressure.
At the same time, as we can see from Figure 3-12, a change in gas
pressure for a fixed current density may be accompanied by a change
in conductivity that exceeds the error of measurement.

     Data that make it possible to relate electrical properties to
the state of the plasma, i.e., to its temperature and the charged-
particle concentration, are of decisive importance for comparing the
theory of plasma conductivity with experiment. Such experimental
data are given for xenon in Figure 3-13. This figure also shows the

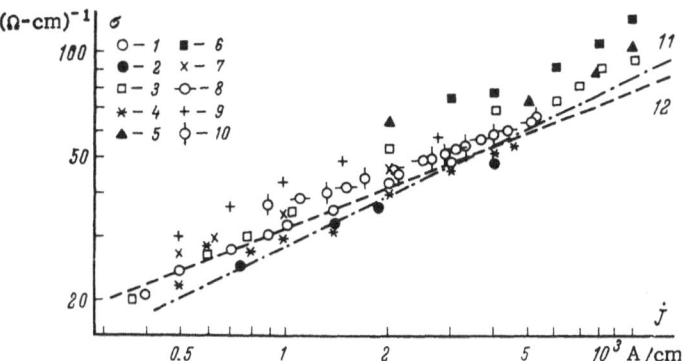

Fig. 3-11. Experimental data on plasma conductivity versus current
density. Xenon: 1, 2) p_0 = 0.053 MPa, d_i = 1.1 and 0.67cm
respectively, l = 13 and 8 cm [3-47]; 3) p_0 = 0.04 MPa,
d_i = 1.1-3.6 cm [3-48]; 4) p_0 = 0.053 MPa, d_i = 0.7 cm,
l = 8 cm [3-49]; 5, 6) p_0 = 0.053 MPa, d_i = 0.74 and
1.55 cm, respectively [3-50]. Krypton, p_0 = 0.053 MPa:
7) d_i = 1.1 cm, l = 13 cm 3-47 ; 8) d_i = 0.7 cm, l = 8 cm
[3-27]. Argon, p_0 = 0.053 MPa: 9) d_i = 1.1 cm, l = 13 cm
[3-47]; 10) d_i = 0.7 cm, l = 8 cm [2-49]. The empirical
relations are 11) $\sigma = 0.885j^{0.5}$ [3-16]; 12) the formula
$\sigma = 2j^{0.4}$ (2-2).

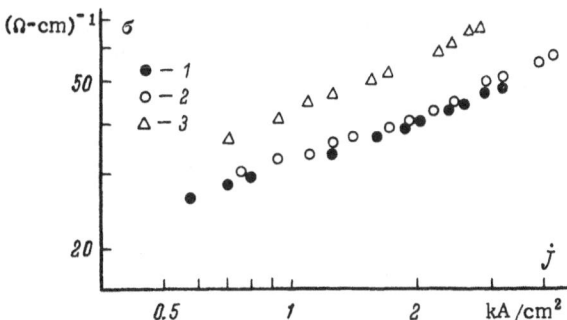

Fig. 3-12. Dependence of conductivity on current density in an
IFP-800 lamp [3-49]. 1) p_0 = 0.053 MPa; 2) 0.08 MPa;
3) 0.013 MPa.

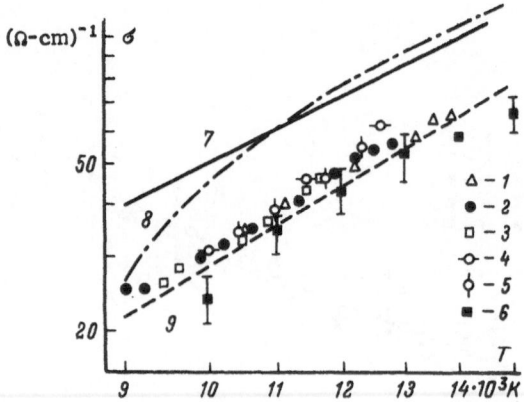

Fig. 3-13. Temperature dependence of the conductivity of a xenon
plasma. Experiment: 1, 2, 3) p_O = 0.013, 0.053, and
0.08 MPa, respectively [3-49]; 4) p_O = 0.04 MPa [3-51];
5) p_O = 0.053 MPa [3-20]; 6) p_O = 0.053 MPa [3-53].
Calculation: 7) the theory of a fully ionized gas [3-39]
(the experimental values of n_e and T were used); 8) the
theory of a partially ionized ideal gas, p = 1 MPa [3-41
and 3-46]; 9) approximation formula (3-33).

results of a calculation of $\sigma(T)$ from Spitzer's formula (3-28) and
from the data of Devoto [3-4] (for p = 1 MPa), which allow for the
contributions of electron-ion and electron-atom interactions in a
partially ionized plasma in the ideal-plasma approximation. A comparison of the two theoretical findings and the experimental data
shows that the contribution of electron-atom collisions is important
only at temperatures below 10,000 K and that the experimental values
of the conductivity are approximately 40% lower than the theoretical
ones.

The results of experimental and computational determinations of
the temperature dependence of the conductivity of krypton and argon
plasmas are presented in Figures 3-14 and 3-15. These figures lead
to the same conclusion, namely that the experimental points are approximately 40% lower than the corresponding theoretical ones. The
experimental values shown in Figures 3-14 and 3-15 for the conductivity
of plasmas of krypton and argon arcs at atmospheric pressure are close
to the theoretical values, exceeding the corresponding values for
pulsed discharges by an average of 30%. The results of all conductivity measurements in the quasi-stationary stage of pulsed discharges
in these three noble gases are summarized by single plots in Figure
3-16. The higher the ionization potential of the gas, the higher the
conductivity at the same temperature (the charged-particle concentration
increases with decreasing ionization potential at a given temperature).

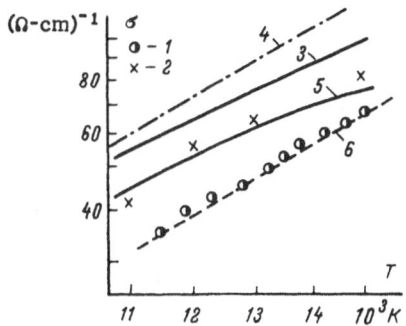

Fig. 3-14. Temperature dependence of the conductivity of a krypton plasma. Experiment: 1) p_O = 0.053 MPa [3-27]; 2) an atmospheric arc [3-54]. Calculation: 3) the theory of a fully ionized gas [3-39]; 4,5) the theory of a partially ionized gas for p = 1 and 0.1 MPa [3-41]; 6) the curve according to relation (3-33).

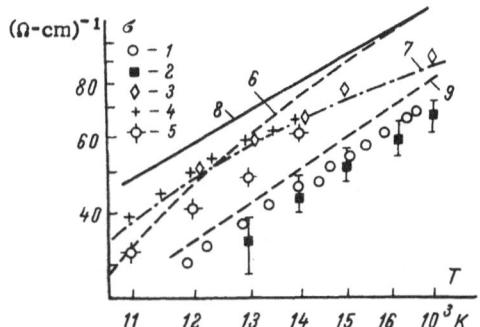

Fig. 3-15. Temperature dependence of the conductivity of an argon plasma. Experiment: 1) p_O = 0.053 MPa [2-49]; 2) p_O = 0.053 MPa [3-53]; 3,4,5) atmospheric arcs according to [3-54 to 3-56]. Calculation: 6,7) the theory of a partially ionized gas [3-44] for p = 1 and 0.1 MPa; 8) the theory of a fully ionized gas [3-39]; 9) the curve according to relation (3-33).

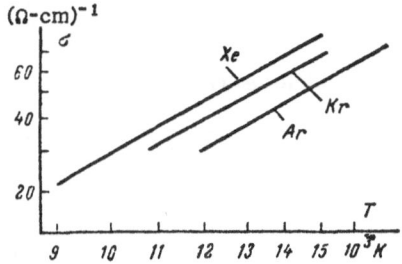

Fig. 3-16. The empirical relations σ(T) for three noble gases [2-49].

The appreciable difference noted above between the measured values of the plasma conductivity of high-pressure pulsed discharges and those of the same gases cannot be explained by the existing theory of charge transport in an ideal plasma with predominant electron-ion interactions. The conditions in a pulsed-discharge plasma differ from those in the plasma of an atmospheric arc in that there are higher pressures in a pulsed discharge (0.3 - 1.8 MPa in the experiments described). As noted above, the conditions in a pulsed-discharge plasma also have a more significant deviation from the ideal and a small number of particles in the Debye sphere (from 1 to 3 particles). Some authors attibute the decrease in conductivity that is observed for pulsed discharges to the increased percentage of electron-atom collisions [3-58 and 3-59] or to the effects of collective Coulomb interactions in a nonideal plasma. Since no methods exist for measuring separately the contributions of various interactions to the conductivity, experiments cannot confirm any of these hypotheses. As for the theoretical approach, calculations of the conductivity using the most reliable values for the cross sections of electron-atom and electron-ion collisions have been made consistently within the framework of the kinetic theory of transport in an ideal plasma [3-44 and 3-45]. These calculations indicate the extremely small effect of electron-atom collisions compared with electron-ion collisions throughout nearly the entire range of temperatures and pressures that are reached in low-temperature, high-pressure pulsed discharges.*

The authors of [2-5] proposed the introduction of a functional correction into theoretical relations (3-27) and (3-28). The correction is determined by comparing the experimental values of the collision frequency in a plasma, as obtained from the relation

$$\nu_{exp} = \frac{e^2}{m_e} \frac{n_e}{\sigma} \tag{3-30}$$

for measured values of n_e and σ, with the theoretical values calculated from relation (3-27) when the experimental values of T and n_e are inserted. The correction, $\beta = \nu_{exp}/\nu_{theor}$, is meaningful if its dependence on the plasma parameters can be determined.

Since β is a factor in relation (3-27), it may be considered a correction to the ion charge Z if we introduce the concept of the effective charge Z_{eff}. (The effective charge is interpreted as a quantity which characterizes the effective field of many ions which interacts with an electron under conditions where the additivity of charged-particle interactions is violated.) For single ionization,

---

*The hypothesis that electron-atom interactions make a significant contribution to the conductivity of a high-pressure plasma is supported by calculations based on approximate interpolation methods which are inferior in accuracy and consistency to the calculations made in [3-44 and 3-45].

the more the plasma state differs from the ideal, the more Z_{eff} differs from unity.

The value of the effective charge based on this interpretation should depend on the average distance between ions, which is proportional to $n_e^{-1/3}$, and on the average velocity of the particles. The average particle velocity is a function of the plasma temperature under conditions where LTE is reached (e.g., the velocity may be proportional to T^α). The hypothesized dependence of Z_{eff} on the plasma parameters may be represented in the form

$$Z_{eff} = 1 + f \left(n_e^{1/3}, \ T^\alpha \right). \tag{3-31}$$

The analysis made in [2-5] of the copious experimental data pertaining to both superatmospheric and atmospheric arcs in various gases has made it possible to express relation (3-31) in explicit form:

$$Z_{eff} = 1 + 53 \, n_e^{1/3}/T^2 \ , \tag{3-31'}$$

where n_e is in cm^{-3} and T is in degrees Kelvin. The experimental data and relation (3-31') are presented in Figure 3-17a.

Allowing for correction (3-31'), the conductivity of the plasma is determined by the relation

$$\sigma = 1.55 \cdot 10^{-4} \, T^{3/2}/Z_{eff}^2 \ln \Lambda_1, \tag{3-32}$$

where $\Lambda_1 = \Lambda/Z_{eff}$. Relation (3-32) matches Spitzer's formula for the conductivity of a fully ionized plasma if $Z_{eff} = 1$, i.e., for an ideal plasma.

It is not difficult to satisfy ourselves that the quantity

$$Z_{eff}^2 \ln \Lambda_1 = Z_{eff}^2 \ln \left(1.25 \cdot 10^4 \, T^{3/2}/Z_{eff} \, n_e^{1/2} \right)$$

turns out to be proportional to the dimensionless factor eU_i/kT after the value of n_e is inserted from Saha's formula. Hence expression (3-32) may be replaced by the approximate relation

$$\sigma \approx C T^{5/2}/U_i, \tag{3-33}$$

where C is a constant to the accuracy of its constituent quantities, which are slightly temperature-dependent. This constant can be determined from empirical data. This factor actually may be considered constant throughout the entire range of temperatures from 9000 to 17,000 K, as we can see from Figure 3-17b, where the empirical temperature dependence of C is shown for three noble gases. Here $C = 3.4 \times 10^{-8}$ Ω^{-1} cm^{-1} K$^{-5/2}$ with an error of less than 12%.

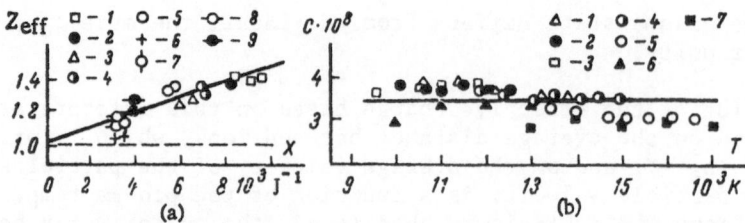

Fig. 3-17. Experimental data on the dependence of the effective
ion charge Z_{eff} on the parameter $X = 1.2 \times 10^{13} n_e^{-1/3} T^2$
[2-5] (a), and data on the temperature dependence of the
coefficient C in relation (3-33) (b).
a) 1,2,3) xenon, p_0 = 0.08, 0.053, and 0.013 MPa; 4,5)
krypton and argon, p_0 = 0.053 MPa [2-49 and 3-27]; 6,7)
argon, p_0 = 0.1 MPa [3-55 and 3-56]; 8,9) nitrogen,
p = 0.1 and 0.2 MPa [3-56]; the solid line represents
relation (3-31').
b) 1,2,3) xenon, p_0 = 0.013, 0.053, and 0.08 MPa [3-49];
4,5) krypton and argon, p_0 = 0.053 MPa [2-49 and 3-27];
6,7) xenon and argon, p_0 = 0.053 MPa [3-53]; the line
represents the average value.

3-4. Radiative and Absorptive Properties of a Quasi-Stationary Plasma

As stated above, the emission mechanism of a low-temperature
gas-discharge plasma is determined by the free-free transitions of
electrons in the field produced by ions or atoms with a loss of part
of the kinetic energy, and by the free-bound and bound-bound transitions of electrons to lower-lying states of the atoms and ions.
The absorption of electromagnetic energy in the plasma is governed
by the corresponding inverse processes. This mechanism leads to the
formation of both continuous and discrete emission and absorption
spectra.

The radiative properties of a plasma are characterized by the
spectral radiation coefficient ε_ν, which determines the amount of
radiant energy emitted by a unit volume of the plasma in a unit spectral interval per unit time per unit solid angle. The radiation
absorption in the given spectral region is determined by the absorption coefficient κ_ν, which is expressed in inverse centimeters. The
quantity $1/\kappa_\nu$ is equal to the ray path of light of frequency ν in a
uniform plasma, on which path the luminous intensity decreases by a
factor e because of absorption.

Under LTE the radiation coefficient and the absorption coefficient are related to each other by Kirchhoff's law (3-4). This
circumstance creates the impression that the two parameters are of

equal importance and that one of them can be used to characterize
the radiative properties of the plasma (provided that the plasma
temperature is known). However, these parameters are only equivalent
for practical purposes in the case of an optically transparent plasma
for which the optical thickness $\kappa_\nu l$ is small. But if the reabsorption of radiation is appreciable and $\kappa_\nu l$ is not small, then difficulties due to the necessity of allowing for radiation transfer are
encountered during the experimental determination of the radiation
coefficient or when its calculated value is used. For example, the
radiation coefficient ε_ν is not equal to Φ_ν/l for a measured radiant
flux Φ_ν from a plasma layer of size l, since the radiation coefficient
is a function of not only the temperature and pressure, but the coordinate as well. Conversely, is the calculated value of ε_ν is used,
the radiant flux Φ_ν cannot be determined without allowing for radiation
transfer. Such difficulties do not arise in the determination of the
absorption coefficient κ_ν. It therefore proves more convenient to
use this parameter.

In addition to the local spectral radiation and absorption
characteristics, it is important in practical problems to know the
overall characteristics of a gas-discharge plasma having certain
geometric configurations: the total radiation of the entire plasma
volume in a specific spectral region or throughout the entire spectral
band. Finding the relations between overall characteristics and the
physical parameters of radiation and absorption is a highly complex
undertaking that can be accomplished for only a few particular cases.

In experiments one usually tries to measure the continuous or
line radiation or absorption separately in view of the fact that the
radiation and absorption are discrete and continuous and the computational methods used to determine quantitities of this type vary.
The methods used to calculate the absorption coefficient (or the
radiation coefficient) in spectral lines have been treated in detail
in [3-59 and 3-60]. As a first approximation, the radiant flux of
a line is proportional to the difference between the populations of
the upper and lower levels of the transition and to the probability
A_{ki} of this transition. The population of a specific level in a
plasma in LTE at a specified temperature and pressure is determined
by Boltzmann equation (3-2). Therefore, the task of calculating the
intensity of discrete radiation (or absorption) reduces to the determination of the corresponding transition probabilities. This quantity
has been found experimentally for many lines of various gases. If it
must be determined by computation, one should keep in mind that an
exact calculation of A_{ki} is possible only for a one-electron system.
There are approximate quantum-mechanical methods of determining A_{ki}
for a multielectron system, but using them requires additional information which is provided by experiment. For example, using the
well-known Hartree-Fock methods, which are covered in [3-5 and 3-59
to 3-62], to calculate the transition probabilities for a large
number of lines turns out to be practically impossible. In such

cases use is made of the method of Bates and Damgaart [3-63], in which experimental data on the energy levels of the given transition are used for the calculation. The accuracy of this method is low for calculations of A_{ki} for a single transition, but the accuracy of the final result proves entirely acceptable for the solution of practical problems involving calculations of the overall radiation coefficient (or absorption coefficient) for a large number of lines.

The continuous absorption (or radiation) in intense high-temperature discharges is commensurate with line absorption (or radiation) in intense high-temperature discharges, and predominates, for example, in a high-pressure xenon discharge. The basic mechanism of continuous absorption in a low-temperature plasma is the process of photo-ionization, which is examined in detail in [3-61, 3-64, and 3-65]. The theoretical calculation of continuous absorption includes a determination of the photoionization cross sections for all states from which a photoelectric effect is possible at a frequency ν, determination of the number of particles in each such state, and summation of the contributions of the individual states [2-23 and 3-6]. Allowing for the contribution of braking processes (free-free transitions) to absorption (or radiation) does not pose fundamental difficulties. The bremsstrahlung contribution in pulsed discharges having a plasma temperature of about 10,000 K is appreciable only in the infrared for $\lambda \geq 2$ μm. This contribution is small compared with the emission in the shorter-wave region of the spectrum.

The central link in the calculation of continuous absorption is the determination of the photoionization cross section. The quantum defect method proposed by Burges and Seaton [3-66 to 3-68] can be used to do this successfully. A substantiation of the method is given in [3-69]. A method of calculating the continuous absorption (or radiation) of multielectron atomic systems has been developed by Biberman and Norman, using the quantum defect method with some simplifying assumptions. The following relations in particular are convenient for practical use [2-23 and 3-70 to 3-72]:

$$\kappa(\nu, T) = 4.3\, n_e\, n_i\, (kT)^{-1/2}\, Z^2 \exp\left[\frac{h(\nu + \Delta\nu) - e\,\Delta U_i}{kT}\right]\xi(\nu)\,\nu^{-3}, \qquad (3\text{-}34)$$

$$\varepsilon(\nu, T) = 6.36 \cdot 10^{-54}\, n_e\, n_i\, (kT)^{-1/2}\, Z^2 \exp\left[\frac{h\,\Delta\nu - e\Delta U_i}{kT}\right]\xi(\nu), \qquad (3\text{-}35)$$

where $\kappa(\nu, T)$ is in cm^{-1}, $\varepsilon(\nu, T)$ is in $J \cdot cm^{-3} \cdot s^{-1} \cdot sr^{-1}$, and T is in degrees Kelvin; ΔU_i is the decrease in the ionization potential, as determined by relation (3-10), for example (in volts); and $\Delta\nu$ is the shift of the threshold frequency of the boundary of the series, as determined in the first approximation by the Ingliss-Teller relation (in s^{-1}) [3-73]:

$$\log h\Delta\nu = 5.069 + 0.267 \log n_e, \qquad (3\text{-}36)$$

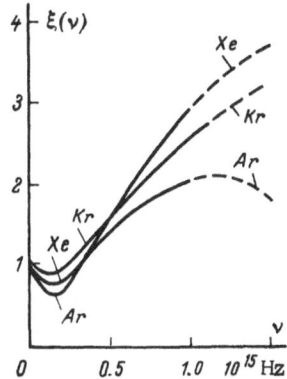

Fig. 3-18.   Spectral dependence of the factor $\xi(\nu)$ for three noble
gases.

where $\xi(\nu)$ is a dimensionless factor that can be determined by a
quantum-mechanical calculation (the calculated values of $\xi(\nu)$ from
[2-23] are shown in Figure 3-18).

Relations (3-34) and (3-35) are valid for frequencies ν below
the threshold frequency ν_{th}, which determines the boundary of the
dense group of terms for a specific atom photoionization from which
can be taken into account on an integral basis. For xenon, for
example, $\nu_{th} \approx 7 \times 10^{14}$ s^{-1}. The contributions of absorption and
radiation at frequencies $\nu > \nu_{th}$ are determined by the relations

$$\kappa(\nu, T) = \kappa(\nu_{th}, T)\frac{\xi(\nu)}{\xi(\nu_{th})}\exp\left[\frac{h(\nu - \nu_{th})}{kT}\right]\left(\frac{\nu_{th}}{\nu}\right)^3 + \Sigma\, \kappa_{n,l}(\nu, T),$$
(3-37)

$$\varepsilon(\nu, T) = \varepsilon(\nu_{th}, T)\frac{\xi(\nu)}{\xi(\nu_{th})}\exp\left[\frac{h(\nu - \nu_{th})}{kT}\right] + \Sigma\, \varepsilon_{n,l}(\nu, T).$$
(3-38)

The sum of the absorption coefficients (or radiation coefficients) in the right-hand side of these relations takes into account
the contributions of the individual, isolated levels of an atom whose
frequencies exceed ν_{th}. The number of such levels is small in noble
gases.

The calculation of the intensities of continuous radiation and
radiation at individual lines for a uniform, optically transparent
plasma makes it possible to determine the total radiation from the
region occupied by the plasma if the geometric dimensions of the
region are known. However, the approximation of an optically transparent plasma is unacceptable in most cases involving pulsed discharges: calculations of the local values of the continuum and

line radiation must be supplemented by allowing for radiation transfer. This requires that the temperature field be known throughout the entire space occupied by the plasma.

A method of calculating the integral radiation of an absorbing plasma was proposed in [3-74] and applied to argon in [3-75 and 3-76]. The total radiant flux passing through a unit area at the center of a uniformly heated hemisphere was determined in the calculation. As was shown in [3-77], the result of such a determination of the radiant flux also remains valid for a discharge with a cylindrical configuration for which the diameter of the plasma cylinder is the characteristic dimension. This makes it possible to extend all computational findings obtained by the method of [3-74] to the case of pulsed discharges having a uniform cylindrical column.

Calculations of the total radiation of an argon plasma made in [3-75 and 3-76] for a broad range of temperatures, pressures, and characteristic plasma dimension suggest a number of interesting qualitative conclusions. First, the radiant flux density is practically independent of the characteristic plasma dimension, at least up to 10 cm inclusively. Hence a plasma that has appreciable absorption exhibits "quasi-transparent" properties. Second, the radiant flux density remains proportional to the pressure in the range from 0.01 to 10 MPa. This proportionality is broken only at low temperatures (around 6000 K), mainly in the high-pressure region (on the order of 10 MPa). Third, the temperature dependence of the total radiation obeys an exponential law quite well if the degree of ionization is not too high (below 50%).

Experimental methods of determining the line absorption coefficient are covered in [3-60, 3-78, and 3-79]. Measurements of the continuous absorption coefficients are based on the same principles but differ somewhat in the techniques of execution.

One method of determining the absorption coefficient, which involves measuring the angular distribution of the radiation intensity in a uniform axisymmetric plasma (the indicatrix method), is based on the theoretical premises in [3-80] (see Section 5-5). This method is quite difficult to implement and requires both high accuracy in the computation of theoretical curves which must be corrected for the distortion of the angular distribution of the luminous intensity because of transmission of radiation through the walls of the tube [3-81] (see Section 5-5), and an extremely precise experiment. The use of Kirchhoff's law in the form

$$I_\nu = I_{p\nu}\left(1 - e^{-\varkappa_\nu l}\right), \tag{3-39}$$

where $I_{p\nu}$ is the intensity of blackbody radiation at a temperature T, may be considered a variant of this method.

A more effective method is to determine the absorption coefficient by illuminating the plasma with its own radiation, as
reflected by a mirror. If a plane mirror having a reflection coefficient r is installed as the radiation source (e.g., a discharge
tube), then a detector positioned in front of the tube will record
an increase in the luminous intensity I compared with the luminous
intensity I_0 when the mirror is either absent or closed. The increase will correspond to the expression:

$$I = I_0 \left(1 + r\delta^4 e^{-\kappa l}\right), \tag{3-40}$$

where δ is the coefficient of transmission of radiation by the walls,
and l is the thickness of the illuminated layer. If the values of r,
δ, and l are known, then the ratio $I_\nu/I_{0\nu}$ must be measured in order
to find the absorption coefficient κ_ν. It was demonstrated in [3-82]
that measurements using a concave mirror are more effective than
measurements using a plane mirror. Successful measurements have been
made of the absorption coefficients in a pulsed discharge in xenon
by means of a parabolic mirror [3-83]. The shortcomings of the method
are the need to determine the values of r and δ exactly at each wavelength and the decrease in the accuracy of the measurements as the
absorption coefficient increases, since the ratio I/I_0 approaches
unity.

Finally, the most effective method of determining the absorption
coefficient is based on illuminating the plasma under study with the
radiation of an auxiliary pulsed source having a flash intensity and
length such that there is a reliable measurement of the probing signal
with adequate time resolution. The method does not require allowance
for the transparency of the walls of the discharge lamp under study,
since this parameter is automatically compensated for during absorption measurements. The appearance of lasers having tremendous
spectral luminance has substantially broadened the capabilities of
the method by extending it to the region of extremely high values of
the absorption coefficient (it now is possible to perform laser
probing of a plasma at various wavelengths over a broad region of
the spectrum: from the ultraviolet to the far infrared). However,
a significant interaction between coherent radiation and the plasma
occurs if the laser beam has a high power density. The interaction
causes a change in the properties of the plasma, determining the
upper limit on the measureable value of the absorption coefficient.

The dependences of radiation absorption on current density were
measured during the first applications of the method to the study of
flashlamps [3-84 to 3-87]. This had a definite practical benefit,
particularly in the selection of parameters of lighting installations
using a specific lamp as a light source. In [3-87] these measurements
were supplemented with measurements of the plasma temperature, but
none of these papers contains data on the concentration of neutral

atoms. The differences in the designs of the lamps used created
conditions for different displacement of the gas into the ballast
volumes. Hence a reliable comparison of the absolute values of κ_λ
obtained by different authors cannot be made. More general information has been provided by research in which both the temperature and
the plasma composition were determined in addition to the absorption
coefficient, since the exact physical characteristic of .the absorption
properties of a plasma is the effective absorption cross section,
which is equal to the ratio κ_λ/n (where n is the concentration of
absorbing particles), rather than the absorption coefficient κ_λ *per se*.
A correct comparison of the various experimental data and a comparison of experiment with theory are possible only for known values of
the effective absorption cross sections that are assigned to a specific temperature. Neutral atoms are the absorption centers in low-
temperature pulsed discharges. Therefore, measurements of the absorption coefficient should be accompanied by a determination of the
neutral-atom concentration.

A typical schematic diagram of the experimental setup for
measuring continuous radiation absorption in the plasma of a pulsed
discharge is shown in Figure 3-19. The lamp under study is illuminated with a narrow beam of radiation from a probing lamp whose pulse
length is approximately an order of magnitude shorter than that of
the lamp under study. This makes it possible to probe the main discharge in various stages of its occurrence. Measuring the intensity
of the probe beam at a wavelength λ in the absence of the main discharge ($I_{\lambda b}$) and when the beam passes through the plasma ($I'_{\lambda b}$) makes
it possible to determine the transparency of the plasma, $\alpha_\lambda = I'_{\lambda b}/I_{\lambda b}$,
which is related to the absorption coefficient κ_λ by the equation

$$\alpha_\lambda = e^{-\kappa_\lambda l}. \tag{3-41}$$

If the absorption is relatively small, then $\kappa_\lambda \approx (1 - \alpha_\lambda)/l$.

Figure 3-20 shows the results of detailed measurements [3-89]
of the wavelength dependence of the spectral absorption coefficient
of the plasma of a pulsed xenon discharge in the spectral region
300-1100 nm with steps of from 5 to 1 nm at a temperature of 11,700 K.
These results include both the continuous and line components of the
absorption coefficient. The curve of the dependences of κ_λ on λ in
relative units, as derived from [3-84, 3-88, and 3-90], is shown for
comparison (the data from these papers are converted to the same scale
at a wavelength of 600 nm). The wavelength dependences of the effective absorption cross section in xenon are shown in Figure 3-21,
and the corresponding data on the composition and temperature of the
plasma at the time when the absorption coefficient was measured are
presented in Table 3-2. The absorption cross sections, recomputed
on the basis of the results of [3-87] by using the indirect data
presented there on the neutral-atom concentration, also are shown in
Figure 3-21. The data of the two papers are in satisfactory agreement

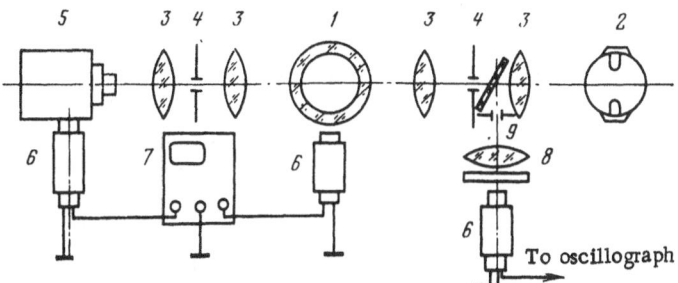

Fig. 3-19.   Optical diagram of a setup for determining the absorption
              coefficient [3-88]. 1) The lamp under study; 2) the
              probing lamp; 3,4) lenses and diaphragms which shape the
              probe beam and direct it to the monochromator slit; 5)
              monochromator; 6) photorecorders; 7) oscillograph with
              a differential amplifier; 8, 9) optical monitoring train.

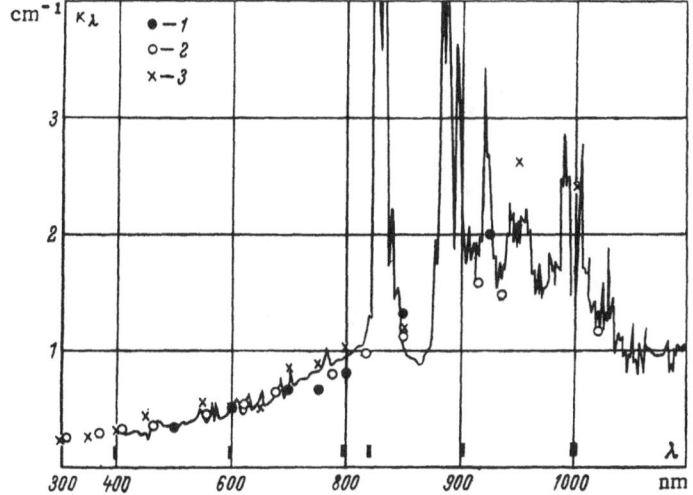

Fig. 3-20.   The spectral absorption coefficient of a xenon plasma in
              an IFP-1200 lamp for $j = 2.6 \times 10^3$ A/cm^2 [3-89]. The
              experimental points from other papers are: 1) [3-88];
              2) [3-90]; 3) [3-86].

at a temperature of about 11,500 K, but at higher temperatures there
is qualitative agreement of the behavior of the curve, with significant divergence of the quantitative values. The discrepancy may be
ascribed to the inaccuracy of the estimate of the neutral-particle
concentration in [3-86], and to some overstatement of the temperature
in the discharge. The empirical dependences of the effective continuous-absorption cross section on the wavelength and temperature

Table 3-2. Discharge Parameters from Measurements of the Absorption
            Coefficient. An IFP-800 Lamp with an inside Tube Diameter
            of 0.7 cm and an Electrode Spacing of 8 cm.

In Xenon, $p_0 = 0.053$ MPa (400 mm Hg).

Regime No. (Fig. 3-21)	j, 10^3 A/cm^2	n_e, 10^{18} cm^{-3}	n_a, 10^{18} cm^{-3}	T, K
1	1.44	0.55	6.3	10,700
2	2.0	0.8	4.3	11,500
3	2.7	1.0	4.0	12,000
4	3.1	1.12	3.8	12,350
5	3.4	1.25	3.7	12,600
6	3.9	1.4	3.6	12,900
7	4.2	1.47	3.5	13,050

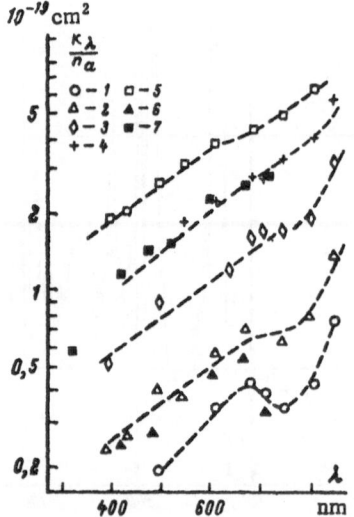

Fig. 3-21. Spectral dependence of the effective absorption cross
            section in a xenon plasma. 1) to 5) correspond to the
            regimes indicated by the same numbers in Table 3-2 [3-88];
            6,7) recalculation using the data of [3-86], corresponding
            to 11,600 and 14,000 K.

may be compared with the theoretical relations given in [2-23 and
3-91]. In [2-23] transitions from levels that form a compact sequence are taken into account on an integral basis, while in [3-91]
some of these levels were calculated individually. A qualitative
difference between the behavior of the calculated and experimental

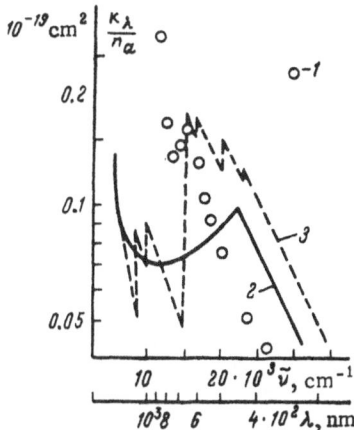

Fig. 3-22. Comparison of the calculated and experimental values of
the effective spectral absorption cross sections in xenon
at T = 10,000 K. 1) Experiment; 2, 3) calculation using
[2-23 and 3-91] respectively.

relations can be seen when λ ≥ 650 nm (Fig. 3-22). This difference
might be smoothed out if allowance were made for the optical decrease
in the ionization potential to the extent indicated in [3-92]. For
example, the continuous absorption in xenon was measured behind the
shock front at T = 10,600 K for λ from 400 to 750 nm [3-93]. The
optical decrease in the ionization potential was 0.57-0.62 V with
allowance for the data of [3-92], according to the authors of [3-93].
This decrease appreciably exceeds the value of 0.5 V calculated from
the Ingliss-Teller formula [3-73]. Appropriately corrected theoretical
data are in satisfactory agreement with experiment throughout the
entire wavelength region indicated.

Analogous measurements of the effective absorption cross sections
were made in [3-27, 3-94, and 3-95] for krypton and argon. The results
are presented in Figures 3-23 and 3-24 (the plasma parameters are
given in Table 3-3).

A comparison with the calculated data for argon [2-23] shows
some qualitative discrepancy at wavelengths shorter than 500 nm and
longer than 700 nm. The quantitative difference does not exceed a
factor of 1.8.

The calculated and empirical temperature dependences of the
effective absorption cross sections are compared in Figure 3-25 for
three noble gases at a wavelength of 500 nm. Entirely satisfactory
agreement is observed here between the calculated and the experimental
values. For example, the current theory of photoionization absorption
for noble gases (xenon, krypton, argon), allowing for all known effects
in a plasma that determine the decrease in ionization potential, allows a satisfactory calculation of the effective continuous-absorption
cross section and of the continuous radiation coefficient of the plasma.

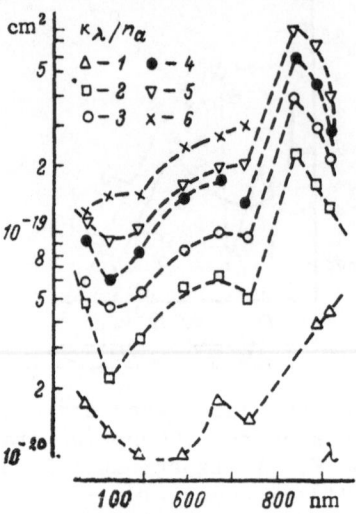

Fig. 3-23. Spectral dependences of effective absorption cross sections
in a krypton plasma [3-27]. Points 1 to 6 correspond to
the regimes indicated by the same numbers in Table 3-3.

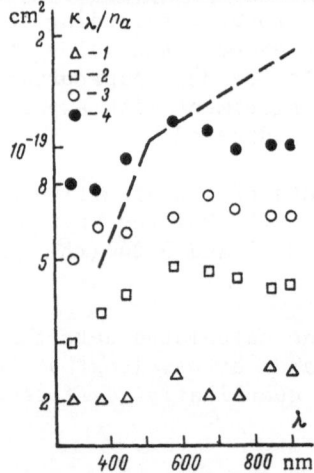

Fig. 3-24. Spectral dependences of the absorption cross section in
an argon plasma [3-94]. Points 1 to 4 correspond to the
regimes indicated by the same numbers in Table 3-3. The
dashed line represents the calculated relation for T =
16,000 K [2-23].

Table 3-3.   Plasma Parameters of Pulsed Discharges in an IFP-800
Lamp Filled with Krypton or Argon at p_0 = 0.053 MPa.
(400 mg Hg)

Regime No.	$j, 10^3$ A/cm^2	$n_e, 10^{18}$ cm^{-3}	$n_a, 10^{18}$ cm^{-3}	T, K
Krypton (Fig. 3-23)				
1	1.5	0.65	1.0	12,400
2	2.2	0.9	4.7	13,200
3	2.8	1.18	4.3	13,750
4	3.7	1.5	4.0	14,400
5	4.4	1.75	3.8	14,800
6	5.1	2.0	3.5	15,200
Argon (Fig. 3-24)				
1	2.1	0.87	5.3	14,300
2	2.75	1.1	5.0	14,900
3	3.3	1.37	4.6	15,400
4	4.0	1.7	4.4	16,000

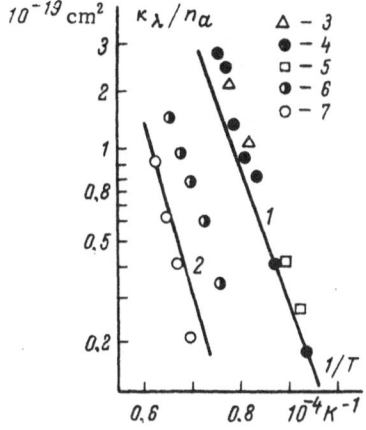

Fig. 3-25.   Temperature dependences of the effective absorption cross
section (λ = 500 nm) [3-27, 3-88, and 3-94]. 1, 2) calculated [2-23]. The experimental points are: 3,4,5)
xenon, p_0 = 0.013, 0.053, and 0.08 MPa, respectively;
6,7) krypton and argon, p_0 = 0.053 MPa.

In order to convert to the overall radiation characteristics of a discharge, allowance must be made for reabsorption and for the presence of line radiation (as well as continuum radiation) that results from the free-bound transitions of electrons. Allowing for reabsorption is unnecessary for an optically thin plasma, but a real plasma is not optically thin throughout the entire spectral band, even at relatively low temperatures. If such an assumption is made nonetheless, the total continuum radiation for an optically transparent plasma is found by integrating relations (3-35) and (3-37) in the range $0 < \nu < \infty$. Then [2-49]:

$$q_{\text{rad}} = 1.38 \cdot 10^{-34} \, \xi_{\text{av}} \left(\frac{h\nu_{\text{th}}}{kT} + 1 \right) T^{1/2} \, n_e^2, \tag{3-42}$$

where ξ_{av} is the average value of the factor $\xi(\nu)$ for the spectrum. The value calculated in this way for the toal emission of xenon $(9000 \le T \le 13{,}000 \text{ K})$ is approximately half the actual value. The missing fraction of the radiation should be made up by allowing for reabsorption and line emission, which makes a significant contribution to the far ultraviolet and the near infrared.

The other extreme case (an optically opaque plasma) is not realized in practice in low-temperature pulsed discharges. A qualitative analysis of the peculiarities of the emission of a xenon discharge was made in [2-49]. The blackbody radiation of an infinite cylinder of radius R was considered as the limiting case. The quantitative discrepancy with the real emission exceeds an order of magnitude, according to indirect data. A method has now been proposed for indirect determination of the total energy radiated by a pulsed discharge at any specified time [3-96]. The method utilizes the fact that under certain conditions practically all of the energy introduced into the discharge in tubular flashlamps during the quasistationary stage is converted to radiation, as we showed in our treatment of the energy balance (Section 3-3). This makes it possible to find the dependence of the total radiation on the basic parameters (gas temperature and pressure) by supplementing measurements of the energy put into the discharge with a determination of the basic parameters.

The results of the experimental determination of the total emission of a xenon discharge are shown in Figure 3-26 for pressures of 0.3, 1, and 1.6 MPa. The temperature dependences of the total emission of xenon, krypton, and argon are shown in Figure 3-27 for a pressure of 1.0 MPa. The experimental points obtained for argon are compared in Figure 3-27 with the calculation made in [3-75 and 3-76] for a 1-cm layer at a pressure of 1 MPa. The fit turns out to be entirely satisfactory.

The following relation has been proposed on the basis of general concepts of the nature of the dependence of the radiation on temperature and pressure and on the basis of experimental data [3-97]:

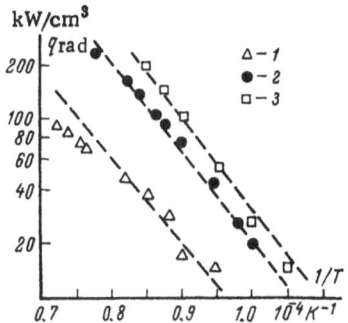

Fig. 3-26. Power density of discharge radiation in xenon at pressures
of 0.3 MPa (1), 1 MPa (2), and 1.6 MPa (3) [3-96]. The
dashed lines represent calculations according to (3-43).

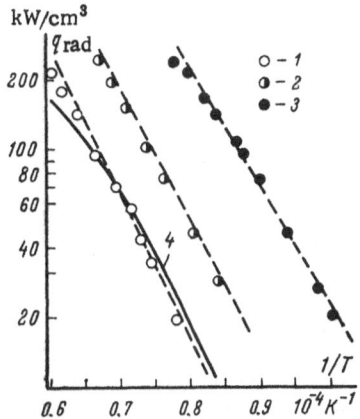

Fig. 3-27. Power density of radiation, calculated from relation
(3-43) for discharges in argon (1), krypton (2), and
xenon (3) at a pressure of 1 MPa [3-96], and as calculated
for discharges in argon (4) (1 MPa, 1 cm) [3-75 and 3-76].

$$q_{rad} = \hat{C}p\exp\left(-aeU_i/kT\right) = 3.7\cdot10^9\,p\exp\left(-10^4\,U_i/T\right) \qquad (3\text{-}43)$$

where q_{rad} is in W/cm^3, p is in megapascals, and T is in Kelvins.
The approximations calculated from relation (3-43) are shown by the
dashed lines in Figures 3-26 and 3-27.

The introduction of the correction factor $a < 1$ into the expo-
nent of relation (3-43) has the purpose of taking bound-bound transitions of electrons. The insertion of the factor a is equivalent to
replacing the ionization potential U_i by some equivalent potential

U_{eq}. It follows from experimental data that $U_{eq} = 0.86U_i$ for noble gases, for which the first excitation level is located relatively close to the ionization limit and the main group of terms forms a compact sequence between the lower excited state and the ionization limit.

Thus, relation (3-43) has a clear physical meaning. It is valid for noble gases under conditions where the predominant emission pro-. cesses are free-bound and bound-bound transitions, i.e., at temperatures of at least 6000 K and no higher than the temperatures at which the bremsstrahlung contribution (free-free transitions) becomes significant. According to the calculation for argon [3-75 and 3-76], the total radiation is proportional to the pressure in the range 0.01-10 MPa (0.1-100 kgf/cm^2). No factors that would substantially alter this range of the linear dependence of the radiation on the pressure can be seen for xenon or krypton. Apart from the condition indicated, the upper temperature limit of applicability of relation (3-43) is determined by the requirement that the exponential growth of radiation with temperature be preserved. This requirement is not satisfied when a high degree of ionization (over 50%) is reached in the plasma. To judge by the theoretical calculation, we may expect that the power density of the radiant flux should be independent of the thickness of the radiating layer up to a characteristic dimension on the order of 10 cm.

New experimental data on the total radiant flux of xenon heated by a shock wave to temperatures of 8500-10,500 K are presented in Table 3-4. The calculated values from relation (3-43) are given in the last column of the table.

A comparison of these values indicates satisfactory agreement between calculation and experiment.

Table 3-4. Total Radiant Flux of Xenon Heated by a Shock Wave [3-98], and the Corresponding Estimates according to Formula (3-43).

T, K	n_e, 10^{16} cm^{-3}	n_a, 10^{17} cm^{-3}	p, MPa	q_{rad}, 10^2 W/cm^3 Exper.	Calc.
8540	2.9	5.08	0.066	1.3	1.7
9025	4.9	5.54	0.08	3.3	4.3
9450	7.5	5.94	0.096	7.2	9.7
9845	1.1	6.26	0.112	21.0	18.5
10,215	5.0	6.5	0.13	33.0	34.0

3-5. Calculation of the Characteristics of a Pulsed Discharge Having a Quasi-Stationary Stage

The aggregate of the processes in a pulsed discharge can be described in general form by a system of equations which includes electrodynamic, gas-kinetic, gas-dynamic, and radiation transport equations for the medium. Solving such a system of nonlinear equations in general form is practically impossible. Therefore, in every case efforts are made to formulate a particular model problem in which allowance is made only for the most important processes in the discharge stage in question. For example, the theoretical approach to studying the channel expansion stage consists in allowing only for gas-dynamic phenomena and in part for electrodynamic effects, which predominate in this period of development of the discharge. On the other hand, gas-dynamic effects are not very pronounced in the quasi-stationary stage of a pulsed discharge and may be ignored. The next simplification stems from the assumption that there is local thermodynamic equilibrium in this stage. As a result, all plasma characteristics can be represented as functions of the temperature T and the pressure p. This makes it possible to convert to a thermodynamic description of discharge and to use the energy balance equation instead of the gas-dynamic equation.

The energy balance equation is written in the following form for a stationary plasma having cylindrical symmetry (this is precisely what we might expect of the plasma of a pulsed discharge in tubular lamps in the quasi-stationary stage):

$$\frac{\partial u}{\partial t} = \frac{1}{\rho} \frac{d}{d\rho} \left(\lambda \rho \frac{dT}{d\rho} \right) - q_{rad} + \sigma E^2, \qquad (3\text{-}44)$$

where u is the internal energy of the plasma, $\rho = r/R$ is the relative radius of the discharge column, and λ and σ are the thermal conductivity coefficient and the electrical conductivity, respectively.

The first term in the right-hand side corresponds to heat losses due to heat conduction, the second term to losses to radiation, and the third term to the power fed into the discharge by the electric field E.

The internal energy of the plasma varies slightly ($\partial u / \partial t \approx 0$) near the current maximum, where the change in temperature is relatively small, and instead of relation (3-44) we get the well-known form of the stationary energy-balance equation, which is called the Elenbaas-Geller equation:

$$\frac{1}{\rho} \frac{d}{d\rho} \left(\lambda \rho \frac{dT}{d\rho} \right) + \sigma E^2 - q_{rad} = 0. \qquad (3\text{-}45)$$

The coefficients $\sigma(T, p)$ and $\lambda(T, p)$ found in this equation can be calculated within the framework of the kinetic theory of gases or can be determined experimentally as indicated in Section 3-3.

For a stationary plasma the equations of electrodynamics degenerate to the condition $E(r) = \text{const}$. We thus achieve a very substantial simplification of the original system of equations. If the dependences of σ and λ on the temperature and pressure are known and if the function $q_{rad}(T, p, r)$ can be expressed in explicit form, then the temperature distribution in a high-pressure stationary discharge can be found by solving equation (3-45) with the boundary conditions

$$T\Big|_{\rho=1} = T_R, \; \frac{dT}{d\rho}\Big|_{\rho=0} = 0, \tag{3-46}$$

where T_R is the temperature of the wall of the gas-discharge tube.

The relatively well-developed state of the methods used to solve equation (3-45) for a stationary cylindrical discharge also makes it possible to use them for the quasi-stationary stage of a pulsed discharge. However, one must be confident that the properties of a pulsed quasi-stationary plasma are in no way affected by the preceding, purely nonstationary phase of development. It can be asserted beyond question that gas-dynamic processes and thermal inertia have no effect by the time the quasi-stationary state becomes established. However, the question whether thermal equilibrium is established between the plasma and the tube wall [2-7] requires perhaps a bit more clarification.

The experimental study of the elements of the energy balance in the quasi-stationary stage of a pulsed discharge (Section 3-3) shows that the overwhelming majority of the external electric field energy released in the discharge is converted to radiation. This fact was used in [2-49] to consider simplified energy-balance equation (3-45) without allowance for heat conduction:

$$\sigma E^2 = q_{rad} \text{ or } j^2 = q_{rad}\sigma. \tag{3-47}$$

Since the conductivity and the radiant flux density q_{rad} are functions of temperature and pressure, the current density j (or the electric field strength E) can be expressed in terms of the same arguments by using relations (3-33) and (3-34). Instead of (3-47) we get

$$j = 11 \, p^{1/2} \, T^{5/4} \, U_i^{-1/2} \exp\left(-5000 \, U_i/T\right), \tag{3-48}$$

where j is in A/cm^2, p is in megapascals, T in Kelvins, and U_i in volts.

Relation (3-48) determines the relation of the temperature and pressure to the current density and the type of gas. The empirical relations j(T, p) are compared with relation (3-48) in Figure 3-28. The experimental values of the working gas pressures were used in the calculations.

Relations (3-33) and (3-43) also make it possible to write the analytic dependence of the field strength E (V/cm) on the temperature, pressure and type of gas:

$$E = 3.3 \cdot 10^8 \, p^{1/2} \, T^{-5/4} \, U_i^{1/2} \exp\left(- 5000 \, U_i/T\right). \tag{3-49}$$

The calculated relations E(T, p) for xenon, as obtained from relation (3-49), are shown in Figure 3-29.

The relations j(T) shown in Figure 3-30 for fixed values of the working pressure are useful for practical purposes. The curves obtained from formula (3-48) are compared with the corresponding calculated relations according to the data of [3-58 and 3-99]. The authors of these papers solved equation (3-45) with boundary conditions (3-46) simultaneously with the radiation transport equations (a diffusion-approximation equation was used in the region of medium optical densities) or with Planck's formula on the assumption that the plasma emits as a blackbody. Since the results of the calculations are highly sensitive to the selection of the functional dependence of integrated radiation temperature, data obtained via the experimental dependence of q_{rad} are of basic importance. The calculations made in [3-58] for "axial" temperature are in satisfactory agreement with these data, especially at a pressure of 1 MPa. For the temperatures in question, the "blackbody emitter" model gives results that are substantially at odds with the model of a real discharge, as a comparison with the data of [3-99] shows. The difference is just as significant for the corresponding calculated dependence of E on T in Figure 3-29.

The pattern presented for the quasi-stationary stage of a pulsed discharge and the equations obtained, which establish the interconnections between the basic plasma parameters, make it possible to outline a scheme for calculating the electrical and radiative characteristics for those tubular flashlamps in which this stage is decisive.

The engineering design of a tubular flashlamp begins with the flash parameters, which are determined by the lamp's intended purpose, the radiant flux, pulse length, spectral composition, the size of the glow volume, and so forth. Such design must take into consideration both the characteristics of the emitter (the column of gas-discharge plasma) and the properties of the structural materials (glass, metals, the filling gas, etc.) as they interact with the plasma and its radiation. The properties of the structural materials include heat resistance (allowing for the duration and periodicity of exposure),

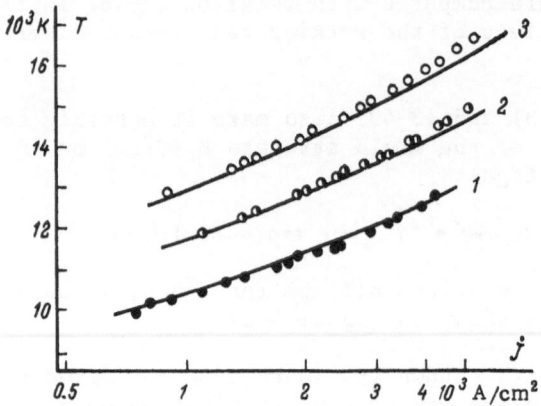

Fig. 3-28.   Temperature versus current density in xenon (1), krypton
             (2), and argon (3) discharges for p_O = 0.053 MPa [2-49].
             The lines represent the calculated values from relation
             (3-48) for the experimental values of the working pressure.

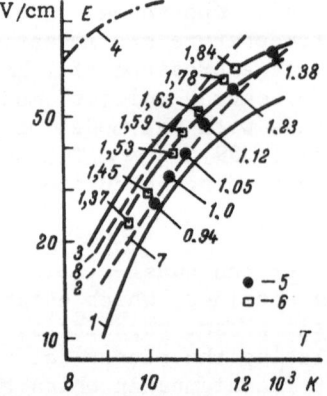

Fig. 3-29.   Temperature dependence of the electric field strength
             in a xenon discharge. 1, 2, 3) calculated results in
             [3-58] for p = 1, 2, and 3 MPa respectively; 4) calcu-
             lated curve on the assumption of blackbody radiation
             [3-99] ; 5, 6) experimental data [2-5] (the numbers
             beside the points indicate the working pressure, MPa);
             7, 8) calculated from (3-49) for p = 1 and 2 MPa.

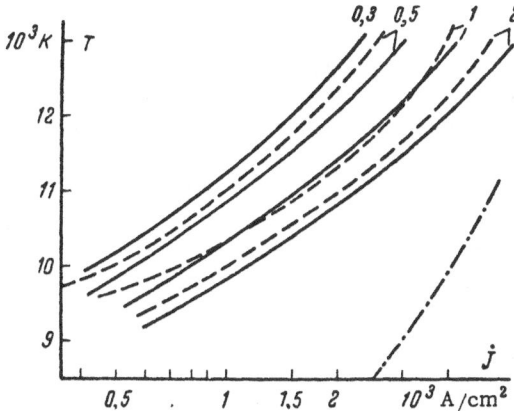

Fig. 3-30. Temperature versus current density in a xenon discharge for fixed values of the gas pressure. The solid curves represent values calculated from (3-48). The dashed lines represent the calculated data of [3-58] (the numbers beside the lines indicate the pressure, MPa). The dot-dash line represents calculated values on the assumption of blackbody radiation [3-99].

chemical resistance, optical transparency of the shell material, sputterability of the electrode material, the possibility of treatment, etc. These properties are examined in the second part of the book. The concepts developed there are used as the basis for first-approximation selection of the geometric dimensions of the lamp and its components, the type of gas, the initial pressure, current density, and the law governing the time dependence of the current density. The parameters obtained are adopted as the initial values for a re-fined calculation of the electrical and optical properties of a quasi-stationary discharge. The results make it possible to correct the initial parameters in order to ensure that the characteristics of the lamp match its function.

The interrelationships among the parameters of a quasi-stationary plasma are determined by the energy balance equation in form (3-47), by the approximate relation for the plasma conductivity (3-33), and by the expression for the radiation density (3-43). Saha's formula (3-3), equation of state (3-15), and formula (2-29), which relates the displacement of gas into the volumes beyond the electrodes to the design parameters of the lamp, are added in order to obtain a closed system of analytic relations. The computational scheme, which uses the system of relations listed, goes as follows.

1. Formula (2-29) is used to estimate the fraction of atoms displaced into the volumes beyond the electrodes. This makes it possible to determine the heavy-particle concentration in the plasma.

2.   Empirical relation (3-24) and

$$j = 2.6 \cdot 10^{-15} n_e \qquad (3\text{-}50)$$

are used to determine the electron concentration n_e (in principle this relation can be obtained from the closed system of equations indicated above by inserting relations (3-48) and (3-3) in expression (3-23)).

3.   The temperature and working pressure which correspond to a given current density are determined by solving equation (3-48) and equation of state (3-15) simultaneously:

$$p = kT (n_0 + n_e). \qquad (3\text{-}51)$$

Graphic solutions of this system of equations are given in Figures 3-31 to 3-33.

4.   The current-voltage characteristic of the discharge is found:

$$j = \sigma E = 3.4 \cdot 10^{-8} T^{5/2} E/U_i. \qquad (3\text{-}52)$$

5.   The radiation density of the discharge is calculated from relation (3-43) for the values found for p and T.

6.   All quantities sought for the entire quasi-stationary stage are found for a specified time dependence of the current density. This is done all the way to regions where the relations obtained no longer are valid.

7.   Given the geometric dimensions of the working region of the lamp, the total current i, the voltage u across the lamp, and the total radiation Q_{rad} are determined.

Whether a calculation can be made is determined by assumptions which restrict the applicability of the relations used. They were considered in detail previously, and briefly reduce the following:

1. Local thermodynamic equilibrium (LTE) is present in the plasma.

2. The discharge takes place under conditions which ensure that the plasma fills the entire cross section of the tube.

3. The dependence of the mean drift velocity of electrons on the temperature and type of gas is negligible.

4. The energy fed into the discharge does not cause appreciable vaporization of the wall material.

5. The working pressure lies in the range $0.01 < p < 10$ MPa.

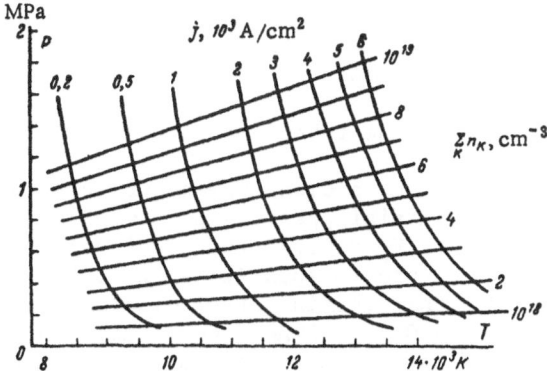

Fig. 3-31. Diagram for determining the values of p and T in a xenon discharge.

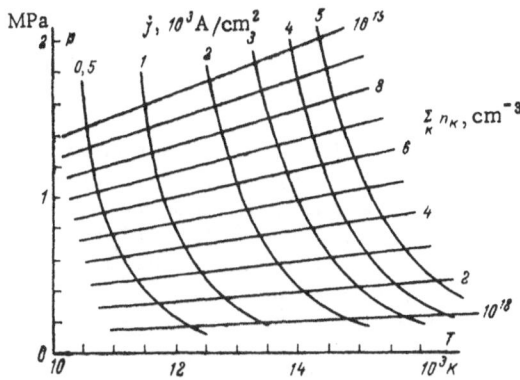

Fig. 3-32. Diagram for determining the values of p and T in a krypton discharge.

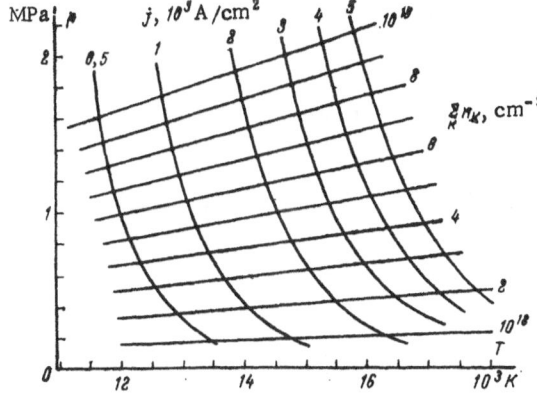

Fig. 3-33. Diagram for determining the values of p and T in an argon discharge.

6. The temperature range is bounded from below by values of 7000 K
    for xenon and 8500 K for argon, and from above by values for which
    the degree of ionization does not exceed approximately 50%.

7. The inside diameter of the tube lies in the range $5 \leq 2R \leq 30$ mm.

    Sample Calculation. Let a discharge occur in a tubular lamp
filled with xenon at an initial pressure $p_O = 0.053$ MPa (400 mm Hg)
with a design parameter $V_e/V_l = 0.17$ (the usual value for a type
IFP-800 quartz lamp). We use formula (2-29) to determine the corresponding fraction of gas that remains after displacement into the
ballast volumes: $n_\infty/n_O = 0.45$. Since the initial concentration of
atoms in a cold lamp is $n_O = 1.42 \times 10^{19}$ cm^{-3}, the number of heavy
particles per unit discharge volume is $n_\infty = 6.4 \times 10^{18}$ cm^{-3}.

    The calculation then follows the scheme indicated above using
the graphic solution of system of equations (3-15) and (3-48) shown
in Figure 3-31. (Intermediate values are found by interpolating
between the corresponding horizontal lines. This interpolation is
facilitated by the fact that the pressure p is proportional to Σn
for a given temperature.) The results are given in Table 3-5.

    The values of n_O (1.6 x 10^{18} cm^{-3} for xenon at $p_O = 0.013$ MPa
and 9.6 x 10^{18} cm^{-3} at 0.08 MPa, for example) change at other initial
gas pressure. This leads to a shift in temperature and pressure for
a specified current density, as follows from Figure 3-31. For example,
a decrease in p_O for a given j is accompanied by an increase in temperature and a decrease in the working pressure. Conversely, an increase in p_O causes a decrease in the temperature and an increase in
the working pressure.

    The curves for given j shift to the right as the ionization
potential U_i of the gas increases (Figures 3-32 and 3-33). For a
fixed current density, this corresponds to an increase in the temperature and pressure, compared with xenon. The results of calculations

Table 3-5. Calculation of the Characteristics of a Quasi-Stationary
            Discharge in an IFP-800 Flashlamp

j, 10^3 A/cm^2	0.5	1	2	3	4	5
n_e, 10^{18} cm^{-3}	0.19	0.38	0.76	1.12	1.53	1.92
Σn, 10^{18} cm^{-3}	6.6	6.8	7.16	7.52	7.93	8.4
T, 10^3 K	9.55	10.35	11.4	12	12.5	13
p, MPa	0.88	1.0	1.11	1.23	1.35	1.45
E, V/cm	20.3	32.3	51.3	67.5	83	92
q_{rad}, 10^3 W/cm^3	10.2	32.3	100	190	310	490

of discharge parameters in various gases are in good agreement with the experimental data, as we can see from Figures 3-7, 3-10, and 3-28.

The calculated plasma parameters are specific, i.e., they apply to a unit volume, a unit cross section, or a unit length of the plasma. The values of the electrode spacing l and the inside radius R of the discharge tube can be used to determine the instantaneous values of the voltage across the lamp, $U = lE$ (ignoring the voltage drop on the electrodes), the current strength $i = \pi R^2 j$, and the radiant flux $Q_{rad} = \pi R^2 l q_{rad}$. Since approximately 15-20% of the radiant flux is retained in the tube walls, the radiation that reaches the outside is $Q'_{rad} = 0.8 Q_{rad} = 2.5 R l q_{rad}$. Thus, the complete characteristics of the quasi-stationary stage of a lamp can be calculated if the new law governing the time dependence of the current density is known, as this law determines the selection of parameters for the entire discharge circuit.

Because of the limitations stipulated, the proposed computational method does not encompass the entire diversity of the cases encountered in practice. For example, tubular flashlamps with initial pressures below 0.013 MPa (100 mm Hg) have come into use recently. They exhibit some redistribution of the energy balance, particularly a noticeable increase in the fraction of losses on the walls, and the discharge conditions do not completely meet the criteria for LTE. Capillary flashlamps having a tube diameter of less than 5 mm form a special class that has its own specific features. Flashlamps which operate in regimes involving appreciable vaporization of the wall material also are not of interest. Finally, lamps having an extremely short flash length comparable to the channel expansion time for which transition of the discharge to a quite prolonged quasi-stationary stage is not assured also lie beyond the framework of the method. Nonetheless, the remaining range of subjects to which the method of computation is applicable is quite extensive from a practical standpoint.

CHAPTER 4

PHENOMENA AT ELECTRODES

4-1. General Information

As yet no one has created a general theoretical concept of the
physical phenomena in the regions near the electrodes during a pulsed
discharge from which regularities suitable for engineering calculations
could be extracted. However, the inadequacy of the early development
of the physics of phenomena at electrodes was not felt very much
because of the comparative isolation of processes near electrodes
from processes in the discharge channel. This also was due to the less-
than-primary importance of such processes in many practical situations
(the main external manifestations of a discharge are determined
by the discharge channel). Only the progress in the technical application of pulsed discharges has drawn the attention of many investigators to phenomena at the electrodes during such a discharge.

The particular complexity of these phenomena is due to the fact
that radial nonuniformity of the discharge channel in time also is
accompanied by longitudinal nonuniformity, in addition to a number
of other interrelated physical processes in the "gas-metal" boundary
layer and in the body of the metal. We may list the following processes above all:

(a) ionization of the gas in the regions around the electrodes, the
     formation of space charges around the electrodes, and electron
     emission from the cathode. These phenomena determine the energy
     supplied to the electrodes (losses at the electrodes which
     affect the efficiency of the discharge) and the conditions under
     which the electrodes are bombarded by plasma particles;

137

(b) heat propagation in the electrodes, and fusion and vaporization
    of the metal of the electrodes. Such vaporization directly deter-
    mines the second most important external manifestation of all
    processes at the electrodes (after the efficiency) from a technical
    standpoint: electrode sputtering and the resultant blackening
    of the flashlamp envelope (this usually limits the life of the
    flashlamp).

    All these processes are so complicated that as yet no generally
accepted theoretical interpretation has been worked out, even for
stationary arc discharges. Moreover, the spatially and temporally
discontinuous nature of the phenomena on the cathode of stationary arcs
has led to a situation in which the study of the arc regions near
electrodes has required experiments involving pulsed discharges.
Studies of the electrode regions of stationary and pulsed discharges
presently are being done on a united front, as it were.

    Taking into account the extremely incomplete state of this re-
search, we can give a brief summary of the main data available on
phenomena at electrodes. We do this so that we may examine some of
the empirical data regarding the influence of the electrodes on the
efficiency and life of devices that are most important in practice.
These data will be examined in Part 2, which is devoted to the technical characteristics of flashlamps.

4-2. The Cathode Spot

    Observations of the cathode spot using an electrooptical shutter,
ultrahigh-speed filming, image converter tubes, the marks left on
the cathodes, and others have showed that the spot has a multiple
structure (Frum, Sommerville, Blevin, Mandel'shtam, Raiskii, Khermokh,
Zhizhki, Kobin and others, and [4-1 to 4-6]). The structural element
of the spot is a microregion having a diameter of 5-10 μm, which produces a current strength of 0.5-5 A in the discharge. As experiment
has demonstrated, the following threshold current strength for the
formation of a single microspot corresponds to every material for
low-pressure high-current arcs [4-7, p. 118]:

$$I_{th} = 0.26 \cdot 10^{-3} T_{boil} \lambda^{1/2}, \qquad\qquad (4-1)$$

where T_{boil} is the boiling point of the cathode material, and λ is
the thermal conductivity coefficient [W/(cm·K)]. Such a microspot,
the current density in which ranges from the order of 10^5 to 6×10^7
A/cm^2, exists at a specific site for an extremely short time (≤ 0.1 μs).
There is a dark space about 0.1 mm thick between the luminous zone
of the microspot and the cathode. Therefore, the current density on
the cathode itself can be even higher. For a uniform surface (a
mercury cathode) the spot moves evenly at a rate of 10^3–10^6 cm/s,

stretching out into an arc segment as the current rises. As this occurs, the current density remains constant, and the small arc approaches a circle (or semicircle if the original spot was formed at the edge of the cathode) when the current steepness di/dt is sufficient. If di/dt is so large that the increase in active surface due to the increase in the radius of the small arc becomes insufficient, then new microspots appear at the side. Similar small annular arcs then race away from the new microspots. If the surface is irregular (due to contamination or a crystalline structure), the microspots move in jumps, focusing on nonuniformities. The average rate of motion of the spots along the surface of the cathode is on the order of 10^5 cm/s. The number of microspots in this case obviously is equal to the quotient from the division of the total current strength in the discharge by the current strength of about 5 A in a single spot. On the whole, the set of spots encompasses an extremely sizeable zone of the cathode surface as the current rises and the channel expands (cf. Figure 2-22a).

Movement of the general boundary of the glow region with a velocity of $(4-18) \times 10^3$ cm/s as di/dt simultaneously varied over the range 4-20 MA/s has been observed by means of high-speed photography of the phenomena on a plane cathode (tungsten or thoriated tungsten) in a tubular xenon flashlamp having a resolution insufficient to detect individual microspots at the start of the discharge [4-6]. The area S of the glow region is proportional to the current:

$$S = (20 - 5) \cdot 10^{-4} i. \qquad (4-2)$$

Less-mobile distinct spots of the second type appear after di/dt falls below 4 MA/s (after 240-150 µs for thoriated tungsten or 250-350 µs for pure tungsten). These spots have a current of 10-50 A per spot, a diameter of 0.1 mm (at the start) and 0.5 mm (at the end), a current density of 3×10^5 (at the start) and 3×10^4 A/cm^2 (at the end), and a velocity of 2×10^3 (at the start) and 10^2 cm/s (at the end) up until the spots come to a complete stop. Spots of the first type do not leave noticeable traces on a tungsten surface, while spots of the second type leave erosion tracks on pure tungsten that have a width close to the diameter of a spot.

High-speed photography of the phenomena on a hemispherical cathode made of highly thoriated tungsten heated to 1200°C by the frequent flashes of a tubular xenon lamp (at a frequency of about 4 Hz) has been used to observe stationary spots of the third type from the very start. These spots have a current of 100 A (at the start) and 200 A (at the end) per spot, a diameter nearly an order of magnitude larger and a current density lower by a factor of 1-1.5 than those of spots of the second type on a cold cathode. The lifetime of such spots is about 1 ms, and they leave practically no partially fused tracks.

The formation of microspots is related to the formation of emission centers. The emission may be explained, for example, by the combined action of the thermionic mechanism (T emission) and the field-emission mechanism (F emission) (allowing for local order-of-magnitude excesses above the average value of the electric field due to the roughness of the cathode [4-8] and the statistical distribution of ions at the cathode [4-9 and 4-10]). A small region on the cathode is heated by the energy supplied, and the population of high energy levels increases in the electron gas of the metal. This makes it easier to overcome the potential barrier for an electric field of about 2×10^7 V/cm at the cathode, which can be produced by the space charge at a current density of about 10^6 A/cm^2. Field emission having the current density indicated also becomes possible. A corresponding estimate of the temperature of the briefly heated region (3000 K allowing for the roughness factor, and 3500 K, ignoring it) seems entirely feasible for cathodes made of copper or more refractory metals.

The transition observed in [4-6] from spots of the first type to spots of the second having a lower velocity and a larger diameter probably is associated with a transition from pure F emission to combined T-F emission. This is confirmed by the decrease in current density; the earlier appearance of spots of the second type on thoriated tungsten, which has a lower work function at which a lower temperature is required for T-F emission; and by the less-pronounced fusion of thoriated tungsten. T-F emission apparently occurs from the very start on preheated highly thoriated tungsten, producing large stationary spots of the third type having a relatively low current density.

However, another mechanism seems more feasible for a mercury cathode (the originators of the mechanism also believe it applicable to copper). The atoms are excited in the metal-vapor cloud formed by local heating. On returning to the cathode (in a short period of time during which they are unable to emit a photon), the atoms induce electron emission which is similar to the emission due to metastables in a Townsend discharge (the ε process; see Chapter 1). Such a mechanism is confirmed by an approximate calculation based on an estimate of the number of returning atoms: 10^{25} atoms/(cm^2.s) (experiment gives the amount of evaporated mercury as 3×10^{-4} g/K1,* i.e., 10^{24} atoms/(cm^2·s); the ratio of the number of returning atoms to the number that left the cloud may be 10:1). These atoms produce an emission current density of about 10^6 A/cm^2 when $\varepsilon \approx 1$. The unconvincingness of such a high estimate of ε led to a number of other hypothetical explanations of electron emission from the cathode: the penetration of ions into the interior of the hot cathode [4-12]; and the formation of a region of superheated vapor 0.1-2 μm thick in front of the cathode. Metal electrodes are converted into such a region because of the overlap of energy levels and the formation of conduction bands (hence such electrodes already emit under the action of the temperature and field [4-13] and other factors [4-14]).

*This is in agreement with the data of [4-11] for low-melting cathodes, according to which the amount of lead or bismuth vaporized is 4×10^{-4} g/K1.

4-3. Electrode Sputtering

The liberation of a significant amount of energy in a microspot leads to instantaneous vaporization of the metal, accompanied by the explosive propagation of the vapors. Experiment reveals the discrete structure of the jets from solid cathodes. Such a structure is consistent with the discrete structure of cathode microspots (the ejection time of a single jet, on the order of 10^{-7} s, is equal to the lifetime of a single spot).

The explosion of a microspot is analogous in a sense to the explosion of a fine metal wire acted on by a condensed electric discharge. Just as the electric current is interrupted at the initial instant because of the high vapor density during the explosion of a wire (the current is restored only after the vapor expands), when a microregion explodes on the cathode further flow of the current at the given point of the cathode becomes impossible. Precisely this fact should explain the movement of the emission center onto adjacent sections of the cathode. This mechanism is confirmed, for example, by the correspondence between the propagation velocity of the shock wave, as calculated from hydrodynamic theory (see Section 2-2), and the average velocity that was obtained experimentally by Boyle for the motion of the pulsed microdischarges of a relaxation circuit having a 100-pF capacitance and a 1-kΩ charging resistance along a filamentary electrode. (Boyle relates the appearance of charges to the movement of the reduced-pressure region behind the shock front. The distance between extremely closely positioned electrodes in this region corresponds to the minimum of Paschen's curve. It should be stipulated that the movement of the microspots also has been linked to other mechanisms: the breaking of equilibrium between the magnetic and the gas-kinetic pressures, fluctuation of the transverse temperature gradient [4-14], the blowing-away of positive ions by the vapor jet [4-15], etc.).

The explosive nature of the ejection of vapors from the cathode is confirmed by spectroscopic studies of the jets formed on cathodes made of alloys, and by absolute measurements of flow velocities. The spectroscopic studies showed that the velocities of various atoms in vapor jets are the same when the cathode is made from a metal alloy, while the absolute measurements of flow velocities showed that the jet velocities are 10^5–10^6 cm/s. These velocities vary according to the change in the current in the discharge and are inversely proportional to the atomic mass of the cathode metal. Thus, the kinetic energy of atoms in jets from cathodes made of different metals turns out to be the same, given the same discharge conditions.

The literature contains reports of several attempts to calculate the process of vaporization of the metal present in a vapor jet [4-1, 4-2, 4-11, etc.]. These calculations are based on some estimate of the energy supplied to the metal from the adjacent region of the

channel. For example, an obviously low estimate of the ion current
is used in [4-11], and the assumption is made that every ion is able
to recombine via the tunnel effect and to impart the ionization energy
to the metal, ignoring the kinetic energy. Mandel'shtam, Sommerville,
and others estimate the energy as the product of the current density,
as determined from oscillograms of the current and from the size of
the microspot, the rough value of the cathode drop (10 V), and the time
interval between ejections of individual jets (1 μs). Such a long
interval is obtained when the inductance of the discharge circuit is
significant (L $\gtrless$ 100 μH). Individual jets are superposed on each
other at smaller L. On the other hand, the thermal conductivity
equation can be used to estimate the depth of the metal zone around
the cathode spot within which the temperature reaches the melting point
or the boiling point. These calculations, which differ in their initial
data, give the same order of magnitude for the rate of vaporization
of the metal (about 10^{-2} g/s). This value is in agreement with experiment. As yet, however, the differences in the initial premises
(the multiple structure of the cathode spot is not even taken into
consideration in [4-11]) and in experimental conditions prevent us
from considering this agreement to be a confirmation of theory on
whose basis the computational methods could be used for practical
purposes. Thus, the results of the calculations more nearly illustrate the acceptability of the underlying physical hypothesis. Another
such illustration is the estimate given by Mandel'shtam and others of
the energy balance at the cathode spot (of the 40 J/cm^2 per spot 4 J/cm^2
is dissipated in the metal by heat conduction, 12 J/cm^2 goes to vaporization, and 24 J/cm^2 is converted into the kinetic energy of the jet)
and of the vapor pressure of the metal ejected from the cathode (about
10 MPa). This pressure determines the formation of the explosive
wave. The hydrodynamic nature of the jet, the vapor temperature in
which is ~(2-3) x 10^3 K, is consistent with the observed thin layer
of dark space between the cathode surface and the brightly glowing
zone of the microspot. This zone evidently is a vapor cloud heated
by the discharge after expulsion (along a path of about 0.1 mm) to
the plasma temperature of (20-40) x 10^3 K.

The relative values of the rates of vaporization of various
metals, as calculated in [4-11], seem more reliable. The following
expression is obtained by solving the thermal conductivity equation for
a mass M of metal surrounded by a melting zone upon the instantaneous
release of an amount of energy W at a point on the surface:

$$M = 0.3\,W/\gamma T_0, \qquad\qquad (4\text{-}3)$$

where γ is the specific heat of the metal, and T_0 its melting point.

The use of this formula would be justified if the energy released
on the cathode were independent of the cathode material (this obviously

should not be the case, given the large difference in the work function) and if the amount of metal ejected were independent of the latent
heat of fusion and of the heating of the metal to vaporization. The
satisfactory agreement of the relative values for the rate of vaporization obtained for different pure metals from formula (4-3) and
from experiment (Table 4-1) demonstrates the feasibility of these
assumptions. (It should be stipulated that the structure of the
material has a significant influence on the rate, as [4-16] for example, shows: a fine-grained cathode sputters more slowly than a
macrocrystalline one does).

Formula (4-3) cannot be used as the basis for an estimate of
the cathode sputtering rate for activated cathodes. This is evident,
for example, from Table 4-2,* which gives data on the amount of sputtered material per discharge in a tubular xenon lamp having a flash
energy of 500 J (a flash length of 1 ms, a current amplitude of 1500 A,
a flash rate of 2-4 Hz, and a cathode temperature of 800-1200°C) for
cathodes containing various activating additives.

Strict proportionality was observed between the vaporized matter
and the amount of electricity that flowed through the discharge under
the specific conditions of the experiment in which measurements were
made of the rate of cathode vaporization in [4-11] (as the capacitance
of the supply capacitor, the working voltage, and the ballast resistance were varied (the ballast resistance went as high as 7 Ω)).
However, such proportionality is not observed for most conditions
encountered in practice, and the dependence of the sputtering rate
on the parameters listed is quite complex. This may be linked to the
fact that, first, formula (4-3), which was derived for the case of
instantaneous energy release at a spot on the cathode, does not extend
to discharges that have a comparatively large pulse length and a large
zone of the cathode all parts of which are in operation at once.
Second, we may not expect that proportionality between the energy
released on the cathode and the amount of electricity that flows
through the discharge should be observed when discharge conditions
are varied significantly.

We have outlined in general features the existing picture of the
processes on a cathode producing a pulsed discharge. Phenomena on
the anode are of less practical importance, since the darkening of
flashlamp envelopes is due mostly to cathode sputtering,** and since

---

*Several other versions of complicated cathodes have been published
 recently. These cathodes use granulated tantalum [4-17], a tungsten-
 vanadium alloy [4-18], and such additives as barium aluminate [4-19],
 samarium dioxide [4-20], barium baryllide [4-21], and others [4-22].
**The anode in flashlamps undergoes practically no sputtering if it
 has dimensions large enough that the anode as a whole does not over-
 heat to the temperature of intense sublimation of the metal.

Table 4-1. Average Rate of Vaporization of Cathodes Made of Different Pure Metals in Pulsed Discharges [4-11]. The Discharges Were Produced in Argon at $p = 0.065$ MPa (500 mm Hg), $l = 4$ mm, the Cathode Diameter was 2.5 mm, $C = 3$ μF, $U_o = 520$ V, the Ballast Resistance was 1 Ω, and the Frequency was 50 Hz

Cathode material	Be	Al	Mo	W	Cu	Ag	Zn	Cd	Sb	Sn	Pb	Bl
Average value of specific heat used in calculation, 10^3 J/(kg·K)	2.4	1.3	0.32	0.17	0.6	0.34	0.56	0.32	0.4	0.32	0.16	0.24
Calculated rate, * 10^{-7} g/s	4.3	16	16	23	20	40	56	130	54	180	260	210
Observed rate, ** 10^{-7} g/s	3.0	12	14	14	17	27	55	91	91	200	340	380

* Absolute calculations were made in [4-11] by inserting in formula (4-3) the following expression for the total energy dissipated on the cathode per second: $W = fCU_0U_i\sqrt{m_e/m_i}$ (where f is the flash rate, U_i the ionization potential of the gas, and m_e and m_i the mass of the electron and ion, respectively), as obtained under the assumptions indicated above.

** If we assume that the off-duty factor of the current pulses in [4-11] was 5×10^3 (corresponding to the actual effective current-pulse length of 4 μs), then a rate of 10^{-2} g/s during the current pulse corresponds to the average rate of vaporization given in the table: 20×10^{-7} g/s.

Table 4-2. Average Amount of Cathode Material Sputtered (g/pulse)
            for Cathodes Containing Various Activating Additives
            [4-16] (discharges in xenon, 0.053 MPa (400 mm Hg),
            l = 80 mm, the inside diameter of the tube is 7 mm, and
            the diameter of the cathode is 6 mm).

Additive	---	4% Hf	5% Ni + 2% Ba	1.5% ThO$_2$	5% ThO$_2$	Impregnated aluminosilicate or aluminate cathodes
Amount sputtered	1.4×10^{-4}	1.8×10^{-5}	1.6×10^{-7}	2.5×10^{-8}	1.2×10^{-8}	2.1×10^{-9}

an estimate of the energy losses around the electrodes in pulsed discharges shows that losses on the cathode are predominant in most cases
(the cathode voltage drop may amount to a few tens of volts) (see
Section 8-2 and [4-23]). Anode effects briefly reduce to the constriction of the channel into a single general spot at which the metal
is vaporized continuously and more smoothly (without explosions and
with lower ejection velocities than on the cathode) under ordinary
conditions. The anode spot also expands as the discharge channel
expands.

The current density at the anode spot is about 10^5 A/cm^2. The
anode voltage drop is 2-9 V. This estimate is based on the thermal
effect [4-16]. (We use the heating of the anode after the flash, or
the thickness of the foil which melts during the discharge; then the
heat propagation is calculated by a procedure analogous to the one
used for the cathode). The anode voltage drop obviously decreases
as the work function of the anode material decreases. This is due
to the decrease in the energy gained by the anode upon the absorption
of electrons. The phenomena on an anode covered with a very thin
insulating film are of particular interest as a means that can be used
to study the conduction zone of a discharge. The anode spot has a
multiple structure when such a film is used. This structure may be
explained as follows. The ion-produced distribution of potential
around the "hole" focuses the electron stream into the hole after the
breakdown of the film which occurs near the axis of the channel at
the start of discharge. The "hole" is used by the expanding discharge
until the buildup of surface electron charges in more peripheral
regions of the insulating film exceeds an electrical gradient in the
film such that the film breaks down and a new "hole" is formed. If the
electrical conductivity of the channel has decreased in the region of
the first "hole," the "hole" is abandoned and the charge is transferred to the peripheral "hole." The longer the discharge is retained

on the original holes, the more pronounced the fusion tracks remaining
on them. The new holes that continually appear during the discharge
are located in concentric annular zones which move away from the center
with a velocity equal to the rate of channel expansion. A method
using extremely short double current pulses separated by various time
intervals enabled the authors of [4-24] to elucidate the pattern
described above. It also allowed them to show that the channel expansion velocity that follows from this pattern is in agreement with
the hydrodynamic theory of expansion (Section 2-2) and to determine
that the conductivity on the axis of an expanding discharge column is
low.

PART 2

GAS-DISCHARGE FLASHLAMPS

CHAPTER 5

RADIATION CHARACTERISTICS OF FLASHLAMPS

5-1.   General Information on Flashlamps

A flashlamp is a gas-discharge device with two primary current-carrying electrodes and a gas-filled gap between them. Such a device is designed to produce high-power pulsed (spark) electric discharges with intense luminous radiation in the gas-filled gap at specified times. Flashlamps have a hermetically tight, usually sealed outer envelope made of glass or quartz which is filled with a chemically inactive gas (usually a noble gas). This is in contrast to open air-filled spark gaps, which are characterized by unstable performance due to wear on the electrodes, a fluctuating dependence on air pressure and humidity, the necessity of using an extremely high voltage, low luminous efficiency, poor controllability, the noise effect, and other factors. The firing time of the lamp usually is controlled by means of a third electrode, a control electrode (or sometimes several control electrodes) mounted inside the lamp or on the surface of its envelope. A high-voltage pulse which triggers the discharge is fed to the control electrode. In some flashlamps there is no third electrode, and the firing time is controlled by a more or less short-duration increase in the potential difference between the primary electrodes.

A pulsed discharge in a lamp is powered by some electrical source which is capable of providing a high current strength for a short period of time. A capacitor charged by a comparatively low-power DC circuit usually is employed as such a source, but other sources also exist (see Chapter 9).

Like continuous light sources, the available variety of pulsed sources should (whenever possible) provide consumers a selection of

149

the lamps required with different sizes and shapes of the luminous
volume and with different luminous and spectral characteristics.
Two other variable parameters should be added to these three basic
parameters for flashlamps: flash length and flash repetition frequency. Therefore, in time the extensive introduction of flashlamps
in various fields of technology may require a product assortment beyond the extremely broad variety of continuous lamps.

At the same time, allowance should be made for a specific
feature of flashlamps, namely the possibility of using them effectively under widely varying operating conditions: pulsed discharges
in a given gas-filled gap having nearly identical efficiency and
spectral composition of radiation are characterized by different
luminous intensities and durations and may occur with different repetition frequencies. This fact expands, as it were, the assortment
of flashlamps available to consumers when the number of types of
lamps produced industrially is limited.

Optical devices which use the radiation of flashlamps may be
divided into two main classes. The first class includes flashlamps
in which the radiation is concentrated in an extremely narrow beam
(e.g., automatic devices, optical communications devices, and lo-
cating devices). Devices of the second class fill a large solid
angle with radiation (photoilluminators, optical pumping systems
for lasers, light-signaling devices, etc.). It generally is required
of the first class that the luminous volume of the source by minimal
and have maximum luminance, while for the second class the luminous
volume of the source does not necessarily have to be small,* and the
primary requirement is that the radiation of energy be maximized.

It is fortunate for the development of pulsed light sources
that most optical devices of the first class interact with high-speed
receiving or detection systems which operate at a high frequency and
react to the peak radiant flux (without a dependence on the flash
duration, or with a slight dependence). For such systems, increasing
the flash length means only a pointless expenditure of energy and a
reduction of the operating life of the device. At the same time,
optical devices of the second class usually actuate slow-responding
detection systems which react mainly to radiant energy (the time-
integrated radiant flux) and require comparatively infrequent flash
repetition. Therefore, the requirement of minimal volume and maximal
luminance of the pulsed light source usually is coupled with the requirements of short flash duration and high flash frequency, and the
requirement for maximal radiant energy is made when the flash duration
and frequency are of secondary importance.

---

* In some cases it is important that the luminous volume be given a
  special shape, such as a ring, a cylinder, a straight line segment,
  or a sphere.

This classification of the requirements made on pulsed light
sources may be compared with the data on the physical characteristics
of a pulsed discharge in gases that were described in Part 1. Here
we may conclude that devices of the first class generally must be
equipped with lamps having a short spark channel which usually is
not confined by the walls of the discharge tube and is located in
the central part of a broad (spherical) envelope. At the same time,
lamps having an extended discharge channel confined by the discharge
tube generally should be used in devices of the second class. Such
a tube may be given a shape determined by the properties of the required light beam.

Both the physical features described above for spherical and
tubular lamps and their parameters are related to their design and
power-supply parameters by essentially different rules. Their design principles and the processes used in their industrial production
also are different.

5-2. The Main Photometric Characteristics of Pulsed Light Sources

The radiation of sources that are continuous in time may be
considered in two aspects: the spatial and the spectral. The radiation of pulsed light sources also must be characterized temporarily.
The peculiarities of the spectrum-time relationship are determined
by the fact that the spectral composition of the radiation changes
continuously during a pulse. The change occurs at different rates
for different directions of the radiation in space [5-1]. This
greatly complicates the use of photometric parameters in the description of phenomena. The time dependences of instantaneous values
must be considered instead of time-continuous photometric quantities
(GOST* 7932-56) and radiometric quantities (GOST 7601-55). Furthermore, a number of time-integrated photometric quantities that apply
solely to pulsed sources have been introduced (GOST 16803-71).

The main photometric quantities used to describe pulsed radiation are given in Table 5-1. The definitions of these quantities
can be found in the standards mentioned above, in the *Vocabulaire
internationale de l'éclairage* [5-2], and in a number of handbooks
of photometry and illumination engineering, such as [5-3 to 5-5].
In most practical cases it suffices to know the duration of the
radiation pulse and the values of the peak (maximum) and integrated
photometric and radiometric quantities. The period of time for which
the luminous intensity $I(t)$ exceeds a specified fraction (0.35) of
its peak value I_p usually is taken as the length τ of a radiation
pulse. The time integrals of photometric and radiometric quantities

---

* Translator's note: a "GOST" is an All-Union State Standard, which
  is a mandatory technical standard in the USSR.

are specific in character, and only some of them currently have
generally recognized names. For example, the *Vocabulaire Internationale de l'éclairage* [5-2] contains four integrated quantities:
luminous energy, radiant energy, exposure, and radiant exposure.
The terms "integrated luminous intensity"* and "integrated radiant
intensity"* are used in the USSR (GOST 16803-71). The other integrated quantities shown in Table 5-1 do not have special names or
designations.

Pulsed sources usually are characterized spectrally by the
spectral density of the peak radiant intensity and the integrated
spectral radiant intensity. The spectral densities of other radiometric quantities also may be used [5-2]. Such characteristics as
the luminous efficacy (the ratio of the luminous energy to the electric input power), the "luminous efficacy per unit solid angle" (the
ratio of the luminous energy to the input power), and the spectral
luminous efficacy (spectral efficiency), which shows what fraction
of the electric energy is converted to radiant energy per unit spectral interval of wavelengths (GOST 16803-71).

There exist tubular flashlamps of a special type (the coaxial
type) in which the discharge occurs in a cylindrical layer between
two coaxial quartz tubes, and the irradiated object is located in
the cavity of the inside tube. An opaque coating which reflects
radiation inward usually is applied to the outer surface of the
envelope of such lamps to improve their efficiency. In principle,
coaxial-type lamps without this coating could be characterized by
the photometric quantities used for ordinary lamps if the relation
is established between these quantities and the characteristics of
the light field in the cavity of the lamp. Coaxial-type lamps having
an opaque outer envelope cannot be characterized by these photometric
quantities, since they only radiate inside the cavity, and the concept of a point source is inapplicable to them. A group of photometric parameters (GOST 16803-71) based on the concept of the spatial
illuminance E_O has been put together to describe the radiation of
such lamps. Physically, E_O is the sum of the normal illuminances at
a given point of the field, or the luminous flux incident on a sphere
having an equatorial cross-sectional area equal to unity [3-80].
The spatial illuminance E_O is defined by the expression $E_O = \int L_V d\Omega$,
where L_V is the luminance in the direction of a unit solid angle $d\Omega$.
It is measured in luxes in photometric units and in W/m^2 in radiometric units. The following radiation parameters of coaxial-type
flashlamps are considered the main ones [5-6]:

---

* Translator's note: the Russian terms used here are "osvechivanie"
  and "energeticheskoe osvechivanie," respectively. "Osvechivanie"
  is defined in one Russian dictionary as a quantity expressed in
  "candela-seconds." Here it is translated as "integrated (luminous
  or radiant) intensity."

Table 5-1. The Basic Quantities of Pulse Photometry

Instantaneous and peak quantities			Time-integrated radiations		
name	symbol and formula	unit of measurement	name	symbol and formula	unit of measurement
Luminous flux	Φ_v	lm	Luminous energy	$Q_v = \int \Phi_v dt$	lm-s
Luminous intensity	$I_v = d\Phi_v/d\Omega$	cd	Integrated intensity	$\Theta_v = \int I_v dt$	cd-s
Luminance	$L_v = d^2\Phi_v/(d\Omega dA\cos\varphi)$	cd/m^2	Integral of luminance pulse	$\int L_v dt$	$cd\text{-}s/m^2$
Luminous exitance	$M_v = d\Phi_v/dA$	lm/m^2	Integral of luminous exitance pulse	$\int M_v dt$	$lm\text{-}s/m^2$
Illuminance	$E_v = d\Phi_v/dA$	lx	Exposure	$H_v = \int E_v dt$	lx-s
Radiant flux	Φ_e	W	Radiant energy	$Q_e = \int \Phi_e dt$	J
Radiant intensity	$I_e = d\Phi_e/d\Omega$	W/sr	Integrated radiant intensity	$\Theta_e = \int I_e dt$	J/sr
Radiance	$L_e = d^2\Phi_e/(d\Omega dA\cos\varphi)$	$W/sr\text{-}m^2$	Integral of radiance pulse	$\int L_e dt$	J/m^2
Radiant exitance	$M_e = d\Phi_e/dA$	W/m^2	Integral of radiant exitance pulse	$\int M_e dt$	J/m^2
Irradiance	$E_e = d\Phi_e/dA$	W/m^2	Energy exposure	$H_e = \int E_e dt$	J/m^2
Spectral density of radiant intensity	$I_{e\lambda} = dI_e/d\lambda$	W/sr-nm	Spectral density of integrated radiant intensity	$\Theta_{e\lambda} = d\Theta_e/d\lambda$	J/sr-nm

Note: dA is the area of a surface element (m^2); $d\Omega$ is the unit solid angle (sr); φ is the angle between the ray and the normal to dA; λ is the wavelength (nm); t is the time; e is the subscript for a radiometric quantity; and v is the subscript for a photometric quantity (it is omitted in obvious cases).

(1) the spatial exposure on the axis of the lamp at the middle of
    the cavity, H_O (the integral of the spatial illuminance over the
    emission time):

$$H_O = \int\limits_{0}^{\infty} E_O(t)\, dt = \int\limits_{0}^{\infty} \int\limits_{(4\pi)} L_v(t, \Omega)\, dt\, d\Omega;$$

(2) the peak spatial illuminance E_{Oc} on the axis of the lamp at the
    center of the cavity;
(3) the pulse length of the spatial illuminance τ_O at the level of
    $0.35 E_{Oc}$;
(4) in spatial distribution of the radiation of a coaxial-type lamp,
    $H_O(x)$: the dependence of the spatial exposure H_O on the axis of
    the lamp on the distance x to the center of the cavity;
(5) the spectral density of the radiant spatial exposure $H_{O\lambda}$ on the
    axis of the lamp at the center of the cavity.

The spatial exposure H_O at the center of the cavity characterizes
the radiation of a lamp of a specific type quite unambiguously, and
can serve as an analogue of the integrated intensity.

Pulsed sources usually are not used for illumination (for use
by the human eye). Nonetheless, their radiation is characterized in
terms of photometric quantities in the reference, scientific, and
technical literature. This is explained by the fact that the system
of photometric measurements using physical photometers has been worked
out in considerable detail and introduced on an international scale.
In many cases flashlamps actuate detectors that are sensitive in the
visible region of the spectrum. Photometric quantities can be used
to evaluate the radiation parameters in various spectral regions and
outside the visible region for a flashlamp that has some sufficiently
invariant spectral distribution under certain conditions of power
supply.

Flashlamps of each type should be considered to be general-
purpose instruments with an extremely broad range of parameters of
the radiation pulses generated, depending on the specific parameters
of the power supply. At the same time, the radiation characteristics
of tubular and spherical flashlamps differ considerably.

5-3. Light Characteristics of Tubular Flashlamps

The nature of the curve of the time dependence of the luminous
intensity $I(t)$ of a pulsed discharge is determined by the parameters
of the lamp and the discharge circuit. Luminous intensity pulses
$I(t)$ usually are characterized by three parameters: the integrated
intensity θ, the peak luminous intensity I_p, and the flash duration
τ. The parameters of radiation pulses can be varied over a wide
range by varying the shape and duration of the pulse of electric

power that is released in a lamp. Luminous intensity pulses that
are nearly rectangular can be obtained when the lamp is powered by
a long line (see Chapter 9). However, the curve of I(t) has a
characteristic form that is common to all conditions (Figure 5-1)
when a more extended power supply is provided from a capacitor and
the discharge circuit has inductances of up to 10 μH. The scales
along the abscissa and the ordinate can be varied to achieve a practical match among all graphs, to the accuracy of small changes in
the steepness of the front and disruptions of the smooth behavior of
the curve due to fluctuations of gas density as the channel expands.

Hence the ratio $K = \Theta/I_p\tau$ and the ratio K' of the integrated
intensity over a period of time τ (see Figure 5-1) to the integrated
intensity Θ of the entire pulse have practically identical values
for the most varied conditions: $K = 0.86 \pm 0.04$ and $K' = 0.81 \pm 0.04$.
The quantities K and K' also do not change too significantly (by no
more than 10%) upon a significant increase in the inductance of the
power-supply circuit of lamps for which the graph of I(t) approaches
a symmetrical bell shape. Thus, pulses of luminous intensity may be
characterized approximately by just two parameters, such as the integrated intensity and the flash duration. Various functions have
been proposed for describing the characteristic shape of the radiation pulse [5-7, 5-7a, 5-7b, and 5-8]. For example, the luminous
intensity is approximated quite well by the following expression
when there is a broad variation of the discharge energy over the
range 50-14,500 J and of the capacitance of the supply capacitor
over the range 200-2500 μF:

$$I\,(t) = I_p\left[\frac{t}{t_p}\exp\,(1-t/t_p)\right]^b,$$

where t_p is the time when the peak value is reached (this varies
from 30 to 600 μs); and b is a parameter of the function, which is
0.14-1.2 when the inductance of the circuit is low and 3 when it is
high. In some cases the following approximation is more convenient:

$$I\,(t) = \alpha I_p\,(e^{-t/\tau_1} - e^{-t/\tau_2}),$$

where α, τ_1, and τ_2 are parameters of the function, and α depends
on the ratio τ_1/τ_2.

It is convenient to compare the radiation pulses for different
electric discharge energies W by using the luminous efficacy of the
pulsed light source: $\eta = \Theta_0\Omega_{eq}/W = Q/W$. Here, as for continuous
radiation sources, Ω_{eq} is the equivalent solid angle expressed in
steradians (the ratio of the luminous energy Q of a flash to the
integrated intensity Θ_0 in the accepted main direction).

The light characteristics are intricately dependent on the
design data of the lamps (d_i, the inside diameter of the envelope;

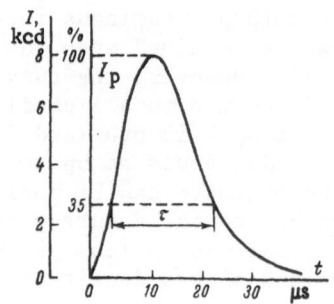

Fig. 5-1. Typical dependence of the luminous intensity $I(t)$ on the
time t for low inductance of the discharge circuit (I_p is
the peak luminous intensity and τ the duration of the
luminous intensity pulse). A xenon capillary flashlamp
($p_0 = 0.08$ MPa, $l = 70$ mm, $d_i = 0.5$ mm), $C = 0.25$ μF,
$U_0 = 1.2$ kV, $L = 1$ μH.

l, the electrode spacing; V_{el}/V_w, the ratio of the ballast volume
around the electrode to the working volume; p_0, the pressure of the
cold gas just before the flash; and the type of filler gas) and on
the power-supply parameters (U_0, the operating voltage; C, the capacitance of the bank of capacitors in the discharge circuit; L, the
inductance of the discharge circuit; and R_b, the active ballast resistance of the circuit). Generalizing the experimental data that
have been amassed is complicated by the fact that in many cases the
data were gathered without sufficient monitoring of these parameters,
e.g., electric power losses in the circuit and the escape of gas into
the volumes around the electrodes.

The dependence of τ on d_i is complex. It goes from being proportional to $d_i^{-1/2}$ to being proportional to d_i^{-2} as l and C increase
when L is low. If the product Cl is constant, then a change in l
produces nearly no change in τ, ignoring the extremely small l (less
than 4 cm) for which a slight increase is observed in τ (10% when
$l = 2$ cm). This increase is due to the fact that the voltage drops
in the discharge near the electrodes become comparable to the voltage
drop in the column. The dependence of τ on U_0 or on the electric
field strength $E = U_0/l$ may be expressed by the formula $\tau_2/\tau_1 \simeq$
$(E_1/E_2)^{0.6}$, where E_1 and E_2 are the two values of the initial strength
and τ_1 and τ_2 are the corresponding flash lengths. The dependences
of τ on C also are complex: a transition from proportionality to
$C^{1/2}$ (when $\tau \lesssim 50$ μs) to proportionality to C is observed as we go
from small values of Cl to large ones. Therefore, the different
empirical formulas for the dependence of τ on E can be used in a
narrow range of C. The left side of Figure 5-2 shows a family of
experimental graphs of the dependence of τ on the product Cl for
xenon lamps, for an initial electric field strength of 100 V/cm and
$L \simeq 1$ μH. Shown at right is the displacement of the ordinate scale

with varying initial strength. Hence Figure 5-2 may be used as a
nomogram for determining the flash duration of various tubular lamps
for different electrical parameters (given small R_b and L). An inspection of Figure 5-2 shows that the slope of the curves and the
vertical distance between them in the region of large τ (over 300 μs)
correspond to the concept that the discharge column is equivalent to
an ohmic resistance R_L, which is determined by the plasma conductivity
σ (which is unique for a given value of E) and by the geometric dimensions of the inner channel of the discharge tube. The calculated
value of the power pulse length ($R_L C/2$) that corresponds to this
concept is in agreement with the experimental data for τ (when $\tau \geq$
300 μs) if we assume $\sigma \approx 80$ $(\Omega\text{-cm})^{-1}$ when E = 100 V/cm.

Based on this concept, we should expect that the length τ should
not have a significant dependence on L or R_b if the inductance of the
discharge circuit is below the critical value $L_{cr} = CR_L^2/4$ and if R_b
$\ll R_L$. This is confirmed, for example, by the fact that the introduction of an ~100-μH inductance into the discharge circuit has
practically no effect on the time dependence of the luminous intensity
for capillary lamps with R_L = 40 Ω and L_{cr} = 100 μH, and does not
result in a noticeable increase in τ. Only the introduction of a
2500-μH inductance increases τ by a factor of 2.2. Analogously, the
additional incorporation of a 100-μH inductance, which exceeds L_{cr}
by a factor of 3, is required to double the τ of the radiation pulse
of a lamp having R_L = 0.4 Ω (xenon, p_0 = 0.013 MPa, d_i = 9 mm, l =
14 cm, C = 800 μF, U_0 = 1 kV). We may assume roughly that the relation $\tau \approx 2.6\sqrt{LC}$ is fulfilled when L is substantially larger than
L_{cr}. The ballast resistance of the circuit, which exceeds the calculated resistance of the lamp, has a qualitatively analogous effect
on the duration of luminous intensity pulses.

The light characteristics have just as intricate a dependence
on the type of gas and on the gas pressure. The corresponding relations for low p_0 and small d_i are most pronounced wherever the
plasma begins to differ significantly from an equilibrium plasma.

Table 5-2 gives the light parameters of capillary flashlamps
containing different fillers. Attention is drawn to the fact that
the peak luminous intensity at low discharge energies has a maximum
at a pressure of 0.013-0.04 MPa, decreases by 30% at large p_0, in
contrast to the flash duration and the integrated intensity, which
initially increase monotonically with increasing gas pressure p_0 and
then stop changing. This is due to the more rapid increase in integrated intensity with increasing pressure, compared with the increase
in τ. It also can be seen from Table 5-2 that capillary discharges
in xenon has a severalfold higher luminous efficacy than discharges
in other noble gases. Analogous but much less pronounced relations
of the light parameters also are observed for flashlamps having
larger inside diameters.

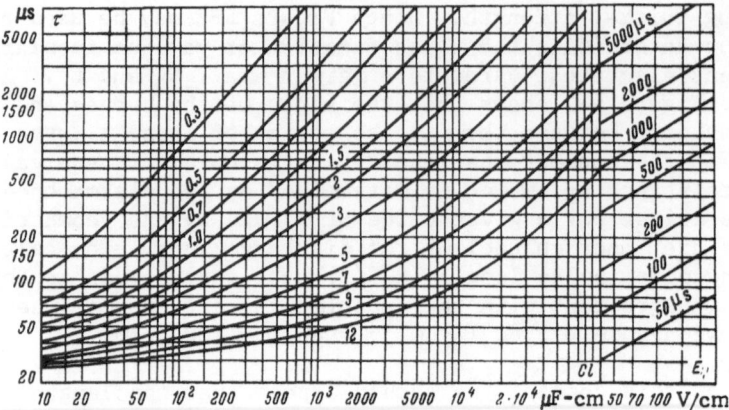

Fig. 5-2. Dependences of τ on Cl for lamps having different inside
diameters of the discharge tube (d_i in mm are given above
the curves) for an initial electric field strength E =
100 V/cm (Xe, p_0 = 0.02-0.08 MPa, l = 2-130 cm [5-10].
The displacement of the ordinate scale with varying E is
shown at right.

Table 5-2. Dependences of τ, I_p, and Θ on the Type of Gas
and on the Pressure (d_i = 0.5 mm, l = 70 mm,
C = 0.25 μF, U_0 = 1.2 kV)

Type of gas	P_0,MPa	τ,μs	I_p, kcd	Θ,cd-s
Argon	0.08	12.6	2.2	0.025
Krypton	0.08	19.4	5.1	0.08
Xenon	0.08	22.2	8.2	0.15
"	0.04	17.9	12.3	0.15
"	0.013	10.6	12.3	0.1
"	0.006	7.0	9.8	0.04

The peak luminous intensity I_p of capillary xenon flashlamps
(d_i = 0.5-3.5 mm, l = 10-80 mm, U_0 = 0.5-3 kV, C = 0.25-2 μF, L =
0.5-10 μH, p_0 = 0.003-0.09 MPa) is linked (approximately linearly)
by an increasing relation to the capacitance C of the working capacitor and the square of the operating voltage U_0 [5-9]. The value
of I_p increases with decreasing C at a constant discharge energy.
The peak luminous intensity I_p is practically independent of the
xenon pressure in the range 0.03-0.1 MPa. It has a maximum when d_i
lies in the range 0.5-2 mm, and decreases appreciably with increasing
inductance of the discharge circuit if the resistance of the lamp
is relatively low.

A useful empirical expression for the optimal pressure has been derived for the characteristic dependences of the luminous efficacy η_v on the gas pressure (Figure 5-3) for τ = 600-900 μs [5-11]: $p_{0.95}$ = 0.3/d_i (where $p_{0.95}$ is the pressure in MPa at which the luminous efficacy reaches 95% of its maximum values, and d_i is the inside diameter of the envelope in mm). This expression has been confirmed for lamps with d_i ranging from 7 to 34 mm. According to available data, the absolute values of the luminous efficacy per unit solid angle $\eta_v' = \Theta_0/W$ (where Θ_0 is the integrated intensity in the direction of the normal to the axis of the lamp, and W is the electric energy of the discharge) range from 4 to 6 cd-s/J in straight tubular xenon flashlamps having high discharge energies at optimal xenon pressures when the energy losses in the discharge circuit are small (not over 10%) for the characteristic ranges of the geometrical parameters (d_i = 7-94 mm, l = 50-1000 mm) and electric power densities in the discharge ((50-500) x 10^3 W/cm^3) [5-11 to 5-13].

The significant influence of the ballast volumes around the electrodes on the luminous efficacy of flashlamps can be seen from Figure 5-4. The decrease in the luminous efficacy of the lamps with increasing V_{el}/V_w may be attributed to the relations shown in Figure 5-3. The escape of a significant fraction of the neutral atoms from the discharge column (in the working volume V_w) into the relatively cold regions around the electrodes that are not occupied by the discharge is characterized by formula (2-29) [2-34, 3-26, and 5-14]. The escape is equivalent to a decrease in p_0. The rated value of V_{el}/V_w lies in the range 0.04-0.16 for most series-produced tubular flashlamps (the value is 0.85 for the ISP-1000 lamp), but the actual value sometimes may deviate from the rating by a factor of 2-4 in either direction [5-15].

The influence of design and electrical parameters on the light characteristics of coaxial-type flashlamps has not been studied in as much detail [5-16 and 5-17], but we may assume that it is analogous. The luminous efficacy of coaxial-type lamps is severalfold lower than that of ordinary lamps because of the increase in energy losses due to the inside tube of the envelope. For example, the luminous efficacy per unit solid angle is just $\eta_v' = 3.9$ cd-s/J for an IFPP-7000 lamp without an external reflective coating if the discharge energy is sufficiently high (W = 6760 J) [5-12]. The duration of illuminance pulses is practically the same at different points in the cavity [5-6], and measurements of the spatial distribution of the radiation inside the cavity (Figure 5-5) are in good agreement with calculations [5-18]. The spatial exposure $H_0(x)$ no longer depends on the axial coordinate x if the radiation detector is immersed comparatively little in the cavity (5 cm for long lamps having an internal diameter of 4.5 cm (sic)). The spatial exposure increases by a factor of about 6 if a diffusely reflective coating having a reflection coefficient of 0.9 is applied to the outer surface of a coaxial-type lamp.

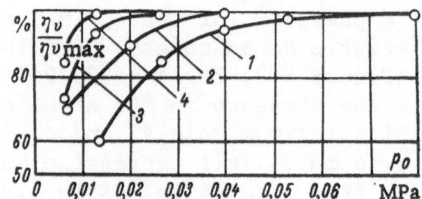

Fig. 5-3. Dependences of relative luminous efficacy η_v/η_{v-max} of tubular flashlamps on xenon pressure p_0 for the following d_i and τ: 1) 7 mm, 600-900 μs; 2) 16 mm, 600-900 μs; 3) 34 mm, 600-900 μs; 4) 34 mm, 150 μs   [5-11].

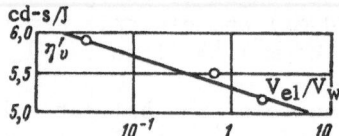

Fig. 5-4. Experimental dependence of the luminous efficacy per unit solid angle of a xenon flashlamp on the volume ratio V_{el}/V_w (d_i = 7 mm, l = 80 mm, p_0 = 0.053 MPa, W = 650 J, C = 600 μF, L = 100 μH)   [2-35].

As we turn to brightness characteristics, let us note first that, because of refraction in the glass of the capillary wall, the diameter (observed along the normal to the axis of the lamp) of the glow tube of a capillary is equal to the true diameter of the ca- pillary, i.e., there is no increase.   Indeed, an examination of the path of rays through a cylindrical envelope in a plane perpendicular to the axis of the lamp [3-17, 5-19, and 5-20] and of rays that sub- tend any angle with the axis [5-21] shows that although a ray emerging from the lamp is rotated with respect to the corresponding ray inside the envelope because of refractions on the surfaces of the envelope, the distance between this ray and the axis of the lamp remains un- changed.   (This is true for any number of coaxial cylindrical shells, provided that the refractive indices of the media are equal inside and outside the shells.   This condition is met exactly for gas-dis- charge tubes in air, since the refractive indices of the gas filler and air differ insignificantly.)   Thus, the envelope does not affect the radial distribution of the luminance of a completely transparent cylindrically symmetric discharge column if the discharge has circular symmetry.   However, different rays may be attenuated selectively if the source (the plasma or the shell) is absorptive.   This can be de- tected from the slight polarization of the radiation.

Allowing for the properties of cylindrical envelopes, we can use the known values of the light parameters to make an extremely simple calculation of the peak overall luminance $L_{vp} = I_p/ld_i$ and

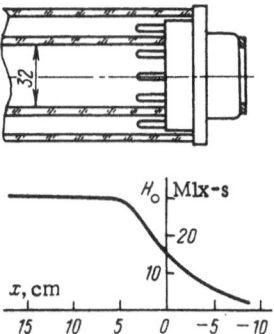

Fig. 5-5. Variation of the spatial exposure H_0 along the axis of a coaxial-type lamp without an external coating. The diameter of the cavity is 32 mm [5-6]. (A cross section of the lamp is shown at top.)

the overall integral of the luminance $\int L_v dt = \Theta/l d_i$ for comparatively long discharge durations for which the channel-expansion stage is insignificant and the plasma fills the entire inside cross section of the discharge tube nearly uniformly. (Analogous expressions which include the corresponding areas of the projection of the luminous volume are used for lamps with curved tubes.) For example, in view of the small diameter of capillary lamps we may assume that τ = 15-30 μs greatly exceeds the time required for expansion of the discharge channel. Magnified photographs of discharges in capillaries with d_i = 0.3-1 mm show [1-67] that filling of a capillary with plasma in a way that ensures uniform luminance is achieved at energies per centimeter length of the discharge which exceed some minimum value $W_{min}/l = (CU_0^2/2l)_{min}$ that depends on d_i (0.01 J/cm for d_i = 0.5 mm, and 0.07 J/cm for d_i = 1 mm; given equal W/l, discharges with a higher voltage and a lower capacitance of the supply capacitor fill the capillary slightly worse than do discharges having the reverse relation between these parameters; therefore, W_{min} can be determined with an error of about 30%).

The time integral of the luminance is of the greatest practical importance for tubular lamps, which generally are used together with an integrating radiation detector (such as a photographic material, the active element of a laser, or the eye). The maximum value of this quantity depends mainly on the product Cl (which determines the pulse length) and on the diameter and material of the discharge tube. Any attempt to surpass this boundary by increasing the peak discharge power for a given Cl, a given d_i, and a given material will result in breakage of the tube (see Chapter 6). Figure 5-6 shows graphs of the dependence of the limiting integral of luminance on the product Cl for different diameters and different materials of the discharge tube. Attention is drawn to the fact that the limiting integral of luminance is practically independent of d_i for quartz lamps having tube inside diameters of 4-12 mm. This can be seen from Table 5-3.

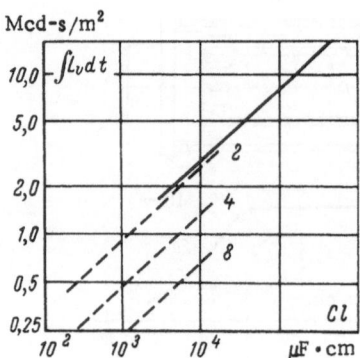

Fig. 5-6. Behavior of the limiting integral of luminance for
tubular xenon lamps. The solid line represents quartz
tubes (d_i = 4-12 mm). The dashed lines represent tubes
made of S-52 glass (formerly called ZS-5K glass), with
the diameters in millimeters indicated in the figure.

For different d_i the same limiting values of the luminance
integral correspond to different pulse lengths. The graphs of the
dependence of the limiting luminance integral on d_i that are shown
in Figure 5-7 can be constructed from data on pulse length and from
Figure 5-6 if the pulse length τ is specified.

Data on the luminance distribution are less definite for extremely short-duration discharges whose lengths are comparable to
the channel expansion time. The formulas presented above for L_{vp}
and $\int L_p$ dt give the average values of the peak luminance and the luminance integral over the diameter of the tube. These average values
may be much lower than the values that correspond to the most luminous section. However, the difference in the luminance over the width
of the channel is comparatively small at values of $CU_0^2/\pi l d_i$ larger
than 2 J/cm². This can be seen, for example, from instantaneous
photographs of the expanding channel or from experimental data on
the instantaneous radiances at various distances from the axis.
Examples are given in Table 5-4.

This conclusion is consistent with composite data on the brightness characteristics of discharges in capillaries. The data were
culled from various papers and are presented in Table 5-5 and Figure
5-8.

The brightness characteristics of various flashlamps are given
in the last three columns of Table 5-5. For the most part, these
values were determined by dividing the luminous intensity and the
integrated intensity by the area of a side projection of the discharge-tube channel. The next-to-last column gives the value of the
peak luminance, divided by an empirically determined coefficient

Table 5-3. Limiting Values of the Luminance Integral $\int L_c dt$ (Mcd-s/m^2) for Quartz Xenon Lamps with p_0 = 0.02 MPa (150 mm Hg) Using Natural Cooling, Interpulse Intervals of 15 s, and an Inductance of Up to 10 μH

Cl,μF-cm	d_i,mm				
	4	6	9	12	15
12,000	3	3	3	–	–
36,000	4.6	4.8	4.5	4.4	3.2
120,000	7	8.7	8.6	8.1	–
324,000	11.5	11.5	10.6	10.3	8

Table 5-4. Peak Radiances of a Capillary Discharge at Various Distances from the Axis in Two Regions of the Visible Spectrum. Air, p_0 = 0.1 MPa; d_i = 0.8 mm; l = 1 cm; C = 0.05 μF; U_0 = 18 kV; the Radiance Is Given in Relative Units

Distance from axis as a fraction of radius of capillary	0	0.25	0.5	0.75
Luminance } λ=570 nm at } λ=434 nm	1 1	1 0.9	0.9 0.7	0.6 0.3

which is equal to the square of the function η/η_{max} (see Section 5-6). This coefficient is a measure of the decrease in the peak luminance of discharges which are characterized by a value of $CU_0^2/\pi l d_i$ below 2 J/cm^2, compared with the maximum value L_{vp} for a given electric field strength. Physically, the coefficient may be related to the optical density of the column of radiating plasma.

As we see from Figure 5-8, the available data on the peak luminance of discharges having a limited channel can be represented satisfactorily by a single graph that corresponds to the formula:

$$L_{vp}/\left(\eta/\eta_{max}\right)^2 = E_0^{0.9}/30, \qquad (5-1)$$

where L_{vp} is in Gcd/m^2 and E_0 in V/cm.

The deviation of points 3 and 5 from the general graph is due to systematic errors in photometric measurements using photocells

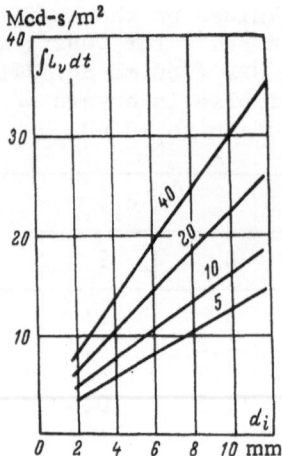

Fig. 5-7. Dependence of the limiting luminance integral of quartz
          lamps on the inside diameter of the envelope for certain
          pulse lengths, as indicated in milliseconds in the figure.

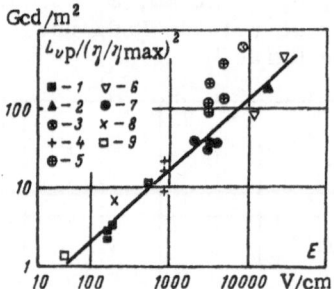

Fig. 5-8. Dependence of the peak luminance of a capillary pulsed
          discharge, divided by the coefficient $(\eta/\eta_{max})^2$, on the
          electric field strength. The coefficient allows for the
          inadequate optical density of the radiating discharge
          column (see Table 5-5 for the meanings of the symbols).

without correcting light filters. The spread of the other points
seems small if we allow for differences in the methods of measurement.
It is especially significant that the graph in Figure 5-8 encompasses
data which apply to different gases, pressures, discharge-circuit
inductances, and discharge parameters that differ by several orders
of magnitude. The use of this graph together with the graph in
Figure 5-2 and the dependence of η/η_{max} on the energy density
(Section 5-6) makes it possible to estimate all spectrally integrated
photometric characteristics of the radiation from pulsed discharges
in tubular lamps of small diameter with an accuracy that is satisfactory in practice. It should be stipulated that at present sufficient data are not available for determining the exact behavior

Table 5.5. Light Characteristics of Capillary Discharges, According to the Data of Various Authors [Refs. 1-68, 2-44, 2-51, and others — see Ref. 0-1]

Authors	No. in Fig. 5-8	Gas	p_a, MPa	r, cm	l, cm	C, µF	U_0, kV	$\frac{CU_0^2}{2}$, J	E_0, V/cm	$\frac{CU_0^2}{2\pi r_a l}$, J/cm²	τ, µs	I_p, Mcd	Θ, cd·s	η_a, lm/W	l_o, Gcd/m²	$L_o/(\eta/\eta_{max})$, Gcd/m²	$\int L_o\,dt$, kcd·s/m²
Marshak, Shchukin	1	Xe	0.08	0.025	7	4	1.2	2.9	170	5.5	100	0.086	7.4	26	2.5	2.5	210
Same		Xe	0.08	0.025	7	0.12	1.2	0.086	170	0.16	14	0.003	0.04	4.5	0.09	3.1	1.15
"		Xe	0.08	0.025	7	20	1.2	14.4	170	27.5	400	0.11	37.4	26	3.1	3.1	1070
Mak Thackeray	2	Xe	0.1	0.025	7	0.25	4	2	570	3.8	18	0.4	6.2	31	11.5	11.5	180
Edgerton	3	Air	0.1	0.04	1	0.05	19	9	19000	71	1.15	1.45	1.45	1.6	180	180	180
"	4	Ar	0.1	0.05	0.8	0.1	7	2.5	8800	20	1.5	4.8*	5.7*	25*	600*	600*	700*
		Xe+30%H₂	0.08	0.05	9.2	0.01	8	0.32	870	0.22	1.2	0.1	0.12	3.7	1.1	22	1.3
		Xe+30%H₂	0.08	0.05	9.2	0.02	8	0.64	870	0.45	1.6	0.22	0.35	5.5	2.4	17.5	3.8
		Xe+30%H₂	0.08	0.05	9.2	0.04	8	1.28	870	0.89	2.1	0.3	0.62	4.9	3.3	9	7
Edgerton, Katu†	5	Xe	0.1	0.06	0.6	0.03	2	0.06	3800	0.52	0.25	0.15*	0.038*	6.3	21*	120*	5.3*
Same		Xe	0.1	0.06	0.6	0.1	2	0.2	3300	1.73	0.6	0.5*	0.3*	15*	70*	91*	42*
"		Xe	0.1	0.06	0.6	0.1	3	0.45	5000	4	0.6	1	0.6*	13.3*	140*	140*	85*
"		Xe	0.1	0.06	0.6	1	2	2	3300	17.3	2.5	1.5*	3.7*	18.5*	210*	210*	510*
"		Xe	0.1	0.06	0.6	1	3	4.5	5000	40	3	2.6*	8*	17.8*	360*	360*	1100*
		Xe	0.39	0.12	1	0.011	12	0.8	12000	2.1	3	1.7	—	—	.70	87	—
Vanyukov, Mak, Ureš†	6	Air	0.1	0.02	1	0.011	29	9.2	29000	73	1	2	1.5	8	500	500	—
Same		Xe	0.3	0.015	1.7	0.1	6	1.8	3500	2.3	—	1.5	2.5*	13*	31	38	31
Kirsanov et al.	7	Xe	0.3	0.015	1.7	0.1	5.3	1.4	3100	1.7	—	2.5*	—	—	23	30	51*
Same		Xe	0.3	0.015	1.7	0.1	3.7	0.7	2200	0.87	—	1.5	—	—	15	39	—
"		Xe	0.3	0.015	1.7	0.05	6.9	1.2	4100	1.5	—	0.75	—	—	24	37	—
Edgerton et al.	8	Xe	0.02	0.17	7.5	12	1.5	13.5	200	3.2	30	1.2	44	32.5	6.5	6.5	170
Marshak et al.	9	Xe	0.013	0.2	7	2500	0.3	113	43	26	1100	0.42	400	35	1.5	1.5	1430

*Measurements were made with a radiation detector whose photocathode had maximum response in the blue region, without a light filter to correct for the spectral sensitivity of the eye.

of the function η/η_{max} for light gases. We may assume that the value of $CU_0^2/\pi l d_i$ for light gases, below which this function becomes smaller than unity, exceeds a value of 2 J/cm^2, which corresponds to xenon and krypton.

The distribution of luminance throughout the volume of large-diameter flashlamps (10 mm or more in diameter) depends largely on the method of triggering, which influences how the discharge tube is filled and over how long a period. In some cases the tube may not fill at all, even for relatively high energies and durations of the discharge. For example, a plasma formation (which was stable throughout the entire discharge) in the form of a ribbonlike column was observed in an IFP-5000 lamp along the wire trigger electrode wound onto the envelope as the conditions of power supply were varied over an extremely wide range ($\tau = 150$ μs, $W = 0.5\text{-}1.8$ kJ, and $\tau = 600$ μs, $W = 0.9\text{-}3.6$ kJ) [5-22].

Particular importance is attached to a uniform distribution of luminance in flashlamps having a channel of rectangular cross section ("rectangular-bore" lamps) and in coaxial-type lamps when they are used for the optical pumping of lasers, for example. The influence of the shape of the trigger electrode is significant for rectangular-bore lamps having cross sections of 5×30 mm and 10×30 mm when the volume density of the discharge electric energy is relatively low ($W/V_d = 80$ J/cm^3) and the radiation pulse length is $\tau = 800$ μs [5-23]. A more uniform distribution is obtained by using a plate-type trigger electrode or a wire electrode positioned along the narrow face of the tube (compared with a spiral electrode). The distribution of the integral luminance becomes more uniform if W/V_d is increased above 130 J/cm^3, and is practically independent of the shape of the trigger electrode.

High-speed photography shows that the triggering of the IFPP-7000 and IFPP-4000 coaxial-type lamps is characterized by the formation of several discharge channels [5-24]. The coaxial cavity then is filled uniformly with radiating plasma. An analysis of the time distribution of the luminance (Figure 5-9) shows that the more symmetrical development of the discharge in coaxial-type lamps with ring-type electrode assemblies (the IFPP-4000) also leads to more uniform distribution of the luminance over the length of the lamp (the coordinates at 30 and 150 mm correspond to the ends of the electrodes). The dependence of the blackbody temperature T_L (Figure 5-10) on the xenon pressure has a mildly sloping maximum at $p_0 = 0.006$ MPa. The quantity T_L increases with increasing discharge energy or current density. The plasma temperature in the transelectrode volumes is significantly lower than in the central part of the discharge gap. The luminance distribution over the length of the lamp is more uniform in the IFPP-4000 lamp. Thus, the design features of a lamp which promote symmetrical development of the discharge also lead to more uniform distribution of the luminance over the length of the lamp.

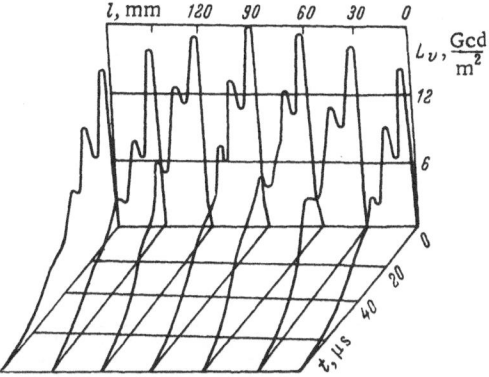

Fig. 5-9.   Time dependence of the luminance distribution over the
length of a coaxial-type IFPP-4000 lamp (the discharge
column, 120 mm long, is bounded by diameters of 19 and
29 mm; $W = 500$ J, $C = 4.7$ μF, $L = 0.4$ μH, $p_0 = 0.0067$
MPa, $\tau \approx 20$ μs; a discharger was connected in series to
the lamp) [5-24].

5-4. Light Characteristics of Spherical Flashlamps

The radiation of discharges in spherical flashlamps, which
generally have a much shorter duration than tubular lamps and a
channel shape that varies from pulse to pulse, is determined not only
by the integrated intensity Θ, the peak luminous intensity I_p, and
the luminous intensity pulse length τ, but also by additional parameters: the rate of increase of the luminous intensity, and the
reproducibility of the characteristics from pulse to pulse (Section
5-9 is devoted to this last factor).

Short radiation pulses are produced in spherical lamps by discharging a low-capacitance condenser across a short discharge gap
that is not confined by walls. Such discharges usually are of the
oscillating type because of the low resistance of the channel. They
have a number of specific features which result in a greatly complicated dependence of their light characteristics on the parameters.
Let us point out the main features.

The ratio of the electric energy W_L released in a spherical
flashlamp to the electric energy $W_0 = CU_0^2/2$ is called the efficiency
η_c of the discharge circuit. This quantity varies over a range
which is broader than the one characteristic of tubular lamps, and
is strongly dependent on the properties of the circuit. If the
ballast resistance R_b is constant, the efficiency η_c essentially
depends on W_0, which determines the diameter and the effective resistance of the discharge channel.

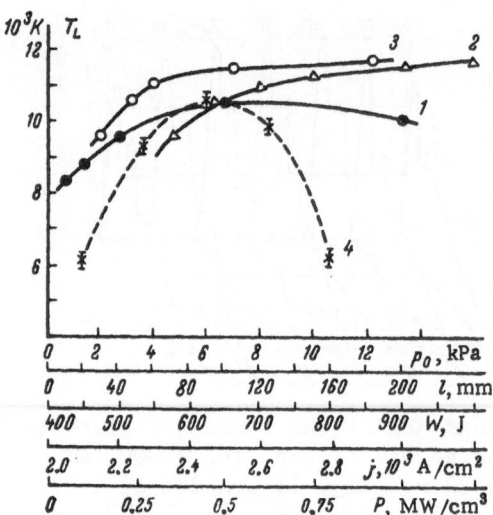

Fig. 5-10. Dependences of the peak blackbody temperature of the
plasma in an IFPP-7000 lamp on the initial xenon pressure
p_0 (1), the discharge energy W, the average power density
P of the discharge (2), and the current density j (3),
and the distribution of the temperature over the length
of the tube (4) [5-24]. (The other discharge parameters
are the same as in Figure 5-9.)

The losses in the ballast resistor, which are proportional to
$R_b/(R_b + R_L)$ (where R_L is the effective resistance of the channel),
can be estimated by determining R_b from data on the circuit design
and R_L from data on the conductance of the column (see formula (2-1))
and on its rate of expansion (see Figure 2-15). The quantity R_b is
determined mainly by losses in the capacitor if the cross section of
the leads is sufficiently large and the contacts in the current circuit are good. On the one hand, the average power in these losses
can be expressed in terms of $CU_{eff}^2 2\pi f \tan \delta$ (where U_{eff} is the effective voltage, equal to $U_0/\sqrt{2}$; $f = 1/2\pi\sqrt{CL}$ is the natural frequency
of the circuit; and $\tan \delta$ is the tangent of the losses in the capacitor). On the other hand, they can be expressed in terms of
$0.85 i_p^2 R_b$ (where i_p is the peak current strength, equal to $0.64 CU_0^2/L$)
by using the concept of the effective resistance R_b. Equating the
two expressions for the power loss, we find:

$$R_b \approx \sqrt{L/C} \tan\delta.$$

For example, according to Figure 2-15 the discharge in an ISSh
100-3 lamp (xenon, 0.3 MPa, $l = 0.5$ cm) has a column radius of about
0.05 cm by the time of maximum current (after a quarter period, i.e.,
0.5 μs) when C = 0.5 μF and L = 0.2 μH. When U_{eff} = 3.2 kV, the
conductivity according to formula (2-1) is 10^{-3} $(\Omega\text{-cm})^{-1}$. Therefore,

R_L = 0.06 Ω. The effective value is $R_b \approx$ 0.02 Ω when an oil-filled paper capacitor with tan δ = 0.03 is used. Thus, an estimate of the losses gives a value of about 25%. An experimental comparison of the integrated radiant intensities of flashes from this lamp, when powered by an oil-filled paper capacitor and a mica capacitor with a negligibly small tan δ, gave a value of about 24% for the oil-filled paper capacitor version of power supply. The same value of R_L would correspond to losses of approximately 83% when the same lamp in the same circuit is powered with an added series resistance of 0.3 Ω. The 62%-decrease in integrated intensity that occurs in this example indicates some increase in R_L due to the significant influence of the high R_b on the column diameter.

A significant fraction of the electric energy supplied to the lamp goes to channel expansion when the discharges are not confined by the walls of the envelope. The corresponding losses can be estimated from the formulas of gas-dynamic theory (in the example given above, they amount to about 20%).

As we pointed out in Chapter 2, a significant fraction of the electric energy input in the channel of a saturated or unsaturated unlimited pulsed discharge may be accumulated by the plasma in the form of internal energy which is released after the discharge current drops. The illumination of the cooling lamp has an appreciable influence on the shape and duration of the radiation pulses. The continuous variation of the channel radius, gas density, electrical resistance, and the power consumed for channel expansion during an unlimited discharge accounts for the large differences in the time dependences of the luminous intensity I(t) and the luminance L_v(t) (Figure 5-11), and complicates the interrelation between the integrated intensity Θ and the integral luminance $\int L_v dt$.

The nonreproducibility of the shape of the discharge channel (which follows a new convoluted path with each occurrence) is especially evident in the regime most widely encountered in spherical lamps: the regime of frequently repeated flashes, beginning with some repetition frequency that depends on W. This nonreproducibility accounts for the greater spread of the light parameters of a lamp from pulse to pulse.

The experimental data accumulated to date enable us to outline (somewhat schematically in some cases) the nature of the dependences of the light characteristics on the parameters in the most-used range of the parameters.

The significant change in the shape of the power pulse for different discharge parameters and the dependence of the deexcitation time of the gas on the type of gas, the gas pressure, and the energy released in the channel lead to the formation of radiation pulses having different ratios between the steepness of the front and that

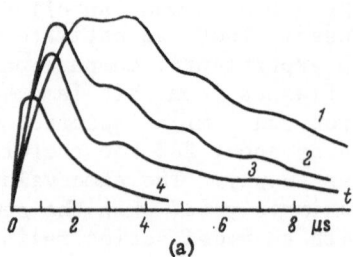 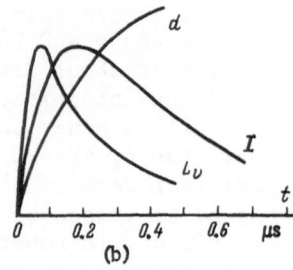

(a)   (b)

Fig. 5-11.   Sample oscillograms.
a) the luminous intensity of discharges in xenon (1,4),
krypton (2), and argon (3): 1, 2, 3) C ≈ 2 µF, L = 0.2
µH, U_0 = 3.75 kV; 4) C = 0.4 µF, L = 0.09 µH, U_0 = 5 kV.
The pressure is 0.3 MPa for all gases. The lamp used
was similar to the ISSh-3, with l = 5 mm.
b) the luminous intensity I, channel luminance L_v, and
discharge-channel diameter d: C = 0.002 µF, L = 0.12 µH,
U_0 = 5.5 kV. A mixture of Xe + 10% H_2 at 0.3 MPa. The
lamp used was similar to the ISSh 300, with l = 6 mm.

of the tail off of intensity. Thus, essentially no similarity is
observed between plots of the time dependences of the luminous intensity of spherical flashlamps. This is illustrated by Figure 5-11.

The spread of the values of the coefficient K (the ratio of the
integrated intensity of spherical flashlamps to the product of the
peak luminous intensity I_p by the duration τ of a luminous intensity
pulse (at the level of $0.35I_p$)) is much larger than for tubular
flashlamps. However, we may assume K = 1 for spherical lamps with
an error that does not exceed ±20% [5-25].

As we see from Figure 5-12, η_v varies by a factor of no more
than 2 as W varies by nearly four orders of magnitude. The quantity
η_v may be assumed constant in the range of W from 0.1 to 10 J. The
curve of η_v versus W in this figure is consistent with the data of
many papers for various ranges of W. The slight increase in the
luminous efficacy and the subsequent constant value with increasing
W that were observed by Glaser and in [5-25] correspond to the left
branch and the middle section of the curve in Figure 5-12, while the
decrease in η_v with increasing W that was observed by Charna relates
to the right branch. We may assume that the decrease in η_v in the
right branch of the graph for W ≥ 5 J is due to a larger extent to
the change in the electrical efficiency of the discharge circuit,
while the increase in the range 0.01-1 J is due to the decrease in
the fraction of energy losses to expansion of the discharge channel,
and to the change in the spectral distribution of the radiation. A
more detailed picture of the dependence of η_v on the discharge parameters can be gained from an examination of Figures 5-13 to 5-15.

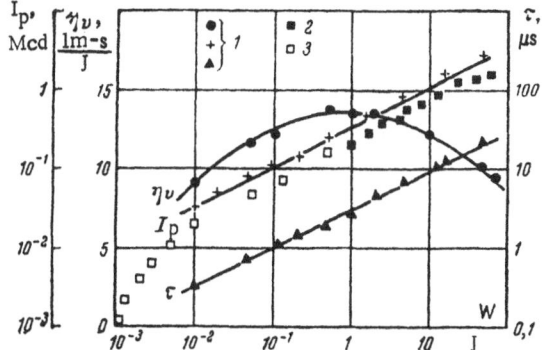

Fig. 5-12. Dependence of the luminous efficacy η_V, the peak luminous
intensity I_p, and the flash duration τ on the discharge
energy of spherical lamps. 1) Xe, 0.2 MPa, l = 3.5 mm,
U_0 = 2-6 kV, L = 0.2 µH, and the lamp used was similar
to the ISSh 400 [5-26]; 2 and 3) Xe, 0.22 MPa, l = 2-2.5
mm, and the lamp used was an ISSh 7 [5-27 and 5-28].

The luminous efficacy changes slightly as the supply voltage,
flash frequency, and capacitance are varied severalfold (Figure 5-13).
The value of η_V is practically independent of the capacitance when
C = 10^{-2}-10 µF (see Figure 5-14), but decreases slowly at first and
then rapidly as the capacitance falls below this range. The value
of C at which η_V no longer depends on the capacitance obviously decreases with increasing atomic number of the gas. Figure 5-15 shows
that reducing the pressure by a factor of 3 or tripling the inductance
of the discharge circuit, or using paper insulation in the capacitor
instead of a film with low dielectric losses leads to an approximately
30% decrease in η_V. A change in the gas density in the discharge
circuit during flashing operation affects the luminous efficacy to
the same extent as a change in the pressure p_0. In particular, decreasing the diameter of the envelope from 60 to 12 mm increases η_V
by 13% for f = 400 Hz and an average power of 90 W (Xe at 0.3 MPa,
l = 2.5 mm, C = 0.05 µF).

An increase in the electrode spacing is accompanied by an approximately linear increase in the luminous efficacy. For discharges
in xenon (0.3 MPa, 0.5 µF, 4.5 kV), an increase in l from 2.5 to 10
mm leads to an increase in η_V from 12 to 21 lm-s/J.

The type of gas has the strongest effect on the luminous efficacy of an unlimited pulsed discharge. As we see from Figure 5-16,
in the region where the luminous efficacy has a slight dependence
on the other parameters, it is directly proportional to the atomic
mass of a gas having a coefficient of about 0.11 lm-s/J per unit
atomic mass. The addition of a light gas (such as hydrogen or nitrogen) to a heavy noble gas which is characterized by high luminous
efficacy reduces η_V approximately in proportion to the ratio of the
light to the heavy gas.

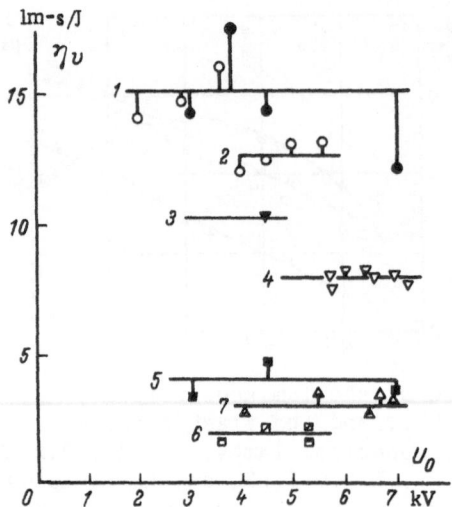

Fig. 5-13. Dependence of luminous efficacy on supply voltage for
            spherical lamps with different fillers for various supply
            capacitances and flash frequencies f. L = 0.2 μH, p_0 =
            0.3-0.4 MPa, l = 5-6 mm. 1) Xe, C = 0.04-0.5 μF; 2) Xe
            + 30% H_2, C = 0.04 μF; 3) Kr, C = 0.5 μF; 4) Kr + 30%
            H_2, C = 0.02-0.05 μF; 5) Ar, C = 0.5 μF; 6) Ar + 30% H_2,
            C = 0.05 μF; 7) Kr, C = 0.01 μF, and a ballast resistance
            R_b = 0.5 Ω is series-connected to the lamp (for the other
            graphs, R_b = 0). Solid symbols represent points obtained
            for single flashes. Unfilled symbols are for f = 400 Hz.
            Symbols that are filled at bottom are for 800 Hz, symbols
            filled at top are for 2000 Hz, and the crossed symbols
            are for 4500 Hz.

        The luminous intensity of spherical flashlamps is equal to the
integral of the continuously varying luminance over the area of the
expanding discharge channel. The maximum of this function determines
the peak luminous intensity I_p. The difference between the rate of
change of the luminance (a rapid rise to a maximum, and a comparatively slow tail off) and the rate of continuous expansion of the
discharge-channel diameter leads to a situation in which the luminous
intensity reaches its highest value during a flash somewhat after
the maximum luminance. A corresponding example of oscillograms of
the luminous intensity I, the luminance L_v, and the time dependence
of the glow-channel diameter d is shown in Figure 5-11b. An analogous pattern also is observed at high discharge energies (over 100 J)
[5-29]. Like the luminance and channel diameter, the peak luminous
intensity I_p depends on the spatial and temporal concentrations of
energy in the discharge, and hence on those parameters of the discharge and spark gap which determine the rate of energy input into
the channel and the expansion process. We can see from the Figure

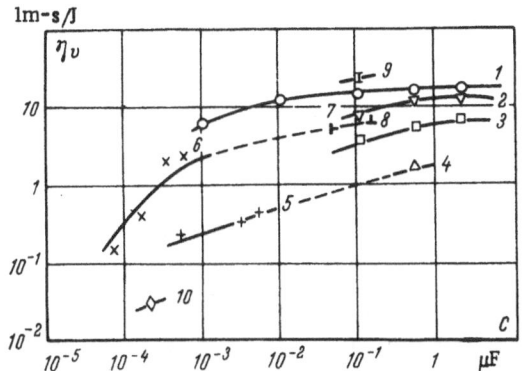

Fig. 5-14. Dependence of the luminous efficacy on the supply capacitance. 1) Xe; 2) Kr; 3) Ar; 4) N_2. For all four graphs: U_0 =
4.5 kV, R_b = 0, l = 5 mm, the envelope has a diameter of 26 mm,
and mica capacitors are used. For curves 1-3: p_0 = 0.4
MPa, L = 0.12-0.2 μH; for curve 4: p_0 = 0.1 MPa, L =
0.02 μH. 5-8) air, p_0 = 0.1 MPa, open discharge, R_b =
0. 5) U_0 = 0.8 kV, l = 3 mm, L = 0.01 μH. 6) U_0 = 8 kV,
l = 3 mm, L = 0.01 μH; 7) U_0 = 20 kV, l = 8 mm, L = 0.03
μH. 8) U_0 = 7 kV, l = 4 mm, L = 0.06 μH. 9) argon, p_0 =
0.1 MPa, open discharge, R_b = 0, U_0 = 7 kV, l = 8 mm,
L = 0.06 μH. The data for plots 6-9 were obtained for
a detector with no filter to correct for its spectral
response to the visibility curve. 10) hydrogen, U_0 = 2.5
kV, R_b = 0, l = 0.4 mm, p_0 = 0.01 MPa.

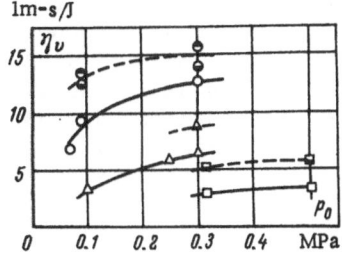

Fig. 5-15. Dependence of luminous efficacy on gas pressure for
various capacitances and inductances of the discharge
circuit. U_0 = 5 kV, l = 5 mm, p_0 = 0.3 MPa, R_b = 0.
0 - Xe; Δ - Kr; □ - Ar. The open symbols are for L =
0.2 μH, C = 0.1 μF, with paper insulation. The symbols
filled at top are for L = 0.07 μH, C = 0.1 μF, with
Hostaphan® (terephthalic ester). The symbols filled at
bottom are for L = 0.07 μH, C = 0.4 μF, with Hostaphan.

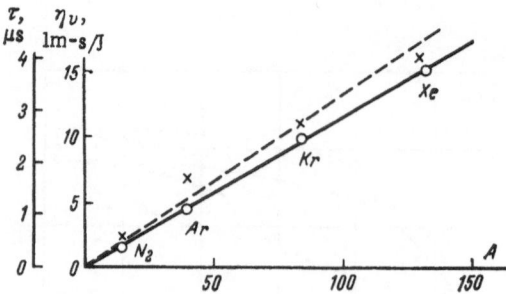

Fig. 5-16. Dependence of the luminous efficacy (solid line) and
            flash duration (dashed line) on the atomic mass of the
            gas. C = 0.5 μF, L = 0.12 μF, R_b = 0, U_0 = 5 kV, l =
            5 mm, p_0 = 0.3 MPa. The points for nitrogen correspond
            to the values for p_0 = 0.1 MPa, corrected according to
            data on the dependence of Θ and τ on p_0.

5-12 that I_p has a linear dependence on the discharge energy W over
a broad range of the energy. The increase in I_p slows only at high
discharge energies at which the luminance of the channel approaches
the maximum value. Under certain conditions the increase in I_p
reaches a limit which is characteristic of a given gas (see Chapter
2). Since the limiting luminance is higher and (all other conditions
being equal) the channel diameter is larger in light gases than in
heavy ones, the peak luminous intensity is higher under limiting
discharge conditions in these gases (Figure 5-17).

     This advantage of light gases is lost in the region of the
maximum attainable luminances for the discharge parameters that are
found in industrial types of spherical lamps (moderate pressures and
supply voltages). Therefore, mainly pure xenon or xenon containing
a small impurity of hydrogen or nitrogen is used to fill spherical
lamps. In this case the peak luminous intensity I_p has an insignificant dependence on U_0 and C at a constant discharge energy over a
broad range of the parameters (Figure 5-18), increasing approximately
in proportion to the square root of the energy. This is essentially
true regardless whether U_0 or C is a variable (Figure 5-19). The
quantity I_p is approximately proportional to $C^{0.8}$ for small C, at
which η_v decreases with decreasing capacitance (Figure 5-14).

     Changes in p_0 and l are accompanied by a change in I_p that is
approximately proportional to $\sqrt{p_0}$ (Figure 5-20) and to $l^{0.7}$. An
order-of-magnitude decrease in the inductance (from 0.2 to 0.02 μH)
increases the I_p of xenon lamps by 15-20% and the I_p of lamps filled
with a mixture of xenon with hydrogen or nitrogen by up to 50%. A
change in the Q factor of the discharge circuit (e.g., the series
connection of a resistance $R_b \leq \sqrt{L/C}$), which has a significant effect
on η_v, has practically no influence on I_p.

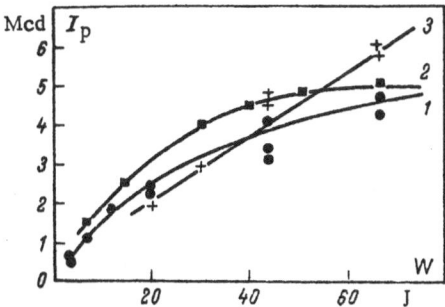

Fig. 5-17. Dependence of the peak luminous intensity on the discharge energy. C = 0.4 µF (a type KBGP paper capacitor).
l = 10 mm. 1) argon, 1 MPa; 2) xenon, 1 MPa; 3) nitrogen,
0.25 MPa [5-29].

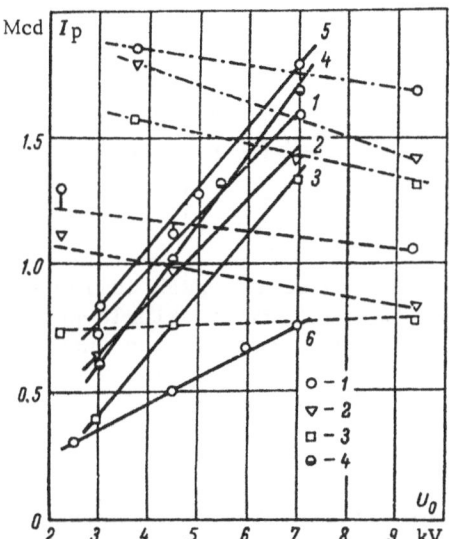

Fig. 5-18. Dependence of the peak luminous intensity on the supply
voltage. l = 5 mm, R_b = 0, and the envelope diameter
is 26 mm. The solid lines are for constant capacitance.
1-4) C = 0.5 µF, L = 0.2 µH; 5) C = 0.4 µF, L = 0.08 µH;
6) C = 0.1 µF, L = 0.2 µH. The dashed and dot-dash lines
are for constant flash energies (5 and 14 J, respectively). 1) Xe; 2) Kr; 3) Ar (for all three gases, p_0 =
0.3 MPa); 4) 0.066 MPa Xe + 0.026 MPa N_2 [5-25].

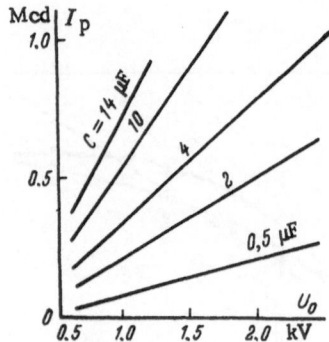

Fig. 5-19. Dependence of peak luminous intensity I_p of an ISShO-1 lamp on the supply voltage U_0 for different capacitances C of the storage capacitor (Xe, p_0 = 0.22 MPa, l = 2.5 mm) [5-27].

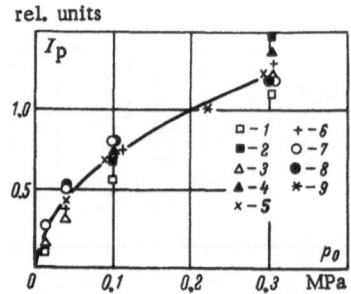

Fig. 5-20. Reduced dependence of the peak luminous intensity of an ISSh-7 lamp (l = 2.5 mm) in single-pulse operation on the xenon presure p_0 for various energies W (in mJ).
1) 0.8 mJ; 2) 1.3 mJ; 3) 2.2 mJ; 4) 3.4 mJ; 5) 14.5 mJ; 6) 22.5 mJ; 7) 640 mJ; 8) 995 mJ; 9) point of reduction to a single ordinate scale [5-28].

A roughly analogous pattern of the dependence of I_p on the discharge parameters also is observed in spherical flashlamps having a submicrosecond flash duration which are filled with light gases and mixtures thereof [5-30 and 5-31] and during discharges in air [5-32]. The peak radiant intensity, as measured with a detector whose maximum spectral sensitivity is shifted toward the blue (e.g., a photomultiplier with an antimony-cesium photocathode that has maximum response around 400 nm and 35% of maximum response at 310 and 570 nm), is of definite interest in the practical use of spherical flashlamps. The peak radiant intensity I_{ep} obtained from the reaction of this detector is higher by a factor of 1.5-2 than the peak luminous intensity for an extremely wide range of the discharge parameters and spark gap (when the detector is calibrated to a type-A source) [5-25].

The duration τ of the luminous intensity pulse is determined by the duration of the processes of electric energy release in the discharge channel and by the deexcitation time of the hot gas. References [1-68 and 5-33] contain data on the duration of the electrical pulse and the light flash under discharge conditions for extremely low inductance of the circuit.

The flash duration τ is approximately proportional to $\sqrt{W}$ (Figure 5-12), regardless of the selection of the variable C or U_0. A more detailed picture of the variation of τ is shown in Figures 5-16, 5-21, and 5-22. Figure 5-21 shows that τ is proportional to $\sqrt{C}$ over an extremely broad range of the parameters. This is analogous to the dependence of τ on C that has been found for tubular lamps in the region $\tau \lesssim 50$ μs (Section 5-2).

The supply voltage and the type of gas have a significant effect on τ, but the other factors have a less-pronounced effect. These factors include the diameter of the envelope (confining the discharge in a capillary with d = 1 mm changes τ by no more than 30%, all other conditions being equal), gas pressure (increasing p_0 by a factor of 5 no more than doubles τ), the discharge-circuit inductance (τ is approximately proportional to $L^{1/6}$), and the electrode spacing (increasing l by a factor of 3 approximately halves τ). Including a small ballast resistance in the discharge circuit approximately halves τ [1-68].

As we see from Figure 5-22, τ is proportional to U_0. It is important to note that the nature of the dependence of τ on U_0 for short unlimited discharges is opposite to the behavior observed during prolonged discharges in tubular lamps, for which τ decreases noticeably with increasing U_0 (Section 5-2).

The dependence of τ on the type of gas is shown in Figure 5-16. It follows from this figure that τ is approximately proportional to the atomic mass of the gas, all other conditions being equal. The addition of molecular-gas impurities (nitrogen and hydrogen) to a heavy noble gas reduces the flash duration approximately in proportion to the ratio of the light-gas content to the heavy-gas content. The general pattern of the dependence of the flash duration on the parameters makes it possible to outline the lower boundary of the attainable flash durations by using practically feasible gases (from the standpoint of electric strength, luminous efficacy, etc.) and power-supply parameters. This boundary is shown by the dot-dash line in Figure 5-21 and can be expressed by the simple relation $\tau = \sqrt{C}$, where τ is given in microseconds and C in microfarads.

The dependence of the growth rate of the luminous intensity dI/dt on the discharge parameters have been studied least thoroughly. The available data, which are summarized briefly in Table 5-6, enable us to form a general qualitative picture of the dependence of dI/dt

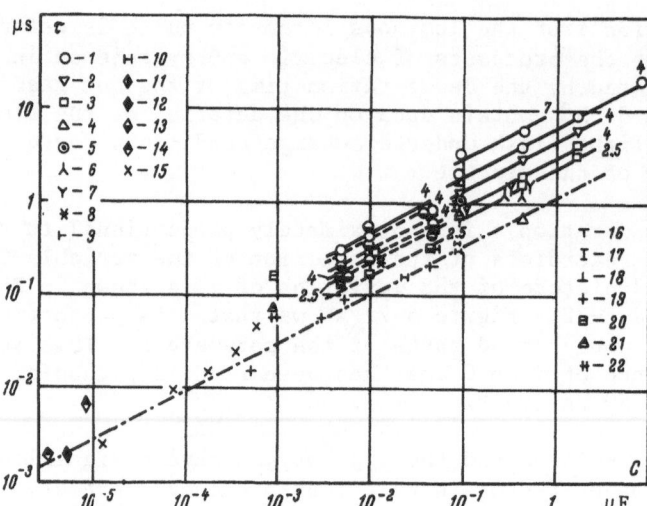

Fig. 5-21. Dependence of the flash duration on the capacitance of
the supply capacitor. The data are from [5-25]: $l =$
5 mm, p_0 = 0.3 MPa, R_b = 0, the diameter of the envelope
is 26 mm, and the solid lines are for 0.15 μH. 1) Xe;
2) Kr; 3) Ar (the crossed points are the same gases containing a 30%-impurity of hydrogen; the supply voltage
in kV is indicated on the plots); 4) N_2, 0.3 MPa, 4.5 kV,
0.02 μH, and the other parameters are the same; 5) Xe,
l = 17 mm, R_b = 0, a capillary with d_i = 3 mm, 0.2 μH,
6 kV. The other points were taken from other papers:
6) Xe, 0.02 MPa; 7) Ar, 0.07 MPa, 3 kV, 12 mm, 0.2 μH;
8) air, 0.1 MPa, 10 kV, 4 mm (a capillary with d_i = 1 mm),
0.01 μH; 9) air, 0.1 MPa, 6.5 kV, 8 mm, 0.03 μH; 10) air,
0.1 MPa, 3-6 kV, 0.8-1.5 mm, 0.001-0.004 μH; 11) air,
0.1 MPa, 17 kV, 10 mm, 0.01 μH; 12) H_2, 0.015 MPa, 2.5 kV,
0.4 mm, 0.005 μH; 13) Hg, 5 kV; 14) air, 0.1 MPa, 1 kV,
0.14 mm, 0.01 μH; 15) air, 0.1-0.17 MPa, 5-8 kV, 1.3-3 mm,
0.01 μH; 16) air, 0.1 MPa, 7 kV, 4 mm, 0.06 μH; 17) Ar,
0.1 MPa, 7 kV, 8 mm, 0.06 μH; 18) the same, but in a
capillary with d_i = 1 mm; 19) air, 0.1 MPa, 8.5 kV, 3 mm,
0.01 μH; 20) Ar; 21) H_2; 22) He (for points 20 to 22 the
other parameters are 1.4 MPa, 8 kV, 1.5 mm, 0.007 μH).

on the main characteristics of the discharge. The values of the
maximum steepness of the front $(dI/dt)_{max}$ and of the rise time $\Delta\tau$
of the luminous intensity also have been determined for pulses of
radiant intensity recorded by a detector with an antimony-cesium photocathode (without correcting absorbers). The last column of the table
gives the corresponding growth rates of the channel radius, as obtained by using an RKS-1 scanning motion-picture camera. These data
confirm that the growth rate of the luminous intensity is related to

Table 5-6. Data of the Growth Rate of the Luminous Intensity

Ref. cited	Gas	p_e, MPa	l, mm	Envelope diam., mm	c, μF	U_e, kV	L, μH	$(dI/dt)_{max}$, Tcd/s		$\Delta\tau$, μs		$\left(\dfrac{dr}{dt}\right)_{max}$, km/s
								true	uncorrected	true	uncorrected	
[Ref. 5-34]	Air	0.1	4	—	0.1	7	0.06	—	—	—	0.33	—
	Ar	0.1	8	—	0.1	7	0.06	—	—	—	0.24	—
	Ar	0.1	8	1	0.1	7	0.06	—	2.1	—	0.27	—
	Xe	0.3	5.3	26	0.5	4.5	0.2	1.1	—	1	0.7	1
	Xe	0.3	5.3	26	0.4	5	0.08	1.35	—	0.75	—	—
	Kr	0.3	5	26	0.5	4.5	0.2	1	3.3	0.8	—	—
	Kr	0.3	5	26	0.4	4.5	0.08	1.37	—	0.63	—	—
	Ar	0.3	5.3	26	0.5	5	0.2	1.37	—	0.46	—	1.3
	Ar	0.3	5.3	26	0.4	4.5	0.08	2.15	4.3	0.35	—	—
	Ar	0.32	5	26	0.5	4.5	0.02	—	—	—	0.4	1.8
[Ref. 5-25]	Xe}	0.09	5.3	26	0.5	5	0.08	1.5	—	0.55	—	1.5
	Xe	0.09	5.3	26	0.4	5	0.2	4.5	—	0.25	0.4	—
	Xe} *	0.09	7.5	26	0.2	7	0.2	—	5.6	—	0.23	—
	Xe	0.09	7.5	26	0.2	5	0.2	—	6.7	—	0.3	—
	Xe}	0.09	7.5	26	0.4	—	0.02	—	5.7	—	0.35	2.5
	Xe	0.09	5	26	0.5	5	0.2	—	—	—	—	—
[Ref. 5-25].	Ar}	0.1	7.5	26	0.2	7	0.2	—	4.5	—	0.15	—
	Ar} *	0.1	7.5	26	0.2	5	0.2	—	4.7	—	0.22	—
	Ar}	0.1	7.5	26	0.4	5	0.2	—	4.7	—	0.23	—
	Ar}	0.1	5	26	0.5	—	0.02	—	—	—	—	3

*+40% nitrogen.

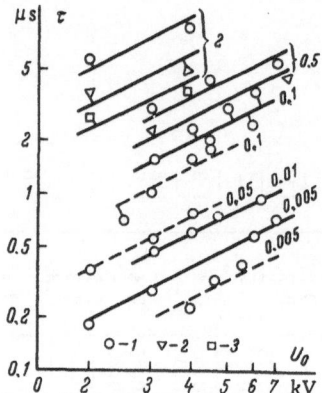

Fig. 5-22. Dependence of the flash duration on the supply voltage.
l = 5 mm, p_0 = 0.3 MPa, R_b = 0, and the diameter of the
envelope is 26 mm. The solid lines are for L = 0.2 µH,
and the dashed lines for L = 0.01 µH. 1) Xe; 2) Kr;
3) Ar; the capacitances in µF are indicated on the plots
[5-25].

the rate of channel expansion. The quantity dI/dt is strongly de-
dependent on the inductance of the discharge circuit, increasing by
a factor of 1.5-2.5 in some cases (argon, a xenon—nitrogen mixture)
as L decreases from 0.2 to 0.08 µH. The voltage across the capacitor
has a significant effect on dI/dt. The type of gas has a very strong
effect on it. For pure noble gases dI/dt increases with decreasing
atomic mass. However, an even steeper luminous—intensity rise front
is found for mixtures of noble gases with nitrogen. Obviously, other
factors also affect the growth rate of the luminous intensity to a
much lesser extent.

5-5. The Spatial Distribution and Polarization of the Radiation
      from Flashlamps

     The spatial distribution of the radiation from flashlamps is
determined mainly by the shape of the plasma glow volume and its
luminance distribution. Reflection and refraction on the surfaces
of the lamp envelope produce some changes in the spatial distribution
of the radiation of the plasma volume. Reflections on the envelope
surfaces are responsible for the polarization of flashlamp radiation,
since the radiation of a plasma is unpolarized.

     The distribution of the plasma luminance in flashlamps changes
continuously during the discharge. Therefore, the spatial distribution of the radiation also changes continuously, strictly speaking. However, these changes are so large that for practical purposes
we may consider the spatial distribution of integrated photometric

parameters: the luminous energy or radiant energy in various spectral
regions. The graphic dependences (in polar coordinates) of the
integrated luminous intensity $\int I_v(t)dt$ or the integrated radiant
intensity $\int I_e(t)dt$ on direction are called their solids of light
distribution [5-4]. The intersections of the solids of light distribution by planes that pass through the origin of the polar coordinates are called the indicatrices of the integrated luminous intensity Θ or the integrated luminous intensity Θ_e. These indicatrices
usually pass through the lamp's axis of symmetry.

A useful quantity for practical purposes which is related to
the spatial distribution of the radiation is the equivalent solid
angle Ω_{eq}. This is the ratio of the luminous energy Q (or the radiant
energy Q_e) to the integrated luminous intensity Θ (or the integrated
radiant intensity Θ_e) in the direction of radiation that is taken as
the main direction. This direction is the normal to the axis of the
lamp for straight tubular flashlamps. The angular argument α of the
indicatrices $\Theta(\alpha)$ and $\Theta_e(\alpha)$ usually is counted from the axis of the
tube. To compare the indicatrices, it is convenient to assume the
value of the integrated luminous intensity in the main direction to
be equal to unity, i.e., to represent the relation $\Theta(\alpha)/\Theta(\pi/2)$.

The calculation made by Gershun [3-80] of the indicatrices of
the luminous intensity $I(\alpha)/I(\pi/2)$ of a uniform cylindrical plasma
column of diameter d having a self-absorption coefficient κ gives

$$\frac{I(\alpha)}{I(\pi/2)} = \sin\alpha \; \frac{F(\kappa d/\sin\alpha)}{F(\kappa d)} \; , \tag{5-2}$$

where

$$F(x) = 1 - \int_0^{\pi/2} \exp(-x\cos\beta)\cos\beta \, d\beta = \frac{\pi}{2}\,[I_1(x) - L_1(x)]$$

(I_1 and L_1 are the first-order Bessel and Struve functions of the
imaginary argument). The calculated indicatrix of a straight tubular
gas-discharge lamp (ignoring screening by the electrodes and bases)
is obtained by multiplying function (5-2) by the transmission coefficient $\tau(\alpha)$ of the lamp [3-81]. This coefficient is calculated
(Figure 5-23) for a transparent circular cylindrical envelope on the
condition that the radiation reflected from the envelope be absorbed
completely by the plasma column. This condition is satisfied quite
well for flashlamps, which are characterized by a rather high optical
density $\kappa d/\sin\alpha$ (especially for small angles α). As we see from
the figure, the transmission coefficient of a cylindrical envelope
decreases somewhat more rapidly with decreasing angle α than the
transmission coefficient of a flat sheet. The indicatrices calculated for different κd with allowance for $\tau(\alpha)$ (Figure 5-24) differ
significantly from the indicatrices given in [3-80] for a plasma
column with no shell, but differ less among themselves. For this
reason the extreme values of Ω_{eq} become equal to 11.7 and 9.6 sr
instead of 4π and π^2 (Table 5-7).

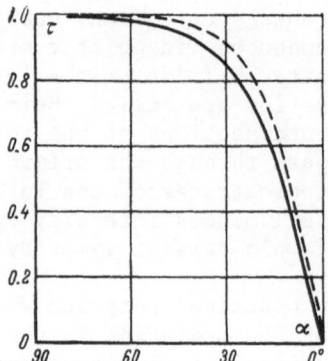

Fig. 5-23. Relative transmission coefficients of a cylindrical
            envelope (solid line) and a flat sheet (dashed line)
            for various angles.

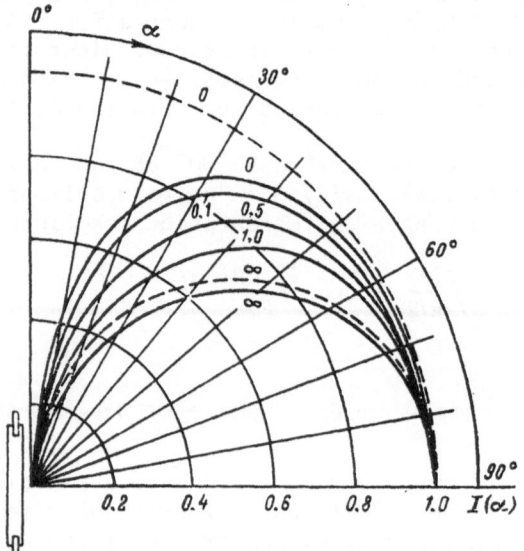

Fig. 5-24. Theoretical indicators (sic) of the radiation of straight
            tubular lamps with (solid lines) and without (dashed
            lines) allowance for the transmission coefficient $\tau(\alpha)$
            of the lamp. The numbers beside the curves indicate the
            values of the product κd.

     The experimental spatial distribution of the light from straight
tubular flashlamps has been investigated in [3-81 and 5-35 to 5-37].
The data are presented in generalized form in Figures 5-25 to 5-28.
The good fit between the calculated and experimental plots indicates
the insignificant effect of radiation screening by the electrodes
and bases (which the calculation did not take into consideration).

Table 5-7.   Equivalent Solid Angles Ω_{eq} for Different
Optical   Densities κd of the Plasma Column

κd	Ω_{eq}, sr	
	per formula (5-2)	allowing for $\tau(\alpha)$
0	12,56	11.7
0.1	12.3	11.5
0.2	12.0	11.3
0.3	11.7	11.2
0.4	11.5	11.1
0.5	11.3	11.0
0.6	11.2	10.9
0.7	11.1	10.8
0.8	11.0	10.7
1.0	10.9	10.6
1.2	10.7	10.4
1.4	10.6	10.3
1.6	10.5	10.2
2.0	10.4	10.1
2.5	10.3	10.0
3.0	10.2	9.9
3.5	10.1	9.8
4.0	9.9	9.7
∞	9.86	9.6

The equivalent solid angles, as calculated from the indicatrices of
the integrated luminous intensity (in the spectral region 400-700 nm),
lie in the range 10.4-11.6 sr. The shape of all indicatrices obtained for the integrated luminous intensity in different spectral
regions varies over the calculated narrow range (the inner and outer
curves in Figure 5-26) according to changes in the absorption coefficient of the plasma: the indicatrix for the infrared region,
where κ is higher, runs inside the indicatrix for the visible region,
where the plasma absorbs more weakly. This accounts for the slight
change in the relative distribution of radiant energy over the
spectral ranges as the angle between the direction of observation
and the axis of the lamp varies.

It is possible to calculate Ω_{eq} for specific values of d as a
function of the density of the electric power P released in a discharge (Figure 5-28) by using Table 5-7 and the data on the absorption
coefficient and conductivity of the plasma that were presented in
Chapter 3. The theoretically predicted trend of decreasing Ω_{eq} with
increasing absorption coefficient and power density thus is confirmed,
as we see. The quantity Ω_{eq} decreases as we go from the visible to
the infrared.

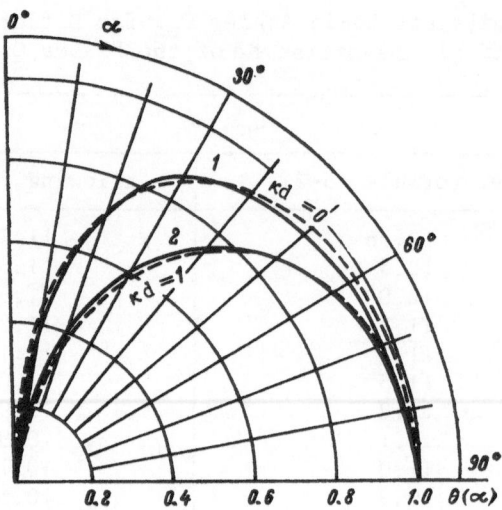

Fig. 5-25.   The indicatrices that differ most in shape for the integrated luminous intensity of lamps studied under rated
conditions: the IFP-800, IFP-1200, IFP-2000, IFP-5000,
IPF-8000, IFP-20000, and ISP-1000. 1) IFP-1200, U_0 =
1100 V, C = 200 µF, L = 10 µH, τ = 350 µs; 2) IPF-2000,
U_0 = 1500 V, C = 1800 µF, L = 50 µH, τ = 840 µs. The
dashed lines represent the calculated indicatrices.

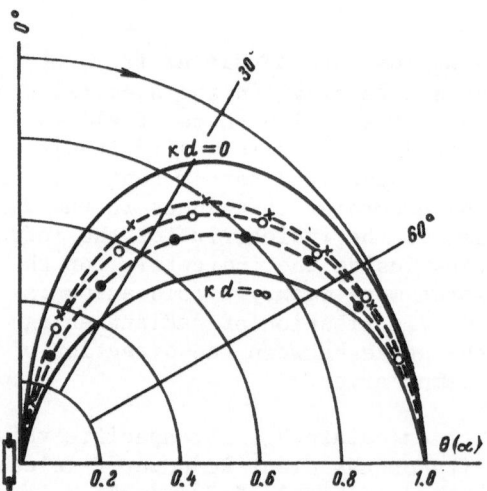

Fig. 5-26.   Experimental indicatrices of the integrated radiant intensity of an IFP-1200 lamp (U_0 = 1450 V, C = 590 µF,
L = 100 µH, τ = 470 µs) in the spectral regions 400–700 nm
(×), 700–2700 nm (●), and 180–4500 nm (O)). The solid
lines represent the calculated indicatrices.

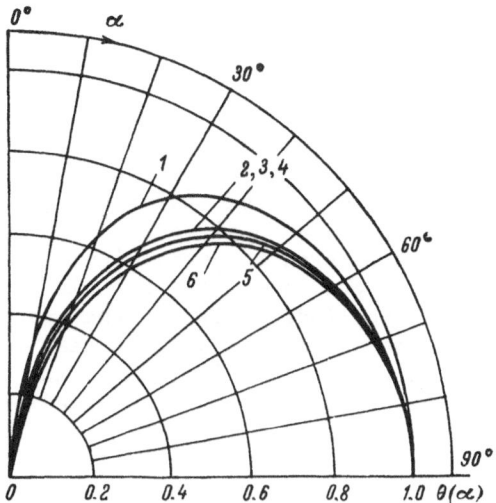

Fig. 5-27. Indicatrices of the integrated radiant intensity of an
IFP-1200 lamp in the spectral region 180–4500 nm for
different U_0, C, and L. 1) 1.1 kV, 220 µF, 100 µH
(τ = 260 µs); 2) 1.5 kV, 1000 µF, 30 µH (430 µs); 3) 1.3
kV, 1000 µF, 16 µH (380 µs); 4) 1.45 kV, 590 µF, 100 µH
(470 µs); 5) 1.8 kV, 300 µF, 16 µH (160 µs); 6) 2.4 kV,
300 µF, 30 µH (190 µs).

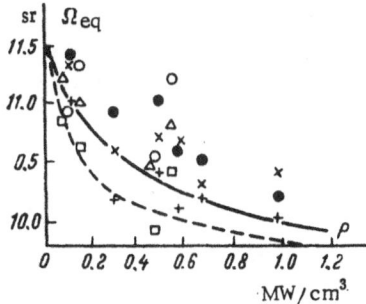

Fig. 5-28. Equivalent solid angles Ω_{eq}, as calculated for the visible
light of lamps with d_i = 7 mm (solid line) and 11 mm
(dashed line), and the experimental values for IPF-1200
and IFP-800 lamps (● represents the visible region, +
the infrared, and × 180–4500 nm) and for IFP-2000 lamps
(○ represents the visible region, □ the infrared, and
△ 180–4500 nm).

According to published data, for IFK lamps Ω_{eq} = 10.1-12.1 sr, and for ISSh lamps Ω_{eq} = 10.2-10.8 sr [5-36]. We may recommend Ω_{eq} = 11 sr for approximate calculations.

The results of calculations and measurement of the degree of polarization of the radiation due to reflection on the surfaces of the envelope are presented in Figure 5-29. The angular dependence of the degree of polarization of radiation transmitted through a flat sheet is shown in the figure for comparison. In contrast to a sheet, the plane of polarization of the radiation from a cylindrical tube has two mutually perpendicular positions, depending on the angle α. The plane of polarization is perpendicular to the axis of the lamp in the direction of the normal to the axis (the polarization is determined solely by the cylindricality of the envelope). As α decreases, the gentle dip of the rays that pass near the axial plane makes an increasingly significant contribution to the polarization. The degree of polarization vanishes at some intermediate value of the angle α because of the opposite effect of these factors. The degree of polarization of the radiation in the direction perpendicular to the axis of the lamp ($P(\pi/2)$) and the angle α_0 at which the polarization vanishes depend on the radial distribution of the luminance of the discharge column. The faster the luminance decreases with distance from the axis of the lamp, the smaller the values of $P(\pi/2)$ and α_0. The experimental dependence of the degree of polarization of radiation on the angle α are the same for all flashlamps studied within the limits of accuracy of the measurements (Figure 5-29), and vary slightly as a function of the operating conditions of the lamp (Figure 5-30). The calculated and experimental dependences of the degree of polarization of the radiation on the angle α are in qualitative agreement. The quantitative discrepancies may be attributed to factors that were not taken into account in the calculation: the finite thickness of the wall, the nonuniformity of the discharge column, and multiple reflections of radiation from the peripheral regions of the discharge.

5-6. The Spectral Characteristics and Efficiency of Flashlamps

Discharges in tubular flashlamps are characterized by plasma temperatures of 8000-12,000 K or more. At such temperatures an optically dense radiator is characterized by the spectral distributions of the efficiency of blackbody radiation (Figure 5-31). Before striking the detector, the radiation from the plasma passes through a number of media, which act as optical filters. We can see from Figure 5-32 that the transparency region of quartz envelopes for flashlamps may extend from 155 to 4500 nm, while that of electron-tube glasses may range from 290 to 3000 nm. Thus, the spectral characteristics and spectrally integrated efficiencies of quartz and glass flashlamps may differ significantly, given the same discharge parameters.

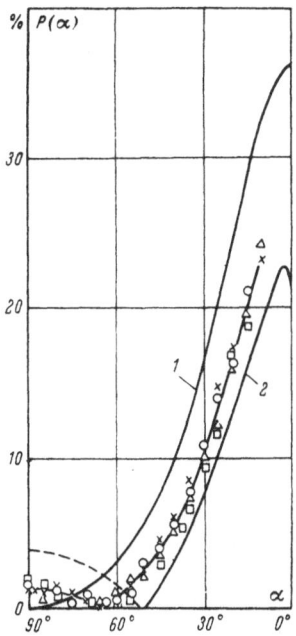

Fig. 5-29. Calculated relations for the degree of polarization
$P(\alpha)$ of the radiation from a flat sheet (1) and a circuit
cylindrical envelope (2). The dashed line represents
radiation polarized mainly in the plane perpendicular
to the axis of the lamp. The solid line is for radiation
polarized in the plane passing through the axis of the
lamp. The experimental points are: $\bigcirc$ for the IFP-1200
(U_0 = 1500 V, C = 975 µF, L = 30 µH, τ = 580 µs); $\times$ for
the IFP-2000 (U_0 = 1500 V, C = 1800 µF, L = 50 µH, τ =
840 µs); $\triangle$ for the IFP-5000 (U_0 = 2250 V, C = 2000 µF,
τ = 620 µs); and $\square$ for the IFP-8000 (U_0 = 2600 V, C =
1800 µF, L = 60 µF, τ = 800 µs) [3-81].

Qualitatively, the picture of emission spectra integrated over
a pulse is characterized by superpositioned lines and the continuous
background. The line spectrum includes the lines of nonionized atoms
(arc lines) and the lines of singly and multiply ionized atoms.
Lines also are observed which are not excited in less intense discharges in the same gases, including lines which correspond to "for-
bidden transitions." The continuous background is explained by the
significant broadening of some lines due to the interaction of atoms,
to recombination, and to free-free transitions, including the "pseu-
docontinuum," which is formed by the coalescence of broadened terms
near the ionization limit. The relationship between the intensities
of the lines and the background depends on the power density in the
discharge, the type of gas, and the gas pressure: the intensity of
the background increases with increasing power, atomic number of the
gas, and gas pressure.

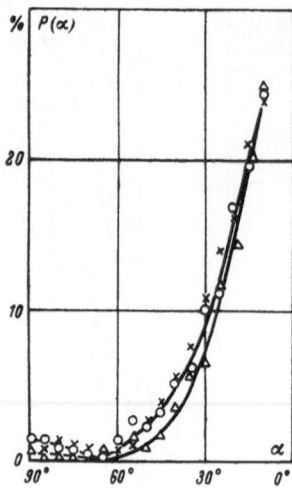

Fig. 5-30. Degree of polarization $P(\alpha)$ of the radiation from an
           IFP-1200 lamp. × is for U_0 = 1500 V, C = 975 μF, L =
           30 μH, τ = 580 μs; Δ for U_0 = 110 V, C = 200 μF, L =
           10 μH, τ = 350 μs; o for U_0 = 1450 V, C = 600 μF, L =
           100 μH, τ = 570 μs [3-81].

The main characteristic required for energy calculations of
various spectrally selective optical systems is the spectral distribution of the time-integrated radiation of lamps. This quantity is
reflected most conveniently by the spectral distribution of the efficiency per unit solid angle.* Quantitative studies of flashlamps
spectra were started as early as the 1940s [2-1]. Many experimental
studies of this distribution for new flashlamps (particularly series-
produced tubular flashlamps having a straight quartz envelope) have
been described in recent years. These flashlamps generally operate
at high peak electric power densities in the discharge [5-15, 5-38
to 5-46a, etc.].

The summarized results of the research are presented for tubular
xenon lamps in Tables 5-8 and 5-9 and in Figures 5-33 to 5-37. The
notations mentioned above are used in these tables and figures. By
$\eta_{\lambda_1-\lambda_2}$ is meant the efficiency of the lamp in the wavelength interval $\lambda_1-\lambda_2$ per unit solid angle (in the direction of the normal to
the axis of a lamp having a straight envelope), and by $P = 2CU_0^2/\pi l d^2 \tau$
is meant the electric power density averaged over the time τ. This

---

* As [5-38] showed, the relative spectral distributions of the integrated radiant intensity and the peak radiant intensity are quite
  close to each other. This is because most of the radiation falls
  in the time interval in which the discharge power differs little
  from the peak value.

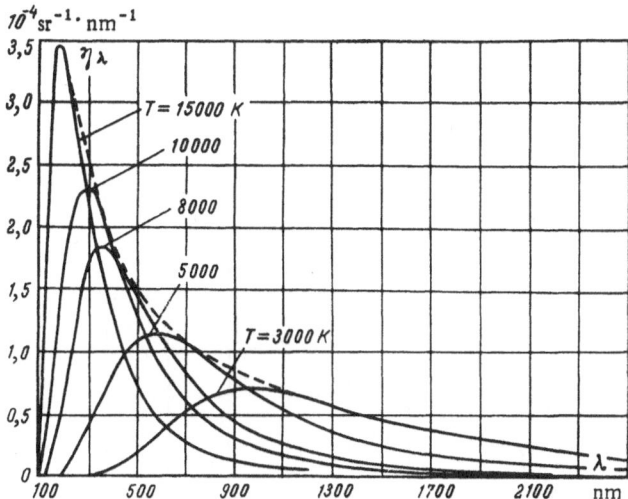

Fig. 5-31. Spectral distributions of the efficiency (per unit solid
            angle) of a cylindrical blackbody at various temperatures
            T. The values of the efficiency were obtained by dividing
            by $\pi\sigma T^4$ (where σ is the Stefan-Boltzmann constant) the
            spectral radiance of a blackbody, as calculated from
            Planck's formula. The dashed line indicates the envelope
            of the maximum possible spectral efficiencies of a black-
            body.

quantity is adopted as a parameter which approximately gives the
temperature and radiation of the xenon plasma in any lamp, on the
basis of the concepts presented in Chapter 3 regarding the physical
pattern of the discharge. (It is assumed that the losses in the discharge circuit do not exceed 10% and may be disregarded.)

As we can see from these tables and figures, the curves of η_λ
are transformed comparatively slowly as P and d_i vary over a broad
range. The values of η_λ in the wavelength interval decrease with
decreasing P in the wavelength interval 300-500 nm, but increase in
the interval 800-1000 nm. This pattern is especially clear if we
examine the curve of η_λ in a series of pulsed discharges with gradually decreasing values of P for a stationary AC discharge in xenon
arc lamps when P $\approx$ 100 W/cm^3 (Figure 5-36). At the same time, attention is drawn to the fact that the values of η_λ go practically unchanged in the interval 650-800 nm as P varies by nearly 4 orders of
magnitude, being about 3 x 10^{-5} sr^{-1}·nm^{-1}.

It should be noted that the spectral distributions of efficiency
differ more significantly in regimes with τ < 0.3 ms because of the
complex time dependence of various factors (the filling of the dis-

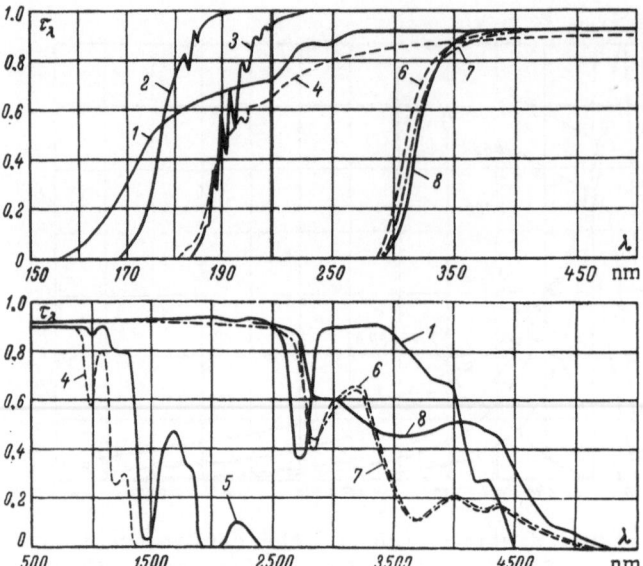

Fig. 5-32. Spectral transmission coefficients. 1) Quartz glass
made from synthetic raw material. The glass is fused
in the presence of hydrogen and oxygen. The gap around
2700 nm is due to absorbed water and can be eliminated
by melting the glass in an inert medium. The curves of
the spectral transmission coefficient of quartz glass
made from natural raw materials run below the curve
shown in the figure in the shortwave region. The thickness of the glass is 2.5 mm. 2) Air, 10 mm. 3) Air,
2200 mm. 4) Distilled water, 10 mm in a quartz vessel.
5) Distilled water, 1 mm in a quartz vessel. 6, 7, 8)
S-40 (ZS-11), S-52 (ZS-5), and S-88 (BD-1) electron-
tube glasses, 1 mm.

charge tube with plasma, energy losses on the walls and in the discharge circuit, vaporization of material, and the reversible opacity
of the quartz envelope [3-33a, 5-11, and 5-48 to 5-52]; see also
Chapter 6). Here the efficiency, integrated over the spectrum, may
be significantly lower than when $\tau > 0.3$ ms (Figure 5-34).* Line
radiation produced by xenon ions and silicon vapor formed upon

_____

* The influence of the electrode spacing and the gas density (due
  to losses around the electrodes and in the transelectrode volumes)
  should be taken into account during compilation of the graphs in
  Figures 5-33 to 5-37, in addition to the factors listed. This influence may explain, for example, the understated values of the
  efficiency in Figure 5-34.

Table 5–8. Power–Supply Parameters and Conditions of Flashlamps for Which the Spectral Distributions Were Obtained

Type of lamps, references cited	Regime No.	d_i, cm	l, cm	P_0, MPa	v_0, V	c, μF	L, μH	τ, μs	w, J
IFK-1500 [5-40]	1	0.4	30.4	0.04	3 200	300	40	500	1500
	2	0.4	30.4	0.04	2 600	600	40	1000	2000
FX-79 [5-41]	—	0.4	15.2	0.026	—	—	—	1900	360
IFK-120	1	0.5	7	0.013	300	2500	—	1000	113
IFK-120, made from uviol glass [5-42]	2	0.5	7	0.013	125	5800	—	2300	42
IFP-1000 [5-43, 5-44]	1	0.5	4.5	0.065	1250	300	30	200	230
	2	0.5	4.5	0.065	900	300	1	84	120
	3	0.5	4.5	0.065	1900	50	1	32	90
	4	0.5	4.5	0.065	1000	375	10	250	188
	5	0.5	4.5	0.065	800	50	0	80	16
IFP-800 [5-43]	1	0.7	8	0.065	1600	600	150	720	770
[5-45]	2	0.7	8	0.006	10 000	0.94	0.2	3	47
IFP-1000 [5-43]	1	0.7	8	0.065	900	2500	0	750	1020
IFP-1200 [5-43, 5-44]	1	0.7	12	0.04	1500	1000	30	580	1100
	2	0.7	12	0.04	1100	200	100	350	120
	3	0.7	12	0.04	1100	200	1000	1000	120
	4	0.7	12	0.04	1450	600	100	570	630
	5	0.7	12	0.04	2400	300	30	230	860
	6	0.7	12	0.04	1300	1000	1	300	840
	7	0.7	12	0.04	1800	300	1	100	490
	8	0.7	12	0.04	3500	50	1	36	310
ISPT-6000 [5-40]	—	0.7	12	—	1450	600	100	570	630
IP-1000 [5-46]	—	0.8	8	0.06	500	1400	—	650	175
IFP-2000 [5-43]	—	1.1	13	0.04	1500	1800	45	840	2030
IFP-5000 [5-43, 5-44]	1	1.1	25	0.04	2250	2000	0	620	5060
	2	1.1	25	0.04	2200	1200	10	360	2900
	3	1.1	25	0.04	3000	600	1	170	2700
IFK-1500 [5-40]	—	1.1	79.5	0.04	5000	973	150	1000	12 000
FX-47A [5-41]	—	1.3	16.5	0.04	—	—	—	750	1000
FX-47A [5-41]	—	1.3	16.5	0.04	—	—	—	750	5000
IFP-5000 [5-43, 5-44]	1	1.6	25	0.04	2600	1200	10	300	4000
	2	1.6	25	0.04	2600	1800	60	800	6760
	3	1.6	25	0.04	2600	2380	150	1320	8030
IFP-20000 [5-43]	1	1.6	58	0.04	4650	1200	70	900	12 960
	2	1.6	58	0.04	4650	1800	140	1400	19 460
Experimental lamps, [3-87]	1	1.8	58	0.04	5000	1115	30	1350	13 900
	2	1.8	58	0.04	4300	2285	80	1300	21 100
	1	2.6	58	0.04	4600	2285	50	1200	2420
	2	2.6	58	0.04	4900	1408	30	950	17 750
	1	3.5	58	0.04	4200	3500	50	620	30 800
	2	3.5	58	0.04	4000	2450	80	540	19 600
DKsTB-2000 [5-47]	—	1.8	17	0.028	220	—	—	4000	20

vaporization of the quartz is observed in the region 185–300 nm (especially around 220 nm) under severe conditions of power supply (see Figures 5–33 and 5–34). It can be seen from a comparison of Figures 5–33 and 5–34 with the curves in Figures 5–35 and 5–36 that the efficiency of flashlamps increases significantly in the ultraviolet under the harsh conditions of short-duration discharges, and the maximum of the curve of η_λ shifts to the shortwave ultraviolet (220–300 nm when $P \approx 0.5$ MW/cm^3 and 170–250 nm when $P > 1$ MW/cm^3) [5-45]. Xenon has a recombinational continuum of emission from 147 to 220 nm [5-53]. Therefore, the shortwave emission spectrum of

Table 5-9.   Spectral Efficiencies of Lamps in Wavelength Intervals (nm) (per unit solid
angle)

Type of lamp	Regime No.	P, MW/cm³	$\eta_{\lambda_1-\lambda_2} \cdot 10^3$, sr⁻¹						
			180—400	220—400	250—400	400—700	700—1100	220—1100	250—1100
IFK-1500	1	0.79	—	9.3	9.1	15.2	5.4	29.9	—
	2	0.52	—	8.9	8.8	15.7	5.5	30.1	—
FX-79	—	0.10	—	—	—	20.2	25.0	—	—
IFK-120	1	0.08	—	3	—	18	—	—	—
	2	0.013	—	—	—	13	—	—	—
ISP-1000	1	1.3	14.5	12.1	—	13.9	4.0	30.0	—
	2	1.6	25.9	20.5	—	14.6	5.7	40.8	—
	3	3.2	15.0	13.3	—	9.8	7.2	30.3	—
	4	0.85	—	9.8	9.0	13.5	8.4	31.7	—
	5	0.23	—	8.3	7.5	14.5	18.8	41.6	—
IFP-800	1	0.35	—	13.8	13.3	27.5	12.3	53.6	—
	2	5.1	45.4	35.2	15.5	6.3	4.0	45.5	25.8
IFP-1000	—	0.45	—	10.4	10.0	19.4	11.9	41.7	—
IFP-1200	1	0.41	—	15.8	14.9	22.4	10.8	49.0	—
	2	0.07	—	7.8	7.5	16.5	17.1	41.4	—
	3	0.03	—	5.8	5.5	13.1	13.1	32.0	—
	4	0.24	—	12.8	12.2	20.1	10.4	43.3	—
	5	0.81	32.0	27.4	—	21.5	6.4	55.3	—
	6	0.61	15.0	13.5	—	16.8	7.1	37.4	—
	7	1.1	26.0	25.2	—	9.0	9.0	53.8	—
	8	1.9	14.0	13.0	—	10.7	3.3	27.0	—
ISPT-6000	—	0.24	—	9.2	8.8	18.2	9.3	36.7	—
IP-1000	—	0.07	—	—	—	15.4	18.9	—	—
IFP-2000	—	0.20	—	11.4	11.0	22.3	8.1	41.8	—
IFP-5000	1	0.34	—	14.2	13.5	21.0	8.2	43.4	—
	2	0.34	16.1	14.8	—	18.1	8.1	41.0	—
	3	0.67	26.2	22.6	—	23.6	10.5	56.7	—
IFK-15000	—	0.16	—	9.3	9.2	18.6	10.4	38.3	—
FX-47A	—	0.06	—	—	—	27.9	22.0	—	—
FX-47A	—	0.30	—	—	—	35.2	17.6	—	—
IFP-8000	1	0.27	24.8	19.6	—	23.2	8.4	51.2	—
	2	0.17	—	15.1	15.0	24.2	10.6	49.9	—
	3	0.12	—	12.2	12.1	24.3	17.3	53.8	—
IFP-20000	1	0.12	—	14.1	13.6	26.7	13.4	54.2	—
	2	0.12	—	13.7	13.4	27.8	17.2	58.7	—
Experimental lamps	1	0.07	—	—	11.2	22.5	11.5	—	45.8
	2	0.11	—	—	9.6	22.6	10.6	—	42.8
	1	0.07	—	—	8.4	21.3	9.8	—	39.5
	2	0.06	—	—	10.3	21.8	10.5	—	42.6
	1	0.09	—	—	6.8	16.7	8.3	—	31.8
	2	0.07	—	—	3.9	15.4	13.4	—	32.7
DKsTB	—	$11 \cdot 10^{-6}$	—	1.0	0.9	8.3	37.9	47.2	—

Note. Flashlamps emit up to 3-5% of their total radiant energy in
far infrared intervals (1100-1800 and 1800-4500 nm) [5-37].

xenon flashlamps is limited by the transmission of the media located
between the discharge plasma and the radiation detector. The shortwave boundary of the emission spectrum may shift within the interval
150-220 nm, depending on the type of quartz from which the envelope
of the lamp is made, the surrounding medium, and the distance from
the lamp to the detector (thin layers of air and water absorb radiation from a lamp at wavelength shorter than 185 nm). However, for
some τ an increase in P beyond some value leads to a lowering of the
ultraviolet efficiency. For example, the efficiency in the wavelength
interval 165-185 nm decreases severalfold (by a factor of up to 10)

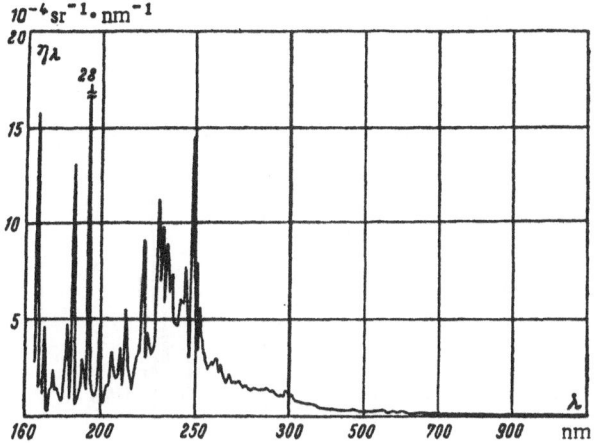

Fig. 5-33. Spectral distribution of efficiency per unit solid angle for the IFP-800 lamp [5-45]. Xe, p_0 = 6.6 kPa, C = 0.94 µF, L = 0.2 µH, U_0 = 10 kV, τ ≈ 3 µs, P = 5 MW/cm^3.

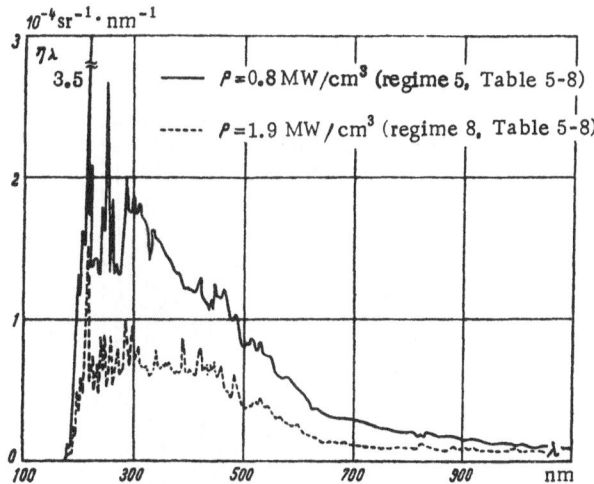

Fig. 5-34. Spectral distributions of efficiency per unit solid angle of an IFP-1200 lamp [5-44].

when τ = 3 µs and P increases from 5 to 20 MW/cm^3 [5-45]. For shorter flashes a decrease in the efficiency begins at much smaller P. This is explained by the heating of the inside surface of the quartz envelope to a temperature at which there is reversible opacity of the quartz, which progresses from short to longer wavelengths [3-33a, 5-11, and 5-48 to 5-52].

The spectra of flashlamps undergo practically no change with changing discharge repetition frequency if the lamp filler reaches

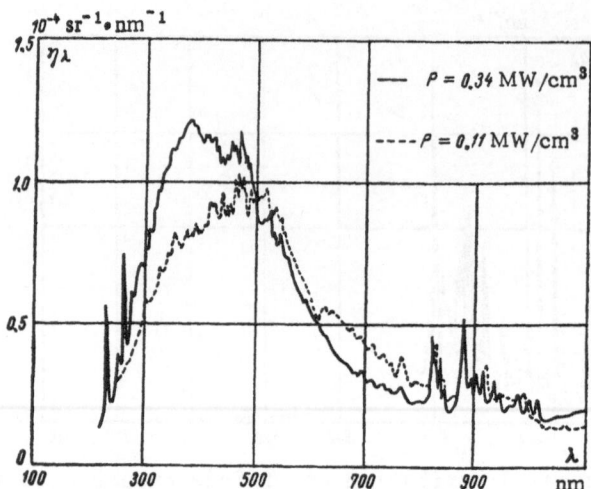

Fig. 5-35.   Spectral distributions of efficiency per unit solid
              angle for an IFP-5000 lamp (solid line) [5-43] and
              an experimental lamp (dashed line) [3-87].

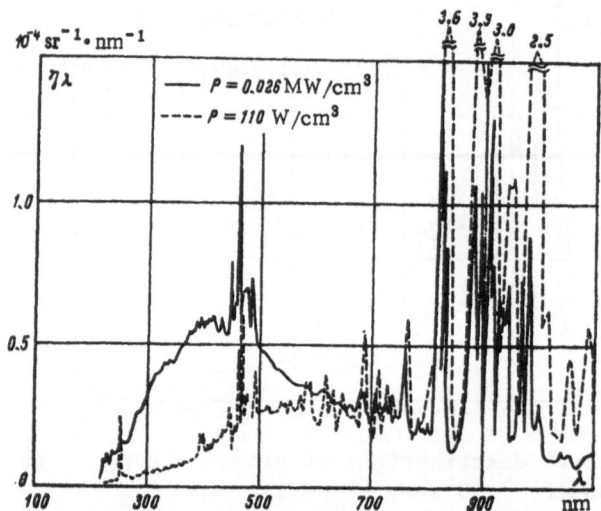

Fig. 5-36.   Spectral distributions of efficiency per unit solid
              angle for an IFP-1200 lamp (solid line) [5-43] and a
              DKsTB2000 lamp (dashed line) [5-47].

a state close to the initial state during the intervals between discharges. However, the noble gas may be contaminated with vapors
from the electrode and wall materials during the investigation of
low-energy, high-frequency flashes. As a result, a much lower number
of intense lines is observed over the continuous background (compared

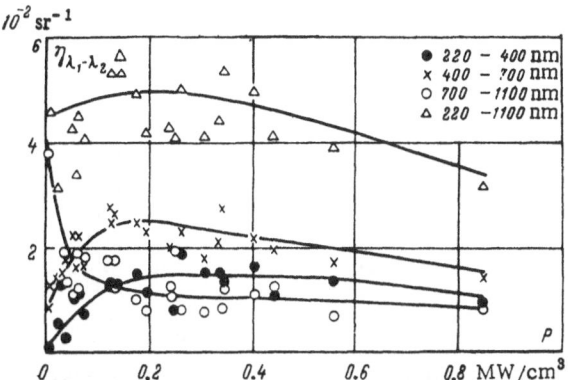

Fig. 5-37. Dependences of efficiency in the spectral intervals
            indicated in the figure (per unit solid angle) on the
            average power density (0.7 cm $\leq d_i \leq 1.8$ cm).

with the high-energy discharges considered above) [5-54]. Lamps with
a quartz or glass envelope have analogous spectral distributions,
but the shortwave emission limit of glass-envelope lamps is located
in the wavelength region near 300 nm, according to the spectral transmission coefficients of the glasses.

    A sample spectral distribution for a stroboscopic capillary
lamp is shown in Figure 5-38. The spectral density of the radiation
gradually decreases in the infrared ($\lambda > 1000$ nm), reaching zero at
the longwave limits of the spectral transmission coefficient of the
envelopes (ignoring the thermal self-radiation of the envelopes).
The relative spectral distribution of the integrated radiant intensity
differs little, despite significant differences in the design and
power-supply conditions of lamps that have low discharge energies.
The luminous efficacy and efficiency of these lamps in the visible
and near-infrared regions (400-1100 nm) are 5-10 lm-s/J and 5-10% in
most cases (see Table 5-10), i.e., they are severalfold lower than
for flashlamps with high energies and long discharge durations.

    As indicated above, the rough value of the peak spectral irradi-
ance $I_{\lambda p}$ (in W/sr-nm) can be obtained by dividing the spectral density
of the integrated radiant intensity θ_λ (J/sr-nm) by the flash duration. However, to make a more exact calculation we must take into
account the slight dependence (which actually does exist) of the
duration of the radiant intensity pulses on wavelength [5-38, 5-46,
and 5-55 to 5-57]. It can be seen from the characteristic dependences of τ on λ that are shown in Figure 5-39 that the duration increases by 20-50% as the wavelength increases from 250 to 750 nm.
The pulse length in the emission lines of xenon (in the near infrared) increases by a factor of up to 1.5 over the background. This
can be explained by the time dependence of the temperature and spectral absorption coefficient of the plasma.

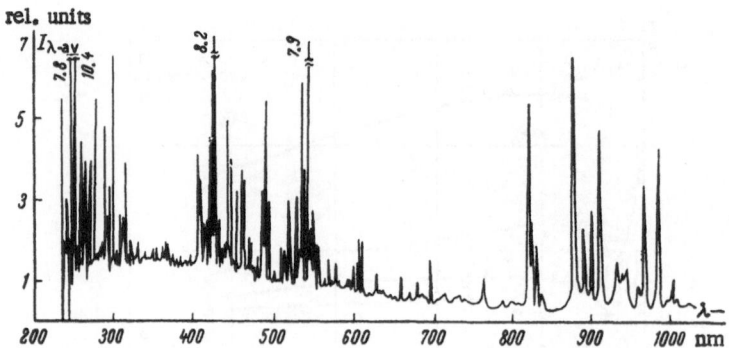

Fig. 5-38. The spectral distribution of the time-averaged radiant
            intensity of an ISK 20-1 quartz lamp [5-54]. d_i = 2 mm,
            l = 22 mm, p_0 = 0.06 MPa, C = 2 μF, U_0 = 300 V, τ ≈ 8 μs,
            P ≈ 0.2 MW/cm^2 (sic), f = 100 Hz.

     The complete time dependence of the emission spectra of tubular
flashlamps [5-38, 5-54, 5-56, and 5-58 to 5-61] is characterized by
the graphs in Figure 5-40. We can see from these graphs that the
start (t = 0.2 ms) and end (t = 2 ms) of the discharge are characterized by a spectral distribution having a background and numerous
less-pronounced lines. The distribution is close to the spectrum
of pulsed discharges for small P and to the spectrum of xenon arc
lamps having a plasma column with a low optical density [2-6, 5-41,
5-47, and 5-62]. At maximum current the energy contribution of the
lines to the total radiant flux decreases sharply, and the spectral
distribution of the radiation approaches blackbody radiation. This
is a result of the significant increase in the optical density of
the plasma column, even against a continuous background, for the
entire spectral interval of intense radiation: 300-1000 nm.

     The curves of the spectral distribution of the integrated radiant
intensity Θ_λ (the circles in Figure 5-40) and the peak spectral irradiance I_{λ_p} (at time t = 0.8 ms) are close to each other in the
wavelength interval 500-800 nm. The discrepancy between the spectral
distributions of Θ_λ and $I_{\lambda p}$ in the region λ < 500 nm is related to
the lower optical density of the plasma column and the accordingly
shorter duration of the radiation pulse, while it is related in the
region of intense arc lines (800-1000 nm) to the different shape of
the radiation pulses of the continuous background and the lines due
to the sharp increase in their absorption coefficient, and hence in
the emissivity. The line emission is prolonged in time and even
exists when the discharge current and the background radiation have
already ceased.

     Xenon is used to fill flashlamps in most cases because of its
superiority over the other noble gases in terms of overall radiation
efficiency. It may prove advisable to use other noble gases to fill

Table 5-10. Parameters of Capillary Discharges, and the Efficiency $\eta_{\lambda_1-\lambda_2}$ in Spectral Intervals $\lambda_1-\lambda_2$, nm

Type of lamp, reference cited	d,cm	l,cm	Gas	p_0,MPa	U_0,kV	C,µF	W,J	τ,µs	P,MW/cm³	$\eta_{\lambda_1-\lambda_2}$,% 230-400	400-700	700-1000
IFK-120 made of uviol glass [5-42]	0.5	7	Xe	0.013	1.6	8	10	–	–	15	11	32
Experimental lamps [5-54]	0.17 0.2	8.5 5	Xe	0.25	2.8	0.25	1	6.6	0.8 1.0	–	6.9	2.6
ISK-20-1 [5-54]	0.2	2,2	Xe	0.06	0.3	2	0.09	8	0.16	3.3	4.1	2.8
ISP-5 [5-54]	0.05	1	Xe	0.066	1 1	0.1 2	0.05 1	1.8 6	14 85	– –	3.2 6.1	1.9 2.2
ISP-70 [5-54]	0.05	7	Xe	0.04	1.25 1	0.25 2	0.2 1	20 90	0.7 0.8	3.5	3.5	5.6
ISP-70 [5-42]	0.05	7	Xe Kr Ar	0.08 0.08 0.08	1.2 1.2 1.2	0.25 0.25 0.25	0.18 0.18 0.18	25 – –	0.5 – –	– – –	7.2 4.2 1.2	15.5 2 2.6

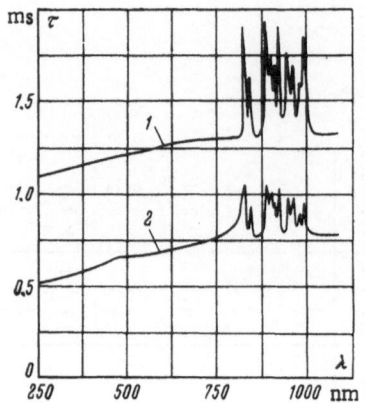

Fig. 5-39. Dependence of the duration τ of radiant intensity pulses
on the wavelength λ of the radiation of a xenon lamp
[5-38] (d_i = 18 mm, l = 580 mm, p_0 = 0.04 MPa). 1) U_0
= 4.3 kV, C = 2285 μF, L = 80 μH; 2) U_0 = 5 kV, C =
1110 μF, L = 30 μH.

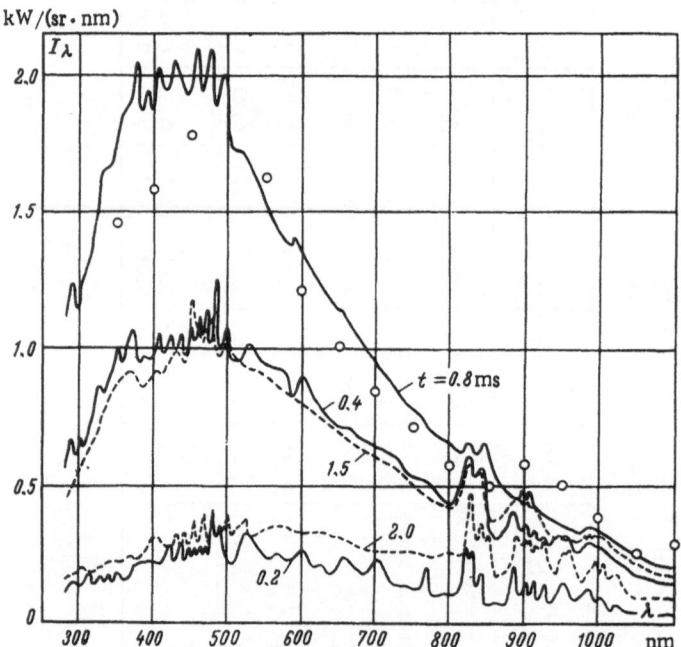

Fig. 5-40. Spectral irradiances I_λ of a xenon flashlamp at different
times [5-38]. d_i = 18 mm, l = 580 mm, p_0 = 0.04 MPa,
U_0 = 4.3 kV, C = 2285 μF, L = 30 μH. The circles indicate
the spectral distribution of integrated radiant intensity
in relative units.

flashlamps [5-63 and 5-64] under comparatively mild power-supply
conditions in order to obtain a higher efficiency in narrow spectral
intervals that contain emission lines. The line selectivity of the
radiation decreases with increasing power density because of the increase in the absorption coefficient of the discharge. A xenon filler
ultimately turns out to be more efficient even for these intervals.

The value of Θ_λ is higher for xenon lamps than for krypton
lamps (Figure 5-41) throughout the entire spectral region in question,
except for the narrow spectral intervals that contain the intense
infrared emission lines of krypton. These lines are shifted toward
shorter wavelengths compared with the analogous group of lines of
xenon, and have a better fit with the infrared absorption bands shown
in the figure for neodymium-activated yttrium-aluminum garnet. Therefore, krypton flashlamps with a diameter d_i = 4 mm turn out to be
more effective sources for pumping lasers containing this active
material to an electric power density of 0.6 x 10^6 W/cm^3 [5-65]. The
selectivity of the radiation in the lines gradually falls off as P
increase, and xenon lamps become more efficient with respect to the
detector indicated.

Comparative data on the spectral distribution of the radiation
of xenon, krypton, argon, and neon lamps [5-46 and 5-66 to 5-69]
lead to analogous conclusions. For example, neon plasma turns out
to be more efficient than xenon plasma as a radiator in the spectral
interval 610-650 nm when P $\leq$ 0.4 MW/cm^3. Nor does the use of mixtures
of noble gases lead to a significant gain in efficiency [5-69 and
5-70a].

Flashlamps containing vapors of various substances are still
under investigation in the laboratory. For example, [5-71 and 5-72a]
report elevated efficiency in the interval 240-300 nm for a pulsed
discharge in a mixture of xenon with zinc and cadmium vapors, compared with a discharge in xenon. References [5-70 and 5-73 to 5-76],
East German patent No.167728, and Japanese patent No.46-77489 note
that flashlamps containing mercury vapor turn out to be efficient
sources for the optical pumping of ruby lasers because of the selectivity of the radiation in the broadened lines of mercury. Analogous
selective advantages over xenon were found in [5-77 and 5-77a] and
in US patent No.3,781,585 for pulsed discharges in vapors of cesium,
sodium, and thallium.

The luminous efficacy and the overall efficiency (over the
entire radiation region) of tubular flashlamps filled with noble
gases increases with increasing atomic number of the gas (Table 5-11)
and are highest for xenon [5-66, 5-78, 5-79, and 5-80]. According
to [1-68], if we take the η_e of xenon as 100, then η_e is close to
80 for krypton and to 60 for argon when d_i = 4-14 mm, l = 15-100 cm,
p_0 = 0.013-0.02 MPa, C = 10-1000 μF, and E_0 = 40-150 V/cm. The value
of η_e is 55 for krypton and 16 for argon when d = 0.5 mm, l = 7 cm,

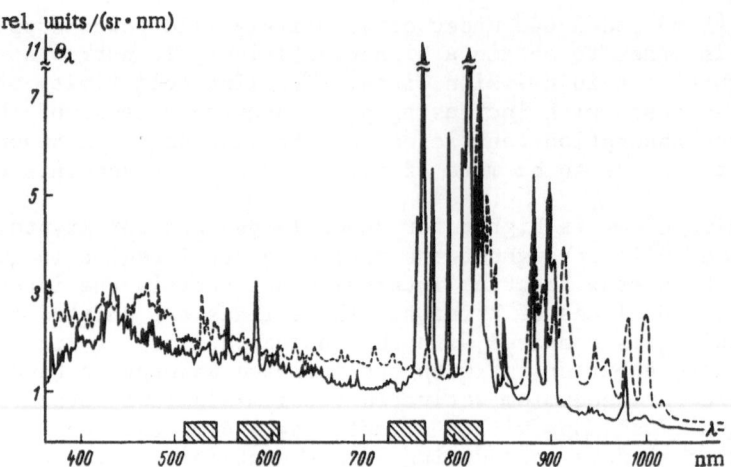

Fig. 5-41. Spectral distributions of the integrated radiant intensity
Θ_λ of krypton lamps (solid line) and xenon lamps (dashed
line) [5-65]. d_i = 4 mm, l = 60 mm, p_0 = 0.08 MPa, C =
290 µF, U_0 = 700 V, L = 13 µH, τ = 250 µs, P = 0.4 MW/cm^3.
The rectangles indicate the main absorption bands of
YAG:Nd^{3+}.

p_0 = 0.08 MPa, C = 0.25 µF, and E_0 = 170 V/cm. The nature of the
dependence of η_e on P_V is shown in Figure 5-42. The overall efficiency η_e increases somewhat as the discharge-tube diameter increases to approximately d_i = 7 mm under optimal conditions with
$P_V \approx 0.5$ MW/cm^3. As d_i increases further, η_e remains practically
constant, reaching about 80%. If the capacitance of the supply capacitor is limited, then the dependence of η_e on d_i have a maximum
that shifts into the region of large d_i with increasing C and decreasing p_0 (Figures 5-43 and 5-44). Apparently because of this,
increasing d_i to 100 mm leads to some gradual decrease of η_{e-max} to
approximately 50% [5-11 and 5-13]. The nature of the dependence of
η_e on p_0 (Figure 5-45) is the same as that of the dependence of the
luminous efficacy on p_0 (see Figure 5-3).

Increasing the electric power density to a few units times 10^6
W/cm^3 (increasing the current density to tens of thousands of A/cm^2)
leads to a decrease in the efficiency to 40% (Figure 5-42b), even if
the discharge tubes have optimal diameters (d_i = 10-20 mm). This is
explained by the significant increase in the fraction of ultraviolet
and vacuum ultraviolet radiation of the discharge column which is
absorbed by the quartz envelope [5-11, 5-44, and 5-55].

The efficiency decreases by approximately 20% as the ratio of
the ballast volume near the electrode to the working volume (V_{el}/V_w)
increases from the minimum value of 0.05 to 10 [2-34, 2-35, and 5-81].

Table 5-11. Concentration of Neutral, Singly Ionized, and Multiply Ionized Nitrogen Atoms at Different Temperatures [5-87]

$T, 10^3 K$	$N, \%$	$N_I, \%$	$N_{II}, \%$	$N_{III}, \%$
20	1	98.7	0.3	–
30	0.03	50	50	–
40	–	4	93	3
50	–	0.5	61	38.6
60	–	0.02	10	90

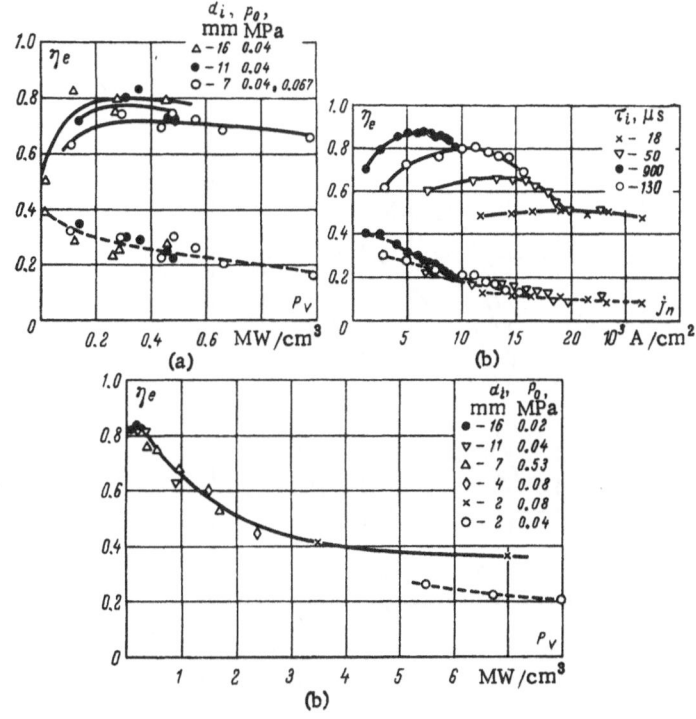

Fig. 5-42. Dependence of the radiant efficiency on various parameters. a) Total efficiency (solid lines) and infrared efficiency (the dashed line, for the spectral interval 700-2700 nm) versus the electric power density P_V [5-39 and 5-37]. b) total efficiency η_e versus the electric power density P_V (τ = 100-900 μs) [5-11]. c) total efficiency (solid line) and infrared efficiency (dashed line, for the spectral interval 700-2700 nm) versus the peak current density j_p for the durations τ_i indicated in the figure for the discharge-current pulses of lamps with d_i = 6.5 mm [5-81].

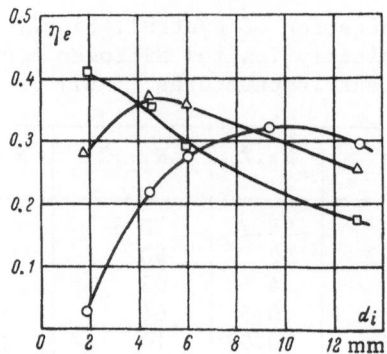

Fig. 5-43. Examples of the dependences of the radiant efficiency
on the discharge-tube diameter for W = 38 J and for
various C and U_0. l = 50 cm, krypton, p_0 = 0.013 MPa.
□ is for 2.1 μF, E_0 = 120 V/cm; △ for 14.5 μF, E_0 =
46 V/cm; ○ for 152 μF, E_0 = 14.1 V/cm [5-82].

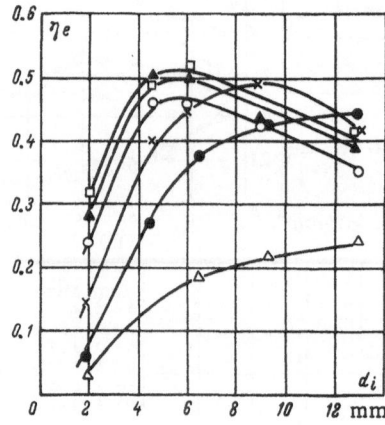

Fig. 5-44. Dependence of the efficiency on the diameter of the
tube for different krypton pressures. l = 50 cm, C =
80 μF, E_0 = 39 V/cm. □ is for p_0 = 0.033 MPa; ▲ for
0.021 MPa; ○ for 0.013 MPa; × for 0.007 MPa; ● for
0.003 MPa; and △ for 0.0013 MPa [5-82].

The efficiency is lower in coaxial-type flashlamps than in ordinary
tubular ones, not exceeding 40% [5-16]. This may be explained by
the increase in energy losses due to the increased contact area between the plasma and the quartz envelope.

In tubular flashlamps the efficiency in the infrared region
(700-2700 nm) reaches 40% under near-arc conditions, 20% as P increases to (0.5-1) x 10^6 MW/cm^3, and about 10% when P is a few units
times 10^6 W/cm^3.

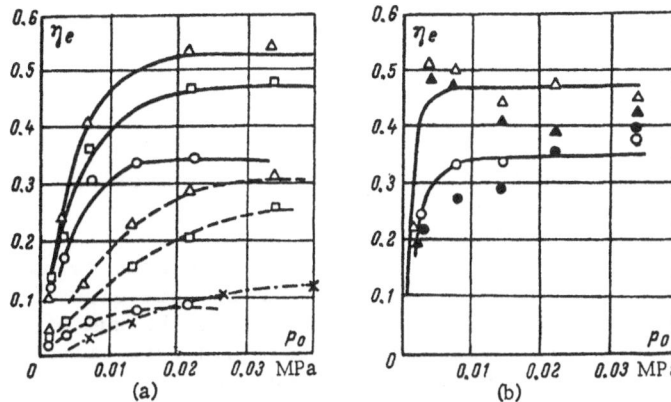

Fig. 5-45. Dependences of the efficiency on the gas pressure p_0 for
various d_i, E_0, and C. a) solid lines: d_i = 5 mm;
dashed line: 2 mm, Kr; dot-dash line: 0.5 mm, Xe; △ is
for C = 152 μF, E_0 = 40 V/cm; □ for 152 μF, 28 V/cm; ○
for 152 μF, 20 V/cm; × for 0.25 μF, 170 V/cm. b) d_i =
13 mm, Kr; △ is for 152 μF, 40 V/cm; ▲ for 48 μF, 71 V/cm;
● for 152 μF, 20 V/cm; and ○ for 48 μF, 35.5 V/cm [5-10
and 5-82].

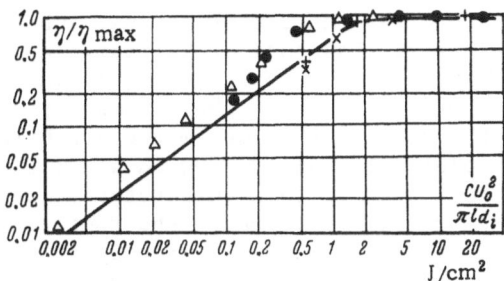

Fig. 5-46. Relative change in efficiency as a function of the capacitance of the supply capacitor for various d_i and l.
● is for d_i = 0.5 mm, l = 7 cm, Xe, p_0 = 0.08 MPa, E_0 =
170 V/cm; + is for d_i = 1.2 mm, l = 0.6 cm, Xe, p_0 =
0.093 MPa, E_0 = 330 V/cm; △ for d_i = 5 mm, l = 7 cm, Xe,
p_0 = 0.013 MPa, E_0 = 50 V/cm; and × for d_i = 13 mm, l =
50 cm, Kr, p_0 = 0.013 MPa, E_0 = 80 V/cm [1-68 and 5-78].

        Figure 5-46 shows that the decreasing efficiency (compared with
the maximum efficiency η_{e-max} for a given E_0 and an extremely large C)
with decreasing C can be represented approximately with a single graph
for different lamps, provided that the quantity $CU_0^2/\pi l d_i$ is taken as
the abscissa. This quantity may be treated as a characteristic of
the optical density of the plasma column. The decrease in efficiency

begins when $CU_0^2/\pi l d_i \approx 2$ J/cm^2 and is described by the approximate empirical relation

$$\eta/\eta_{max} = \left(CU_0^2/\pi l d_i\right)^{0.72}, \qquad\qquad (5-3)$$

where C is given in microfarads, d_i and l in centimeters, and U_0 in volts.

The concept of the discharge channel as a plasma column that is longitudinally uniform with an initial longitudinal electric field strength E_0 equal to $(U_0 - U_{ac})/l$ (where U_{ac} is the sum of the near-electrode voltage drops, which amounts to a small fraction of U_0 for long lamps) leads us to expect a weak dependence of η on the length of the discharge channel for a constant U_0/l (Figure 5-47).

5-7. The Spectral Characteristics of Spherical Flashlamps

The difference between the spectral characteristics of tubular and spherical flashlamps is due to the following peculiarities of the latter: the absence of a quasi-stationary stage due to the continuous expansion of the column as a significant amount of electric power is liberated in it; the electric field strength, which is elevated by an order of magnitude or more (thousands or tens of thousands of V/cm) and which determines the corresponding, substantially higher plasma temperature (tens of thousands of kelvins); and the short duration of the discharge (microseconds or less), which is comparable to the afterglow time of the column. The maximum spectral density of the radiation of the column of an unlimited discharge should lie in the vacuum ultraviolet interval 100-200 nm because of the high optical density of the column at such temperatures (see Figure 5-31). But the observed spectral distribution of the radiation of spherical lamps differs from this because of absorption of the shortwave portion of the radiation from the column by the surrounding media (the gas that fills the envelope and the glass of the envelope). Thus, absorption by quasi-molecules in the region of the resonance lines of the corresponding atoms [5-83] determines the shortwave limit of the spectrum as 160 nm in xenon (ignoring absorption by the envelope) and as 120 nm in argon. The glass envelopes of spherical lamps transmit radiation beginning at 300 nm, while quartz envelopes do so from 160 nm (see Figure 5-32).

The characteristic spectral distributions of radiation efficiency for spherical flashlamps are presented in Figures 5-48 and 5-49. Groups of the spectral lines of xenon are superposed on the continuous background. The steep boundary in the ultraviolet region (Figure 5-48) is explained by absorption by the glass envelope of the lamp. Without such absorption there is a sharp increase in the spectral efficiency when $\lambda < 250$ nm (Figure 5-49) which is due not

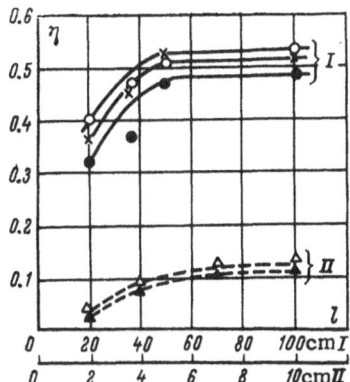

Fig. 5-47. Dependence of η on electrode spacing for constant Cl and U_0/l. 1) Kr, $p_0 = 0.013$ MPa, $d_i = 4.5$ mm, $U_0/l = 80$ V/cm ($\circ$ is for $Cl = 7600$ μF-cm; $\times$ for 4000 μF-cm; $\bullet$ for 2400 μF-cm); II) Xe, $d_i = 0.5$ mm, $U_0/l = 170$ V/cm, $Cl = 1.5$ μF-cm ($\triangle$ is for $p_0 = 0.04$ MPa; $\blacktriangle$ for $p_0 = 0.08$ MPa) [2-34].

only to the continuous background, but also to the group of intense lines of xenon ions. Radiation has been recorded up to 1700 nm in the infrared region.

The study of spectra at different times (Bogdanov and Wulfson; Mandel'shtam et al.; Vanyukov and Mak; and others: see [2-1 and 2-42], and the latest paper [5-54, 5-58, and 5-84]) has showed that the development of individual elements of the spectra (arc lines, the lines of singly and multiply ionized atoms, and the continuous background) which have different excitation energies are mutually shifted in time according to the temperature and optical density of the channel, which change during the discharge. Thus, it is evident from a comparison of the spectral distributions of the peak radiant intensity $I_{\lambda p}$ and the efficiency (Figure 5-49) that the emission lines of xenon in the wavelength interval 380-750 nm are more pronounced (in comparison with the continuous radiation) in the spectrum of $I_{\lambda p}$ than in the spectrum of η_λ. In the interval 800-1100 nm the spectrum of η_λ contains a number of arc lines of xenon that correspond to the 6P-6S transitions of the excited atom, whereas these transitions show up as absorption lines in the spectrum of $I_{\lambda p}$ (the colder outer layer of xenon excited to 6S levels absorbs the continuous spectrum of the hotter axial portion of the column).

In the infrared region of the graph of η_λ lies above that of $I_{\lambda p}$, while these curves have the opposite position in the blue portion of the spectrum. This difference may be attributed to the fact that a higher plasma temperature corresponds to the distribution of $I_{\lambda p}$, and some lower temperature, averaged over the flash, corresponds to the distribution of η_λ.

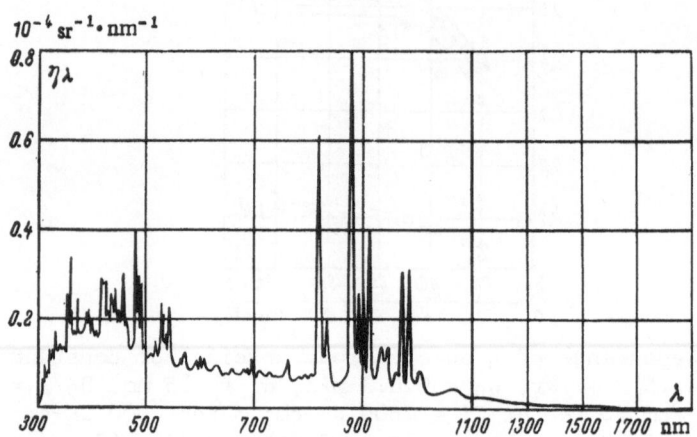

Fig. 5-48.   Spectral distribution of efficiency per unit solid angle
             for an ISSh-7 glass lamp. Xenon, 0.22 MPa, l = 2.5 mm,
             1000 V, 6800 pF, τ = 0.35 µs, f = 2 kHz, I_{av} = 4.8 cd
             [5-54].

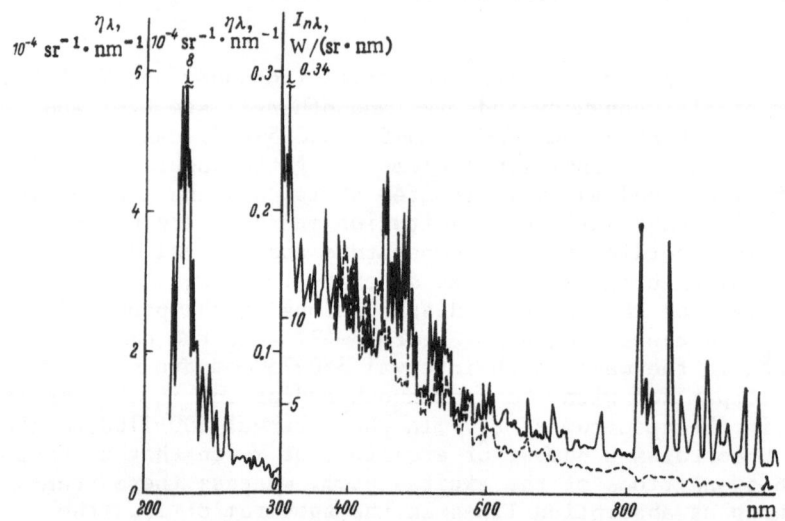

Fig. 5-49.   Spectral distribution of the peak radiant intensity
             (dashed line) and of the efficiency per unit solid angle
             (solid line) for a modified ISShO-1 lamp with a magnesium
             fluoride window. p_0 = 0.22 MPa, l = 2.5 mm, U_0 = 800 V,
             C = 4 µF, τ = 3 µs, I_p = 0.3 Mcd [5-84 and 5-84a].

The different nature of the spectral distributions of $I_{\lambda p}$ and η_λ is clarified by oscillograms of the radiation pulses and close-lying regions of the continuous background (Figure 5-50). The radiation in the line at 823.16 nm has several maxima in time and a much longer duration than the background radiation pulse at $\lambda = 815$ nm. The same pattern also is observed for all 6P-6S lines, and the second maximum exceeds the first for the radiation pulse at 881.94 nm.

Figure 5-50b shows the time dependence of some elements of the spectrum of a short pulsed discharge in nitrogen at reduced pressure. Here the comparatively slow development makes it possible to trace the change in the spectrum not only as the temperature decreases after the maximum electric power of the discharge is reached, but also as the temperature rises. The molecular bands of nitrogen and the weak lines of neutral atoms àre observed at the very start. The first maximum of the spectrum of singly charged ions is reached later, and the maximum for doubly charged ions (the maximum temperature) is reached still later. After this, the maxima alternate in reverse order.* The time at which the maximum intensity is reached decreases with increasing ionization potential U of the upper level of a line (Figure 5-51). The point having an abscissa equal to the time at which the maximum background is reached can be found by using the fact that the experimental points for all lines having different excitation potentials fit on a common curve under given discharge conditions. The ordinate of this point may be taken as the effective excitation potential of the continuous background, which turns out to lie between 55 and 45 V for different discharge intensities. Hence it follows that triply or doubly charged ions play a significant part in the formation of the background, which is produced mainly by bremsstrahlung and recombination.

If we assume that the population of the levels is proportional to $\exp(-eU/kT_e)$ (where T_e is the electron temperature and k the Boltzmann constant), according to Boltzmann equation (3-2), then the intensity of the i-th spectral line of frequency ν_i is equal to**

$$I_i = \frac{g_i}{g_0}\, h\nu_i\, A_i\, N_0 e^{-eU_i/kT_e},\qquad (5\text{-}4)$$

---

* The curves in Figure 5-50 apply to oscillatory discharges. This explains the repeated increase in intensities in the second half-period of the discharge.
** We consider here lines which do not undergo reabsorption. In [5-85] the criterion for the absence of reabsorption was that the relative intensity of the multiplet line be constant throughout the discharge.

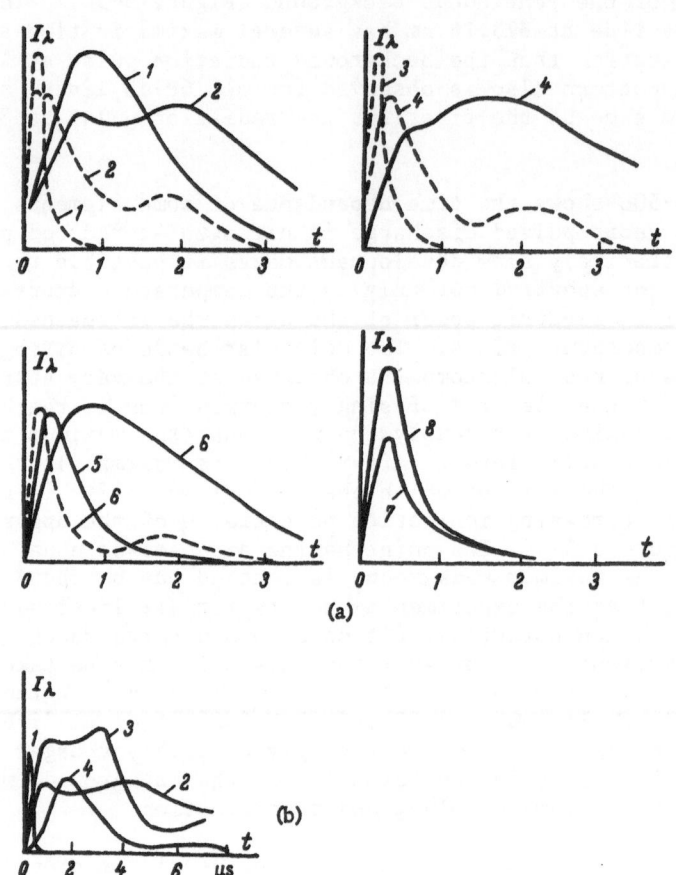

(a)

(b)

Fig. 5-50. Oscillograms of radiation pulses.
a) an ISShO-1 lamp (for the parameters see Figure 5-49)
in lines (the even-numbered curves) and in the background
(the odd-numbered curves). 1) λ = 815 nm; 2) 823.16 nm;
3) 975 nm; 4) 881-94 nm; 5) 962 nm; 6) 916.26 nm; 7)
450 nm; 8) 484.43 nm. The scale division of the t axis
is 4 μs for the solid curves, and 40 μs for the dashed
curves [5-84].
b) a short-duration spark discharge in nitrogen. 0.013
MPa, 10-15 kV, 0.24 μF, 10 μH. 1) molecular band of
nitrogen; 2) H_α line of the hydrogen present as an impurity; 3) the line of a singly ionized atom; 4) the
line of a doubly ionized nitrogen atom [5-85].

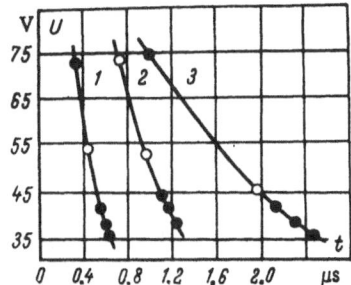

Fig. 5-51. Dependence of the time at which the maximum intensity of
nitrogen lines is reached (solid symbols) on the excitation potential U of their upper level (including the
potentials for single or double ionization, which are
14.5 and 29.5 V, respectively). Air, 0.1 MPa, 10-15 kV.
1) 0.01 μF, 10 μH; 2) 0.25 μF, 2.6 μH; 3) 0.25 μF, 10 μH.
The abscissas of the circles correspond to the time at
which the maximum intensity of the continuous background
is reached.

where g_i and g_0 are the statistical weights of the upper and lower
levels, A_i is the transition probability, and N_0 is the concentration
of atoms or ions in the unexcited state. Hence

$$\frac{1}{e\,(U_i - U_j)} \ln \frac{I_i}{I_j} = B_{ij} - \frac{1}{kT_e} , \qquad (5\text{-}5)$$

where B_{ij} is a constant which is characteristic of a given pair of
lines and which is independent of the temperature.

Thus, the temperature dependence and hence the time dependence
of the quantity $\dfrac{1}{U_i - U_j} \ln \dfrac{I_i}{I_j}$ should be the same for any pair of lines
to the accuracy of some constant. Figure 5-52 confirms this con-
clusion, and thus also confirms the hypothesis that these exists a
specific temperature and that the population of the levels is of the
Boltzmann type.

The electron temperature of the discharge plasma can be determined from measurements of the line intensities during the flash;
from formula (5-4), for which the transition probabilities have been
calculated; and from Saha's formula. The three pairs of lines de-
noted by the numbers 1, 2, and 3 in the caption to Figure 5-52 gave
estimated electron temperatures of 43×10^3 K, 38×10^3 K, and $52 \times
10^3$ K, respectively, for a time t = 0.5 μs under the discharge con-
ditions indicated in the same caption.

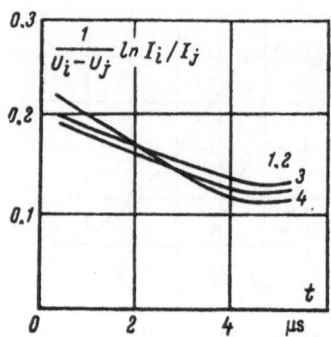

Fig. 5-52.   Time dependences of the logarithm of the ratio of the
ratio of the intensities of various pairs of lines
(superposed for a time t ≈ 2.5 µs). According to formula
(5-5), these plots confirm the Boltzmann distribution of
the atoms over the levels.   1) λ_i = 409.7 nm, U_i = 74 V
(the spectrum of N_{III}), λ_j = 399.5 nm, U_j = 35.5 V (N_{II});
2) λ_i = 517.9 nm, U_i = 44.5 V (N_{II}), λ_j = 504.5 nm,
U_j = 35.5 V (N_{II}); 3) λ_i = 553.5 nm, U_i^j = 41.5 V (N_{II}),
λ_j = 549.5 nm, U_j = 47.5 V (N_{II}); 4) λ_i = 549.5 nm, U_i
= 47.5 V (N_{II}), λ_j = 460.7 V, U_j = 35.5 V (N_{II}) [5-85].
The discharge conditions are: air, 0.1 MPa, 0.25 µF,
10-15 kV, 10 µH.

Temperatures of (26-33) x 10^3 K were obtained in [5-86] for
four pairs of lines under much harsher discharge conditions (0.05 µF,
2-8 kV, 0.1-0.04 µH). The average estimate for the two papers, 35
x 10^3 K, is in satisfactory agreement with the estimate of the gas
temperature T_{gas} = 40,000 K, that was obtained in studies of channel
expansion and the electrical characteristics of the channel, and in
luminance measurements. Experiment thus confirms the theoretical
hypothesis that thermodynamic equilibrium becomes established quite
rapidly in the channel of a pulsed discharge and that the Boltzmann
equation and Saha's formula are applicable to it. (A conservative
estimate gives a value of no more than 10^{-10} s for the time required
for establishment of a stationary distribution of excited atoms and
ions, and a value of no more than 10^{-7} s for the stationary value of
ionization.)

Figure 5-53 shows the variation of the temperature, as calculated in the manner described from records of the change in spectral-
line intensities. Attention is drawn to the comparatively small
range of the temperature as the power and intensity of the discharge
vary significantly (cf. Section 2-5). Table 5-11 gives the results
of calculations [5-87] of the degree of ionization of nitrogen at
several temperatures. This table graphically clarifies the predominance of the N_{III} spectrum at the time of the maximum temperature
and the subsequent increase in the N_{II} spectrum.

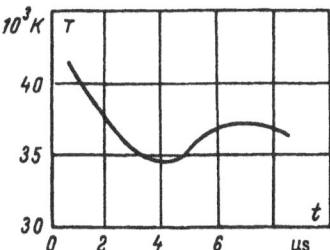

Fig. 5-53. Time dependence of the temperature of the discharge
            channel in air, as calculated from the emission of
            lines at 409.7 nm (N_{III}) and 399.5 nm (N_{II}) [5-87].

     The pattern described for the time dependence of elements of
the spectra of unlimited discharges is supplemented by data on the
spectral distribution of the radiant intensity at different times
during the flash over sizable wavelength intervals (Figures 5-54 and
5-55). Apart from the difference in the phases of appearance and
disappearance of lines having different excitation potentials for
the upper levels, these plots show that the discharge spectrum (including even the continuous background) is dissimilar to the blackbody spectrum, even at the time of maximum power in the discharge.
Attention is drawn to the fact that the spectrum approximates an
equal-energy distribution at times following maximum power, and to
the fact that the continuous background decreases even when maximum
power is reached, given low absolute values of the power (which is
reduced by the high inductance of the discharge circuit).

     Measurement of the spectral irradiance of the continuous back-
ground (Figure 5-56) showed that the spectral distribution $L_{e\lambda}$ agrees
well with the temperature distribution of the luminance of a blackbody at a temperature of 27,500 K (in the case of xenon) for an unlimited discharge and a sufficiently high energy concentration at
which luminance saturation is reached. This indicates the quite
high optical density of the middle of the column, and enables us to
assign the plasma the value obtained for the temperature. A determination of the plasma temperature from the maximum spectral luminances on sections of the continuous spectrum and at the center of
broadened lines (Figure 5-57) gives temperature values which are in
satisfactory agreement with other estimates.

     Table 5-12 gives the values of the efficiency of spherical
flashlamps in three consolidated spectral intervals. The values
were obtained by integrating the spectral characteristics and multiplying by the equivalent solid angle Ω_{eq} = 10 sr. Like the luminous
efficacy, the efficiency of spherical lamps is severalfold lower
than that of tubular lamps, amounting to a few percent in the visible
region. The overall efficiency in the spectral region where the
envelope of the ISSh-7 lamp is transparent (see Figure 5-48) is 9%.

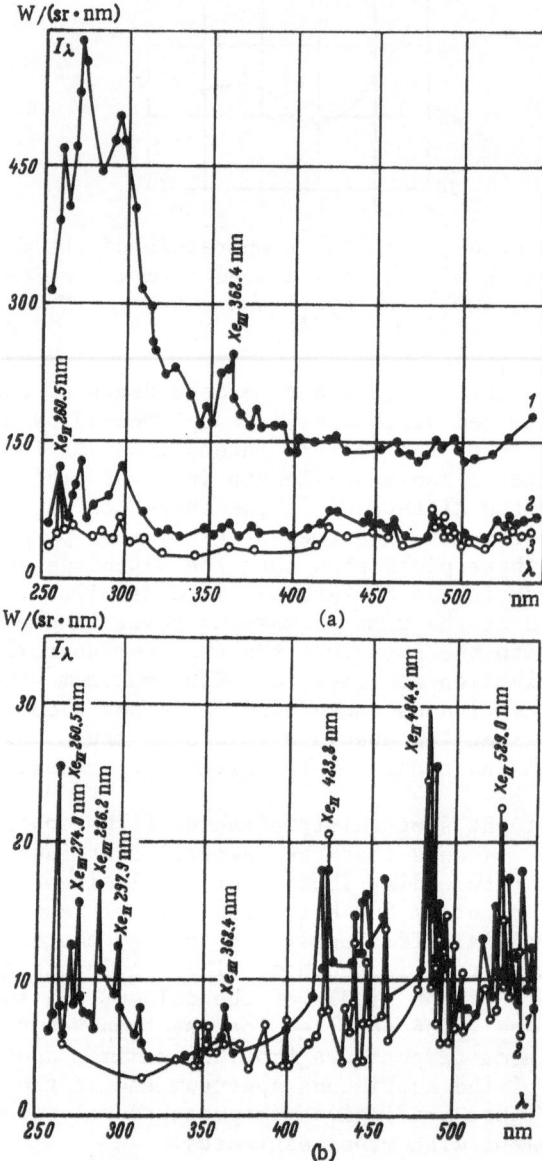

Fig. 5-54. Spectral distribution of the radiant intensity of an
            unlimited discharge in xenon (p_0 = 0.35 MPa, l = 10 mm)
            at different times and for different power-supply param-
            eters [5-88].   a) 0.05 μF, 5 kV, 0.1 μH (t, in μs:
            1) 0.3; 2) 0.8; 3) 1.2).   b) 0.05 μF, 12 kV, 25 μH
            (t, in μs: 1) 3.3; 2) 7.1).

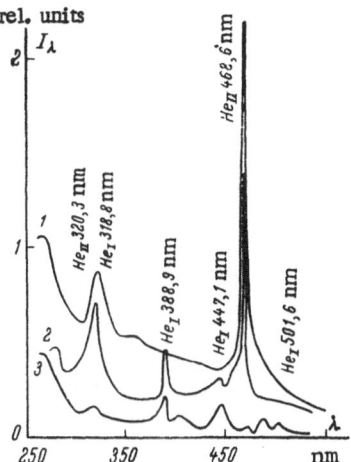

Fig. 5-55. Spectral distribution of the radiant intensity of an
unlimited discharge in helium (0.5 MPa, 0.035 μF, 7 kV,
0.18 μH) at different times [5-89]. t, in μs: 1) 0.1;
2) 0.7; 3) 4.5.

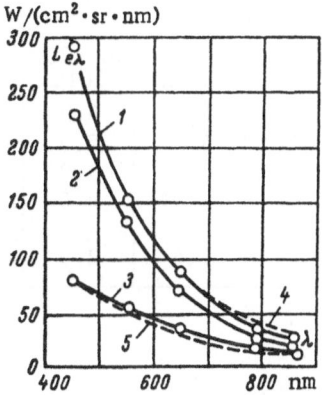

Fig. 5-56. Spectral irradiance of the channel of an unlimited discharge at different times during a flash [2-42]. Xe,
0.5 MPa, 0.011 μF, 12 kV, 0.12 μH. 1) at time t = 0.1
μs; 2) 0.3 μs; 3) 0.8 μs; 4 and 5) curves of the spectral
blackbody luminance at temperatures of 27,500 and
14,500 K, respectively.

We should recall that the efficiency is lowered substantially by the
absorption of a significant fraction of the radiant energy in the
ultraviolet by the envelope. For example, the efficiency in the
interval 220-250 nm is 0.5 the efficiency in the interval 220-1050 nm
(see Figure 5-49). If we take into account even shorter-wave radiation (including vacuum radiation), the overall efficiency of high-
power unlimited discharges probably may reach 0.5 [5-84a].

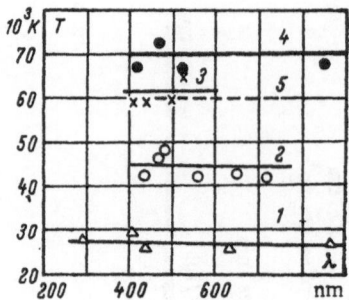

Fig. 5-57. Plasma temperature of unlimited discharges, as calculated
from the maximum values of the spectral radiance $L_{e\lambda}$ in
different spectral regions for continuous and line
emission. 1) xenon; 2) argon; 3) nitrogen; 4) helium
[5-86]; 5) air (revised values according to [5-90]).

Table 5-12. Efficiency $\eta_{\lambda_1-\lambda_2}$ of Spherical Xenon Flashlamps
(in spectral intervals $\lambda_1-\lambda_2$)

Ref.	p_0, MPa	l, mm	U_o, kV	C, µF	$\eta_{\lambda_1-\lambda_2}$, % 230–400 nm	400–700 nm	700–1000 nm	Envelope glass
5-42	0.3	2.5	3	0.27	4.5	5.9	5.5	Uviol
5-42	0.3	2.5	3	0.025	4	5.5	6.5	Uviol
5-54	0.22	2.5	1	0.068	–	3.2	1.7	S-88
5-84a	0.22	2.5	0.4	4	12.9*	2.6	1.3	Magnesium fluoride
5-54	0.22	3	5	0.01	1.4	2.9	1.7	S-52

*In the interval 220-400 nm.

5-8. The Effect of an Optical System on a Flashlamp

When a lamp is placed in some optical device that contains
optically reflective surfaces, part of the self-radiation may be
returned to the discharge column and partially absorbed by the plasma.
Part of the self-radiation that is returned to the lamp also is absorbed by the lamp envelope and the electrodes. The absorbed self-
radiation provides the additional power fed into the discharge which
causes an elevation of the plasma temperature, a corresponding increase in the luminance of the column, and overheating of the lamp
structure. This results in a reduced maximum permissible load (see
Chapter 6) and shorter operating life. The phenomenon of energy
interaction between flashlamps and their own radiation has been investigated in [5-39 and 5-91 to 5-97b].

The effect is most evident in an extremely tight reflector that closely surrounds the lamp envelope (Figure 5-58). The peak discharge current increases by 20% in a tubular lamp and by 10% in a coaxial-type lamp. The duration of the discharge-current pulse decreases somewhat. The effect is less pronounced in a coaxial-type lamp because of the significant energy losses on the additional quartz tube (additional relative to a tubular lamp). The corresponding increase in the luminance of the discharge column can be estimated by using the dependences of the plasma temperature and the absorption coefficient of the plasma on the current density (Chapter 3). For example, the increase in the spectral luminance (λ = 420 nm) observed through a narrow slit in the tight reflector (with a coefficient of diffuse reflection of 90%) reaches 60% for a power density of 0.2 x 10^6 W/cm^3 [5-97a].

It was established in [5-93] that when a tubular lamp (d_i = 4.2 mm, l = 60 mm, containing xenon or krypton) is placed in an illuminator, the radiation that emerges from it in different spectral regions has a front duration 15-20% shorter, and at the 0.35 level a duration 15% longer, than for the same lamp in open space. The differences in the time dependence of the radiation from open lamps and lamps placed in an illuminator decrease with decreasing discharge energy and gas pressure.

The radiation incident on a lamp also can change the nature of the radial dependence of the luminance and temperature of the discharge channel [5-97b]. This can be seen clearly when a tubular flashlamp is placed in a coaxial reflector [5-92].

The pattern of the energy interaction of a flashlamp with its own radiation becomes substantially more complicated if the radiant flux generated is partially extracted from a closed optical system through a spectrally selective element, such as the active medium of a laser. This case has been investigated for a coaxial-type flashlamp used for laser pumping [5-94]. Under these conditions the lamp absorbs the returning radiant flux, whose spectral distribution contains gaps in the region of the absorption bands of the laser's active element. During reradiation, the plasma acts here as a nonlinear element which converts the radiation from one spectral interval to another. Hence, in order to achieve maximum efficiency of the pumping system, we must work not only toward maximum efficiency of the flashlamp in the spectral bands where pumping is done, but also toward the maximum overall efficiency for the entire spectrum. To accomplish this, the radiation losses in the components of a laser illuminator must be minimal over the entire spectral region (even outside the pumping bands). The larger the fraction of the radiant flux that passes through the plasma a second time, the more noticeable the reradiation effect will be. This is true in systems where the plasma occupies a significant fraction of the volume of the pumping system (a coaxial-type lamp, a closely wrapped

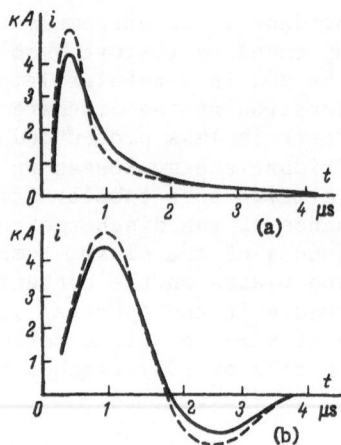

Fig. 5-58. Oscillograms of the discharge current of an IFP-5000
            tubular lamp (a) and an IFPP-7000 coaxial-type lamp
            (b) in open space (solid line), and when surrounded
            from without by a diffusely relecting magnesium oxide
            powder (dashed line), without any change in the param-
            eters of the discharge circuit [5-39].

illuminator, etc.). The effect diminishes with increasing losses in
the illuminator (losses on the reflector, losses on the filter,
losses through the end faces, etc.) and with decreasing total efficiency of the lamp.

    A theoretical calculation [5-95] shows, for example, that if a
selectively reflective coating (e.g., an interference filter which
only transmits radiation in the absorption bands of the active
medium) is applied to the envelope of a tubular flashlamp, then the
efficiency in the pump bands of yttrium-aluminium garnet can be increased by a factor of 2-3. For a xenon lamp 7 mm in diameter, the
maximum effect is achieved at plasma temperatures of 11,000-12,000
K or at electric power densities of $(100-200) \times 10^3$ kW/cm^3 in the
discharge.

    A strong energy interaction with self-radiation can be observed
for spherical flashlamps only in exceptional cases in which a significant fraction of the radiant flux is returned to the discharge
channel. A system consisting of a spherical lamp and a deep reflector which focuses the radiation on the charge channel (sic) itself is an example of such a case.

5-9.   The Stability of Flashlamp Radiation

    In many cases involving the technical application of flashlamps
it is important to know how the photometric parameters are reproduced

in different samples of lamps of a specific type and from pulse to pulse in a specific sample. The first type of nonreproducibility is of primary importance for tubular flashlamps, while the second type is the main factor for spherical flashlamps.

It is important to distinguish between two cases of lamp operation when considering the reproducibility of light parameters from pulse to pulse: single-discharge operation (infrequently repeated discharges), and high-frequency operation. In single-pulse operation each successive discharge occurs under identical initial conditions after the gas-dynamic, thermal, and ionization disturbances in the working volume have disappeared almost entirely. During high-frequency operation of a flashlamp, each successive discharge begins under varying conditions due to the gas-dynamic, thermal, turbulent, and ionization disturbances remaining from the previous discharge. Noticeable instability of the photometric parameters of successive pulses can be observed during high-frequency operation as a result of these factors. By contrast, single flashes have extremely stable emission parameters when the power-supply conditions (the voltage across the working capacitor) are reproduced exactly. This fact was noted long ago, and flashlamps accordingly have even been used as photometric devices [5-98 to 5-101].

The spread in the average values of the photometric characteristics of lamps of one type is due to the spread in their geometric parameters (discharge-tube diameter, electrode spacing, the shape and size of the near-electrode ballast volume, etc.), in the gas pressure, and in the composition of the gas that is allowed during production. The coefficient of variation of lamps with respect to luminous efficacy is about 0.09 for each type and mode of operation of series-produced tubular flashlamps [5-12]. The distribution of lamps with respect to luminous efficacy is close to a normal distribution, hence the effect of the factors may be considered random. As yet such data are not available for spherical lamps.

From a photometric standpoint, the use of flashlamps in various devices reduces to two cases. In the first, the device accepts radiation from all points of the discharge channel (it reacts to the peak luminous intensity or to the integrated luminous intensity), and in the second it accepts radiation from some fixed area that is isolated from the image of the glow volume by means of a diaphragm (it reacts to the peak luminance or to the time-integrated luminance).

Thus, the instability of one of four photometric quantities must be considered, depending on the application of the flashlamp: the peak luminous intensity I_p; the integrated luminous intensity Θ; the peak luminance L_{vp}; and the integral of the luminance pulse, $\int L_v(t)dt$.

The data of numerous measurements [5-100 to 5-103] show that the relative variation of the average value of the integrated luminous intensity $\delta\overline{\Theta}$ and the peak luminous intensity $\delta\overline{I}_p$ does not exceed 1% per 1000 flashes for a number of series-produced tubular and spherical flashlamps operating in the infrequent-flash mode, while the coefficient of variation of the photometric parameters generally does not exceed 0.02. It should be noted that the reproducibility of the photometric parameters of flashes is influenced significantly by the triggering method and the parameters of the trigger voltage of the flashlamps [5-104].

The dependences of the coefficient of variation of the peak luminous intensity V_I on the energy W of individual discharges by an ISSh-7 spherical lamp ($l = 2.5$ mm) for a different xenon pressures (Figure 5-59) show that a pressure of 0.22 MPa (which was used in the design) is optimal. They also show that V_I is practically independent of the capacitance of the supply capacitor and of the working voltage in the pressure region studied for W of over 0.1 J [5-28]. (When W < 0.1 J, a sharp increase is observed in V_I. The higher the working voltage, the sharper the increase.)

The increasing instability of the radiation from spherical lamps (unlimited discharges) with increasing flash frequency [5-28, 5-98, 5-99, 5-102, and 5-104 to 5-109] is characterized by the plot in Figure 5-60. During high-frequency operation the instability of the peak radiant intensity I_{ep} of spherical lamps is exhibited in two qualitatively different forms [5-107]: (1) comparatively small random fluctuations from flash to flash, and (2) infrequent drops of I_{ep} by an order of magnitude or more which occur for a specific sample some time after the lamp is switched on and which persist for an entire series of successive flashes. As I_{ep} decreases, the discharge channel flattens out, and regions of elevated luminance appear. The peak discharge current corresponding to these flashes increases appreciably. The neon and helium impurities reduce the depth and duration of such drops by a factor of approximately 2-3. An almost complete lack of drops is observed when hydrogen is added to xenon, but the spatial stability of the discharge channel is impaired. The combined action of the added hydrogen and the spherical gas-dynamic reflectors located on the electrodes makes it possible to ensure quite stable operation of spherical lamps at frequencies up to 4 kHz.

The instability of the luminous intensity of spherical lamps during high-frequency operation is due mainly to the change in the gas density in the electrode gap. The gas density does not manage to restore itself during the time period between flashes [5-28, 5-107, and 5-108]. The conditions for each successive discharge may be different since the density is restored through the turbulent intermixing of cold and hot gas in the electrode gap. An instability such as the falling-off described is due to the prolonged decrease

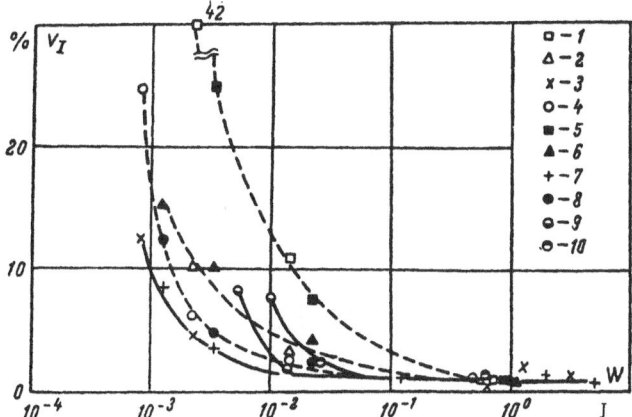

Fig. 5-59. Dependences of the coefficient of variation V_I of the peak luminous intensity of an ISSh-7 lamp on the discharge energy W in single-pulse mode for various xenon pressures p_0 and working voltages U_0 [5-28].

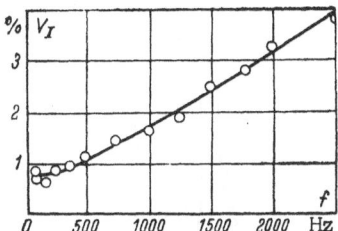

Fig. 5-60. Dependence of the coefficient of variation V_I of the peak luminous intensity of an ISShO-1 lamp on the flash repetition frequency f. Xenon, $p_0 = 0.22$ MPa, $l = 2.5$ mm, C = 6800 pF, $U_0 = 1$ kV, $I_p = 10$ kcd [5-27].

in the gas density in the discharge zone. The discharge apparently is associated with the formation in the envelope of standing sound waves, which sharply slow the process of turbulent intermixing of gas. When this occurs, the gas at the center of the envelope is heated to high temperature, its density decreases, and the pulsed discharges occurring in the highly rarefied gas have a low luminous efficacy. Such inefficient discharges are repeated until the standing wave is broken and the original gas density is restored.

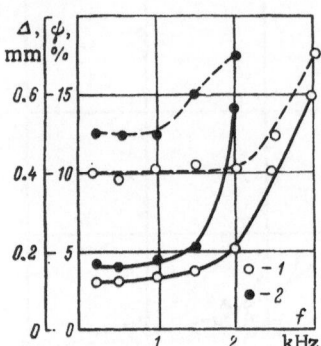

Fig. 5-61. Dependences on the frequency f of the maximum deviation
ψ of the average value of the peak radiant intensity
(———) and of the maximum deviation Δ of the discharge
channel (----) from the axis of the ISSh-7 lamp [5-28].
C = 0.0066 μF, U_0 = 1000 V. 1) series-produced samples;
2) samples containing molecular-gas impurities.

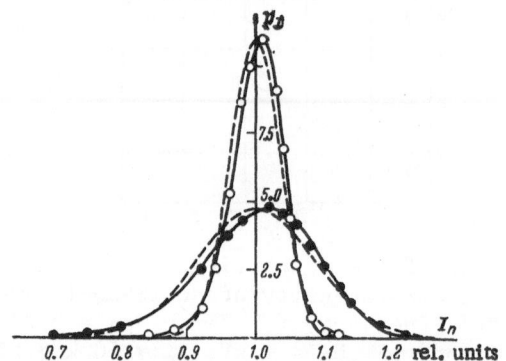

Fig. 5-62. Distribution of the probability density of the peak
luminous intensity of spherical flashlamps. The solid
circles represent values for an ISSh-7 lamp: l = 2.5 mm,
p_0 = 0.02 MPa, C = 0.007 μF, U_0 = 1000 V, τ = 0.5 μs,
f = 2 kHz. The open circles represent values for the
ISSh-7 lamp: l = 3.5 mm, p_0 = 0.2 MPa, C = 0.01 μF,
U_0 = 5000 V, τ = 1 μs, f = 3 kHz. The solid curves
represent the probability density distribution, calculated from the function $\varphi(I)$. The dashed curves represent
the probability density distribution according to the
normal law [5-108].

The stability of the peak luminous intensity is impaired substantially when a few tenths of a percent of impurities of such
molecular gases as O_2, CO, and CO_2 are added to the gas filler of a
spherical lamp [5-28]. The influence of such impurities is especially
pronounced at flash frequencies above 1 kHz (Figure 5-61). The
stability of the peak luminous intensity and location of the discharge channel take a sharp turn for the worse as the limits of the
range of controllability are approached.

The distribution of the peak luminous intensity of a series of
pulses is random, approaching the normal law for the distribution of
a random quantity (Figure 5-62). The difference between the statistical distribution of the peak luminous intensity of spherical
flashlamps operating in a high-frequency mode and a normal distribution is that the maximum is shifted toward large values of I_p
relative to the mathematical expectation. Hence, the maximum deviation from the mathematical expectation is larger in the direction
of smaller values.

The coefficient of variation of the peak luminance and of the
integrated luminance pulse generally exceeds V_I because of the
spatial instability of the discharge channel, and is dependent on
the location of the area between the electrodes that is being
measured photometrically [5-102].

CHAPTER 6

THE LOAD CHARACTERISTICS OF FLASHLAMPS

6-1. A Survey of Load Limits

Apart from the range of controllability with respect to supply
voltage, the possible variations of power-supply conditions for flashlamps are restricted by the limit of the power load on the lamp beyond
which it fails or malfunctions. A malfunction may entail damage to
the electrodes of the lamp (especially the leads), a loss of controllability, or breakage of the envelope.

The load characteristics of flashlamps include the permissible
energy W per pulse which can be dissipated in the lamp during operation
with infrequent, essentially isolated pulses with intervals between,
such that an increase in the intervals does not affect the value of
the maximum energy; and the permissible energy during operation with
pulses W_f that are repeated often with a frequency f, and the associated maximum maximum permissible average power $P_{av} = W_f f$. The values
of all these quantities may approach the corresponding limiting values
W_{lim}, W_{f-lim}, and P_{av-lim} to one extent or another. The extent to
which the limiting values are approached is determined by the required
number of pulses that the lamp can provide with the required reliability. An idea of how this extent of approach varies as a function
of the number of pulses is given by Figure 6-1. This figure illus-
trates the experimental data of various authors on the dependence of
the limiting load W_{lim} on the number of pulses (for lamps operating
in an infrequent-pulse mode) for which the lamps break with equal
probability.

Experience gained in the operation of different types of flashlamps has demonstrated that the permissible energy that ensures 1000
pulses in the infrequent-pulse mode usually is $0.6W_{lim}$ for several

223

flashes*. The value of W_{lim} decreases (by 30% in some cases [5-97])
when lamps are operated in an optical system that returns part of
the self-radiation (or that directs onto the lamp part of the radiation
from other lamps operating synchronously), and W should be correspondingly lower. The operation of lamps in a radiation-absorbing liquid
also may lead to a decrease in W_{lim} [6-5]. (A decrease in the limiting
energy that apparently is due to the external pressure of the flash-
evaporated layer of liquid adjacent to the lamp is observed even when
lamps having a pulse length of 0.5 ms are operated in distilled water.)

The lowest value of the electric energy or power input into the
lamp at which an ordinary pulse (or a short series of pulses during
stroboscopic operation) results in failure of the lamp (with a probability close to unity) due to mechanical breakdown or loss of controllability is considered the load limit. The quantity W_{lim} depends above
all on the pulse length, but P_{av-lim} also depends on the pulse frequency. All other conditions being equal, the limiting values also
depend on a number of other factors: the supply voltage and the
voltage of the trigger pulse, cooling conditions, the return of part
of the self-radiation to the lamp [5-96 and 6-6], and others.

The load limits for electrodes have no practical significance
for glass flashlamps in which metal leads of large cross-sectional
area can be used (this also is true of quartz lamps with leads sealed
into the quartz by means of glass connectors or solder). The calculation of the permissible load on a lead is highly important in the
design of quartz lamps that have thin ribbon-type leads which are
not matched in their thermal expansion coefficients (the leads are
strips of molybdenum foil 0.02-0.1 mm thick).

If the leads are selected properly, then glass and spherical
quartz lamps (in contrast to tubular quartz lamps) experience practically no failures during single-flash operation due to loss of
controllability without visible breakage of the envelope. The trigger
voltage may increase unacceptably as a result of overloading without
breakage in some types of tubular quartz lamps (those having a small
diameter and a large length).

Both failure of the envelope and loss of controllability can serve
as the criterion for failure at a high flash frequency in small tubular
lamps. The flash repetition frequency usually is low for the tubular
quartz lamps of medium and large size that are used in the optical
pumping of lasers, for example, and their working voltage U_o is high
($U_o/l \gtrsim 100$ V/cm). Under these conditions W_{f-lim} and P_{av-lim} are
related mainly to the failure of the envelope.

---

* Exceptions are some special modes of operation (see below) in which
  the spread of W_{lim} increases from 10 to 50%. Hence a value no
  higher than 0.2-0.3 the average value of W_{lim} must be selected.

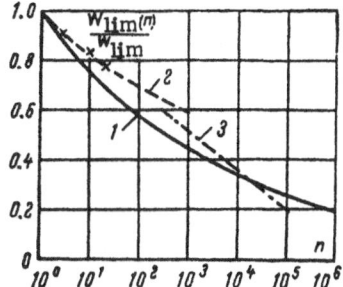

Fig. 6-1. The ratio of the energy $W_{lim}(n)$ which breaks a lamp in n
          pulses to the energy $W_{lim}(1)$ which breaks it after a single
          pulse. 1) corresponds to the function $n^{-0.117}$ [6-1 and
          6-2]; 2) the same, but for $n^{-0.08}$ [6-3]; 3) data from [6-4].

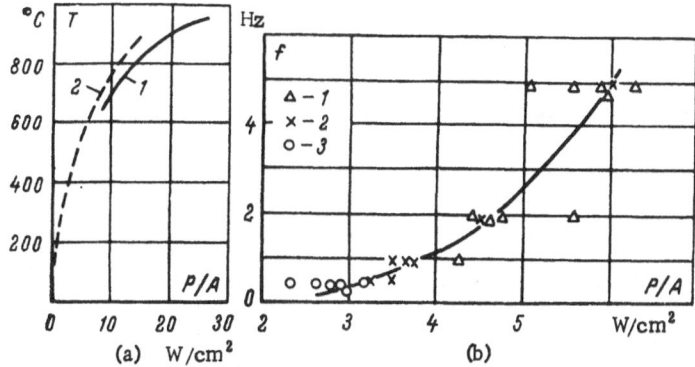

Fig. 6-2. Load characteristics of flashlamps during high-frequency
          pulsing operation.
          a) dependence of the average temperature of the outer
          surface of the discharge tube of a quartz flashlamp on
          the average power P dissipated in a lamp of average power
          P, referred to the outer surface A of the working portion
          of the discharge tube; natural cooling. 1) d_i = 23 mm,
          2) d_i = 1 mm. The experimental points for spherical lamps
          are close to curve 2.
          b) dependence of the load limit of ISSh-300 spherical glass
          flashlamps on the flash-frequency. 1) loss of control-
          lability; 2) explosion of the envelope; 3) failure of the
          trigger electrode.

      Glass lamps (spherical and tubular) often are used in modes
involving a significant pulse repetition rate which cause substantial
heating of the envelope during prolonged operation (Figure 6-2a).
The disruption of controllability, which generally precedes failure,
may occur as a result of overheating of the glass, for example, until

there is a sharp increase in the conductivity of the glass or a
chemical reaction occurs between the glass and the materials used
in the lamp. For most glasses the corresponding average temperature
is about 200°C for tubular lamps and 300-400°C for spherical lamps
(with smaller fluctuations of the instantaneous temperature about
the average values). It is 850°C for quartz (whose temperature limit
is characterized by a sharp speedup of crystallization). It follows
from Figure 6-2a that for periodic operation without forced cooling
the average power dissipated in the lamp per square centimeter of
outside surface area of the working portion of the envelope should not
exceed 0.5-1 W for tubular glass lamps and 1.5-3 W for spherical glass
lamps. (The heat losses on the walls of various lamps during stroboscopic operation may amount to 40-90% of the power dissipated in the
lamp. Therefore, the difference between the abscissas on graphs of
the type shown in Figure 6-2a may not exceed a factor of 2 for different lamps). The considerable experience gained from the development of different types of lamps confirms these standards. An example is the graph in Figure 6-2b, which also shows that the decrease
in the temperature fluctuation with increasing flash frequency
somewhat exceeds the load limit at which the lamp fails. The corresponding load limit is 10-20 W/cm^2 for quartz lamps (it apparently
is the same for spherical and tubular quartz lamps). The standards
indicated may be doubled in cases where a significant fraction of
the power is dissipated in the near-electrode region of the discharge
(e.g., in cases of extremely short tubular lamps).

A significant increase in the average load on the walls of flashlamps operating in the periodic mode is possible when forced cooling
is used. The most intensive cooling is achieved by placing the lamp
in a stream of coolant liquid. (This is true if we ignore systems
using eddy pumping of gas through the lamp [0-2 and 0.2a]). Short
quartz lamps having a working section 10-20 mm long and an outside
diameter of 5 mm can handle an average load of about 300 W/cm^2 per
square centimeter of outside surface area of the tube when they are
cooled with running water flowing at a rate of 4 m/s. The corresponding value is 150 W/cm^2 for long lamps (about 1 m long) having an
outside diameter of 2-3 mm. Combining liquid cooling with turbulent
pumping of gas makes it possible to raise these values another order
of magnitude.

The failure of envelopes during single-flash or infrequent-
flash operation of tubular lamps is linked to a number of processes.
Failure may occur through various mechanisms or combinations thereof,
depending on the material of the envelope, its size, the wall thickness, and the wall temperature, which is determined by heating due
to previous flashes. In the general case, the explosion or cracking
of a lamp envelope occurs when the dynamic stress in the glass of the
envelope exceeds the breaking stress. This may result from the
pressure of the hot gas and glass vapors during the discharge as a
result of thermal stresses in the walls or as a result of shock waves.

The pressure pulse formed in the shock front initiated by the expanding discharge channel is of primary importance for spherical lamps.
Experience shows that failure occurs irregularly in this case, with
a significant spread of the discharge parameters. This is due to
the fact that different samples of lamps of the same design have different static stresses in the glass which remain after flame treatment, different wall thicknesses and different curvature of individual
parts of the envelope walls. Hence the walls experience different
dynamic stresses due to the pressure pulse.

The lower limits of the load under which the envelope explosions
sometimes are observed are similar for spherical glass and quartz
lamps. According to a rough estimate, explosions of lamps having a
"cold" gas pressure of about 0.3 MPa are observed when a specific
pulse energy for a given envelope having a given diameter and wall
thickness is exceeded (the other structural elements and power-
supply parameters obviously are of secondary importance).* These
limiting energies are 70, 150, and 300 J for envelopes having a wall
thickness of 2-3 mm and outside diameters of 30, 45, and 60 mm, respectively. Doubling the gas pressure approximately halves the limiting energy. The limiting energy is reduced by a factor of about 2-3
if the wall thickness is reduced to 1-1.5 mm and we go from a spherical
to a hemispherical or cylindrical envelope.

Spherical lamps are used quite often for short series of pulses
having a total duration ranging from fractions of a second to several
seconds. In such regimes the value of the permissible envelope
temperature is practically independent of the cooling conditions. It
is determined mainly by the heat capacity of the envelope. This fact
makes it possible to increase the permissable average power for a series
significantly (severalfold) over the average power during prolonged operation.

It would be advisable to supplement our brief review of the types
of failures of tubular and spherical lamps having glass or quartz
envelopes during single-flash and infrequent-flash operation with
more detailed data concerning the load characteristics that are of
the greatest practical importance, specifically:

(a)   the load limits on the leads of tubular and spherical lamps
      (during single-flash and infrequent-flash operation);

(b)   the load limits on tubular lamps during single-flash operation.

---

* This relationship is an indirect confirmation of the fact that
  failure is due to the shock wave, since the shock velocity also
  depends mainly on the pulse energy, as we demonstrated in Chapter 2.

6-2. The Load Limits on Leads

The failure of metal leads is due either to the instantaneous liberation of the critical energy per unit volume of the metal (this is similar to the electrical explosion of a wire), or to the prolonged overheating of the metal and oxidation of the portion of the metal in contact with the air when the lamp is operated at the critical average power. Data on the critical energy [6-7] have been used to obtain the following relation among the maximum permissible energy of a single pulse $CU_0^2/2$ (J), the cross section A of a molybdenum lead (cm^2), and the effective resistance R of the discharge (ohms):

$$A = \sqrt{\frac{CU_0^2/2}{\xi_{cr} R} + \frac{\rho_M^2 l_M^2}{4R^2} - \frac{\rho_M l_M}{2R}} \,, \tag{6-1}$$

where ξ_{cr} is the ratio of the critical energy released per unit volume of the molybdenum to the effective resistivity ρ_M and l_M is the total length of the leads (cm).

Experiment gives values ranging from 4×10^7 to 23×10^7 J/(Ω-cm^4) for the ξ_{cr} at which the leads fail under various conditions (a different structure of the metal, a nonuniform cross section, different sizes of the lamps and leads, different power-supply conditions).

An expression based on the concept of a quasi-stationary discharge (Chapter 3) may be used to estimate R for tubular lamps:

$$R = \rho l / \pi r^2, \tag{6-2}$$

where l is the electrode spacing, r is the inside radius of the tube, and ρ is the resistivity of the discharge plasma. (The value of the discharge plasma resistivity may be set at approximately 0.02 Ω-cm. The data on the channel radius that are given in Chapter 2 may be used for more exact calculations in cases where the discharge channel falls noticeably short of filling the cross section of the tube.) For tubular lamps, the effective resistance of the channel can be estimated by using formulas (2-1) to estimate the plasma resistivity and (2-31) to estimate the maximum column radius (we assume that the effective resistance of the channel is determined by the radius of the column, which is equal to θr_{max}, where θ is a proper fraction:

$$R = \frac{40 l^{5/3}}{\theta^2 U_0^{2/3} (CU_0^2/2)^{0.8}} \,. \tag{6-3}$$

The fraction θ can be determined experimentally by using expressions (6-1) and (6-3) for spherical lamps, for which the luminous efficacy at pulse energies close to the limiting energy has a normal value and the term $\rho_M l_M/2R$ in (6-1), which is related to the energy loss in the leads, thus may be ignored.

Experiment shows that the pulse energy which can cause failure of a lead exceeds 200 J for spherical xenon lamps with a xenon pressure of 0.3–0.4 MPa, l = 0.8 cm, U_0 = 6 kV and A = 1.2 x 10^{-2} cm^2. The luminous efficacy of such lamps usually is 15 lm-s/J at an energy of 200 J. Inserting the values indicated and the calculated value ξ_{cr} = 2 x 10^7 J/(Ω-cm^4) for tubular lamps into (6-3) and into simplified expression (6-1), we find that θ is equal to 0.42. This value can be used for calculations (with some margin) of the minimum cross section of a lead that is required under analogous conditions. These conditions thus can be fulfilled by using the approximation formula

$$A = 10^{-5} \left(C U_0^2 / 2 \right)^{0.9} U_0^{1/3} \, l^{-5/6}, \qquad (6\text{-}4)$$

where C is given in farads, U_0 in volts, l in centimeters, and A in square centimeters.

The value obtained for θ and the fact that the lead resistance has practically no effect on the luminous efficacy under the experimental conditions indicated enable us to adopt 10^{-4} Ω-cm for the upper limit on the possible values of the effective resisitivity ρ_M (according to reference data, ρ_M = 0.3 x 10^{-4} Ω-cm for molybdenum at 1200°C). It follows from this estimate that a reliable design of a lead can be made from formula (6-4) if the second term under the radical turns out to be equal to or less than the first when the value ρ_M = 10^{-4} Ω-cm, which is known to be too high, is inserted into (6-1) under certain experimental conditions. The cross section obtained for the lead is sufficient to ensure that the resistance of the lead will have practically no affect on the luminous efficacy of the lamp.

Analogous methods can be used to design the leads of lamps intended to operate in the frequent-pulse mode, but the pulse energy $C U_0^2 / 2$ must be replaced by the average power P_{av} that is dissipated in the lamp, and the constant ξ_{cr} must be replaced by the square of the maximum permissible density of the direct current j_{cr} through the molybdenum lead. We know from data obtained from the development of a number of types of quartz lamps that the operating life of a lead is on the order of 1000 hr. Such a life can be ensured if the effective density of the alternating current in the lead does not exceed j_{eff} ≈ 5 x 10^4 A/cm^2.* Therefore, j_{cr}^2 = $2 j_{eff}^2$ ≈ 5 x 10^9 A^2/cm^4, and

$$A = 1.4 \cdot 10^5 \sqrt{P_{av}/R}. \qquad (6\text{-}5)$$

---

* These data apply to lamps operating without forced cooling. If the lead is cooled intensively (with water, for example), then j_{cr} has a much higher value.

The resistance R of a lamp may be assumed to be equal to $\rho l/\pi r^2$ for tubular lamps (if the tube is filled by the discharge channel), but R is calculated from formula (6-3) for spherical lamps.

It should be noted that the values of A calculated from formula (6-5) by using values of P_{av} that are close to the limiting power lamp envelopes always turn out to be smaller than those calculated from formulas (6-1) or (6-4) when values of the pulse energy that also are comparable to the limit that the lamp envelope can withstand are inserted in the formulas. Therefore, formula (6-5) may prove useful only in special cases of lamp design in which the lamp is known to be intended only for frequent-pulse operation, and it is advisable to use a lead with a minimal cross section because of manufacturing considerations.

6-3. The Load Limits of Tubular Lamps During Single-Flash Operation

The load limit for single-pulse operation usually is determined from successive series of a specified number of pulses (hereafter, 10, unless otherwise specified). The pulse energy of each successive series is increased by a few percent over the previous series. The intervals between pulses are made quite long (usually a few minutes) so that the lamp has time to cool down before the next discharge.

The pulse energy at which the lamp fails is taken as the limiting value W_{lim}.

In a glass flashlamp a network of annular hairline cracks appears on the inside surface of the discharge tube as the pulse energy increase to some value that is well-defined (to an accuracy of 5-10%) for lamps of a given design and for a given pulse length. If the overloads are small, these cracks are so slight and infrequent that they can be seen only with difficulty (only by careful inspection of a lamp illuminated with an intense light source). The cracks become more noticeable and denser as the supply voltage is increased further. At length, the development of the cracks leads to loss of the vacuum tightness of the envelope and to complete failure of the lamp. (If the cracks are slight, then the lamp may retain its performance, for example, at an overload equal to 20-30% the minimum energy at which the first isolated cracks appear, and may do so for thousands of pulses.)

An analogous development of hairline cracks occasionally is observed when quartz lamps having an inside diameter of over 0.5 cm are overloaded. For quartz lamps, the probability that the envelope

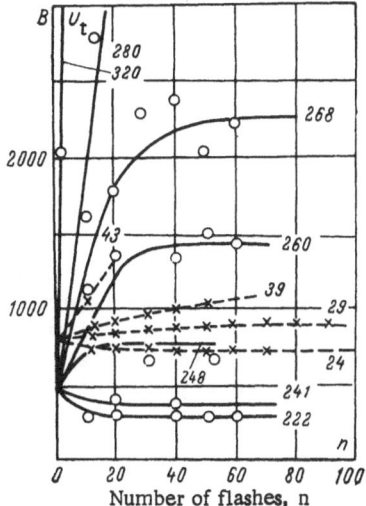

Number of flashes, n

Fig. 6-3. The nature of the dependence of the trigger voltage U_t of
quartz capillary lamps on the number n of pulses for the
various energies indicated in the figure (J) [1-68 and
6-7]. The solid lines are for l = 50 cm, d = 2 mm, Kr,
0.013 MPa, C = 60 µF, with a 1.5-s interval. The dashed
line is for l = 7 cm, d_i = 0.53 mm, Xe, 0.08 MPa, C =
40 µF, with a 2-s interval. In both cases the lamps were
soldered onto the exhaust unit, and the heating due to a
series of comparatively frequent pulses should have lowered
the gas density significantly.

will explode increases roughly simultaneously with the development
of cracks.*

As the pulse energy gradually increases, the trigger voltage
eventually begins to increase in quartz lamps of smaller diameter as
a result of the increase in voltage across the capacitor. (This
phenomenon also is observed in a less-pronounced form in lamps with
wide tubes.) The nature of the variation of the trigger voltage of
such lamps over a series of several tens of pulses with a pulse
energy that is constant for each series is shown in Figure 6-3. In
this figure the number n of pulses having a given energy is plotted
along the abscissa. Then the next measurement of the trigger voltage
is made. (The trigger voltage of a lamp that has not been severely
overloaded can be lowered by briefly aging the lamp with discharges

* A white film of vaporized and condensed silicon monoxide (or sometimes a brown film of metallic silicon) often appears at energies
   equal to 60-90% the limiting energy in quartz lamps (especially in
   the cold zones: near the electrodes and in the stem). Hairline
   cracks sometimes are observed in quartz lamps long after the pulses
   are halted.

of reduced energy. The trigger voltage partially decreases if the
lamp is not operated for a few tens of minutes.) The conditions under
which an increase in U_t changes from reversible to irreversible (pulse
energies of 320 and 43 J for the cases shown in Figure 6-3) usually
are the limiting conditions for small-diameter quartz lamps that have
a comparatively low supply voltage ($U_0/l \approx 60$ V/cm) and low-power
triggering.

Some data on the limiting loads for glass and quartz lamps
operating in a low-inductance circuit (1 μH, $R_L \gg 2\sqrt{L/C}$) are given
in Figure 6-4. (The data for quartz lamps correspond to comparatively
long flash durations of 0.3-20 ms.) It follows from the figure that
the average values of the load limits for large-diameter quartz lamps
which are associated with different kinds of failure practically match.
(The limit associated with an increase in U_t for small-diameter lamps
with low-power triggering may be an order of magnitude below the
limit associated with an explosion. The first hairline cracks appear
in the lamps at energies that lie between these two limits.)

Since the slope of the dependence of the limiting energy, referred
to the length l of the discharge section, on Cl is 0.5 on a log scale,
the following expression is valid for both quartz and glass lamps:

$$(CU_0^4/l^3)_{\lim} = K, \tag{6-6}$$

where K is a constant that depends mainly on the tube material, and
to a lesser extent on the type of gas and the gas pressure. (Replacing
xenon with krypton reduces K by 30% for thin lamps, whose efficiency
η is strongly dependent on these parameters. Lowering the pressure
from 35,000 to 2500 Pa (from 250 to 20 mm Hg) decreases K by a factor
of 3. The indirect dependence of K on the diameter of the tube also
is related to this fact.) The quantity K has a very weak dependence
on the thickness of the tube walls, the interval between flashes, and
the cooling conditions. Ref. [6-7] gives a value $K \approx 5$ μF-kV^4-cm^{-3}
for quartz lamps having a high efficiency and sufficiently powerful
triggering. The load on such lamps is limited by explosion.*

---

* The use of quartz flashlamps for the optical pumping of lasers
  usually requires that the pulse energy and power be increased as
  much as possible with a low probability of explosion of the tube.
  At the same time. the requirements for expanding the range of controllability (a decrease in U_t) are eased by the use of relatively
  high supply voltages and powerful triggering during optical pumping.

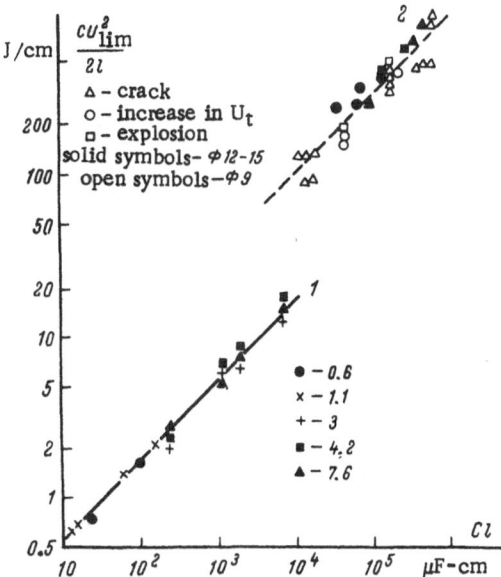

Fig. 6-4. Dependence of the limiting loads per centimeter length of the tube on Cl for glass (1) and quartz (2) xenon lamps with the d_i (mm) indicated in the figure. The appearance of cracks is the criterion for the load limit for glass lamps; the criteria for quartz lamps are indicated in the figure [1-67 and 6-7].

Thus, the experimental data indicate that the limiting loading conditions for a given lamp with a low-inductance discharge circuit are determined by the constant CU_0^4, which is the characteristic load parameter. This parameter sometimes is called the "load factor."

In view of the fact that the relation $\tau \approx RC/2 = \rho Cl/2\pi r_i^2$ (Section 5-3) is fulfilled for a prolonged flash with $\tau \geq 300$ μs when the inductance of the circuit is low, we can derive the following relation from (6-6) by transforming CU_0^4 into

$$\frac{4W^2}{C} = \frac{4W^2 \rho l}{2\pi r_i^2 \tau},$$

$$W_{\lim} = 1000 \sqrt{\frac{\pi K}{2\rho}} r_i l \sqrt{\tau} \, ,$$

If we insert the value of K indicated above for quartz lamps, and the value $\rho = 0.02$ Ω-cm, then this expression assumes the form

$$W_{\lim} \approx 20\,000 \, r_i l \sqrt{\tau}. \qquad (6\text{-}7)$$

In this form, the dependence of the limiting energy on the power-supply parameters can be applied not only to a low-inductance circuit, but also to a circuit having supercritical inductance.

As will be evident from what follows, the proportionality implied by relation (6-7) between the load limit and the product of the square root of the pulse length by the inside surface area of the tube is confirmed by extensive experimental data over a broad range of conditions. The relation of $W_{\lim}$ to the parameters of tubular lamps makes it possible to outline some regularities which are analogous to formula (6-7) for "coaxial-type" lamps, in which the discharge occurs between the walls of coaxial tubes [5-16], and for lamps with a discharge cavity of rectangular cross section [5-23]. To judge by the limited available data [6-3], experiment is in satisfactory agreement with the calculated load limits for such lamps when formulas similar to (6-7) are used with correction factors that are not too far from unity, specifically:

for coaxial-type lamps,

$$W_{\lim} \approx 20\,000 \, (r_{\text{in}} + r_0) \sqrt{\frac{r_{\text{in}} - r_0}{r_{\text{in}}}} l \sqrt{\tau} \, , \qquad (6\text{-}8)$$

where r_{in} and r_0 are the inside and outside radii of the gas-discharge annulus (cm):

for lamps of rectangular cross section,

$$W_{\lim} \approx 6500 \, (b + d) \sqrt[4]{d/b} \, l \sqrt{\tau}, \qquad (6\text{-}9)$$

where b and d are the width and thickness of the gas-discharge cavity (cm).

Expressing the effective discharge power P_{eff} in the form $P_{\text{eff}} = U_0^2/2R = U_0^2 \pi r_i^2/2\rho l$ and using expression (2-1) to estimate ρ, we can employ relation (6-7) to obtain the formula

$$(P_{\text{eff}}/2\pi r_i l)_{\text{lim}} \approx 1260\, r_i\, (U_0/l)^{1/4}\, \tau^{-1/2}, \tag{6-10}$$

which has been confirmed experimentally for pulse durations of 0.3-4 s when tubular quartz xenon lamps briefly connected to a high-power AC line [0-6].

Finally, the concepts of a lamp failure mechanism related to vaporization of the walls (see below), which are consistent with experimental formula (6-7) for single flashes, have made it possible to derive a formula for the limiting average power $P_{\text{av-lim}}$ of lamps operating with a pulse frequency f (with forced water cooling) [3-47]. The limit is expressed in terms of the limiting energy W_{lim} for single pulses:

$$P_{\text{av-lim}} \approx \frac{0.85 W_{\text{lim}} f}{1 + 100\,(1 - \eta_e)\, r_i \ln\!\left(\dfrac{r_i + \delta}{r_i}\right) f \sqrt{\tau}}, \tag{6-11}$$

where W_{lim} is given in joules, f in hertz, τ in seconds, and r_i and the wall thickness δ in centimeters (η_e is the radiant efficacy of the lamp).

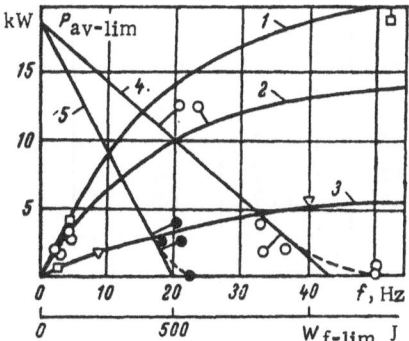

Fig. 6-5. Dependence of the limiting average power of a water-cooled tubular flashlamp on the pulse frequency f(1-3) and on the limiting flash energy at a given frequency $W_{\text{f-lim}} = P_{\text{av-lim}}/f$ (4 and 5). The solid lines correspond to relation (6-11), and the dashed lines to the segments where the computational formulas are invalid. 1) $r_i = 0.35$ cm, $l = 12$ cm, $\tau = 400$ µs; 2 and 4) $r_i = 0.35$ cm, $l = 8$ cm, $\tau = 500$ µs; 3) $r_i = 0.25$, $l = 4$ cm, $\tau = 150$ µs; 5) $r_i = 0.35$ cm, $l = 8$ cm, $\tau = 100$ µs.

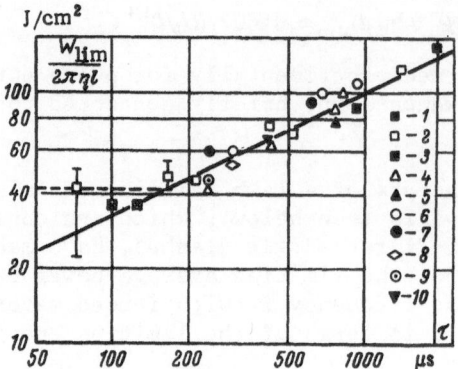

Fig. 6-6. Load limits per square centimeter of inside surface area
of various quartz lamps versus flash duration. The line
represents the graph of relation (6-7) [3-47, 5-16, and
6-8].

	r_i, cm	δ, cm	p_o, MPa	l, cm	Type of lamp
1	0.25	0.1	0.053	4-4.5	ISP-1000, ISK-25
2	0.35	0.15	0.053	8, 12	IFP-800, IFP-1200
3	0.35	0.15	0.007	8	IFP-800
4	0.57	0.2	0.04	3, 25	IFP-2000, IFP-5000
5	0.57	0.2	0.007	25	IFP-5000
6	0.78	0.15	0.04	13	Experimental
7	0.78	0.15	0.007	13	"
8	1.3	0.15	0.007	30	"
9	1.7	0.15	0.007	50	"
10	4.7	0.25	0.007	100	"

As we see from Figure 6-5, this expression is confirmed experimentally for small lamps with an optimal coolant-water flow rate.

It can be seen from a comparison of the experimental points
presented in Figures 6-6 to 6-9 with the graphs of relation (6-7)
that significant deviations are observed in addition to the good
agreement of the experimental results with this formula for large
and short pulses. These deviations run both toward an increase in
the limiting energy for lamps with a greater wall thickness and lower
gas pressure (such an increase seems to indicate an anomalously "mild"
effect of the discharge on the tube), and toward a decrease in the
limiting energy for lamps with normal walls and pressure (this decrease
indicates an anomalously "severe" effect).

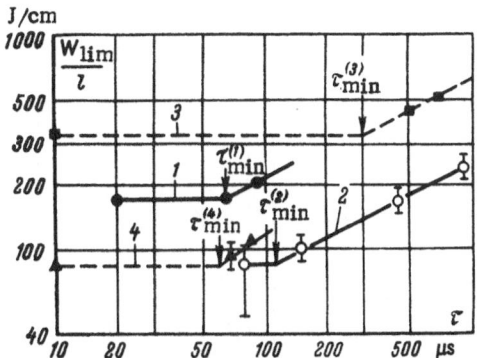

Fig. 6-7. Dependence of load limits per centimeter tube length on
flash duration for lamps with a reduced xenon pressure or
an increased wall thickness. The sloped lines correspond
to relation (6-7) [6-3, 6-4, and 6-9]. 1) r_i = 0.8 cm,
δ/r_i = 0.19, p_0 = 0.007 MPa; 2) 0.35 cm, 0.43, 0.04 MPa;
3) 0.95 cm, 0.32, 0.04 MPa; 4) 0.52 cm, 0.47, 0.04 MPa.

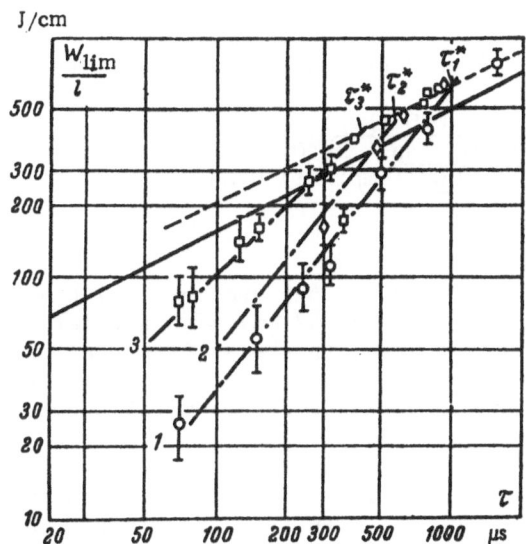

Fig. 6-8. Dependences of limiting loads per centimeter tube length
on flash duration for large lamps with the usual xenon
pressure and wall thickness. The solid line represents the
calculated graph for relation (6-7). The dashed and dot-
dash lines represent the empirical relation for a reduced
number of flashes in a series of pulses (3 instead of 10)
[2-36]. Points τ_1^* to τ_3^* are the boundaries of the transition to the "hard" regime, with r_i = 0.78 cm, δ/r_i = 0.19,
p_0 = 0.04 MPa. 1) l = 58 cm (an IFP-2000 lamp with a foil
lead); 2) l = 87 cm (an IFP-25000 with a cap-type lead);
3) l = 25 cm (an IFP-8000 with a foil lead).

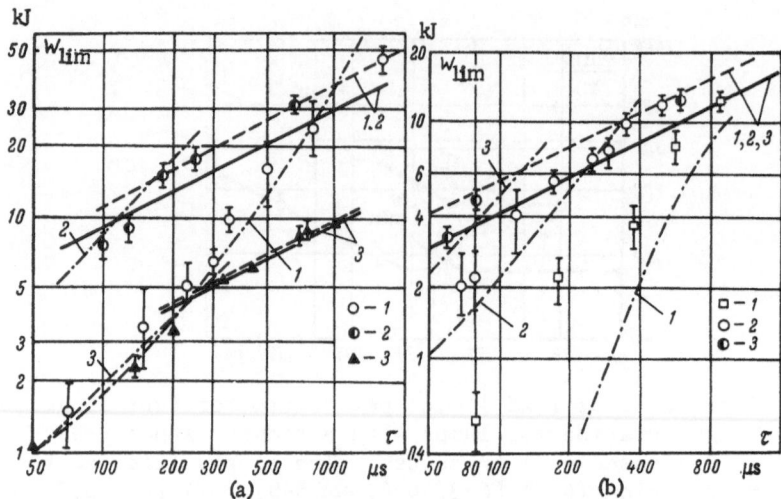

Fig. 6-9.   Dependences of limiting loads on pulse length for different
            xenon pressures. The solid lines represent the calculated
            curves of relation (6-7). The dashed lines represent the
            experimental curves for normal operating conditions. The
            dot-dash lines represent the curves of relation (6-25)
            [2-36].
            a) the IFP-20000 lamp: 1) 0.04 MPa;   2) 0.013 MPa;
            the IFP-5000 lamp: 3) 0.04 MPa.
            b) the IFP-8000 lamp: 1) 0.08 MPa;   2) 0.04 MPa;
            3) 0.027 MPa

It can be seen from Figures 6-8 and 6-9, which give data for
large lamps (r_i = 0.8 cm, l = 25-87 cm) with a relatively low envelope
strength (the ratio of the wall thickness δ to the radius r is about
0.2), that the value of W_{lim} begins to decrease as τ to the first
power, rather than as $\sqrt{\tau}$, when the duration falls below some τ^* that
is specific to each type of lamp. The limiting pulse power W_{lim}/τ
(Figure 6-10), which hitherto has increased in inverse proportion to
$\sqrt{\tau}$ according to relation (6-7), remains constant in this region
with decreasing τ. It can be seen from the figures that the spread
of W_{lim} for a given duration increases from 10-15% (the average spread
of W_{lim} for large τ) to 50% in the region $\tau < \tau^*$. The boundary of the
"hard" regimes of τ^* depends on the design features of the lamps (the
type of leads, the size of the transelectrode volumes, the radius r_i,
the wall thickness δ, and the initial pressure p_0). For example (see
Figure 6-8), for an inside radius of 0.8 cm, $\delta/r_i \approx 0.19$, and p =
0.04 MPa, $\tau^* \approx 1000$ μs for IFP-20000 lamps (l = 58 cm, the diameter
of the foil lead is 18 mm), 600 μs for IFP-25000 lamps (l = 87 cm,
with a cap-type lead), and 350 μs for IFP-8000 lamps (l = 25 cm, with
a foil lead 12 mm in diameter). The boundary of the "hard" regimes
is shifted significantly toward lower ($\tau^* \approx 250$ μs) for the IFP-5000
lamp (r_j = 0.57 cm, δ/r_i = 0.34, l = 25 cm, and the diameter of the

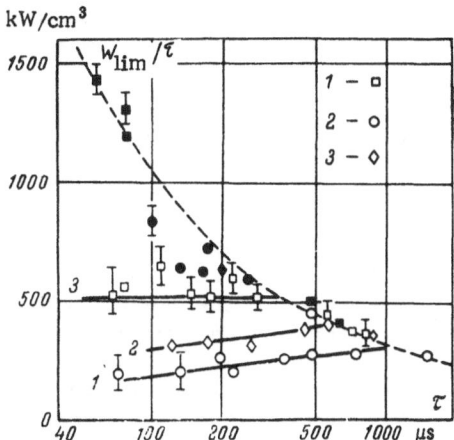

Fig. 6-10. Dependences of the limiting pulse power per cubic centimeter of plasma on the pulse length for lamps with r_i = 0.78 cm and a xenon pressure of 0.04 MPa (open symbols) and with a mixture of 0.013–0.027 MPa xenon + 400–670 Pa nitrogen (solid symbols). 1) IFP-8000 lamp; 2) IFP-20000 lamp; 3) IFP-25000 lamp. The solid lines represent "hard" conditions. The dashed line represents normal conditions.

foil lead is 12 mm), which has a smaller diameter and thicker walls than the IFP-8000. The boundary of the regimes shifts from τ^* = 350 μs to τ^* = 150 μs (6-9b) for IFP-8000 lamps with a reduced initial xenon pressure (0.026 MPa, instead of the usual 0.04 MPa).

The characteristic phenomena observed upon the explosion of an envelope reduce to the following for each of the range of conditions mentioned.

The explosion of an envelope is characterized by the following features in the range of <u>normal</u> conditions in which relation (6-7) is valid:

(a) the experimental values of W_{lim} have a small spread (10%);

(b) W_{lim} is weakly dependent on the wall thickness and on the initial pressure;

(c) at pulse energies close to W_{lim}, each flash leads to an increase in the trigger voltage (see Figure 6-3), a quartz film is formed on the walls, and the absorption lines of silicon appear in the emission spectrum of the lamps [5-11].

(d) one or two flashes before explosion (when the number of pulses in the series is n ≥ 10); the end faces of the electrodes darken and partially melt; the trigger voltage increases (sometimes several fold); and in some types of foil lamps, depressions whose formation requires a force equivalent to a pressure of 10 MPa appear on foil sections that are not confined by quartz when the pulse lengths are long (10^{-3} s) (at the same time the foil burns through more often);

(e) explosion of the envelope leads to the formation of a multiplicity of small fragments (no larger than 1 mm in size); the explosion generally occurs on the trailing edge of the pulse and is accompanied by marked distortions of the normal behaviorof oscillograms of the luminous intensity and discharge current.

The same phenomena are seen to accompany the explosion in the region of anomalously "soft" conditions (W_{lim} does not decrease with decreasing τ when the flashes are short) that is characteristic of stronger lamps with a reduced xenon pressure, but the experimental spread of the values of W_{lim} increases significantly.

Finally, in the region of "hard" conditions (W_{lim} proportional to τ):

(a) the spread of the experimental values of W_{lim} reaches 50%;

(b) the transition to the "hard" regime is determined by the aggregate of the design parameters of the lamps (r_i, δ, p_0, l, etc.);

(c) characteristic features are the absence of a film of vaporized quartz on the walls, the large size of the fragments from the explosion (a few centimeters), and generally the absence of distortions, accompanying the explosion, of oscillograms of the luminous intensity and discharge current.

(d) the effect on W_{lim} of an increase in the transelectrode volumes of the lamps is more noticeable.

The existence of regions having different types of dependences of W_{lim} on τ indicates the existence of various mechanisms of envelope failure.

6-4. Possible Mechanisms of Failure in Tubular Lamps

The following processes which accompany a discharge generally may be involved in the failure of the envelopes of tubular flashlamps:

(a) rapid heating of the thin surface layer of glass (mainly due to absorption of the plasma radiation, which is shortwave in the early stages, but also longwave as the quartz heats up and its transparency limit shifts [3-33a, 5-11, 5-49, 5-51 and 5-52]);

(b) the action of the shock waves formed during expansion of the discharge channel;

(c) the effect of the gas-kinetic pressure pulse (the gas-kinetic
pressure depends on the size of the plasma volume and the initial
gas pressure, which determine the particle concentration in the
working zone of the lamp, and on the operating conditions, which
determine the plasma temperature);

(d) the sputtering of material from the electrodes and walls. In
addition to increasing the pressure in the envelope, sputtering
initiates a number of chemical processes (such as the dissociation
of oxides) which ultimately lead to a decrease in the transparency
of the walls and to an increase in the heating of the walls.

The following mechanisms have been proposed in the literature
to explain the causes of failure:

(1) thermal failure due to the appearance of compressive stresses that
exceed the breaking stress of the material [6-10]. These compressive stresses appear in the thin layer (less than 0.1 mm thick)
near the inside surface of the tube. The stresses (tensile stresses
in this case) also may be related to the short-duration melting and
subsequent rapid hardening of a thin layer of glass (about 30 μm
thick) [6-11] (the appearance of cracks at the depth of the molten
layer contributes to hardening of the tube, since there are no
stresses under the layer);

(2) the appearance of cracks on the inside surface of the tube due
to the splashing of drops of liquid metal as the electrodes
erode [6-12];

(3) the action of the shock waves formed during development of the
discharge [2-36, 6-3, 6-10, and 6-13];

(4) the increase in internal pressure in the lamp due to vaporization
of the wall material [2-37, 6-3, and 6-14] ([6-15] lists another
mechanism of envelope failure due to the recoil impulse resulting
from the intense vaporization of the material, but an estimate of
the analogous effect on a material vaporized by laser light [6-16]
gives radiant fluxes much larger than those which suffice for the
pressure-failure mechanism to operate);

(5) the effect of the pressure pulse of the working gas as its temperature rises during the discharge [2-36, 2-37, 6-4, and 6-14].

The thermal failure mechanism takes into account the short-duration
effect of a pulsed heat flux on the inside surface of a tube filled
with a material having low thermal conductivity. The thermal shock
that appears in the thin layer around the inside surface of the tube

produces significant compressive stresses σ_c with relatively low
tensile stresses in the outside layers of the tube. Failure should
occur when $\sigma_c = \sigma_b$ (where σ_b is the breaking stress). The limiting
energy should be slightly dependent on the wall thickness, and should
be determined mainly by the properties of the envelope material [6-10]:

$$W_{\lim} = \frac{k}{1-\eta_e} \sqrt{\pi} \; \frac{\sigma_b(1-\nu)\lambda}{\alpha_{ex} E \sqrt{a}} V \sqrt{\tau},$$

$$(6-12)$$

where k is a coefficient close to unity that depends on the pulse
shape; $(1 - \eta_e)$ is the fraction of energy that is converted to heat;
ν is the Poisson ratio; λ is the thermal conductivity coefficient;
a is the thermal diffusivity coefficient; α_{ex} is the thermal expansion
coefficient of the glass; E is the modulus of elasticity; and V is
the inside volume of the lamp.

When the constants for quartz glass are inserted in expression
(6-12), it assumes the form

$$W_{\lim} = 8.3 \cdot 10^3 \; \frac{k}{1-\eta_e} r_i l \sqrt{\tau}.$$

$$(6-13)$$

If we conditionally assume $\eta_e = 0.7$ and $k = 0.7$, for example,
then expression (6-13) matches expression (6-7).

However, thermal theory does not explain a number of phenomena
accompanying an explosion which indicate that the increase in gas
pressure in the lamp has a significant effect (deformation of the foil
in the leads, small glass fragments, etc.). Nor does it explain the
transitions to anomalously "soft" and "hard" regimes. Therefore,
thermal theory may explain correctly only the mechanism of formation
of hairline cracks in flashlamp tubes. Such cracks are particularly
characteristic of glass lamps.

The explanation of the explosion of lamps in terms of the effect
of electrode erosion was based on the observation that star-shaped
cracking which gives rise to a system of hairline cracks occurs in
points where drops of molten metal strike the surface of the quartz
tube of a flashlamp (in experimental lamps using inactivated tungsten
electrodes). However, this phenomenon is characteristic mainly of
modes of operation that have a very long duration (10^{-2} s). Heavy
erosion of the electrodes (with the formation of drops of molten
metal) generally is not observed when $\tau \approx 10^{-3}$ s in series-produced
lamps using a cathode material that has a low work function.

The mechanism of explosion due to shock waves appears at first
glance to fit the dependence of $W_{\lim}$ on the parameters of the discharge
circuit (L, U_o) and on the design characteristics of the lamps (r_i,
δ, l, p_o) that is observed when envelopes fail in "hard" regimes.

Using formula (2-4), which relates the pressure p_f in the shock front to the shock velocity and the characteristics of the gas, inserting the value of the rate of channel expansion from relation (2-27) into (2-4), and assuming $\alpha = 0.3$ for xenon, we can estimate the maximum value of p_f for comparison with the bursting pressure p_b of the lamp:

$$p_f \approx 3.2 \left(\frac{p_0}{L}\right)^{0.5} U_0^{0.64}/l^{0.4} \,, \tag{6-14}$$

where p_0 and p_f are given in megapascals, L in microhenrys, U_0 in kilovolts, and l in centimeters.

It follows from (6-14) that the criterion for rupture of the tube $p_f \geq p_b$, is fulfilled only for very high initial voltages and short pulse lengths. For example, for a lamp with $p_b \approx 2$ MPa (58 cm long, with $r_i = 0.8$ cm and $\delta/r_i = 0.19$; see Table 6-1 below), an explosion of the envelope due to the shock wave may be anticipated according to formula (6-14) when $U_0 > 100$ kV on the assumption that $p_0 = 0.04$ MPa and $L = 2$ µH.

The shock velocity does not exceed 5×10^4 cm/s in regimes with $\tau = 500$–1000 µs [2-30], and the maximum value of p_f is known to lie below 0.5 MPa for the gas density corresponding to an initial pressure of 0.04 MPa (300 mm Hg).

The mechanism of lamp failure due to the pressure of the hot vapor of the wall material may be considered on the basis of concepts of the vaporization of an opaque material due to a laser light pulse [6-16]. According to these concepts, when the radiant power absorbed per unit surface area reaches the critical value

$$P^* \approx \delta_c \Omega \sqrt{a/\tau_0} \tag{6-15}$$

(where δ_c is the density, Ω the specific energy of sublimation, and a the thermal diffusivity of the substance), practically all of the radiation incident on the surface after time $t = \tau_0$ begins to go to vaporization.

Thus, providing a set of values of P and τ that correspond to formula (6-15) is a criterion for the start of a rapid increase in the vapor density of the irradiated material. For a discharge in a quartz lamp, this increase may be linked to the explosion of the envelope by vapor pressure. In this case the energy released in a discharge that leads to explosion should be (for a pulse length τ_0)

$$W^* = \frac{\delta_c \Omega \sqrt{a}}{1 - \eta_e'} \, 2\pi r_i l \sqrt{\tau_0} \tag{6-16}$$

$((1 - \eta_e')$ is the fraction of energy absorbed by the inside surface of the envelope; this fraction approaches $(1 - \eta_e)$, where η_e is the radiant efficiency of the lamp, but it is not equal to this quantity because of the possible delay of part of the thermal losses in the discharge).

After the actual values η_e' = 0.5-0.8 and the constants of the material are inserted in relation (6-16), the estimate assumes the form

$$W^* = (33\,000 - 13\,000)\, r_i l \sqrt{\tau_0}. \qquad (6\text{-}17)$$

This relation is close to formula (6-7) in terms of the nature of the dependence on $r_i l$ and τ_0 and the value of the numerical coefficient. Formally, relation (6-17) defines the amount of energy whose release in a discharge should just initiate intense vaporization. However, a number of processes such as sputtering of finely divided, readily vaporized silicon monoxide onto the tube walls and a decrease in the transparency of quartz when heated* may substantially shorten the time required for the onset of the critical regime. (Let us recall that an explosion usually is observed on the trailing edge of a radiation pulse with a delay relative to the time when the maximum power of the radiation pulse is reached.) Because of these processes, a significant fraction θ of the energy W* released in the discharge may turn out to have been expended on vaporization by time τ_0.

The mass of the vaporized quartz when the amount of energy that has gone into its vaporization is $\theta(1 - \eta_e')W^*$ is equal to

$$M = \frac{\theta\left(1 - \eta_e'\right) W^*}{\Omega}. \qquad (6\text{-}18)$$

Hence we can estimate the pressure of the vaporized material when $W^* = W_{lim}$, assuming for example that $1 - \eta_e' = 0.3$, $\theta = 0.7$, Ω = 5900 J/g, a plasma temperature of 14 x 10^3 K, and complete dissociation of the vaporized SiO_2 molecules:

$$p_{lim} = \frac{300 \sqrt{\tau_0}}{r_i}, \qquad (6\text{-}19)$$

where p_{lim} is given in megapascals, τ_0 in seconds, and r_i in centimeters.

---

* See, e.g., Figure 6-11, which indicates increased heat losses in quartz flashlamps as the pulse energy approaches the limiting energy.

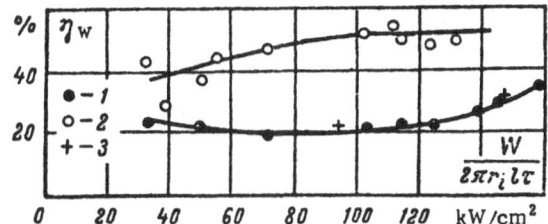

Fig. 6-11.   Dependence of losses on walls (η_w) of an IFP-5000 tubular quartz lamp on the pulse energy as the latter approaches W_{lim} [3-47].

1) p_o = 0.04 MPa, τ = 750 s, $\dfrac{W_{lim}}{2\pi r_i l \tau}$ = 144 kW/cm²;

2) p_o = 0.007 MPa, τ = 720 s, $\dfrac{W_{lim}}{2\pi l r_i \tau}$ = 136 kW/cm²;

3) p_o = 0.04 MPa, τ = 250 s, $\dfrac{W_{lim}}{2\pi r_i l \tau}$ = 162 kW/cm².

Since expression (6-19) is equivalent to the bursting pressure p_b, it enables us to find an estimate of the lower boundary of the flash durations τ_{min} for which this mechanism of lamp failure, like the thermal shock mechanism, should be characterized by a proportionality between W_{lim} and $\sqrt{\tau}$:

$$\tau_{min} \approx 1.1 \cdot 10^{-5} p_b^2 \, r_i^2. \qquad (6\text{-}20)$$

For shorter pulses, achieving a set of values of P^* and τ_0 that correspond to the fulfillment of condition (6-15) no longer should suffice for the lamp to explode, and an explosion via the same mechanism should occur at a constant value of W_{lim} which is independent of τ_0 and which is defined by the expression:

$$W_{lim} = 20\,000 \, r_i l \sqrt{\tau_{min}}, \qquad (6\text{-}21)$$

where τ_{min} is a constant that can be determined from relation (6-20).

The region of anomalously "soft" regimes with a flash duration shorter than some value (see Figures 6-6 and 6-7) that is consistent with estimate (6-20) in its order of magnitude has been noted in a number of experimental papers [2-36, 6-3, 6-7, and 6-9]. This region is characteristic of lamps that have a particularly strong tube or low gas pressure, and it serves to confirm that such a mechanism exists. Experimental data regarding the limiting loads associated

with an increase in the trigger voltage of the lamps (slight vaporization of quartz leads to the contamination of the noble gas with
molecular impurities; see Section 1-3) also are fully consistent with
this mechanism. The limiting loads increase in proportion to $\sqrt{\tau}$,
just as the W^* associated with an explosion increases according to
relation (6-16). Other observations concerning the explosions of
lamps under normal and "soft" conditions also follow naturally from
this mechanism (the deformation of foil leads, the multiplicity of
fragments, the appearance of the absorption lines of silicon in the
spectrum, the sputtering of a silicon monoxide film, and oxidation of
the electrodes by the dissociation products of the silicon monoxide).

Estimates of p_{lim} using formula (6-19) show that for normal
operation with a pulse length of 10^{-3} s the vapor pressure of the wall
material under critical conditions may reach high values (10 MPa) that
greatly exceed the bursting pressure p_b (this pressure is 2.6 MPa for
the IFP-8000 lamp, for example; see Table 6-1). If we also take into
account the steep rise in pressure (which is proportional to the power
to the 3.5 power) that is characteristic of this mechanism when
$\tau > \tau_{min}$ after the critical regime is reached [6-16], then we may
assume that the small spread (50%) of the values of p_b for actual
lamp samples is natural. Here we include the weak dependence of W_{lim}
on the wall thickness which is observed in the range of normal conditions. (The spread and the dependence on wall thickness increase
in the region of anomalously "soft" regimes in which an excess vapor
pressure is not ensured.)

Thus, it is entirely possible to use the mechanism of lamp
failure due to the pressure of hot vapors from the wall material to
explain the entire pattern of normal and anomalously "soft" modes of
lamp failure. However, this mechanism cannot be used to explain the
regularities that are observed in the region of "hard" regimes.

The effect of the pressure pulse upon the heating of the filler
gas has been considered too weak, compared with the breaking point
p_b of ordinary types of lamps, to cause the explosion of lamps, especially for durations of about 10^{-3} s [6-3 and 6-14]. However, the
bursting pressure of a cylindrical cavity may exceed severalfold the
actual strength of the envelope. The bursting pressure can be estimated, for example, from the well-known formula:

$$p_b = \frac{(R/r)^2 - 1}{(R/r)^2 + 1} \sigma_b \quad , \tag{6-22}$$

where R is the outside radius, r the inside radius, and σ_b the breaking
stress. This formula does not take into consideration the reduced
strength of certain regions: for example, at the solder points of a
lead or stem, or at other points where the wall is nonuniform in thickness, curvature, residual stresses, and other factors. Such regions
are formed during the discharge, especially near electrodes at the

Table 6-1. Static Bursting Pressures of Flashlamps, and an Estimate of Pb from Data on W_{lim}

Type of lamp	Type and diameter of lead, cm	r_i, cm	δ/r	l, cm	V_e/V_L	$\dfrac{n_\infty}{n_0}$	Pb, MPa		
							experi-ment	calc. from (6-22)	calc. from (6-25)
IFP-800	Foil, 0.7	0.35	0.43	8	0.15	0.5	6.7	20	≥ 6.0
IFP-5000	Foil, 1.1	0.57	0.35	25	0.09	0.55	3.5	18	3.4
IFP-8000	Foil, 1.1	0.8	0.19	25	0.035	0.64	2.6	10	2.5
IFP-20000	Foil, 1.8	0.8	0.19	58	0.11	0.53	2.1	10	1.8
IFP-25000	Cap-type	0.8	0.19	87	0.1	0.51	≥ 6.0	10	3.6
Sealed quartz tube	—	0.8	0.19	25	—	—	10	10	—

edge of the inside surface heated by the plasma, as a result of thermal stresses [6-10] and as a result of an inaccurate estimate of σ_b, which actually does have a significant dependence of the size of the sample and the condition of its surface (e.g., the value may be lowered as a result of previous discharges).

The importance of factors which reduce the strength of real flash-lamps can be seen from Table 6-1, where the experimental static values of p_b [2-36] are compared with the values calculated from formula (6-22). (The value σ_b = 600 kg/cm² was inserted in formula (6-22). The same formula was used to find this value from the experimental static values of p_b for a sealed uniform quartz tube – see the last column of the table.) The main dimensions are included in the table for each type of lamp. These include the ratio V_{el}/V_L of the ballast and working volumes, and the corresponding estimate of the decrease in density n_∞/n_0 due to its (sic) partial escape into the near-electrode spaces, as determined from formula (2-29).

As we can see from the table, the experimental values of p_b are much lower for lamps with foil leads than for a sealed uniform tube (they are lower by a factor of 3-5). We may assume that the values of the static p_b for lamps are determined mainly by the mechanical strength of the area where the foil is soldered to the quartz tube, which is thinner in this area. (This is consistent with the increase in p_b as the diameter of the lead increases.)

The maximum pressure generated in these lamps during a discharge can be estimated by using experimental data on the plasma temperature [2-5, 2-22, and 5-40] (see also Chapter 2), allowing for the effect of the escape of gas into ballast volumes. The corresponding approximation relation has the form [2-36]:

$$p_{Xe} = 2n_\tau/n_0 \left[16.7\, p_0 + 1.2 \cdot 10^{-4}\, (p_0/0.04)^{0.3}\, j \right], \qquad (6\text{-}23)$$

where p_{Xe} and p_0 are given in MPa and j in A/cm²; n_τ/n_0 is determined from expression (2-30).

The nature of the dependence of j on the pulse energy may be represented in the form [2-36]:

$$j = (p_0/0.04)^{0.15} \left(\frac{W}{V_L \tau} \right)^{2/3}. \qquad (6\text{-}24)$$

If we set p_{Xe} at the limiting energy equal to p_b and assume that this mechanism should account for the anomalously "soft" regimes with a duration shorter than τ^* (Figure 6-8), then from relations (6-23) and (6-24) we find:

$$W_{lim} = \frac{24 \cdot 10^3}{(p_0/0.04)^{0.3}} \left[\frac{p_b}{0.2 n_\tau/n_0} - 167\, p_0 \right]^{3/2} V_L \tau \qquad (6\text{-}25)$$

where V_L is given in cm^3, W_{lim} in joules, and τ in seconds. This expression actually does characterize a linear dependence of W_{lim} on τ.

As we see from Figure 6-9, the values of W_{lim} calculated from formula (6-25) by using the experimental values of p_b (Table 6-1) and the values of n_∞/n_0 and n_τ/n_0 estimated from formulas (2-29) and (2-30) are in satisfactory agreement with the experimental findings for IFP-8000 and IFP-20000 lamps with foil leads at initial pressures of 0.026-0.04 MPa. The worse fit between experiment and the plot for $p_0 = 0.08$ MPa may be due to the fact that the discharge channel contracts somewhat at high initial pressures [2-37].

As we can see from Table 6-1, the experimental value of p_b for a lamp with a cap-type lead (IFP-25000) is approximately twice as high as the calculated value determined by inserting the experimental value of W_{lim} into formula (6-25). This must be attributed to thermal stresses that arise during the discharge process, or to a decrease in σ_b due to disruption of the uniformity of the surface of the wall by previous pulses. Hence, the dynamic strength of lamps with cap-type leads may decrease to the dynamic strength of foil lamps (such as the IFP-5000), despite the greater static strength of the leads in the former.

Thus, the available experimental data indicate that explosions of lamps under anomalously "soft" conditions may be attributed to the pressure of the hot noble gas.

CHAPTER 7

OPERATING CHARACTERISTICS OF FLASHLAMPS

7-1. Starting Characteristics

The controllability properties of flashlamps or the possibility
of producing light pulses at specified times are described by a number
of characteristics which determine the conditions for the triggering
of a pulsed discharge. These characteristics are called the starting
characteristics. They include:

(a) the voltage interval between the primary electrodes of the lamp,
during which controllable triggering of the discharge is possible
(the range of controllability);

(b) the power (or energy) released in the auxiliary discharge channel;

(c) the time lag of the light pulse with respect to the trigger pulse;

(d) the functional relations among these parameters, and the dependences
of triggering parameters on design and operational factors.

The range of voltages between the primary electrodes within which
controlled triggering of lamps is possible is limited by the trigger
voltage and the self-breakdown voltage. These two voltages may turn
out to be substantially different for different lamp samples and different conditions of triggering (such as the parameters of the trigger
pulse, conditions of external ionization, the location of the trigger
electrode and of external conductors that form capacitive couplings
with the plasma column). Therefore, for operating purposes a narrower
range of supply voltages is standardized for flashlamps. This range
is bounded by the minimum and maximum permissible voltages between
which reliable triggering of a discharge is ensured (with some specified high probability), regardless of changes in operating conditions.

251

The trigger voltage depends on the power of the auxiliary discharge ("firing"). However, the effect of the firing power is different in different possible power ranges. It is especially noticeable if there is a low trigger-pulse energy close to the minimum energy required for the formation of the auxiliary discharge itself, and also in the region of high energies where the conductivity of the auxiliary discharge channel increases significantly. The existence of a region where the trigger voltage has a slight dependence on the firing power makes it possible to standardize the voltage required for reliable triggering of lamps and to have a quite broad range of trigger-pulse parameters.

The fundamental difference between the methods used to trigger tubular and spherical flashlamps (an internal trigger electrode is used in the latter) accounts for the different nature of the physical mechanisms underlying the process of discharge triggering, and hence also accounts for the relations among the corresponding parameters.

The triggering process and the basic starting characteristics of tubular flashlamps were examined in Section 1-3. The data given there may be supplemented with the relation between the trigger-lag time and the operating voltage of the lamp. This relation is of practical importance for some flashlamp applications. The delay of a flash relative to the trigger pulse (which is highly dependent on the characteristics of the lamp) is estimated in terms of the time interval t_A. This interval is measured from the time when the auxiliary discharge is formed to the time when there is a sharp increase in the luminous flux or current strength of the lamp (the latter usually is determined from a level 0.1 the maximum value of the corresponding parameter). The other components of the lag (the rise time of the trigger voltage or of the luminous intensity pulse) are more dependent on the parameters of the circuit controlling the lamps and are not treated here.

Figure 7-1 shows the dependences of discharge lag on the voltage U_0. The dashed plot, which is based on the same data reconstructed relative to the argument U_0/U_t, shows that a change in gas pressure and length of tube has the same effect on the lag as the trigger voltage does (provided that the trigger-pulse parameters are the same).

Controlled breakdown of a short discharge gap in pulsed light sources that have a spark channel which is not confined by walls (spherical flashlamps) is accomplished by means of one or several auxiliary trigger electrodes whose ends are inserted directly into the region of the gap between the primary electrodes (see Figure 1-2b). Breakdown occurs at a voltage lower than U_{self} (the self-breakdown voltage). A low-power high-voltage (U_a) auxiliary pulse applied to the trigger electrode forms a fine spark channel between the trigger electrode and at least one of the primary electrodes. This channel induces breakdown of the primary gap if a sufficient potential difference is produced between the main electrodes. Typical

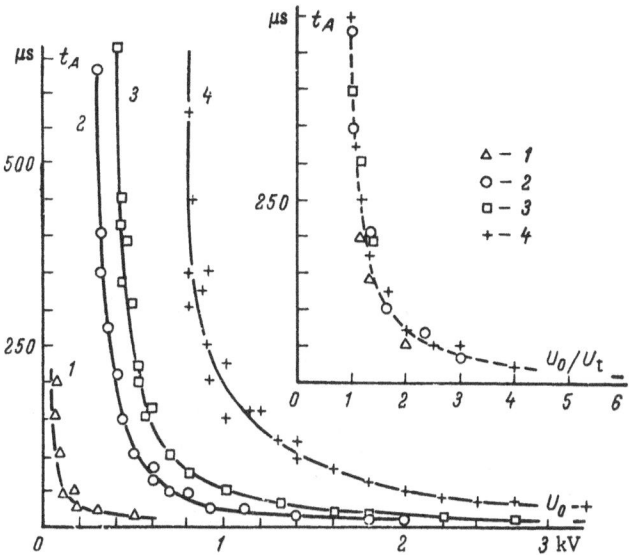

Fig. 7-1. Lag time t_A versus the supply voltage U_0 (solid lines)
and the ratio U_0/U_t (dashed line) for quartz xenon lamps
with a tube inside diameter of 1.7 cm, with series connection of the pulse transformer (Figure 1-2c). U_t was
determined for a 0.95 probability. 1, 2, and 4) p_0 = 0.04
MPa; 3) p_0 = 0.093 MPa; 1) l = 1.7 cm; 2 and 3) l = 27 cm;
4) l = 84 cm [7-1].

schematic diagrams of controlled discharge gaps in spherical lamps
are shown in Figure 7-2. Also shown there are examples of characteristic oscillograms of the trigger-pulse voltage and the primary-
discharge current for the corresponding discharge gaps. These
examples illustrate the development of the breakdown process in time.

Controlled breakdown between primary electrodes at voltages
below U_{self} and upon the formation of an auxiliary channel which
overlaps only part of the gap between the primary electrodes may be
linked to such ionizing factors as the photoelectric effect in the
gas and on the surface of the electrodes, which is due to the shortwave radiation of the auxiliary spark; excitation of a discharge due
to the rapid growth of impact ionization in the region of reduced
gas density, which results from the auxiliary spark; the sharp
variation of the electric field in direct proximity to the trigger
electrode at the time when the pulse is fed to it; and the consequent
formation in the discharge gap of regions with a large overvoltage
in which intense ionization processes develop [1-1 and 7-2].

The schematic diagram shown in Figure 7-2a is used extensively
in controlled spark dischargers, which are called trigatrons [7-2

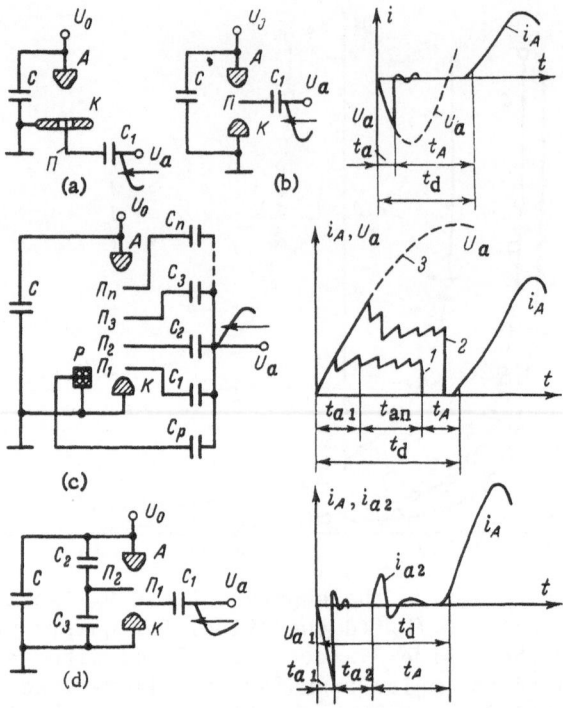

Fig. 7-2.   Typical schematic diagrams of spark gaps with auxiliary
            electrodes, and the corresponding oscillograms of the
            voltage U_a of the auxiliary trigger pulse and of the current
            i_A in the primary discharge gap and i_a in the auxiliary
            gap. A – anode; K – cathode; П – trigger electrode;
            P – electrode of auxiliary discharge; t_a – rise time of
            voltage on the trigger electrode up until formation of
            the auxiliary discharge; t_A – lag time of primary break-
            down relative to auxiliary discharge, as determined for
            the level of 0.1 the peak value of the current i_A in the
            discharge circuit; t_d – total lag time of the primary
            breakdown.

to 7-4] , and in demountable flashlamps [7-5] (it is used comparatively
seldom in sealed lamps). In this scheme the trigger electrode П is
insulated in the opening of one of the primary electrodes (usually
the cathode) K, and the channel of the auxiliary spark is formed between the outer edge of the opening and the end of the trigger electrodes.

     Depending on the relation between the size of the trigatron gap
and the discharge parameters, we can identify three physical mechanisms which play a dominant part in the development of a breakdown in
trigatrons in each specific case [7-2, 7-4, and 7-6].

The sharp increase in the electric field strength at the end of the trigger electrode Π and the rapid growth of the potential difference between electrodes A and Π at the time when the pulse is applied are the main factors responsible for breakdown in the case where the lengths of the auxiliary and primary gaps of trigatrons are approximately equal, the working voltage $U_0 > 0.8U_{self}$, the polarity of the trigger pulse is the reverse of that of electrode A, and the voltage pulse and the steepness of its leading edge are quite high. The distinguishing features of breakdown under these conditions are that it occurs simultaneously with the breakdown of the auxiliary gap (sometimes even leading it somewhat) and that the lag has a weak dependence on the voltage and power of the trigger pulse. Extremely short trigger times can be obtained in the discharger if the edge of the trigger pulse is very steep, there are large overvoltages in the primary discharge gap at the time the pulse is applied, and the inductance of the discharge circuit is low. In this case the time required for development of the discharge becomes commensurate with the growth time of the avalanche up until it is regenerated as a streamer (such a discharge is called a nanosecond pulsed breakdown [7-7]).

The development of the primary breakdown in trigatrons having a primary discharge gap substantially longer than the auxiliary gap is determined by the photoionization due to the auxiliary discharge when the power of the auxiliary discharge is low (the field distortion has a lower value). Here the trigger lag may be as long as tens of microseconds and has a sizeable statistical spread. Satisfactory controllability is achieved at voltages close to U_{self}. A plasma cloud spreads into the primary discharge gap when such a trigatron is triggered by a high-power auxiliary discharge from the gap Π-K. As a result, the breakdown strength of the gap K-A is reduced sharply, and it proves possible to control breakdown at extremely low voltages (all the way to $U_0 \geq 0.05U_{self}$). In this case the delay of breakdown is determined by the time required for the plasma to penetrate the primary gap. The delay essentially depends on the electrical parameters of the trigger circuit and on the length of the gap K-A. The polarity of the trigger pulse has practically no effect on the lag. The lag time usually is a few tens of microseconds and has a sizeable spread. The spatial stabilization of the primary discharge channel (at a flash frequency of up to 20-30 Hz [7-8]) can be improved somewhat by means of a jet-like ejection of plasma, directed toward the anode, from a high-power auxiliary discharge.

The mechanisms of breakdown of multielectrode gaps according to the diagrams in Figures 7-2b to 7-2d have been determined to a lesser extent. However, the available data enable us to present a qualitative picture of the processes [5-106, 7-2, 7-3, and 7-9 to 7-16]. The primary gap generally breaks down with some lag t_A after breakdown of the gap K-Π and formation of the auxiliary discharge. Exceptions are the limiting conditions, under which the trigger-pulse

voltage or power is very high [7-15] or U_o is close to U_{self}. In
such cases the primary discharge occurs almost simultaneously with
the breakdown of the auxiliary gap. The process of development of
breakdown of the gap Π-A is governed mainly by two factors: photoionization due to the auxiliary discharge, and the rapid rise of the
field strength on segment Π-A after the gap K-Π is closed by the channel of the triggering spark. From this time on, nearly all of the
voltage U_o is applied to the gap Π-A, and the field strength on this
segment increases in proportion to the ratio of the lengths of the
gaps K-A and Π-A. If we assume that the minimum trigger voltage U_t
across a multielectrode gap, which has a strictly defined value under
these conditions, is analogous to the static breakdown voltage, then
the ratio U_o/U_t represents the overvoltage. The dependence of the
breakdown lag time t_A on the ratio U_o/U_t for a three-electrode gap
(Figure 7-3) is analogous to the dependence of the breakdown delay
on the overvoltage for a two-electrode gap that was mentioned in
Section 1-2. Here we also observe long delays in breakdown and a
large statistical spread for small overvoltages ($U_o \approx U_t$), as well
as a rapid decrease in t_A with increasing U_o/U_t to values determined
by the time required for formation of the auxiliary discharge channel.
Changing the gas pressure (xenon pressure) from 0.03 to 0.3 MPa has
no effect on t_A for a constant U_o/U_t.

Breakdown control by means of a single auxiliary electrode, as
in the diagram in Figure 7-2b, is very widely used in currently
manufactured spherical flashlamps because of its simplicity and
effectiveness. The starting characteristics of the lamps can be
varied over a wide range by changing the position of the trigger
electrode relative to the primary electrodes. Figure 7-4 gives an
idea of the effect of the position of the trigger electrode Π relative
to the primary electrodes (i.e., the length of the segment K-Π)
on the range of contollability of the anode-pulse voltage (at which
reliable firing of the primary discharge is ensured) during infrequent-
flash operation (at a frequency below 0.1 Hz). It can be seen from
the figure that the largest range of controllability of U_o (the
region enclosed between the plots of the self-breakdown voltage U_{self}
and the trigger voltage U_t is hatched in the figure) and the minimum
t_A are observed when the trigger electrode is placed at equal distances
from the primary electrodes. However, in this case the breakdown
voltage of the auxiliary gap for the trigger pulse is high (the curves
of U_a). This voltage is slightly dependent on the polarity of the
trigger electrode is positioned in the cathode half of the discharge
gap, then more favorable conditions are provided for controlling the
breakdown. This is evident from a comparison of the left and right
branches of the relations. At the maxima of the U_a curves the auxiliary discharge channel is formed on segments K-Π or Π-A with approximately equal probability. As the trigger-pulse energy increases from
1 to 50 mJ, the nature of the dependence of U_t and t_A on the location
of the decrease (by a factor of up to 5-6) in the absolute values of

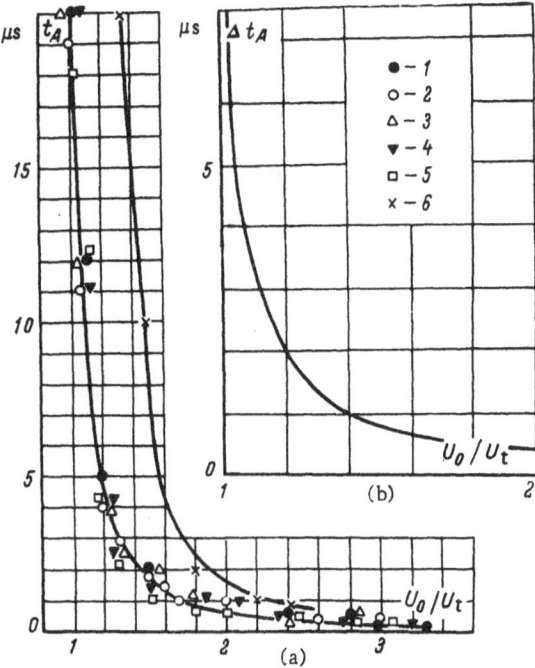

Fig. 7-3. Dependence of the lag time t_A of the primary breakdown relative to the auxiliary discharge (a) and of the maximum
absolute spread Δt_A of the lag time (b) on the ratio U_0/U_t
for a three-electrode discharge gap (the schematic diagram
in Figure 7-2b). The gap length is $l_{K-A} = 4.5$ mm, $l_{K-\Pi} =$
1.5 mm, xenon. 1) 0.033 MPa; 2) 0.067 MPa; 3) 0.1 MPa;
4) 0.2 MPa; 5) 0.3 MPa; 6) $l_{K-A} = 5.5$ mm, $l_{K-\Pi} = 0.5$ mm,
0.3 MPa. In determining U_t it was assumed that the probability of breakdown was 0.95 [7-14].

U_t and t_A is observed as the firing energy is increased from 1 to
15-20 mJ. A further increase in energy does not lead to an appreciable
decrease in U_t or t_A [5-106 and 7-16].

Triggering of flashlamps by means of several auxiliary electrodes
included in the discharge gap according to the diagrams in Figure
7-2c is used mainly in low-power strobotrons [7-11 to 7-13]. This
procedure makes it possible to control the time of the primary breakdown of elongated spark gaps at anode voltages much lower than in
the previous case. Positioning the ends of the trigger electrodes
on the axis of the discharge gap improves the spatial stability of
the discharge channel. The formation of the auxiliary discharge
by the trigger pulse in such lamps consists in the successive breakdown of the separate discharge gaps formed by the electrodes, mov-

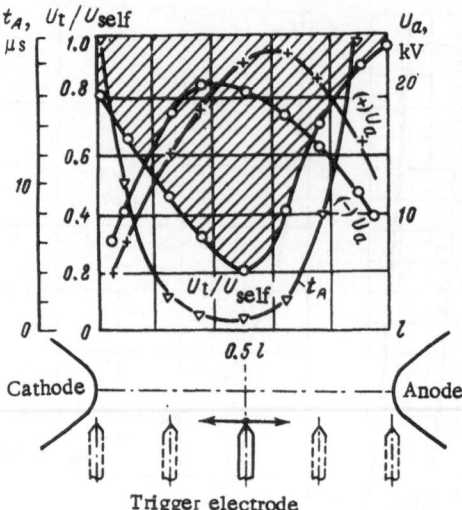

Fig. 7-4. Dependence of the range of controllability of a three-
electrode discharge gap (the hatched region between U_{self}
and the curve of U_t/U_{self}), the breakdown voltage of the
auxiliary gap for a trigger pulse of different polarity
(+) U_a and (-) U_a, and the primary-breakdown lag time t_A
on different positionings of the trigger electrode. Xenon,
0.3 MPa, $l = 4.5$ mm, the distance from the apex of the
trigger electrode to the axis of the primary electrodes
is 1 mm, $U_{self} = 7.5$ kV, the probability of breakdown is
0.95, and the illuminance of the discharge gap just before
breakdown is 100 lx. The plot of t_A was obtained for
$U_o/U_t \approx 1.5$ and $U_a = -22$ to -24 kV [7-14].

ing from the cathode to the anode. As a result, the auxiliary channel
overlaps the electrode gap on the segment $K-\Pi_n$. The interval Π_n-A is
broken down by the anode voltage (when $U_o \geq U_t$). This concludes the initial
stage of primary breakdown. Such breakdown occurs if the breakdown
voltage of the gap $K-\Pi_1$ is much lower than that of each of the other
gaps. For this to be true, the length of the gap $K-\Pi_1$ must be shorter
by a factor of 3-10 than the length of any of the other gaps. The
segment $K-\Pi_1$ is the first to break down if the trigger-pulse voltage
is supplied simultaneously to all auxiliary electrodes through coupling
capacitors C_1, C_2, ..., C_n. The potential of electrode Π_1 practically drops to zero at the time of breakdown. The potential differences between electrodes Π_1 and Π_2 and the ionizing effect of the
spark in the gap $K-\Pi_1$ leads to breakdown of the gap $\Pi_1-\Pi_2$, and so
forth. The capacitances of the coupling capacitors C_1, C_2, ...,
C_n must exceed the capacitance of the trigger electrodes relative to
ground by at least an order of magnitude in order for the voltage
drop observed on the other electrodes at the time of breakdown of

the previous gap (the "teeth" on the oscillograms in Figure 7-2c)
to have no effect on the subsequent development of the breakdown
[7-3 and 7-11]. The capacitance of the coupling capacitators in the
trigger circuits of industrial strobotrons usually is 10-25 pF.

The limiting case of controllable triggering of a multielectrode
discharge gap sometimes is described by the dependence of the trigger
voltage U_t on the voltage on the auxiliary pulse on the auxiliary electrodes. This is called the lower limit of the range of controllability with respect to the working and control voltages. Examples of
such characteristics are shown in Figure 7-5 for a breakdown probability of at least 0.95. Curves 2, 3, and 4 illustrate the dependence
of the lower limit of controllability on the xenon pressure.

The breakdown of multielectrode discharge gaps at small overvoltages (near the lower limit of the range of controllability) is
extremely sensitive to external ionization factors: the radiation
background, illumination, and the discharge repetition frequency
[7-12 and 7-13]. Preillumination of the cathode gap, which with the
radiation produced in the flashlamp envelope itself between the electrodes of the auxiliary discharger is used to reduce and regulate the
breakdown voltage of the cathode gap [7-12]. This is analogous to
regulation by means of spark relays. The auxiliary gap is powered by
an additional coupling capacitor, usually with the same trigger pulse
that is fed to the auxiliary electrodes. The effectiveness of illuminating the gap with a spark is evident from a comparison of the oscillograms of the voltage on the trigger electrodes that are shown in
Figure 7-2c with (curve 1) and without (curve 2) the discharger in
operation, and of the lower limits of the range of controllability
in Figure 7-5 (the dashed and solid lines).

When the minimum voltage of the auxiliary pulse is used, the
lowest trigger voltage for a lamp can be obtained through the successive breakdown of individual parts of the discharge gap in Figure
7-2d. In some regards this is analogous to the breakdown of cascade
dischargers [7-3]. The breakdown of the shortest gap $K-\Pi_1$ (0.1-
0.4 mm) promotes the formation of the auxiliary channel on the segment
$K-\Pi_2$, which in turn stimulates the breakdown of the gap K-A under the
anode voltage. The widest range of controllability is achieved
when electrode Π_2 has positive polarity, corresponding to maximum
uniformity of the electric field between the primary electrodes.
Given such uniformity, U_{self} increases with decreasing U_t [5-106].
The ratio of the potentials on the electrodes Π_2 and A, which are proportional to the self-breakdown voltages of the respective gaps, is
optimal. This ratio is fulfilled when $U_{\Pi 2} \approx 0.7-0.85U_0$ in industrial-
type lamps with two trigger electrodes, such as the ISSh-5 and the
ISSh-7. Several auxiliary electrodes also may be used to regulate
the breakdown of long discharge gaps. The equalizing potentials
on each of these electrodes are determined on the basis of the same
considerations as for a single electrode. The lag time t_A of primary

breakdown relative to the time of breakdown of the control gap K-Π_1 in the diagram in Figure 7-2d depends mainly on U_0 and U_{Π_2}. When $U_{\Pi 2}/U_0 \approx 0.8$ and $U_0/U_t \approx 1.5$-2, we usually have $t_A < 1$ μs, but t_A increases sharply with decreasing U_0/U_t, reaching tens of microseconds.

7-2. Frequency Response

The possibility of using a flashlamp over a broad range of frequencies is determined by the design features of the lamp itself, on the one hand, and by its power-supply parameters on the other: the time dependence of the voltage across the storage capacitor and the capacitance of the storage capacitor, the methods used to charge the capacitor, the methods used to initiate and terminate the current in the discharge gap, and other factors.

The frequency response of a lamp depends on many physical mechanisms which are at work in the gap in the post-discharge period: the cooling of the plasma column, deionization of the gas (due to volume recombination and recombination on the walls of the envelope and on the surfaces of the electrodes), the leveling-out of the gas density and temperature in the inside volume of the lamp following disturbance by the spark discharge, the cooling of the electrodes and envelope walls (which is accompanied by restoration of the low conductivity of the walls), and a number of other processes. In aggregate these mechanisms lead to a situation in which the frequency capabilities of lamps are limited by the time required for recovery of the electric breakdown strength of the spark gap in the post-discharge period; by the stability of the parameters of the light flashes at frequencies at which recovery processes have not yet had time to end; and finally, by the possibility of modulating the luminous flux of the lamp at flash repetition frequencies such that the processes of plasma deexcitation between two current pulses have not been completed. The first constraint is the most significant from the standpoint of the commercial application of flashlamps.

There are three types of malfunctions or disruptions of the normal operation of flashlamps in the high-frequency (stroboscopic) mode.

During a pulsed discharge the voltage across the storage capacitor decreases to the extinction voltage, at which the current in the discharge gap falls to zero. If the storage capacitor is charged up somewhat just before this happens, holding at a level above the extinction voltage, then the discharge will not stop. If this happens the lamp loses its controllability and ceases to operate as a pulsed light source. This corresponds to the first type of disruption of regular operation: a transition to continuous-burn mode.

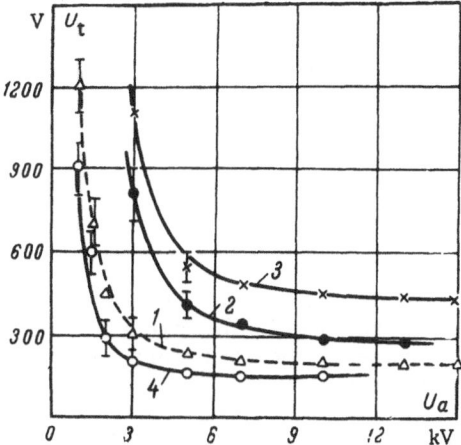

Fig. 7-5. Lower limits of the range of controllability of ISSh-2
multielectrode flashlamps (for a schematic diagram see
Figure 7-2c). The dashed line represents operation with
the auxiliary discharger connected. The solid lines
represent operation without the discharger. The gas is
xenon. 1 and 2) 0.044 MPa; 3) 0.08 MPa; 4) 0.013 MPa
[7-13].

     If the current is completely cut off after a flash, then the
discharge gap of a lamp undergoes a substantial change in its electrical properties within just a few tens to hundreds of microseconds.
The voltage that can be applied to the lamp with recovery of its current increases quickly with elapsed time after the flash: the electric breakdown strength of the lamp is restored. Upon the completion
of this process, the breakdown strength of the gas-filled gap reaches
its original value (the self-breakdown potential of the lamp cold).
The time dependence of the maximum voltage that can be applied to
a lamp in the post-discharge period without restoration of the current is called the breakdown strength recovery characteristic. Spontaneous breakdown ("self-breakdown") of a lamp, corresponding to the
second type of disruption of the normal operation of flashlamps,
occurs in the case of a storage capacitor (connected directly to the
electrodes of the lamp) with a rate of voltage rise such that at some
time the voltage exceeds the breakdown strength of the discharge gap.
This strength increases during breakdown strength recovery. Self-
breakdowns usually occur randomly, accompanied by flashes of reduced,
irregular intensity.

     Skipped flashes, which usually are observed in a lamp at high
frequencies and low voltages, are the third type of disruption of the
regularity of flashlamp flashes during stroboscopic operation. One
possible cause of missed flashes may be severe heating of the glass

in the lamp envelope. When this occurs, the conductivity of the glass
increases and the discharge gap is partially screened by the conductive walls from the auxiliary voltage pulse fed to the external trigger electrode. Skips sometimes are due to contamination of the lamp
filler by molecular-gas impurities formed upon the heating of structural elements, to gas-dynamic processes caused by the reflection
of compression waves from the envelope or electrodes, or to a decrease
in the power of auxiliary voltage pulses.

The first two types of disruptions of regularity, which are the
most significant, involve disruption of the normal process of breakdown strength recovery in the gap. Therefore, the breakdown strength
recovery characteristics are the most important of all the frequency
characteristics of flashlamps.

Strictly speaking, the breakdown strength recovery characteristic
for specified operating conditions (without disconnecting the lamp
from the storage capacitor) should be determined experimentally by
varying the time dependence of the voltage across the capacitor during
the intervals between flashes and by selecting the fastest possible
rise in voltage for which complete controllability of the lamp would
be preserved but slightly beyond which individual self-breakdowns
would start at any point of the charging curve. However, circuits in
which main voltage pulses of a regulated value are fed to the lamp
at different times during the interval between flashes are used to
measure breakdown strength recovery characteristics because of the
complexity of building appropriate apparatus. The corresponding
breakdown strength recovery characteristic curves can be constructed
from the voltages of the pulses which result in the first breakdowns
of the gas-filled gap during different phases of charging.

For example, in [7-17 to 7-20] the breakdown strength recovery
characteristic curves of flashlamps and of the air-filled spark gap
were determined by means of a pulse generator assembled from high-
power electron tubes. The generator charged a storage capacitor
with short pulses immediately before a flash. The main pulse was
fed to the lamp with a regular lag in the interval between two
schedule pulses following a burst of 32 regular pulses. The pulse
delay and the voltage for which the probability of main-pulse breakdown was 5, 50, or 95% were recorded. These data were used to construct breakdown strength recovery characteristic curves for 5%,
50%, and 95% probability of breakdown, respectively.

In [7-21 to 7-24] the breakdown strength recovery characteristic
curves were measured as the lamp was fed a probing voltage "step"
(Figure 7-6) in the intervals between flashes. This step was shifted
by some time interval relative to the previous discharge. During
stroboscopic operation the lamp was supplied by charging a storage
capacitor to operating voltage during the period of 1-3 μs before the
next flash. The amplitude of the probing "step" was regulated over

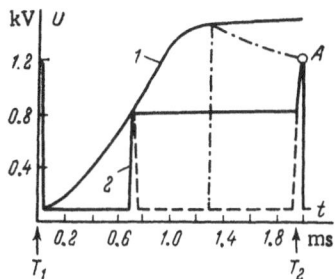

Fig. 7-6. The breakdown strength recovery characteristic curve (1)
of an ISSh-15 lamp (with a flash energy of 0.03 J, a
frequency of 500 Hz, and an operating voltage of 1200 V)
and the probing voltage "step" (2). T_1 and T_2 are the
times at which flashes begin. The dashed line represents
the potential across the lamp in the case of self-breakdown during the main pulse. The dot-dash line represents
the probing "step" with a potential above the operating
voltage of the lamp.

a broad range, and could be higher or lower than the operating voltage
of the lamp (point A in Figure 7-6). The values for which only isolated breakdowns (about one per second) of the gap due to the main-
pulse voltage occurred were recorded by varying the amplitude and
lag time of the "step" and these data were used to construct the breakdown strength recovery characteristic curves. The voltage curves
corresponding to a nearly 100% probability of gap breakdown were determined. Some typical curves of both types are shown in Figure 7-7 for
some tubular and spherical flashlamps. In the figure, zones with a
breakdown probability of approximately 1% to 99% are hatched for both
types of lamp. Assuming a normal distribution of the probabilities
of self-breakdown in this interval, we can use these data to estimate
the probability of disruption of the regular operation of lamps when
the capacitor is charged according to any relation in use.

An analysis of the breakdown strength recovery characteristic
curves obtained in the bibliographic references mentioned for different
types of flashlamps enable us to draw the following general qualitative
conclusions regarding the behavior of breakdown strength recovery
curves.

(a) Breakdown strength recovery curves usually consist of three
segments that have pronounced transitions from segment to segment: the
breakdown strength of the discharge gap goes practically unchanged
during the first few tens to hundreds of microseconds after the end of
a discharge, when a high-temperature plasma still exists in the gap.
The breakdown strength does not exceed the extinction voltage of the

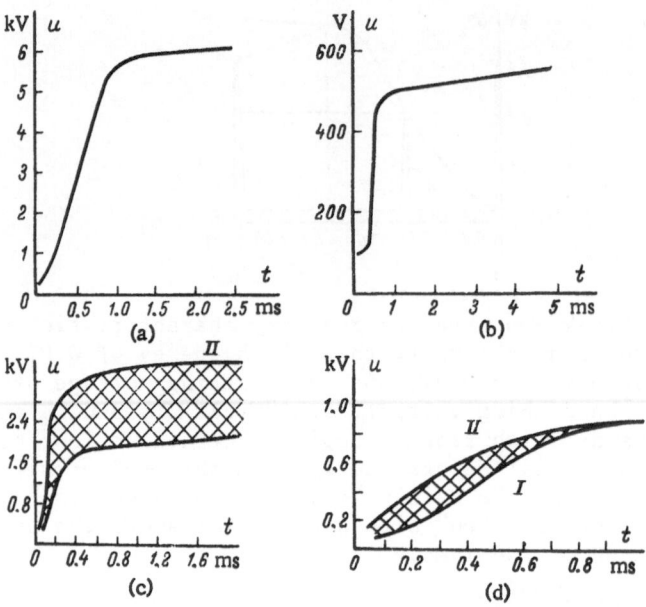

Fig. 7-7. Breakdown strength recovery characteristics. a) For the
ISSh-100-2 spherical lamp (0.12 J, 400 Hz); b) for the ISK-
25 tubular lamp (0.045 J, 200 Hz); c) for the ISP-70 capillary lamp (0.088 J, 400 Hz) (I) and voltage curve (for the
same lamp) for a nearly 100% probability of breakdown (II);
d) for the ISSh-15 spherical lamp (0.03 J, 500 Hz) (I) and
the curve (for the same lamp) for a nearly 100% probability
of breakdown (II).

discharge (which usually is a few tens of volts to a few hundred volts).
The steep second segment is characterized by a rapid increase in the
electric breakdown strength of the lamp, and obviously is associated
with destruction of the current-carrying channel as a result of plasma
deionization and the intermixing and cooling of the gas in the discharge channel. The gentle third segment of the breakdown strength
recovery curves apparently is determined by thermal processes in the
spark gap during which the gas density evens out throughout the entire
volume of the lamp. For spherical lamps the transition from the second
to the third segment usually occurs at voltages close to the self-breakdown potential in the cold state, while for tubular lamps it occurs at
voltages severalfold lower than the self-breakdown voltage (Figure 7-7).

(b) The behavior of breakdown strength recovery curves for each
type of lamp is determined mainly by the discharge energy and has very
little dependence on the capacitance of the storage capacitor and the
voltage across the capacitor just before the flash.

(c) A decrease in flash energy for a constant average power leads to displacement of the breakdown strength recovery curves upward and to the left.

(d) The effect of the average power dissipated in the lamp on the behavior of the breakdown strength recovery curves is much weaker than that of the flash energy.

(e) The inclusion in the discharge circuit of components that absorb part of the discharge energy has the same kind of effect on the breakdown strength recovery curves as a decrease in the energy of the storage capacitor.

(f) There usually is no sharp change in the nature of the breakdown strength recovery curve on the first and second segments during the operating life of a lamp. The level of the breakdown strength recovery curve sometimes changes somewhat on the third segment as the static breakdown potential of the lamp changes.

(g) The presence of voltage across the electrodes of the lamp in the post-discharge period does not affect the behavior of breakdown strength recovery curves as long as this voltage does not cause a current to flow through the lamp, i.e., as long as its value at every instant is lower than the voltage values determined by the breakdown strength recovery curve.

(h) The breakdown strength recovery curves for lamps of a given kind may differ significantly from sample to sample.

(i) The breakdown strength recovery curves are probabilistic in nature. Therefore, in theory there always exists some probability of self-breakdown of a lamp even if the voltage across it increases along a curve that lies below the breakdown strength recovery curve (the larger the difference between these curves, the lower this probability).

(j) The use of power-supply circuits with a very high charge on the storage capacitor just before the flash (for a period of time commensurate with the lag time of the discharge) prevents spontaneous breakdowns in the lamp and makes it possible to obtain stable operation of lamps at the limiting voltages and frequencies.

(k) The nature of the breakdown strength recovery curves depends on the design of the lamp: for equal flash energies, breakdown strength recovery occurs most slowly in spherical lamps with a short spacing between cathode and anode, and most quickly in capillary lamps that have a very long discharge of small cross section. Breakdown strength recovery usually occurs faster in low-pressure lamps than in high-pressure ones. Breakdown strength recovery is accelerated by the presence of cold surfaces (the surfaces of the envelope and

the electrodes) in direct proximity to the discharge channel, on
which ion recombination processes can occur. Breakdown strength
recovery processes occur more rapidly when light molecular gases
such as hydrogen or air are added to the lamp filler.

The standard breakdown strength recovery characteristic of the
ISK-10 flashlamp can be used to characterize and compare the frequency
response of several widely used power-supply circuits for stroboscopic
flashlamps. These are shown in Figure 7-8. The flashlamp itself
performs as the discharge-current commutator in these circuits, and
the frequency capabilities of these circuits therefore depends solely
on the rate of breakdown strength recovery processes in the lamp
and on the charging conditions of the storage capacitor.

The potential across the capacitor increases most steeply when
the capacitor is charged through a charging resistance R to the power-
supply voltage $U_{x.x}$. As we can see from Figure 7-8a (curve I), in
this case the current i_0 at the initial instant of the discharge is
nonzero, the voltage immediately after the flash should increase
faster than the breakdown strength recovery process can occur, and
the lamp should go into the continuous-burn mode. The charging
resistance must be increased sharply in order to reduce the initial
charging current i_0 to the value allowed by breakdown strength
recovery, but then the capacitor is unable to charge to the specified
voltage U_0 by the time of the next flash. In this case the flash
frequency must be reduced severalfold or the emf $U_{x.x}$ (curve II) must
be approximately tripled by charging the capacitor to just $U_{x.x}/3$ by
the time of the flash. For this reason either a capacitor designed
for a voltage 3 times as high as the rated voltage, or reliable
circuits must be used to cut off the charge. A charging circuit
operating through a resistance is characterized by the lowest upper
limit on the flash frequency: usually tens or hundreds of hertz.

A charging circuit using a ballast choke (Figure 7-8b) provides
a slower voltage rise in the initial stage with an improved shape for
the entire charging curve, as well as resonant charging of the storage
capacitor to nearly twice the voltage of the power supply. Diode D
was used in the circuit to prevent reverse leakage of charge from
the storage capacitor. The efficiency of such a charging circuit
may reach 85-95%, and the minimum interval T_M between flashes (with
respect to the ISK-10 lamp in this mode) may be 500-700 μs. The use
of a nonlinear choke with a saturated magnetic circuit additionally
slows the initial charging rate of the storage capacitor, and raises
the maximum flash frequency somewhat. The circuit has two shortcomings.
First, if the lamp enters continuous-burn operation accidentally, the
current in the charging circuit may rise to values that are dangerous
to the lamp. Hence the circuit needs protection with respect to the
maximum current in the charging circuit. Additionally, if resonant
charging of the capacitor does not occur for some reason (e.g., upon
a smooth increase in $U_{x.x}$), then the first firing of the lamp may be

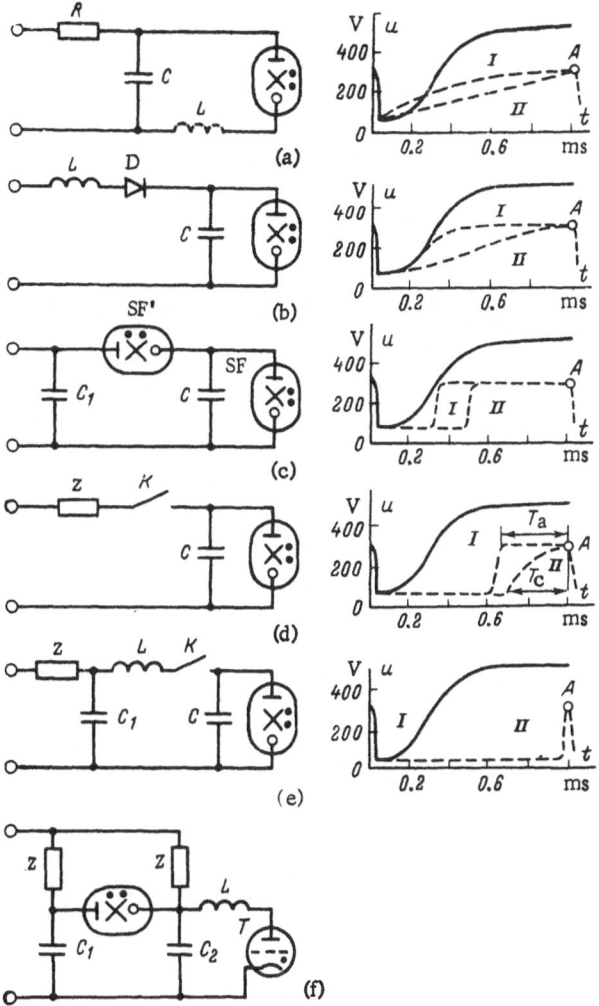

Fig. 7-8.   Some schematic diagrams of the charging of the storage
capacitor for power supply to a flashlamp during stroboscopic operation. In the graphs the solid lines represent
the standard breakdown strength recovery characteristic
of the ISK-10 lamp under the conditions W = 0.01 J,
f = 1000 Hz, U_0 = 315 V, C = 0.2 μF (according to test
data for 10 samples). The dashed lines represent the
voltage curves for the lamps. A is the voltage at the time
of a flash.

a) a circuit with a charging resistance R (a small inductance L which lowers the potential across the lamp when

Fig. 7-8.   Ctd. the discharge is extinguished may be included in the
            discharge circuit).   I) $U_0 \approx U_{x.x}$, $R = 1.67$ kΩ, $1/f = 3RC$,
            $\eta \approx 50\%$, $i_0 = 140$ mA, $T_M > 20$ ms; II) $U_0 = \frac{1}{3} U_{x.x}$, $R = 15$ kΩ,
            $1/f = RC/3$, $\eta = 20\%$, $i_0 \approx 55$ mA, $T_M > 1$ ms.

            b) charging circuit working through a choke L and a diode.
            $U_0 \approx 2U_{x.x}$, $i_0 = 0$ (relay protection is required against
            continuous-burn operation), $\eta = 90\%$, $T_M \approx 0.5$ s   (I) L =
            0.12 H; II) L = 0.5 H).

            c) "two-cycle" circuit with two flashlamps in the charging
            and discharging circuits of the capacitor C. $C_1 \gg C$,
            $U_0 \approx U_{x.x}$, $i_0 = 0$, $\eta = 95\%$ (relay protection is required
            against a series discharge across two lamps) (I) $T_M = 0.3/f$;
            II) $T_M = 0.5/f \approx 0.2$-0.5 ms).

            d) a scheme with a switch K in the charging circuit.
            $i_0 = 0$, $U_0 = U_{x.x}$, $\eta = 50\%$.   I) a thyristor, trigatron,
            or discharger as K, and T_B is the breakdown strength
            recovery time of the switch; II) an electron tube, trans-
            istor, or bidirectional thyristor as K, and T_C is the
            charging time of capacitor C.

            e) a circuit with an intermediate capacitor $C_1 = C$, a
            switch K (a thyratron, thyristor, or discharger), and an
            inductance L which determines the rate of charge transfer
            from C_1 to C. $i_0 = 0$, $U_0 = U_{x.x}$, and the efficiency is
            determined by the method used to charge C_1.

            f) a circuit with a "flip-flop leg": a resonant recharger
            of capacitor C_2 through a thyratron T (or a thyristor
            or discharger) and an inductance L. $C_1 = C_2 = 2C$, $i_0 = 0$,
            $U_0 = 2U_{x.x}$, and the efficiency is determined by the method
            used to charge C_1 and C_2.

difficult because the voltage across the capacitor is nearly half
the rated value.   To correct this situation, an additional resistor
must be series-connected to the inductor, but then charging occurs
through a combined RL circuit that has properties of both the first
and second circuits (see Section 9-5).

     Some improvement of frequency properties is provided in the
circuits shown in Figures 7-8a and 7-8b by the series connection of
a low-inductance choke (shown by the dashed line in Figure 7-8a) to
the flashlamp in the discharge circuit.   This may correspond, for
example, to the limiting case of aperiodic discharge of the capacitor
across the lamp.   The residual voltage across the capacitor turns out
to be close to zero.   As a result, during the initial stage the
charging curve runs below the breakdown strength recovery curve.

In a "two-cycle" circuit (Figure 7-8c) two identical flashlamps
flash alternately with a time difference T/2. The pulsed charge of
working capacitor C from high-capacitance filter capacitor C_1 passes
through lamp SF', while the discharge of the working capacitor passes
through lamp SF [7-25]. In this circuit the voltage across each of
the lamps does not increase at all during the first half-period after
the flash, but increases stepwise to the full working value when the
other lamp flashes. The circuit works reliably during aperiodic
charging and discharging of a storage capacitor (the voltage fluctuations that sometimes occur during the operation of spherical flashlamps can cause breakdown of a nonburning lamp, with discharge of
capacitor C_1 through the two series-connected lamps, and also can cause
disruption of the normal operation of the circuit). As we can see from
the breakdown strength recovery curve, the minimum time between flashes
of two ISK-10 lamps in this circuit* may decrease to 250-500 μs when
W = 0.01 J (depending on the working voltage). Some commutator other
than the lamp SF' may be included in the circuit. Such a commutator
would have a quite high rate of breakdown strength recovery (e.g.,
a trigatron, thyristor, or discharger [5-78 and 7-26, US patent No.
3,019,371, and French patent No. 1,227,228]) and would be series-
connected to a low ballast resistance (Figure 7-8d; curve I is the
corresponding voltage curve). If the breakdown strength recovery time
T_B of the commutator is much shorter than the period between flashes,
then it also is possible to obtain a flash frequency close to the
limiting frequency in this scheme. If an electron tube [2-39, 7-17,
and 7-19. and US patent No. 3,525,016] or a transistor (US patent No.
3,828,222) is used as the key, the charging of the capacitor can be
ended just before the flash, the tube or transistor then can be
blanked, and the next flash can be produced. However, the maximum
current strength of such keys usually is low and the charging time
T_C (curve II in the same figure) turns out to be comparatively long:
tens to hundreds of microseconds (e.g., see the specific circuit
in Figure 7-9, in which three transistors act as the key).

The highest flash frequency can be achieved by using the circuits
shown in Figure 7-8e [7-27 to 7-29] and Figure 7-8f [7-30], in which
the capacitor is charged throughout the entire interval between flashes
but the voltage across the lamp does not appear until just before flash.
This regime is approximated by the circuit shown in Figure 7-10c, in
which intermediate energy storage is provided at PT. The commutator
in such schemes is a high-frequency, high-current device: a thyristor,
thyratron, or discharger. In the circuit in Figure 7-8e, resonant recharging of capacitor C_1 to capacitor C of equal capacitance occurs
when the control pulse is supplied to key K a few microseconds before

---

* This circuit is convenient for determining the limiting flash
  frequency of a lamp operating with specified C and U, since it
  generates a probing voltage "step," and each disruption of regular
  operation can readily be seen visually as a high-power flash and
  a disruption of cyclicality.

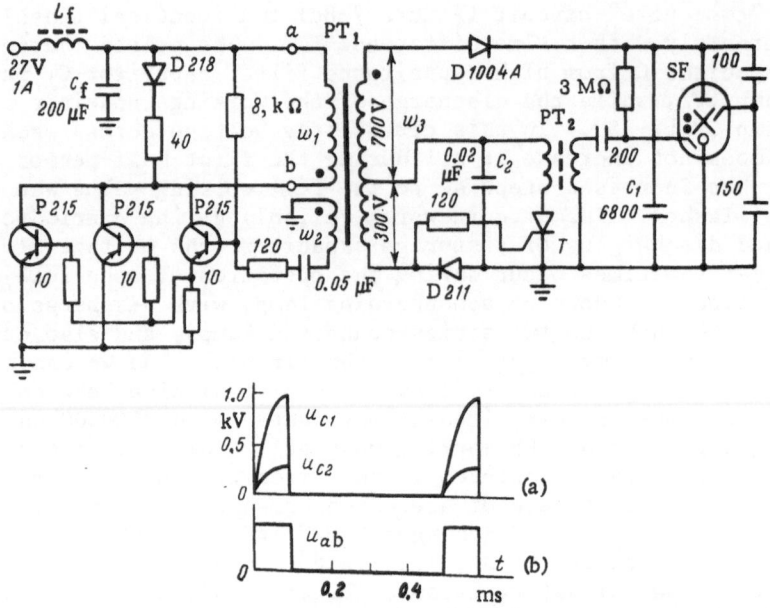

Fig. 7-9. Power supply circuit of a spherical flashlamp (SF) operating at 2 kHz and using a 27-V DC source (this is an example of an implementation of the circuit in Figure 7-8d, with $T_c =$ 100 μs). a) curves of voltage across capacitors C_1 and C_2; b) voltage across the primary winding of pulse transformer PT_1 (w_1 = 190 turns, φ 0.5 mm, w_2 = 20 turns, φ 0.1 mm, w_2 = 11,000 turns, φ 0.05 mm, and the cross section of the magnetic circuit is 0.5 cm^2). After C_2 is charged, the reverse voltage surge across PT_1 triggers a type KU202L thyristor (T in the figure), generating a pulse to fire the lamp by means of pulse transformer PT_2 (w_1 = 6 turns, φ 0.3 mm, w_2 = 160 turns, φ 0.1 mm, the cross section of the magnetic circuit is 0.1 cm^2, F2000 ferrite).

the flash. On completion of recharging, a trigger pulse is fed to the lamp and both capacitors turn out to be discharged. In the circuit in Figure 7-8f capacitors C_1 and C_2 are charged simultaneously to a voltage equal to $U_o/2$, a control pulse is supplied to the trigatron a few microseconds before the flash, and capacitor C_1 discharges through the lamp to capacitor C_2 when a trigger pulse is supplied to the lamp after resonant recharging of C_2 from $+U_o/2$ to $-U_o/2$. The second circuit is somewhat less stable when the flashlamp misses flashes. In these circuits the potential is eliminated almost entirely after a flash from both the lamp and the commutator, in which breakdown strength recovery processes proceed in parallel (the time required for breakdown strength recovery usually is an order of magnitude shorter in the highest-frequency commutators than in most flashlamps).

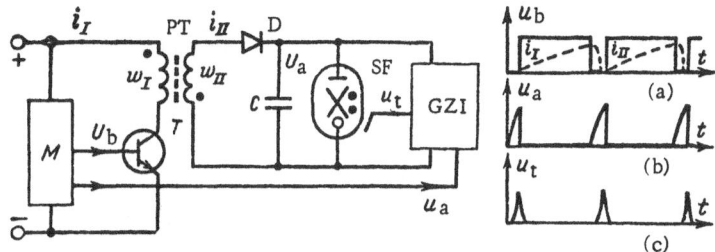

Fig. 7-10. Schematic diagram of the power supply for a stroboscopic
lamp with an intermediate inductive energy storage in the
form of a gapped transformer [7-29 and US patent No.
3,339,108]. When transformer T is opened, a quantity of
energy is stored in PT (i_I is the current in coil w_I).
This energy is dumped to C after T closes (i_{II} is the
current in coil w_{II}). When T opens, a control pulse u_a
is supplied to GZI. a, b, c) plots of the voltages across
the base of the transistor, the capacitor, and the trigger
electrode of the lamp.

When flashlamps are powered by capacitors, a further increase in
the flash frequency can be produced only by the series connection in
the discharge circuit of yet another additional commutator which has
high electrical conductivity in the open state and high electric
breakdown strength in the closed stage. This commutator allows
charging of the storage capacitor before the lamp recovers its breakdown strength (Figure 7-11). The limiting flash frequency for such
circuits is determined not only by the capabilities of the commutator,
but also by the capability of the lamp to generate light pulses at
such frequencies with sufficiently stable photometric parameters. In
addition to high-speed breakdown strength recovery (tens of microseconds), the discharge-circuit commutator in these circuits must
allow generation of current pulses of up to a few kiloamperes with
a minimal voltage drop (tens to hundreds of volts), and must pass
current in both directions when spherical strobotrons having a
characteristic oscillatory discharge are used.

The running time of lamps in such circuits usually is short,
since a high average power, leading to rapid heating, is dissipated
at high frequencies, even when the flash energy is low. Since the
lamp should not recover its breakdown strength, in tubular lamps
it is convenient to provide a low-current stationary ("keep-alive")
discharge as a continuous triggering factor. Such a discharge is
powered by an external DC supply and maintains continuous ionization
of the gas-filled gap. In some cases the lamp may be connected to
the discharge circuit through a matching pulse transformer.

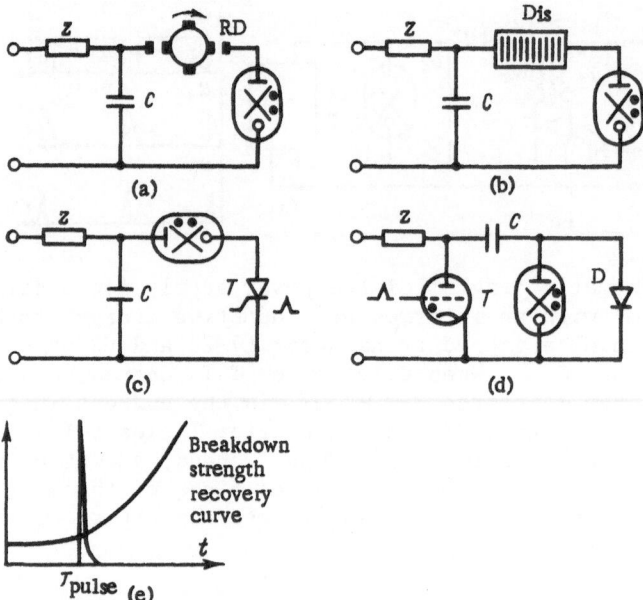

Fig. 7-11. Power supply circuits of flashlamps using a commutator in
            the discharge circuit. a) With a rotating discharger RD
            [7-31]; b) with a multichamber hydrogen discharge Dis [2-
            45, 7-5, and 7-32 to 7-34, and French patent No. 1,227,228];
            c) with a thyristor; d) with a hydrogen trigatron (D is
            the diode of the charging circuit) [7-35 to 7-37].
            In these circuits a discharge-current pulse having a
            length Tpulse passes through the lamp when its process
            of breakdown strength recovery is in the initial stage
            (e).

         Tubular lamps with a small internal volume [2-45 and 7-38]
operate quite satisfactorily in such circuits. The decrease in their
peak luminous intensity upon successive flashes usually does not
exceed 20-40% compared with the first circuit. This evidently is
due to the fact that upon repeated discharges the current flows immediately through the entire ionized gas volume inside the lamp.
Therefore, the channel has a larger cross section and lower resistance
than it does during the first breakdown, when the spark channel
develops with an expansion phase. The effective resistance of the
lamp turns out to be higher during the first pulse. Hence, the
fraction of energy losses is lower in the commutator. The decrease
in gas density due to the escape of part of the gas into the transelectrode volumes of the lamp also has an effect.

The use of spherical lamps in such circuits sometimes is complicated by the development of strong instability of the position of
the discharge channel between the electrodes and by a strong decrease
in the luminous intensity of successive flashes (see Section 5-4).

The limiting flash repetition frequencies (on the order of 10^5
Hz) can be obtained in circuits in which energy enters the lamp
directly from a DC power supply (Figure 7-12a and 7-12b) [7-40],
and the pulse frequency and duration are determined by the commutator.
Similar conditions also are created in the other two circuits shown
in Figure 7-12. In the circuit in Figure 7-12c a packet of successive
flashes is generated by the lamp upon the successive triggering of
the dischargers, which discharge the capacitors onto the lamp (such
an apparatus, containing 20 dischargers, provided a flash frequency
of 50 kHz in a capillary lamp, with a flash energy of 1.5 J [7-39]).
In the circuit in Figure 7-12d [7-24], a transformer with a saturated
core transforms sinusoidal oscillations into rectangular voltage
pulses which power the discharges in a tubular lamp connected to the
transformer secondary.

In these circuits the flashlamp experiences practically no
recovery of its breakdown strength during the flash "packet" and is
constantly on the first gently segment of the breakdown strength
recovery curve, entering a continuous burn when there is a ripple
current. If this happens, the plasma is unable to become completely
deexcited in the interpulse intervals, and a permanent luminous
background is formed in its radiation. The percentage modulation of
the luminous flux of the background decreases with increasing pulse
frequency and energy (Figure 7-12e).

As the pressure of the filler gas and the concentration of energy
released in the discharge channel decrease, the rate of deexcitation
of the plasma at high frequencies may be increased significantly (see
e.g., [7-13]). (Under these conditions of power supply, the high-
pressure spherical lamp described in the previous example would have
a percentage modulation that would not exceed a few percent.) For
lamps with a xenon filler, a frequency of a few hundred kilohertz
apparently is a fundamental physical limit which is defined by the
rate of deexcitation of the gas.

The use of light gases (hydrogen, helium, an argon-hydrogen
mixture) at low pressures in discharge gaps bounded by narrow capillary walls and at low flash energies makes it possible to create even
higher-frequency pulsed light sources with a very short light flash,
a short discharge-deexcitation time, and a high rate of breakdown
strength recovery. However, the amplitude of the luminous intensity
and the luminous efficacy of such light sources turn out to be much
lower than in xenon flashlamps. For example, in the circuit shown
in Figure 7-8a, a flash frequency of up to 400 kHz was obtained in
a laboratory setup using a spark light source [7-42 and 7-43] for

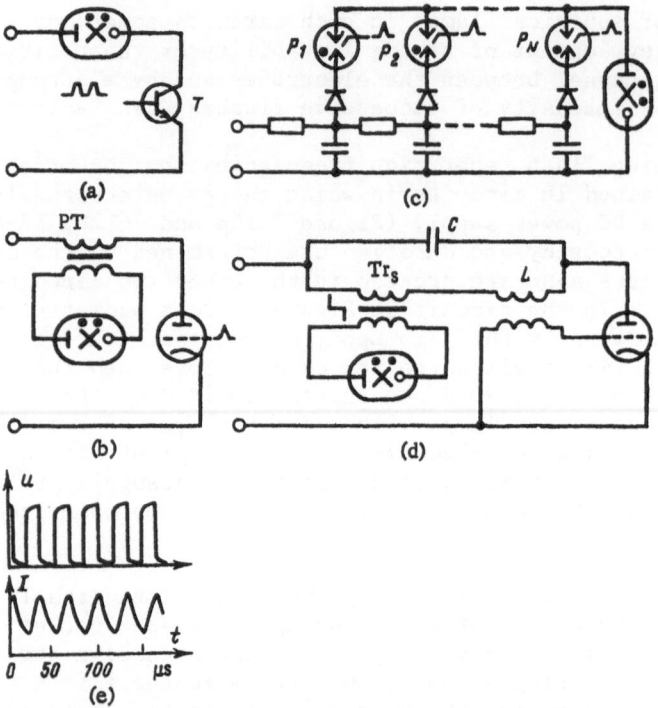

Fig. 7-12.   Schematic diagrams of the power supplies of flashlamps
             in which current pulses through the lamps are formed
             without periodic charging and discharging of the storage
             capacitor. a, b) Circuits with a DC power supply directly
             through transistor T or a generator flashlamp L, such as
             a type GMI-2B lamp; c) a multicircuit power supply con-
             sisting of N circuits connected to the lamp through dis-
             chargers P_1, P_2, ..., P_N successively; d) a circuit con-
             taining an oscillatory circuit consisting of C, L, and
             a transformer Tr_s with a saturated core; e) oscillograms
             of the voltage on the electrodes of a spherical lamp,
             and of its luminous intensity at a frequency of 30 kHz
             [7-40]. Xenon, 0.8 MPa, $l \approx 1$ mm, W $\approx$ 0.01 J. The current
             pulse length is 15 µs, the peak luminous intensity is
             500 kcd, and the luminous intensity is 70%-modulated.
             The decrease in the perecentage modulation limits the
             frequency to tens of kilohertz for an average power of a
             few hundred watts.

an electrode spacing of 23-28 mm, a hydrogen pressure of 0.1-0.2 MPa,
and a supply voltage of about 20 kV. The first flash took place at
an energy of a few tenths of a joule, and subsequent flashes occurred
at 0.04 J. The luminous intensity and duration of the first flash
were 8 kcd and 1.8 µs, respectively; those of subsequent flashes

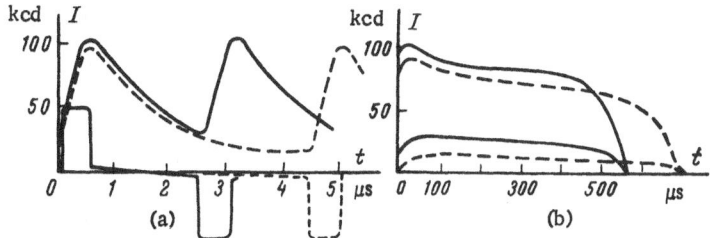

Fig. 7-13. Graphs of the luminous intensity of a tubular xenon lamp
(with a diameter of 3.5 mm and a pressure of 1.3 kPa) at
frequencies of 400 kHz (solid lines) and 230 kHz (dashed
lines) [7-24]. a) Instantaneous values (the rectangles
are the corresponding voltage pulses); b) envelopes of
the packet with respect to the maximum and minimum luminous intensity.

were 1.5 kcd and 0.7 μs. The gas-discharge gap operated at frequency
(sic) in self-breakdown mode at a voltage of about 4 kV. The frequency
was regulated by changing the charging resistance. The percentage
modulation of the luminous flux was 100%.

The overwhelming majority of stroboscopic flashlamps (strobotrons) produced by industry are intended for use in devices that have
an average power ranging from a few watts to hundreds of watts and
comparatively low flash rates (1-10 kHz; generally, the higher the
frequency, the lower the average power). Such flashlamps have simple
circuits that contain ordinary electronic components.

During the design of these devices, if the data on the frequency
characteristics of the flashlamp being used are incomplete, it is
most convenient to determine them under laboratory conditions by
using a very simple power supply circuit, such as a two-cycle circuit*
(see Figure 7-8c) or a circuit with fast charging of the storage
capacitor (see Figure 7-8d). In these circuits a voltage pulse which
inevitably intersects the breakdown strength recovery curve of the
lamp with increasing frequency is produced on the storage capacitor.
Given the flash irregularity noted, this makes it possible to determine one of the points on the breakdown strength recovery curve. In
order to find the other points, the capacitance of the storage capacitor and the working voltage across it can be varied (while holding

---

* In this case there should be an oscillograph, a two-channel pulse
generator (e.g., a type G5-7A), and two trigger-pulse generators to
control the flashlamps, in addition to a variable-voltage power
supply.

the flash energy constant) and the experiment for finding the limiting frequency can be repeated.

The results of the measurements can be represented graphically,
either as the breakdown strength recovery curve, or as a diagram of
the range of controllability of the flashlamp when it is powered by a
capacitor: the maximum working voltage across the lamp at which self-
breakdowns begin to appear is determined for some frequency and constant capacitance of the supply capacitor, and the minimum voltage at
which flashes begin to be missed is determined. The set of points
obtained for different frequencies is used to construct curves of
the upper and lower ranges of controllability of the lamp as a function of frequency, as shown in Figure 7-14. The dashed lines in this
figure show the curves for the upper and lower limits of the power
that can be produced with this lamp when a 2000-pF capacitor is used.

Studies of the range of controllability of various lamps [7-20
and 7-44] have made possible the following conclusions, which are
consistent with our concepts regarding breakdown strength recovery
processes:

(a) as the flash frequency increases, the breakdown potential of the
    lamp decreases and its range of controllability grows narrower.
    Here there is some limiting frequency (5.5 kHz for Figure 7-14)
    beyond which both missed discharges and self-breakdowns may occur;

(b) the limiting frequency is characterized by either a transition of
    the discharge to stationary behavior, or the development of un-
    stable operation with a large number of missed flashes and self-
    breakdowns;

(c) there is some frequency, dependent on the capacitance, for which
    the highest average power is dissipated in the lamp;

(d) for a specified discharge repetition frequency, the power released
    in the spark gap increases with increasing gas pressure and in-
    creasing capacitance of the working capacitor.

The study of the ranges of controllability and limiting loads
of the spark gap in air [7-19], which has a better deionization
capability than xenon, shows that an air gap retains its controllability to frequencies 4-5 times higher than a xenon lamp does (e.g.
20 and 4 kHz, respectively), all other conditions being equal.

The ranges of controllability, the limiting frequencies, and the
average power for which a strobotron can operate are related to the
specific power-supply circuit and may change if a different circuit
is used [7-45]. The ranges of controllability and the limiting
frequencies that have been determined experimentally for lamps in
circuits using fast charging of the storage capacitor just before the

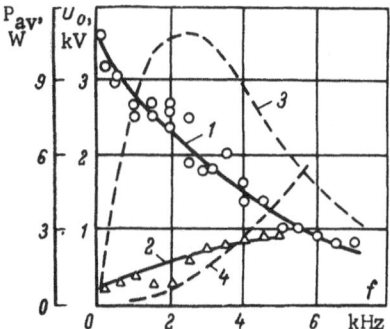

Fig. 7-14. Boundaries of the range of controllability (1, 2) and the
load characteristics (3, 4) of the ISSh-15 lamp as a function of flash frequency (C = 2000 pF) [7-44].

flash are maximal for circuits that do not have a commutator in the
discharge circuit. These frequencies are approximately half as high
for a two-cycle circuit as for fast-charging circuits. The graph
of the ranges of controllability of a lamp in a circuit with known
charging of the storage capacitor usually can be used to estimate the
frequency capabilities of the same lamp in another circuit as well.
Here one always should keep in mind that the breakdown strength recovery curve, like the range of controllability of each specific lamp,
is an individual curve. To make a final selection of operating conditions, one must have relations obtained from studies of a sufficient
number of like strobotrons (the spread of breakdown strength recovery
characteristics for lamps of the same type generally does not exceed
a factor of 2, while the variation of these characteristics during
the operating life of the lamp usually lies in the range 20-40%).

7-3. Operating Life

By the operating life of flashlamps we mean the service life over
which the parameters of the lamps remain constant within prescribed
limits. Operating life is computed in terms of accrued operating
time: the number of flashes is used for lamps with infrequent pulses,
while the number of hours or cycles under fixed discharge conditions
is used for stroboscopic lamps.

Processes of aging of the components and structural assemblies
of flashlamps (electrodes, envelopes, leads) alter some illumination-
engineering and electrical parameters of the lamps and cause random
failures in the form of breakage of the lamp envelope and loss of
controllability.

The diversity of the commercial applications of flashlamps make it difficult to set a single criterion that would define their operating life. For example, the criterion for the operating life of photoflash bulbs and light-warning lamps is a 30% decrease (in the Soviet Union) or a 50% decrease (abroad) in the luminous flux by the end of the lamp's service life. In laser technology the same parameter is limited to a decrease of 10 or 20%. The permissible variation of the electrical parameters of lamps usually is standardized with respect to the lowest power supply voltage of the lamp, with consideration for its specific function and operating conditions. Other characteristics of lamps (spectral composition of radiation, maximum supply voltage, flash delay, etc.) generally change little during operation.

The main factor that causes a decrease in the luminous flux of flashlamps is the gradual blackening of the inside surface of the envelope by the products of electrode sputtering (mainly cathode sputtering), which is associated with the physical mechanisms described in Chapter 4. A thick film on the envelope walls also may cause an increased probability of lamp failure. In tubular lamps the film helps to increase the energy dissipated on the walls of the tube, and accordingly promotes an increase in the temperature in the surface layer, leading to the buildup of internal stresses in the glass and to the growth of microdefects on its surface which can cause rupture of the tube. In spherical lamps the film additionally reduces the surface electric strength of the glass, contributing to self-breakdowns over the surface between electrodes and to a loss of controllability of the lamp.

Some idea of the cathode sputtering in spherical and tubular flashlamp designs that are used in practice can be gained from the following data. Essentially no blackening of the envelope is observed after 5000 hr of use in spherical xenon lamps having a pressure of 0.08 MPa (600 mm Hg), an electrode spacing of 2.5 mm, and a cathode made of grade VNB-3 tungsten–nickel–barium alloy (a design counterpart of the ISSh-15 lamp) at an extremely low average power (1 W) and a flash frequency of 50 Hz. If the same lamps are operated at 15 W, slight darkening of the envelope (15% absorption) occurs after 300 hr of use. At high power levels (hundreds of watts) the material VNB-3 is totally unusable, since heating above 1200°C causes instantaneous vaporization of the nickel fraction of the alloy. This alloy also turns out to be unsuitable for operation with an oscillatory discharge or short (microsecond) high-power (kiloampere) current pulses. Twenty percent blackening of the envelope of xenon lamps (60 mm in diameter, with an electrode spacing of 5 mm and p_0 = 0.3–0.4 MPa) for electrodes made of grade BT-15 thoriated tungsten (1.5% thorium) 5 mm in diameter occurs after 30 min of operation at an average power of 1 kW and a frequency of 400 Hz, after 2 hr at 500 W, and after 10 hr at 2000 W (sic). At an average power of 50 W, the same degree of blackening of the envelope of xenon lamps (p_0 ≈ 0.3 MPa) having a 30-mm diameter and VT-15 tungsten electrodes 2.5 mm

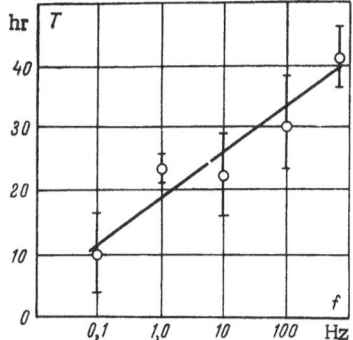

Fig. 7-15. Dependence of operating life of IFK-120 lamps with a film-
type potassium cathode on flash frequency [7-46]. The
supply voltage is 300 V and the average power is 12 W.

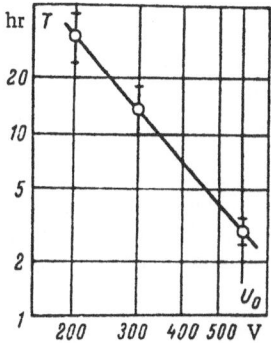

Fig. 7-16. Dependence of operating life of IFK-120 lamps on supply
voltage [7-46]. The frequency is 0.4 Hz and the average
power is 12 W.

in diameter is reached after 5 hr. The sputtering of pure tungsten
electrodes occurs approximately twice as fast under analogous conditions (except for regimes involving maximum power, in which no
significant difference is observed in the rate of sputtering). Some
data on the sputtering of cathodes made of pure metals in spherical
argon lamps at low pressure are given in [4-11].

The variation of the operating life of tubular glass lamps using
a film-type potassium cathode as a function of flash repetition frequency (Figure 7-15) and supply voltage (Figure 7-16) at constant
average power indicates the high sensitivity of a film-coated cathode
to discharge conditions. (The running time of lamps over which the
luminous flux decreases by 25% is taken as the operating life in the

graphs. The average values of the operating life and twice the average absolute spread are plotted in the figures.)

The stability of cathodes made of different grades of tungsten
alloys [7-47] has been estimated from the transmission coefficient of
the envelope and the integrated radiance (in the spectral interval
200-400 nm) for quartz xenon lamps with an inside diameter of 11 mm
and an electrode spacing of 130 mm during infrequent-flash operation
with a load close to the limiting load ($0.9W_{lim}$, $W_{lim} \approx 3000$ J,
$C = 560$ µF, $L = 60$ µH, a current pulse length of 250 µs, with 100
flashes in the series). This estimate showed that the alloy VNB-3
is most stable, while tungsten containing various impurities of lan-
thanum oxide (VL-15), thorium (VT-15), and yttrium (VI-15) is inferior to VNB-3 by a factor of 10-15 (based on the estimated radiation
loss at a wavelength of 300 nm). Thus, the sputtering of electrode
material has a particularly significant effect on the gradual reduction of the radiance of lamps in the ultraviolet region.

Summarizing the available data on comparative tests of tubular
xenon lamps using electrodes made of different materials, we can
put together a rough scale of the comparative service life, as estimated from equal decreases in visible radiation under rated (non-
intensive) operating conditions. The scale is presented in Table
7-1.

An evaluation of the effect of cathode size shows that the
operating life of lamps initially increases with increasing area of
the working surface of the cathode, but then becomes independent of
it. Increasing the xenon pressure roughly quadruples the operating
life.

Another important aging process which develops during the operation of tubular flashlamps is the erosion of the envelope material
and the concomitant decrease in the mechanical strength of the envelope. This is most typical of tubular lamps with a quartz-glass
envelope, which usually operate under high-energy loads. In such
lamps the temperature of the discharge plasma causes gradual vaporization of the wall material. Quartz vapor precipitates in the form
of a white film on the cold regions of the inside surface of the tube.
The rate of film formation increases with increasing discharge energy.
The luminous efficacy is changed little by this film, but the periodic
effect on the tube of thermal shocks that are not uniform over the
surface of the tube leads to thermal stresses and hence to cracking.
The vaporization of quartz due to the discharge has a stronger effect
in intricately shaped flashlamps, such as helical or U-shaped lamps.
Under some conditions vaporization proceeeds so intensely that it
leads to an appreciable local decrease in the thickness of the tube
wall (in capillary lamps, it leads to a significant increase in the
diameter of the capillary). For example, in the IFK-2000 lamp the
wall thickness decreases by 1-1.5 mm at the point where the tube

Table 7-1. A Rough Scale of the Comparative Service Life of Tubular Lamps with Different Cathodes [7-46]

Cathode material	MRN molyb-denum	Alu-minum	VCh forged tung-sten	VT-15 tungsten forged	VT-15 tungsten semi-sintered	VT-50 tungsten, semi-sintered	VT-150 tungsten, semi-sintered	Film potas-sium	Film ces-ium	VNB-3
Test lamps	IFK-2000	IFP-500	IFP-500	IFK-120 IFP-500	IFP-500	IFK-120 IFP-500	IFP-500	IFK-120 IFK-500	IFK-500	IFK-120 IFK-2000
Relative operating life	1	5	5-15	5-15	20	400	600	20	100	800

bends with respect to its inside radius after 10^5 discharges when the lamp is operated with parameters of 1 kJ, 1 kV, 50 μH, and 1 ms. In the ISP-70 lamp the capillary diameter increases by 20% after 500 flashes of operation at 35 J.

In many cases the decrease in the mechanical strength and luminous efficacy of tubular quartz lamps is caused by the formation of a light yellowish-brown film in the discharge region on the inside surface of the tube. The film usually appears during high-frequency operation of lamps under a high unit load (150-250 W/cm^2) with intensive forced cooling. A study of the composition of such films in lamps using electrodes made of VT-15 showed that they do not result from the sputtering of electrode materials (tungsten and thorium are not found in the films), but are dissociation products of silica which result from the interaction of the discharge plasma with the quartz tube [7-48].

Three-dimensional violet coloration of the tube glass, which reduces the light transmission of the envelope by 3-8%, sometimes is observed under certain operating conditions of tubular quartz lamps with forced cooling (when cooling is not halted just before a series of flashes, or under underloaded conditions). This coloration is reversible, vanishing after the glass is heated (thermal decolorization) [7-49]. Three-dimensional coloration of quartz glass due to shortwave radiation [7-49 and 7-50], which also is characteristic of a xenon discharge, occurs most often in tubes made of a glass that contains impurities. Coloration is insignificant under the standard operating conditions of lamps because of the intense heating of the envelope [7-48].

The increase in trigger voltage that sometimes occurs during prolonged operation of lamps is due mainly to the release of gas (as a result of discharges) from the envelope walls and the electrodes. Under conditions far from the limiting conditions, this phenomenon can be largely prevented by appropriate aging of the lamps with pulsed discharges of somewhat higher energy just before they are finally filled, and by using suitable getters in the lamp design. Under very large loads the phenomenon is due to the vaporization and dissociation of silica (see Chapter 6) and cannot be eliminated entirely.

A general idea of the operating life of flashlamps is given by the curves of its dependence on power supply conditions (Figures 7-17 and 7-18). Estimation of the expected operating life of tubular lamps during infrequent-flash operation by means of such plots with a reference point corresponding to the limiting energy W_{lim} is used extensively in practice.

The operating life of flashlamps is at least as dependent on the parameters that characterize the discharge-current pulse as it is on the flash energy. All other conditions being equal (flash

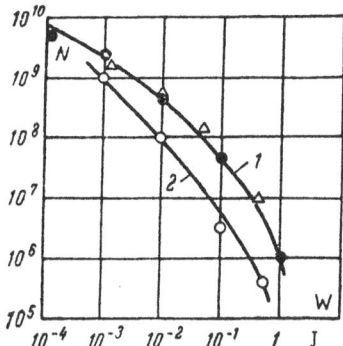

Fig. 7-17. Dependence of the operating life (the number N of flashes)
            of spherical flashlamps on the discharge energy. The
            circles represent the ISSh-4 and ISSh-4-1 lamps (1) and
            the ISSh-100-3M lamp (2) [7-51]. The triangles represent
            data on the operating life of FX-108 and FX-137 lamps with
            a series-connected diode to cut off the negative half-
            wave of the current [7-52].

energy and duration, average power, type of lamp, etc.), the longest
operating life of a lamp can be obtained for aperiodic discharge with
a current pulse shape approaching a half-sinusoid [7-56]. For example,
given the same discharge energy, the operating life of the IFP-800
lamp is nearly halved if the amplitude of the current in the second
half-period. The data presented in Table 7-2 provide some idea of
the quantitative variation of the operating life of lamps (the criterion for the operating life is a 30% decrease in luminous efficacy)
as a function of the discharge-circuit parameters. As we can see
from this table (columns 4 and 5), the triggering method has just as
significant an effect on the operating life of lamps [7-57 and 7-58].
In particular, a trigger circuit using a "keep-alive arc" makes it
possible to increase the operating life of tubular lamps during stroboscopic operation by a factor of approximately 5 over pulse triggering.
(The convenience of a "keep-alive arc" also is related to the possibility of supplying the lamp at minimum voltage. This makes it
easier to select a pulse shape that is optimal in terms of increasing
the operating life of the lamp.)

    Mechanical, climatic, and other environmental factors have no
practical effect on the operating life of lamps if they do not exceed
the values stipulated by the specifications for a given type of lamp.
An exception is the effect exerted on the service life of lamps by
certain coolants which are used in laser equipment to provide forced
liquid cooling of pump lamps and which simulataneously act as ultraviolet filters (French patent No. 1,589,478). Gradually decomposing,
such a coolant forms a dark film on the outside surface of the tube

Table 7-2. Effect of Power Supply Conditions and Discharge Triggering Conditions on Operating Life

Item No.	Type of lamp or size of discharge gap	Type of discharge: A - aperiodic; O - oscillatory	Triggering: P - pulsed; K - keep-alive	Discharge parameters				No. of flashes
				C, μF	L, μH	U_0, kV	W, J	
1	IFK-2000	A	P	8000	1	0.32	400	4×10^5
2	IFK-2000	A	P	800	1	1.0	400	5×10^5
3	IFK-2000	A	P	800	700	1.0	400	2×10^6
4	Counterpart of IFK-800	A	P	100	100	1.45	100	10^5
5	Same	A	K	100	100	1.45	100	5×10^5
6	ISSh-100-3M	O	P	0.25	0.5	3.5	3	0.3×10^5
7	ISSh-100-3M	A	P	0.5	0.5	4.5	5	0.3×10^6

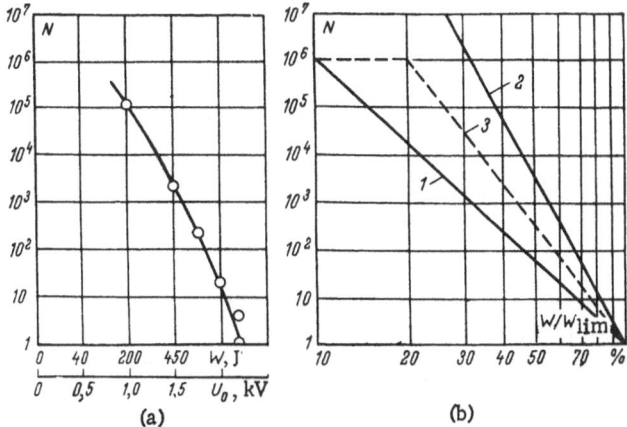

Fig. 7-18. Dependence of the operating life of lamps (the number N of flashes) on the discharge energy. a) the FX-38A lamp, with an inside diameter of 4 mm, a length of 76 mm, C = 400 F, L = 300 H, and a current pulse length of 0.9 ms (a 50% decrease in integrated luminous intensity is the operating-life criterion) [6-1]. b) tubular quartz pump lamps produced by Edgerton, Germeshausen and Greer and by ILC Technology (United States). 1, 2) the limits of the minimum and maximum operating life, respectively; 3) the standard curve given in company catalogs [7-53 to 7-55].

which can be eliminated by special cleaning (tap water also leaves a rust film). Coolants which have an aggressive effect on the quartz envelope of a lamp may not be used, since erosion of the envelope surface lowers its strength and creates the prerequisites for failure of the lamp.

7-4. Mechanical and Climatic Properties, and Other Operating Features

The overwhelming majority of flashlamp designs are extremely heavy-duty (the metal parts outside the seal are short and have high normal-mode frequencies). Having no external fittings, they are quite durable and remain operable under considerable mechanical overloads (vibration, shocks, constant acceleration).

In quartz lamps cyclindrical foil leads [7-60] have significant advantages over cap-type leads [5-72] or leads based on transition glasses (see, e.g., US patent No. 3,742,117) in terms of mechanical strength. This fact is used extensively in the design of Soviet flashlamps, most of which meet the mechanical strength requirements for the heavy-duty ratings of GOST 16962-71 (vibratory loads of 1-600 Hz, vibratory accelerations of 50-100 m/s^2, shocks of 450-1500 m/s^2, and linear accelerations of 500-1500 m/s^2). For example,

IFK-2000, IFK-75, and ISK-25 flashlamps operate successfully in aircraft light-signaling devices without any special shock-absorbing measures. Shock-absorbing accessories usually are required for large lamps when they are used in devices that are subjected to significant mechanical loads.

As we examine the effect of the surroundings on the performance of a lamp, we should keep in mind above all how the surrounding medium affects two characteristics of the lamp: the load characteristic and the range of controllability. A low-ambient-air temperature does not affect either of these two characteristics all the way down to the boiling point of the gas filler (-108°C for xenon), but a high temperature reduces the maximum average power that can be dissipated by the lamp as the difference decreases between the ambient temperature and the maximum permissible temperature of the lamp glass or of the seals of the current-carrying leads. When lamps are operated in closed chambers of limited volume, allowance should be made for the overall effect on the lamp of the ambient temperature and the temperature produced by the power dissipated in the lamp, which must not exceed the limiting values.

A decrease in ambient air pressure affects the triggering of lamps above all: at pressures below 4 kPa (30 mm Hg) but above 10^{-2} Pa, the high-voltage control pulse supplied to the trigger electrode is shunted over the surface by the discharge. As a result, the lamp may lose its controllability. Trigger circuits without an external control electrode, similar to the ones shown in Figure 1-2 (c to f) with high-voltage leads that are sealed (e.g., by pouring on an insulating mass), usually are used to ensure the performance of lamps at such pressures. Lamps can be triggered quite reliably in a deep vacuum by using an external electrode. However, the permissible average load on the lamp is reduced considerably in this case, since energy is dissipated into space only by radiation. The maximum average power dissipated in this manner can be estimated for quartz lamps whose permissible temperature is 850°C by using the power of a "graybody" emitter (having an integrated radiation coefficient of ~0.5) that has dimensions equal to those of the working portion of the envelope, for heat losses of 0.2-0.9 in the discharge (depending on the luminous efficacy of the lamp). The thermal radiation of the envelope makes up a much smaller fraction of the blackbody radiation for glass lamps, which can only be operated at low temperatures.

Elevated pressure of the ambient air or liquid has a practical effect on the characteristics of lamps, up to values which exceed their mechanical strength.

Immersing a lamp into a liquid medium that has good electrical insulation properties (e.g., carbon tetrachloride) makes it possible to increase the average load on the lamp without affecting its range of controllability. If the liquid is electrically conductive (e.g.,

ordinary water), then the trigger voltage of the lamp decreases
appreciable and the self-breakdown potential is approximately halved.
The circuit shown in Figure 1-2c is preferable for the triggering of
tubular lamps. The distribution of the electric potential on the
walls of the chamber surrounding the lamp has a somewhat smaller effect
on the range of controllability, though the effect is nearly the same.
Finally, a reduction of the gas ionization background by external
agents (light, cosmic rays) may affect the voltage and the firing lag.

The external portions of the current-carrying leads and the base
are the structural components of flashlamps that are exposed to
humidity. Foil leads are the most vulnerable in this regard. Lamps
designed for operation at high humidity employ various methods to
protect these assemblies from moisture. These involve the use of
sealants, the application of a corrosion-resistant coating to the lead,
or special designs which use a complex combination of a foil lead with
a cap-type lead made of transition glasses ([7-61] and US patents
Nos. 3,420,944, 3,675,068, 3,785,019 and 3,793,615).

The level of the electromagnetic noise that appears during the
firing and operation of flashlamps is of considerable importance in
some cases in which flashlamps are used in laser technology and automatic equipment. The noise is generated by rapid changes in time
of the voltages and currents due to trigger pulses and the primary
discharge in the lamp. The time when a noise signal appears and the
maximum noise level occurs coincide in time with the breakdown of the
discharge gap of the lamp by the trigger-pulse voltage, i.e., it
coincides with the formation of the auxiliary discharge. All subsequent variations of the voltages and currents during a pulsed
discharge are accompanied by an interference level that is at least an
order of magnitude lower.

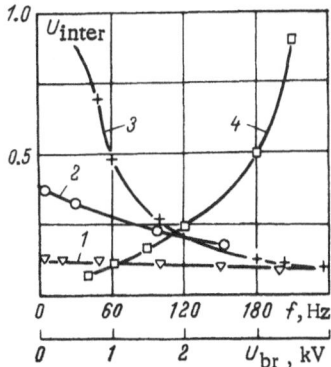

Fig. 7-19. Relative level of radio interference U_{inter} due to the
operation of an ISSh-2 lamp at various flash frequencies
f (curves 1-3) and various breakdown voltages U_{br} of the
cathode spark gap (curve 4). 1) The cathode gap was illuminated by an auxiliary discharger; 2) no illumination
from the discharger; 3) the same lamp without a getter
[7-62].

As we can see from Figure 7-19, the noise level is approximately proportional to the square of the breakdown potential of the cathode gap, decreasing with increasing flash repetition frequency. The lowest noise level and frequency independence of the noise level are observed in lamps which use preillumination of the cathode gap with a spark from an auxiliary discharger (curve 1). In such lamps this gap breaks down at the lowest voltage. A lowering of the noise level which is just as effective occurs in tubular pump flashlamps when firing by means of a pulse transformer is replaced by triggering by a circuit using a "keep-alive arc."

CHAPTER 8

INDUSTRIAL PRODUCTION OF FLASHLAMPS

8-1. The Main Features Used to Classify Lamps

The main design and operating features of a flashlamp and its photometric parameters are determined by its field of application and the nature of the corresponding optical problem or task. First a selection is made of the type of pulsed discharge (a wall-confined or a freely expanding discharge) and hence of the basic design of the lamp (tubular or spherical), as the design elements differ particularly significantly for these.

Spherical lamps, which can dissipate much more power and have a much longer operating life than capillary lamps with the same electrode spacing, have indisputable advantages in cases where the main requirement is to obtain a minimal glow volume (not necessarily strictly localized) with a high peak luminance and a short flash duration (Table 8-1). The use of a tubular lamp to produce short flashes from a small glow volume can be justified only when the operating life and average power are secondary factors compared with the positional stability of the discharge channel (such stability is especially important, for example, when a narrow slit or some other small object must be illuminated) or compared with the compactness of the lamp as a whole (when the lamp is used in narrow cavities). It also is preferable to use spherical lamps in most cases where a comparatively high flash frequency is required with a low energy per flash. In these cases the luminous efficacy of spherical lamps, which is weakly dependent on the flash energy, becomes greater than that of tubular lamps. By contrast, except where strict localization of the glow volume is required, tubular lamps are more advantageous in cases in which the product of the channel radius by the flash energy per cubic centimeter of the channel exceeds 2 J/cm^2, the flash duration is tens

289

Table 8-1. Main Areas of Application of Flashlamps

Applications*	Most important features	Class of lamp
Most tasks in general and special photography (except for superhigh-speed photography and microphotography); light signaling and warning, printing	High luminous energy. Special shape of glow volume	Tubular
Optical pumping, photochemistry	High luminous energy. Special shape of glow volume. Maximum matching of emission spectrum of source to absorption spectrum of detector	Tubular
Illumination of narrow optical slits in optoelectronic devices, high-speed photography, etc.	Special shape and strict localization of glow volume	Tubular
Equipment with highly directional data and command channels (communications, optical location, control); stroboscopy, superhigh-speed photography	High frequency and low flash energy	Spherical
Photolithography (e.g., in the production of microelectronic components), microphotography	Matching of emission spectrum of source to absorption spectrum of detector. Small glow volume and short flash duration	Spherical

*The classification is somewhat arbitrary, as some applications go beyond its framework. For example, lamps of both classes are used in photometry: the ISK-25 tubular lamp is used in the FM89 pulse photometer, and the spherical ISShO-1 is used in the FM120 photometer.

of microseconds or longer, and at least one dimension of the glow
volume is tens of millimeters or larger.

The design of the lamp is the second classification feature.
Depending on the requirements for spatial distribution of the luminous flux, tubular lamps are subdivided into lamps with a straight
cylindrical discharge cavity and lamps with a cavity having a complex
configuration helical, U-shaped, rectangular-bore, coaxial, and so
forth. Given the same type of glow body (small volume and linear
dimensions), spherical lamps may differ significantly in structural
elements (the number and arrangement of the trigger electrodes in
the discharge gap, the design of the current-carrying leads, shell,
bases, etc.).

The next classification feature is the type of operation for
which the lamp is designed: high-frequency (stroboscopic) or single-
flash operation. Despite the fact that most types of flashlamps can
be operated in both modes, this feature is important because the tasks
of increasing the economy and efficiency of lamps for a specific application with allowance for operating conditions often determine the
selection of special materials and design decisions which make lamps
of a given type significantly different from others.

The three main features mentioned above are reflected in the
system of conventional nomenclature for Soviet flashlamps that was
established by GOST 19685-75 [8-1]. According to the standard, the
first elements of the designation (the letters) characterize the function of the lamp, while subsequent elements define the shape of the
discharge region or the design of the lamp. Numbers are used for
the serial number of the lamp type and for the dimensions of the discharge region (for pump lamps), the discharge energy (for photoilluminators), or the average power (for lamps operating at high
frequency). For example, INP2-6/240 represents a flashlamp (I),
pump type (N), straight (P), with a glow body 6 mm in diameter and
240 mm long; SK-25 represents a light-signaling flashlamp with a
glow body of complex configuration having a rated discharge energy
of 25 J (before the introduction of GOST 19685-75, the designation was
ISK 25).

Apart from the main classification features listed, flashlamps
also may be characterized by some additional features. There are
lamp types which are characterized by a low self-inductance and a
low-level magnetic field generated by the discharge; there are lamps
designated for low and high working voltages which differ in the
material used in the envelope (quartz or glass) and accordingly in
the spectral composition of their radiation; and so forth. The
continuing expansion of the areas of application of flashlamps and
the novelty of this field help promote the development of new features
in flashlamps and new ideas for implementing them in designs.

8-2. Design Priciples and Production Processes

   The main structural elements of a flashlamp are:

(a) the envelope and the gas filler of the lamp;
(b) the electrode assemblies: the cathode assembly (including the
     cathode, the getter, the lead going into the glass, and the lead-
     out); the anode assembly (the anode, the lead-in, and the lead-
     out); and the control electrode (the trigger electrode);
(c) external components: the base, the protective envelope, forced-
     cooling fittings, positioners, etc.

   For different types of lamps all these elements differ not only
in size, but often in design concept. Individual components sometimes
may be combined (e.g., the cathode mount may be combined with the
trigger electrode or even with the anode mount in some spherical
lamps) or omitted (baseless lamps; lamps with no control electrodes,
designed for operation in the circuits shown in Figure 1-2c-e; lamps
with no getter; etc.). There also may be additional structural
elements: additional control or current-carrying electrodes, a
reflector, regular electronic components that are built into the
design of the lamp, etc.). Therefore, in a brief description of the
design of flashlamps and their production processes it is wise to
consider only the most widely used technical solutions that are
employed in series-produced designs.

   The Envelope and Gas Filler. The envelopes of light-signaling
and photoilluminating lamps are made from a glass or quartz tube
whose length and diameter are selected mainly on the basis of data
on permissible loads during infrequent-flash operation. Depending on
the optical task to be handled, flame working can be used to give
the tube the proper configuration (a ring, a U shape, a cylindrical
spiral, a flat spiral, or a spiral inscribed in a sphere, etc.). The
production volume determines the processes used in manufacturing the
envelopes: manual labor, simple stamps and mandrels, or a mechanized
method. Examples of lamps with different kinds of envelopes are lamps
with a straight tube for photographing the tracks of nuclear particles
in cloud or diffusion chambers; annular lamps which are put around
the lens of a camera for shadowless photography; lamps having a tube
in the form of a flat zigzag for uniform illumination of the surface
of a negative during reproduction; lamps with an end-on radiation
exit for some medical photodiagnostic devices; and others.

   As we noted in Chapter 6, the permissible load on a lamp increases significantly with increasing flash duration, e.g., with
increasing supply capacitance and decreasing supply voltage. For
this reason, the trigger voltage of the lamp is the second element of
design, after the flash energy (ignoring the illumination-engineering
calculation which determines the size of the glow volume). The
required value of the trigger voltage is determined from the supply

voltage (which is specified according to the optimal capacitor, current source, and other factors), with some margin allowed for the spread of this voltage from sample to sample and for the possible increase in the trigger voltage of lamps during their service life. In other cases the second element may be the requirement that the lamp's flash duration not exceed some value set by the conditions of use (e.g., the requirement that there be no image displacement during the photography of a moving object).

A computational estimate of envelope parameters usually is made in the following manner in the first group of cases. The required luminous energy of the radiation pulse is determined from the required amount of lighting on the object, its distance, and the gain of the optical system. The rough value of the luminous efficacy is used to estimate the discharge energy that must be supplied to the lamp from the capacitor.

The working voltage of a lamp usually is specified according to the voltage of the supply capacitor. The values of C and U_0 make it possible to use formula (6-6) or (6-7) and Figures 6-4 and 6-6 to 6-9 to determine the corresponding electrode spacings l for envelopes made of glass or quartz tubes having different inside diamters d_i. The values obtained for l and d_i are compared with the permissible dimensions of the glow volume and with data on the trigger voltage. The adaptability of the design to manufacture and other economic factors determine the advantage of glass lamps. Therefore, it is preferable to select a glass envelope for photoilluminating and signaling lamps, provided that the required trigger voltages are met. If reliable triggering of the lamps is not provided for the value of l calculated for the glass, then it is wiser to use a higher-voltage capacitor than to use quartz glass. Only the specific requirements of small dimensions of the glow volume, which produce high-energy loads on the tube wall or a high average power, justify the use of a quartz discharge tube.

If the flash duration must be limited, the design procedure is further supplemented by a comparison of the set of values of l and r with data on the dependence of τ on the parameters. Such a calculation becomes particularly simple if the maximum permissible flash duration exceeds 300 µs. In this case

$$\tau \approx \frac{1.2^* RC}{2} = \frac{2.4\rho l C}{\pi d_l^2}. \tag{8-1}$$

*On average, the duration of the light pulse exceeds the duration of the electric power pulse by 20%. Hence the factor 1.2 is introduced here.

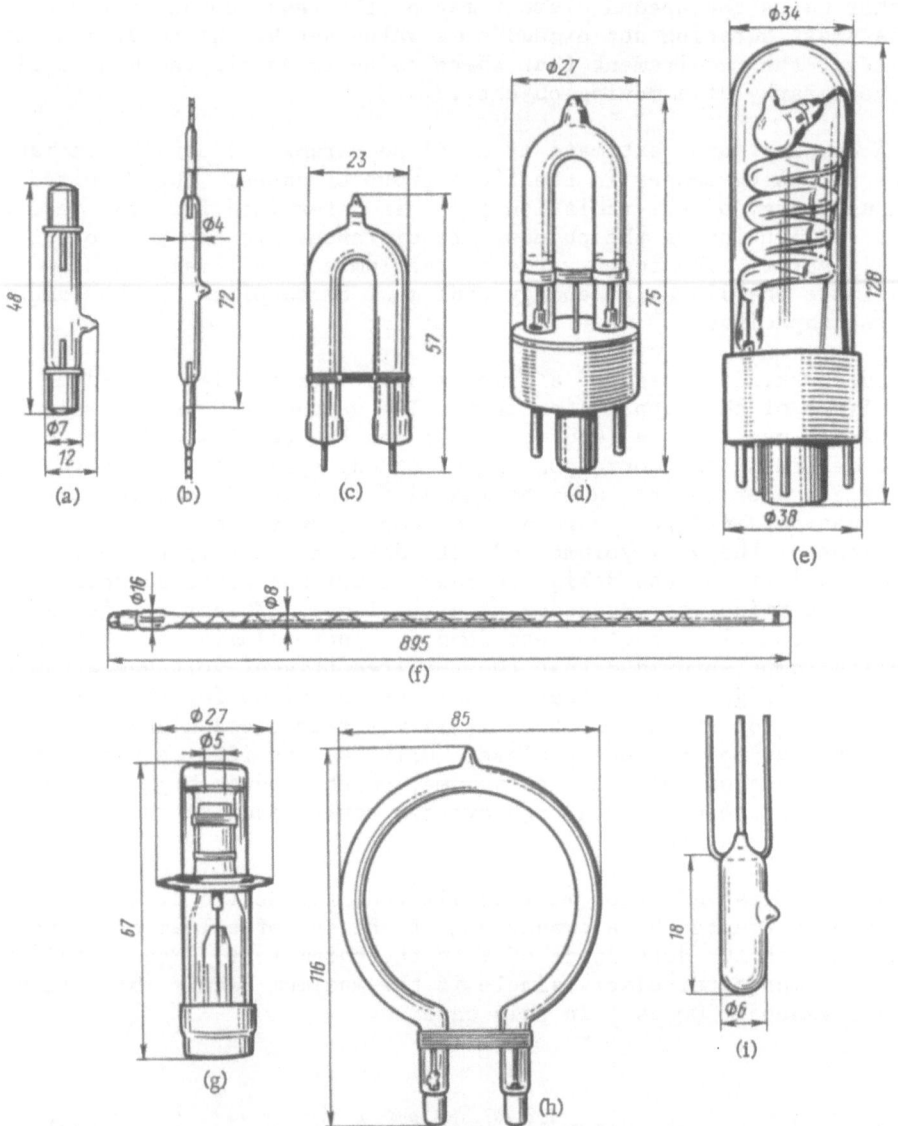

Fig. 8-1.   Tubular glass flashlamps for photography.
a) IFK-20;   b) IFK-20-3;   c) IFK-120;   d) ISK-10;
e) IFK-500;   f) IFP-4000;   g) IFT-200;   h) IFB-300;
i) FX-27 (United States).

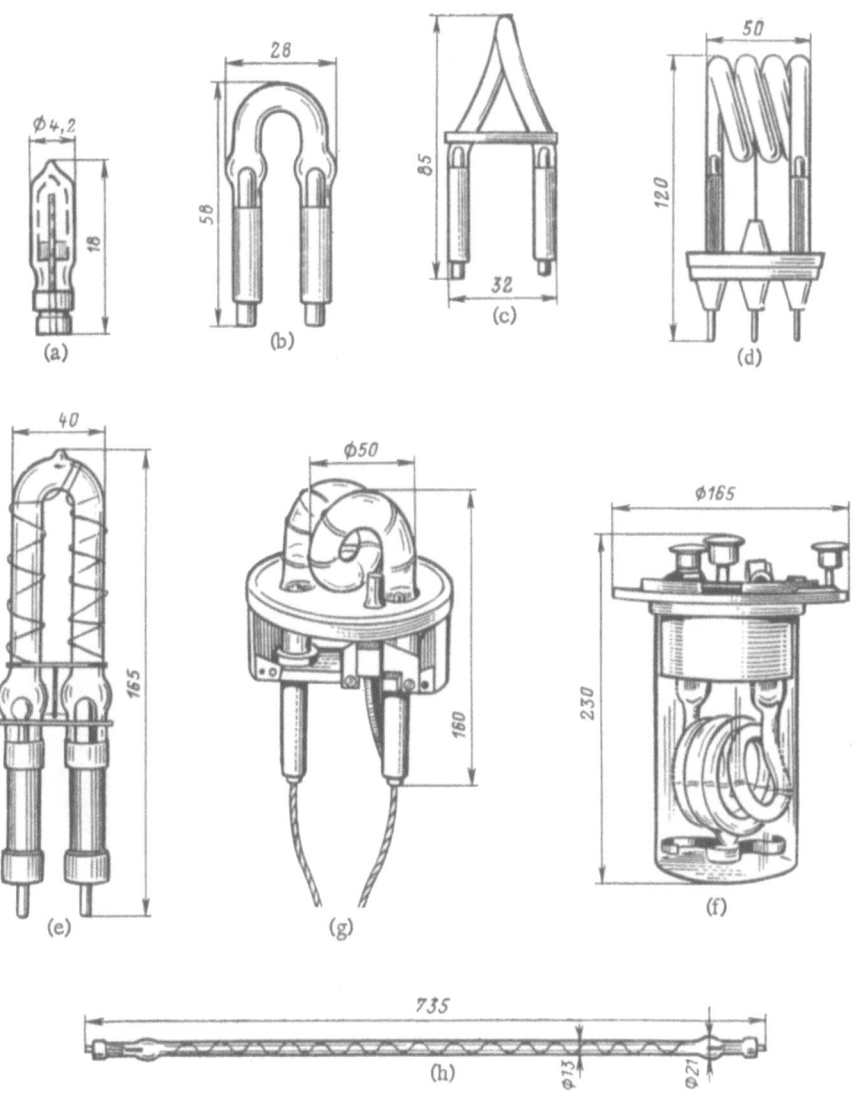

Fig. 8-2. Tubular quartz flashlamps for photography and signaling.
a) IFK-15-2; b) ISK-25; c) IFK-75; d) IFK-150;
e) IFK-2000; f) IFK-2000-3; g) IFK-20000; h) IFP-
15000; i) FP-1000; j) 103723 (Netherlands); k) ISP-
15; 1) ISK-200.

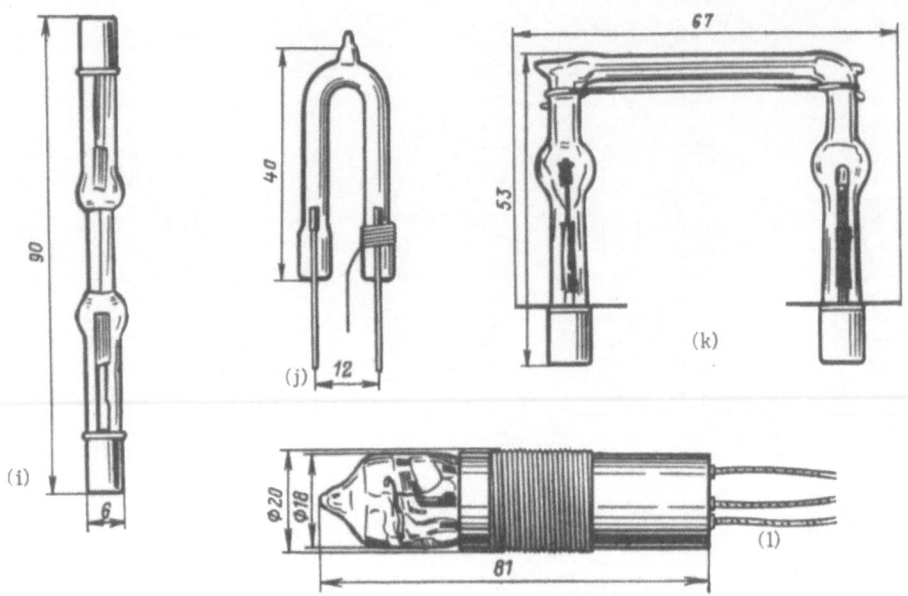

Fig. 8-2, continued.

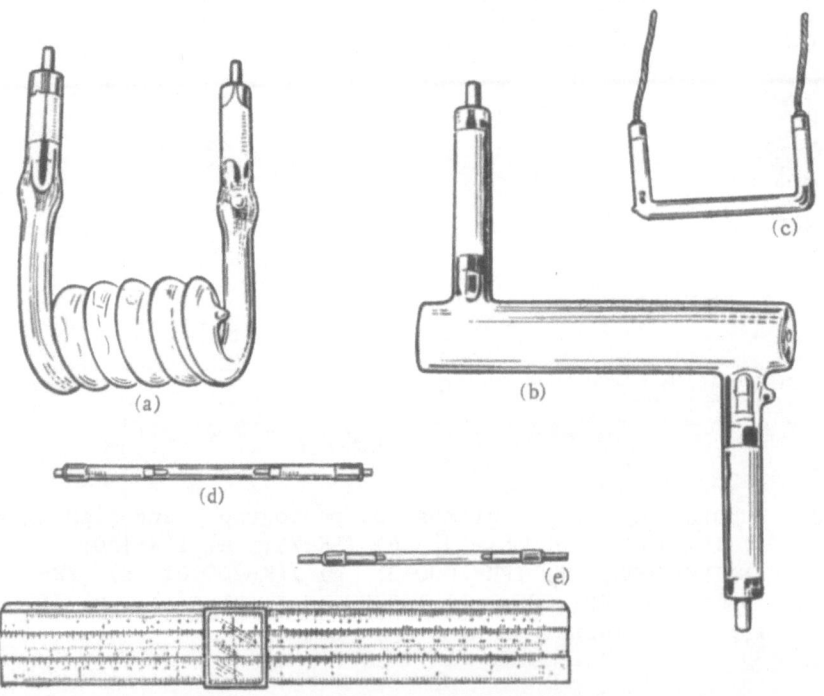

Fig. 8-3. Quartz pump flashlamps. a) IFK-15000; b) IFPP-7000;
c) IFP-600; d) ISP-600 (lamps a through d are foil type);
e) IFP-1000 (with cap-type leads).

Using formula (2-1) (the effective value $U_0/2l$ is taken as the electric field strength) and formula (6-6) (in which we take 0.6, the limiting value of K, as the permissible value of the constant in the righthand side, as follows from the graph in Figure 6-4), we find

$$d_i \approx 0.07\, \tau^{-1/2} (C^7 U_0^7 / K^{5/2})^{1/9}, \qquad (8\text{-}2)$$

$$l = 1.1 U_0^{4/3} (C/K)^{1/3}, \qquad (8\text{-}3)$$

where l and d_i are given in centimeters, C in microfarads, U_0 in kilovolts, τ in microseconds, and K in $\mu F\text{-}kV^4/cm^3$.

For cases in which K is independent of the diameter, formula (8-3) gives the overall length of lamps with different flash durations. According to (8-2), such lamps differ only in their diameter, which varies inversely with $\sqrt{\tau}$.

In order to achieve high-energy loads, the envelopes of lamps for the optical pumping of lasers generally are made of quartz glass. A check calculation of the limiting load is done for such lamps, the size of whose discharge region is rigidly linked to the size of the active elements of lasers, and the expected operating life is estimated from the ratio of the planned operating conditions to the limiting conditions, GOST 17399-72 has established a standardized series of geometric sizes, discharge energies, and flash durations for Soviet quartz pump lamps designed to operate in single-pulse mode [8-2].

The size of the envelope of lamps intended for high-speed photography and the illumination of narrow slits in optoelectronic devices is determined entirely by the nature of the optical task or problem. The diameter of the tube is selected so that the discharge channel fills the cross section of the tube, and the length corresponds to the size of the slit diaphragm or reflector. The envelopes of such lamps usually are made of a glass or quartz capillary to which sections of a wide tube are soldered. These sections are designed to hold the electrodes.

The selection of the envelope for spherical flashlamps, which normally are intended for high-frequency operation, is based on data on the permissible average power per square centimeter of surface area of the envelope (both for prolonged operation of the lamp under steady thermal conditions, and for operation in short series of flashes during which stationary thermal conditions do not become established in the lamp). If the requirements placed on illumination characteristics make it necessary to fill the lamps to high pressure, then the envelope is made near-spherical. The wall thickness usually is

selected in the range 1.5-3 mm, ensuring maximum strength of the envelope. It is inadvisable to increase the wall thickness beyond this
range because of the increase in inhomogeneous internal stresses in
the glass. In general, the residual stresses in envelope walls that
remain after flame working of the glass have a decisive effect on
the strength of the envelope, and much attention is paid to relieving
these stresses by annealing the glass during the production of

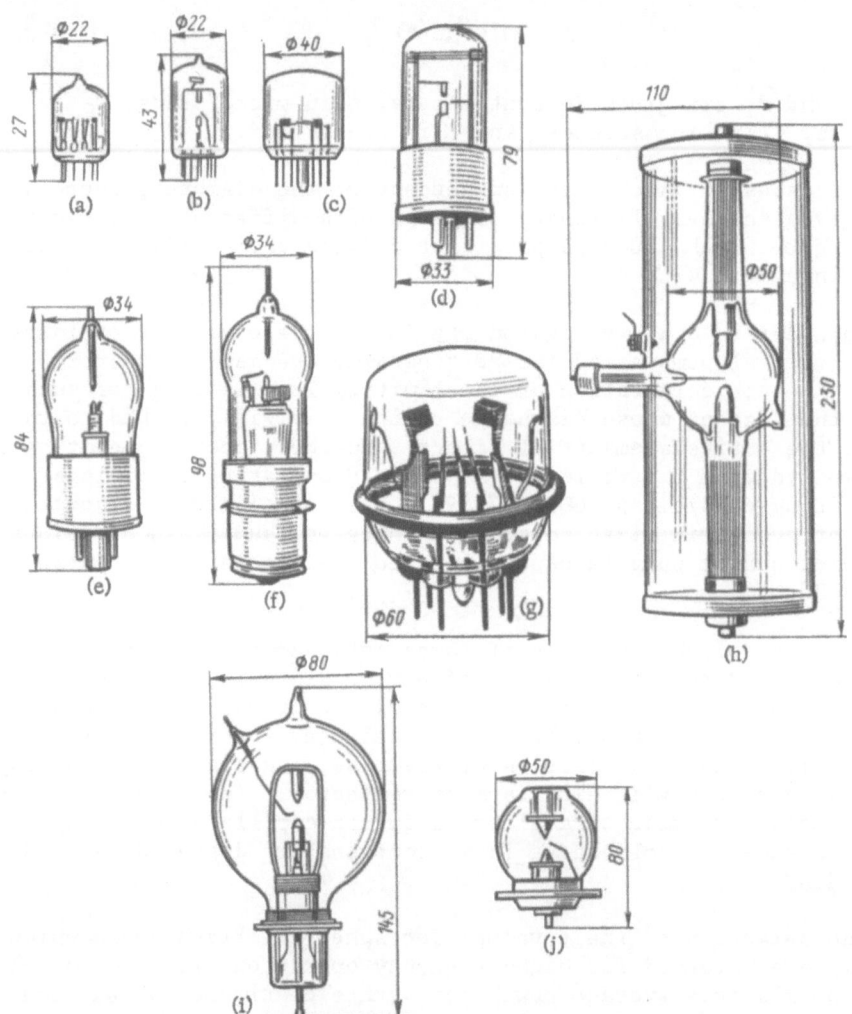

Fig. 8-4.   Spherical flashlamps. a) ISSh-4; b) ISSh-5; c) SSh-
            12; d) ISSh-15; e) ISSh-100-5; f) ISSh-100-3M;
            g) FX-132 (United States); h) ISSh-500; i) ISSh-300;
            j) ISSh-400-3.

high-pressure lamps. If the optical task can be handled at a filler
gas pressure below 0.1 MPa, then the lamp envelope may be cylindrical:
apart from simplifying the production of the envelope, low pressure
is advantageous because of the simplicity of the process of filling
the lamps, because of their explosion hazard, and because of the
decrease in the supply and control-pulse voltages.

If the lamp must have enhanced emission in the ultraviolet
(e.g., lamps for photochemistry) or by contrast must lack ultraviolet
radiation (some pump lamps), then the lamp envelope is made from
special grades of glass or quartz with a specific spectral passband
[5-72 and French patent No. 1,589,478], or a window made of such glass
is sealed into the envelope (lamps for photolithography).

Most types of flashlamps are filled with xenon, which provides
maximum values of spectral efficiency throughout virtually the entire
range of wavelengths. It also provides the best load characteristic
and the longest operating life (at equal pressures, the rate of
sputtering of electrode materials is lowest in xenon). The selection
of the xenon pressure usually is based on a compromise between the
desire to ensure maximum luminous efficacy of the lamp and a minimum
trigger voltage (ordinary tubular lamps) or the minimum probability
of explosion of the envelope (spherical lamps and high-voltage tubular
lamps for photochemistry with an ultrashort flash duration). For
example, designers usually stop at a pressure of 0.013-0.04 MPa
(100-300 mm Hg) for most tubular photoilluminating and warning lamps
having a tube inside diameter of 5-20 mm, since there is practically
no increase in the luminous efficacy above this pressure. The xenon
pressure in pump lamps and capillary lamps, which do not require a
particularly low trigger voltage, is placed in the range 0.04-0.08
MPa to increase their operating life. The xenon pressure sometimes
is reduced to a minimum (e.g., to 0.006 MPa) if ensuring a minimal
trigger voltage is one of the main requirements, and the impairment
of other parameters is ignored. The use of krypton, neon [5-63 to
5-69] and mixtures of xenon with krypton, mercury vapor, or the vapor
of alkali, alkaline-earth, or other metals to fill pump lamps has
not yet found commercial application, since the resultant small increase in pump efficiency does not compensate for the worsening of
other characteristics of the lamps, especially the operating life.

Spherical flashlamps usually are filled with xenon or a mixture
of xenon with light molecular gases (hydrogen, nitrogen). Argon or
nitrogen is used less often. The bases for the selection of a filler
are data on the luminous intensity, luminance, and flash duration of
the lamps, as well as data on their frequency characteristics. At
high energy concentrations (a high supply voltage, low inductance of
the discharge circuit, sufficient capacitance, and a short service
life), the best solution may be to fill the lamp with a xenon-
nitrogen mixture at an overall pressure of about 0.09 MPa and a

nitrogen partial pressure of about 0.03 MPa. For low energy concentrations, xenon retains its advantage, as it ensures the longest
operating life under all conditions, and in this case also provides
better illumination characteristics. The natural shortcoming of a
xenon filler (the increase in flash duration during an oscillatory
discharge) can easily be compensated for by including small ballast
resistors (a fraction of an ohm) or diodes in the discharge circuit.
These components damp current oscillations, approximately halving
the flash duration, and at the same time slightly reduce the peak
luminous intensity and luminance while simultaneously producing a
significant increase in the operating life of the lamps.

    Electrode Assemblies. Metal rod leads made from bars of tungsten,
molybdenum, or special alloys are used for practically all types of
tubular glass lamps. The selection of the lead material is governed
by the requirement that the thermal expansion coefficients of the metal
and glass be matched. Selecting the cross section of a lead that will
ensure its integrity under the most severe operating conditions of
the lamp poses no difficulties: a rod lead 1-1.5 mm in diameter
usually is sufficient for even the most powerful glass lamps. It
usually is advisable to reduce the diameter even further for reasons
related to the mechanical stiffness of the electrodes. Therefore,
bar leads are used for tubular glass lamps with different power
ratings. The manufacture of such leads during mass production of
lamps can be mechanized by stamping one or two metal rods directly
into the neck of the discharge tube. Disk-type leads or multirod
leads in the form of flat bases or stem presses are used in spherical
glass lamps with an end-on radiation exit. Here allowance must be
made for the method used to mount the lamp in the apparatus, which
must have extremely low inductance, be compact, and have good mechanical strength. The design of such leads, which are used widely in
electric vacuum devices, and the production processes have been described in detail in the corresponding literature [8-3 and 8-4].

    Leads having the following types of designs are used most widely
in industrial quartz flashlamps: flat foil leads, cylindrical foil
leads, rod leads based on transition glasses, and cap-type leads
with a soft-solder seal (so-called "glued" cap-type or rod-type leads
with a polymer-compound seal [8-6] and dismountable leads with a
mechanical seal based on gaskets made of special rubber or soft metals
such as lead or copper [8-7]). Each of these types of leads has
specific operating and technological features which are taken into
consideration during the design of each specific type of lamp.

    The main advantages of foil leads are [7-60]:

(a) reliable vacuum tightness of the lead;
(b) heat resistance, which makes possible the high-temperature
    cleaning of parts. This is essential to the operating life of
    the lamps [8-8];

(c) the mechanical strength of the lead and the entire electrode
    assembly. Combined with electrical insulation properties, the
    mechanical strength makes it possible to use very simple designs
    in the fittings for mounting the lamp in the illuminator (this
    is particularly important for the illuminators of lasers, liquid-
    cooled systems, and small equipment).

The main negative features of foil leads are the limitation on
current strength that was described in Section 6-2; comparatively
large length; and low corrosion resistance of the outer unsealed areas
of molybdenum foil when exposed to moisture and the ambient temperature over a prolonged period of time. The last shortcoming can be
eliminated by using the sealants or molybdenum anticorrosion coatings
mentioned in Section 7-4. Another significant drawback of foil leads
of large cross section (a cylindrical design) is the labor-intensive-
ness of their production and the difficulty of mechanizing their
manufacture.

The advantage of leads of the other two types lies in the possibility of transmitting an extremely high current through a much
shorter lead. A lead based on transition glasses [8-9] is extremely
sensitive to sharp temperature drops and has much lower mechanical
strength and reliability compared with a foil lead. The most convenient leads for production mechnization are cap-type leads (US
patent No. 2,756,361 and [8-10 to 8-14]), which are based on a metal
cap or disk soldered to the metallized surface of a quartz tube.
Soldering usually is done with a comparatively low-melting solder
made of a titanium-tin alloy (high-temperature solders also have been
mentioned [8-15]). Leads based on low-melting solders, like leads
based on transition glasses, do not allow high-temperature treatment
of lamps during their manufacture.

Thus, it is advisable to use cylindrical foil leads in stroboscopic lamps with a high average power load, and flat leads in low-
power stroboscopic lamps. Cap-type leads and leads based on transition glasses are preferable for lamps designed for single-flash
operation (cap-type leads are used for mass production). Such leads
can be used in stroboscopic lamps if they have adequate forced cooling
(generally with a liquid).

Cylindrical foil leads (being the most universal type) are now
used most widely in Soviet flashlamps, while cap-type leads (in the
United States) and leads based on transition glasses (in the FRG
and France) are used abroad.

Cylindrical leads are made from molybdenum foil that has been
etched until a bladelike edge is obtained. At first a foil 20 μm
thick was used. It became possible to increase the thickness of
foil sealed into quartz to 50-100 μm without exfoliation upon cooling

of the lead as a result of the application of a thin film of tran-
sition glass having a thermal expansion coefficient of $2 \times 10^{-6} {}^{\circ}C^{-1}$ to
the quartz surface adjacent to the molybdenum (the coating is applied
by steeping the quartz in an aqueous suspension of the powdered transition glass, drying the suspension, and melting it in a flame). This
film acts as a "buffer" between the quartz and the glass, ensuring
hermetic tightness of the joint between them over a large range of
temperatures (from the softening point of the quartz to the temperature
of liquid nitrogen). In addition to increasing the thickness of the
foil, the area of the working cross section of the metal in these
leads can be increased by using extremely wide strips of foil (up
to 50-60 mm wide) rolled into a cylinder 30 mm long (for small diameters) or 60 mm long (for large diameters). The edges of this
cylinder which forms its longitudinal cross section are etched in
an electrolytic bath (the cylinder is one electrode, and a metal rod
of equal length positioned parallel to the cross section of the
cylinder is the second electrode). During this process the edges
of the cross section are given a blade-like profile.

Because of the high sensitivity of the trigger voltage of flashlamps to gas impurities, the small quantities of molecular gases that
are released from the internal components of a new lamp during flashing
cause a sharp increase in U_t and may lead to failure of the lamp.
High-temperature treatment of the electrodes and lamp envelope under
a vacuum, discharge aging, and various getters are used to prevent
an increase in U_t. Getters in the form of films of alkali or alkaline-
earth metals are used in tubular glass lamps, and titanium is used in
quartz lamps.

External Components. Any metal part that is adequately insulated
from the primary electrodes and located on or a few millimeters away
from the surface of the discharge tube can serve as the control electrode in tubular lamps. The part should be extended so that it can
be positioned a considerable distance along the tube or that it approaches the tube in several areas. In lamps with low average power,
a narrow strip of an electrically conductive compound applied to the
outside surface of the tube often is used as the trigger electrode
(the strip stops 10-15 mm short of the seals of the primary electrodes).

For lamps whose discharge tube heats up severely, the trigger
electrode may be made in the form of a nickel wire wound in several
turns along the tube or stretched out parallel to the tube (2-3 mm
from the surface) between clamps on the cooler sections of the tube.
A type of monoblock design for a quartz capillary lamp with a compact
glow volume has been proposed for devices that are operated under
special conditions (mechanical impacts, moisture, vapors of aggressive
substances, etc.). In this design the trigger electrode is located
inside the envelope wall [8-16].

The base of a flashlamp has no specific features compared with the base of other electric vacuum devices, except for the requirements of adequate dielectric strength of the gaps between the metal parts of the base that are connected to the trigger electrode and parts connected to the other electrodes. The base of a flashlamp also must have a low contact resistance between the current-carrying lead-outs (the base pins) and the corresponding parts of the tube socket.

For safety reasons (protection against high voltage), some tubular lamps have an outer protective envelope whose cavity communicates with the atmosphere through the base, and also have auxiliary contacts. The protective envelope sometimes is made reflective (in such cases it is sealed, containing a noble gas, or is evacuated to prevent corrosion of the reflecting layer).

8-3. The Product Range of Flashlamps

The product range of current series-produced Soviet flashlamps, of which there are several dozen types, allows the solution of problems in every known area of flashlamp application. The development of many types of series-produced flashlamps for the pumping of lasers has been a characteristic feature of the last 10-15 years. There has been a significant modernization of product variety among the spherical and warning-type lamps. The diversity of flashlamp applications and the design embodiments do not allow the setting of a single scale of standard sizes that would encompass all lamps. Therefore, it is most convenient to group lamps according to the conditional classification of their main areas of application that is given in Table 8-1. The variety of Soviet flashlamps that currently are in series production and their main characteristics are given in Tables 8-2 to 8-6. The most characteristic types of lamps and some of their foreign counterparts are shown in Figures 8-1 to 8-4.

An extensive range of publications and patent materials give descriptions of the most varied design improvements of flashlamps, which contain original technical concepts for individual assemblies of the lamps. One also can find descriptions of entire lamps and integrated pulsed illumination devices which are combinations of a pulsed light source, an incandescent lamp, optical elements (reflectors, lenses, diaphragms), and other components. However, most of these refinements have not found application in commercial models because of their unconfirmed feasibility or the complexity of introducing them.

Table 8-2. Flashlamps for Optical Pumping of Lasers with Infrequent Pulses

Parameter	IFP-250	IFP-40 mode A	IFP-40 mode B	IFP-600 mode A	IFP-600 mode B	IFP-600 mode C	IFP-600-2	IFP-600-3
Discharge energy, J	250	400	200	600	550	250	650	600
Av. power, W	25	13	7	200	550	125	22	120
Flash interval, s	10	30	30	3	1	2	30	5
Working voltage, kV	0.9-1.3	1.9	1.9	1.7	1.65	1.2	1.8	2
Length of luminous intensity pulse,* ms	0.5	0.5		0.5		0.4	0.5	0.25
Peak trigger-pulse voltage, kV	15	20		15		30	30	20
Operating life, 1000 flashes	20	10	20	20			10	25
Cooling**	N	N		N	FL	FA	N	N
Integrated lum. intensity, kcd-s	0.8	1.0	0.5	2.5		1.2	2.2	2.8
Size of luminous volume, mm	Ø5x36	Ø5x40			Ø7x80		Ø7x80	Ø7x120
Size of lamp, mm	Ø8x48x55	Ø8x105			Ø10x170		Ø10x170	Ø10x240
Trigger voltage, kV	0.5	0.7			0.7		0.7	1
Self-breakdown voltage, kV	2	2.2			3		2.5	2.5
Limiting discharge energy (roughly), kJ	0.48	0.52		1.3		1.2	1.35	1.4
Lamp resistance (roughly), ohms	0.3	0.4		0.45			0.4	0.45

*The flash duration is determined by the capacitance and inductance of the discharge circuit.
**Cooling: N - natural cooling; FA - forced air cooling; FL - forced liquid cooling (distilled water).

Table 8-2, continued

Parameter	IFP-800			IFP-1000-2A		IFP-1200-2			IFP-2000		IFP-5000
	mode			mode		mode			mode		
	A	B	C	A	B	A	B	C	A	B	
Discharge energy, J	800	800	400	350	800	800	800	600	2000	2000	5000
Av. power, W	53	800	4000	117	80	100	80	6000	133	2000	165
Flash interval, s	15	1	0.1	3	10	8	10	0.1	15	1	30
Working voltage, kV	2.2	2.2	1.4	1.65	1.5	2	1.5	1.7	1.5	1.5	2.25
Length of lum. intensity pulse,* ms	0.75	0.75	0.6	0.25	0.5	0.5	0.5	0.7	0.75	0.75	0.5
Peak trigger-pulse voltage, kV	25	25		20	20	25	25		25	25	25
Operating life, 1000 flashes	25	25	50	30	5	10	10	50	5	5	10
Cooling**	N	N	FL	N	N	N	N	FL	N	FL	N
Integr. lum. intensity, kcd-s	2.2	2.2	1.4	1.3	2.2	3.3	4.2	2.1	7.5	7.5	18
Size of lum. vol., mm	Ø7x80	Ø7x80	Ø7x80	Ø7x75	Ø7x75	Ø7x120	Ø7x120	Ø7x120	Ø11x130	Ø11x130	Ø11x250
Size of lamp, mm	Ø10x304	Ø10x304	Ø10x304	Ø10x140	Ø10x140	Ø10x345	Ø10x345	Ø10x345	Ø23x290	Ø23x290	Ø22x470
Trigger voltage, kV	0.7	0.7		0.8	0.8	0.8	0.8		0.6	0.6	1.2
Self-breakdown voltage, kV	3	3		3	3	3	3		2	2	3
Limiting discharge energy (rough), kJ	1.55	1.55	1.4	1.0	1.4	1.9	1.9	2.3	4.2	4.2	6.5
Lamp resistance (rough), ohms	0.5	0.5		0.5	0.5	0.6	0.6		0.26	0.26	0.3

Table 8-2, continued

Parameter	IFP-5000-2	IFP-8000 mode		IFP-20000 mode		IFP-25000	IFP-40000 mode		IFPP-7000
		A	B	A	B		A	B	
Discharge energy, J	5000	6000	8000	13000	20000	25000	25000	40000	6000
Av. power, W	500	200	266	440	670	415	800	1300	200
Flash interval, s.	10	30		30		60	30		30
Working voltage, kV	3	2.6		4.65		6	5	5	2.6
Length of lum. intensity pulse,* ms	0.8	0.8	1.5	0.8	1.5	1	1.5		1
Peak trigger-pulse voltage, kV	25	25		25		25	25		25
Operating voltage, 1000 flashes	5	25	20	5		3	5	5	0.5
Cooling**	FL	N		N		N	N		N
Integrated lum. intensity of flash (min.), kcd-s	18	25	30	45	70	100	90	140	12
Size of lum. vol., mm	Ø11x250	Ø16x250		Ø16x585		Ø16.5x870	Ø16.5x1000		(Ø30-12)x130
Lamp size, mm	Ø22x470	Ø24x470		Ø24x815		Ø19.5x1110	Ø19.5x1235		182x262
Trigger voltage, kV	1.5	1.8		2		2	3		1.5
Self-breakdown voltage, kV	3.5	3.5		6		6.5	7.5		3
Limiting discharge energy (rough), kJ	8.5	13	18	31	43	52	74		12
Lamp resistance (rough), ohms	0.4	0.3		0.6		0.8	1		0.2

Table 8-3. Flashlamps for Optical Pumping of Lasers with Frequent Pulses

Parameter	ISP-200	ISP-250	ISP-600 mode					ISP-2000	ISP-2500
			A	B	C	D	E		
Av. power, kW	0.2	0.25	0.6	0.6	0.2	0.4	0.03	2	2.5
Flash freq.,* s^{-1}	2	5	2	2	1	2	0.3	6.6	50
Discharge energy, J	100	50	300	300	200	200	100	300	50
Working voltage, kV	1.4	1	1.2	1.3	1.25	1.25	1.25	1	0.7
Peak trigger-pulse voltage, kV	20	20	25					25	20
Type of triggering**	A	A	A					A	P
Integrated lum. intensity of flash, kcd-s	0.3	0.17	0.9	0.9	0.4	0.4	0.3	0.9	0.5–1 A
Length of lum. intensity pulse, ***ms	0.3	0.1	0.5	0.3	0.2	0.2	0.1	0.65	0.25
Cooling****	FA	FL	FL			N		FL	FL
Operating life, 1000 flashes	300	100	30	30	50	100	40	125	3600
Size of lum. vol., mm	Ø5x75	Ø5x40	Ø5.5x80					Ø7x80	Ø5.5x60
Size of lamp, mm	11x180	Ø8.3x162	11x180					14x300	10.5x182
Trigger voltage, kV	0.3	0.5	0.6					0.5	0.5
Self-breakdown voltage, kV	3	2	2.5					2	3
Limiting power (bursting power), kW	1.3	0.6	1.8	1.5	0.6	1.1	–	5	10
Lamp resistance (rough), ohms	0.5	0.3	0.5					0.45	0.4

*This frequency is ensured by forced switching.
**P – pulsed, A – using a keep-alive arc.
***This quantity is determined by the selection of the capacitance and inductance of the discharge circuit.
****N – natural cooling, FA– forced air cooling, FL – forced liquid cooling.

Table 8-3, continued

Parameter	ISP-3000-2	ISP-5000 (mode)					ISPT-6000
		A	B	C	D	E	
Av. power, kW	3	0.05-0.1	2-3	2-6	3-10	3-10	6
Flash frequency*, s^{-1}	1	0.1	5	10	20	20	10
Discharge energy, J	3000	500-1000	400-600	200-600	150-500	150-500	600
Working voltage, kV	3	1.4-2.2	1.4-2	1.4-2	1.4-2	1.4-2	1.45
Peak trigger-pulse voltage, kV	20	25	25	25	25	25	25
Type of triggering**	P	P			1-2 A, current		P
Integrated lum. intensity of flash, kcd·s	12	2 at 500 J and 0.5 ms					2.5
Length of lum. intensity pulse, *** ms	0.55			0.5		0.3	0.8
Cooling****	FL	N	FL				FL
Operating life, 1000 flashes	15	30	300-200	700-150	1000-70	1000-70	150
Size of lum. vol., mm	Ø11x250 21x470			Ø7x120			Ø7x120
Size of lamp, mm				Ø10x345			Ø10x260
Trigger voltage, kV	0.9			0.7			0.7
Self-breakdown voltage, kV	6			3			2.5
Limiting power (bursting value), kW	7		6.5	13	20	16.5	12
Lamp resistance (rough), ohms	0.45	0.6		0.6		0.7	0.6

Table 8-4. Light-Signaling and Photoilluminating Flashlamps

Parameter	IFK-15-2	IFK-20	IFK-20-3	IFK-50	IFK-75	IFK-75-1
Shape of luminous volume	Straight				Half-ring	
Size of luminous vol., mm	Ø2.6x9	Ø4x14	Ø2.5x30	Ø4x24	Ø32x50	Ø32x50
Discharge energy, J	15	20	20	50	75	55
Flash interval, s	5	10	15	10	2	1
Av. power, W	3	2	1.3	5	37	55
Working voltage, kV	0.25	0.13	0.36	0.2	1.5	0.7
Peak trigger-pulse voltage, kV	5	5	8	5	15	15
Operating life, 1000 flashes	5	30	10	30	20	300
Integrated lum. intensity (min.), kcd-s	0.035	0.025	0.035	0.1	0.18	0.12
Flash duration, ms	3	0.2	0.3	0.4	0.15	0.16
Peak lum. intensity, Mcd	0.1	0.1	0.2	0.2	1.2	2
Peak luminance, Gcd/m^2	2	2.5	2.5	2.2	2	1.6
Lamp resistance, ohms	0.2	0.2	0.4	0.3	0.6	0.6
Trigger voltage, kV	0.15	0.1	0.3	0.14	0.5	0.5
Self-breakdown voltage, kV	0.6	0.7	0.7	1	2	2
Load factor, µF-kV^4	—	1	1	6	200	200

Table 8-4, continued

Parameter	IFK-120	IFK-150 mode A	IFK-150 mode B	IFK-500	IFK-2000 mode A	IFK-2000 mode B	IFK-2000 mode C
Shape of luminous volume	U-shaped	Spiral			U-shaped		
Size of lum. vol., mm	Ø5x23x30	Ø26x30		Ø30x45	Ø9x40x70		
Discharge energy, J	120	150	80	500	2000	400	150
Flash interval, s	10	0.75	0.5	15	15	1.3	1.3
Av. power, W	12	225	160	30	130	300	113
Working voltage, kV	0.3	1		0.5	0.5	0.32	0.7
Peak trigger-pulse voltage, kV	10	25		15		20	
Operating life, 1000 flashes	10	500	750	10	5.0	400	800
Integrated lum. intensity (min.), kcd-s	0.25	0.45	0.25	1.0	6.0	1.2	0.45
Flash duration, ms	1.0	1.0	1.0	8.0	4.0	2.0	0.1
Peak lum. intensity, Mcd	0.25	0.45	0.25	0.13	1.5	0.6	4.0
Peak luminance, Gcd/m^2	0.7	0.8	0.8	0.09	1.3	0.5	0.6
Lamp resistance, ohms	0.8	2		4.0		0.45	
Trigger voltage, kV	0.18	0.5		0.4		0.25	
Self-breakdown voltage, kV	1.0	2		3.5		2.2	
Load factor, µF-kV^4	25	500		250		1000	

Table 8-4, continued

Parameter	IFK-2000-3	IFK-20000 mode			IFT-200** mode		
		A	B	C	A	B	C
Shape of luminous volume	Spiral	Spherical spiral			Disk		
Size of lum. vol., mm	Ø50x60	Ø85			Ø6		
Discharge energy, J	2000	20000*	10000*	2000	40	120	200
Flash interval, s	4	20	1.8	0.5	15	15	15
Av. power, W	425	1000	5500	4000	2.7	8	13
Working voltage, kV	1.7		6			0.2	
Peak trigger-pulse voltage, kV	25		25			10	
Operating life, 1000 flashes	75	–	7.0	–	5	2	1
Integrated lum. intensity (min.), kcd-s	9	60	34	6	0.05	0.15	0.2
Flash duration, ms	1	2	1	0.6	3	9	10
Peak lum. intensity, Mcd	–	30	30	10	0.06	0.8	4.5
Peak luminance, Gcd/m^2	–	5.5	5.5	0.5	0.05	4	8
Lamp resistance, ohms	0.5		3.5			0.15	
Trigger voltage, kV	1.2		2			0.15	
Self-breakdown voltage, kV	3		20			0.6	
Load factor, µF-kV^4			1.5×10^6			250	

* Forced blower cooling.
** Discharge across a 0.8 mH inductance.

Table 8-4, continued

Parameter	IFB-300	IFP-200	IFP-500	IFP-1500	IFP-4000	IFP-15000	IPO-75	ISh0-1
Shape of luminous volume	Ring				Straight			
Size of lum. vol., mm	Ø8x85	Ø5x200	Ø5x350	Ø5x600	Ø6x800	Ø10x580	Ø7x40	Ø1x2
Discharge energy, J	300	200	500	1500	4000	15000	75	1.3
Flash interval, s	7.5	7.5	7.5	15	15	12	30	1
Av. power, W	40	27	65	100	270	1250	2.5	1.3
Working voltage, kV	0.3	0.5	0.5	1	1.4	2.4	0.7	0.8
Peak trigger-pulse voltage, kV	15	10	10	15	15	25	10	5
Operating voltage, 1000 flashes	10	10	10	10	10	10	5	5
Integrated lum. intensity (min.), kcd·s	0.5	0.4	1	4	12	50	0.3	0.001
Flash duration, ms	40	1.6	7	9	16	4.5	0.3	0.003
Peak lum. intensity, Mcd	10	0.25	0.14	0.45	0.75	10	0.001	0.25
Peak luminance, Gcd/m2	1.6	0.23	0.08	0.13	0.16	1.6	–	–
Lamp resistance, ohms	2.5	2	3.5	8	8	1.5	0.2	0.01
Trigger voltage, kV	0.2	0.45	0.5	0.9	1.3	1.6	0.5	0.6
Self-breakdown voltage, kV	1.6	2	3	4	5	5	1.2	2
Load factor, µF-kV4	1000	600	3200	17500	32000	2×10^5	300	–

Table 8-5. Tubular Stroboscopic Lamps

Parameter	ISP-15 mode A	ISP-15 mode B	ISP-50	ISP-70-1 mode A	ISP-70-1 mode B	ISK-10 mode A	ISK-10 mode B
Shape of luminous volume Size of lim. vol., mm	Ø0.5x35		Straight Ø3x30	Ø0.5x70		U-shaped Ø5x2 3x30	
Av. power, W	8	15	10	18	4	10	
Flash frequency, s⁻¹	100	100	100	100	0.2	1	200
Discharge energy, J	0.07	0.15	0.1	0.18	20	10	0.05
Working voltage, kV	0.8		0.4	1.2		0.3	
Peak trigger-pulse voltage, kV	12		9	18		10	
Operating life, 1000 flashes	500	100	40	40	–	0.18	360
Peak lum. intensity, kcd	3	6	8	6	40	40	5
Flash duration, μs	15	15	13	25	300	200	25
Integrated lum. intensity, cd·s	0.03	0.07	0.1	0.17	20	8	0.008
Peak luminance, Gcd/m2	0.3	0.4	0.3	0.2	1.5	0.1	0.005
Lamp resistance, ohms	35		2	100		0.8	
Trigger voltage, kV	0.6		0.3	0.9		0.18	
Self-breakdown voltage, kV	2.5		1.5	3		1	

Table 8-5, continued

Parameter	FP-1500* mode A	B	C	D	ISK-20-1	ISK-25 mode A	B	ISK-200**
Shape of luminous volume Size of lum. vol., mm	Straight Ø1x10				Ø3x20x20	U-shaped Ø5x21x20		Spiral Ø12x15
Av. power, W	1600	840	420	400	9	25	30	200
Flash frequency, s^{-1}	5000	2000	1000	50	100	1	0.05	200
Discharge energy, J	0.32	0.42	0.42	0.8	0.01	25	600	1
Working voltage, kV	8				0.3	0.3		3
Peak trigger-pulse voltage, kV	15				18	15		18
Operating life, million flashes	0.025				360	0.36	0.01	0.1
Peak luminous intensity, kcd	50	100	60	120	9	300	300	250
Flash duration, μs	2	3	5	5	8	150	6.10^3	5
Integrated lum. intensity of flash, cd-s	0.1	0.3	0.3	0.6	0.07	60	1200	1.25
Peak luminance, Gcd/m^2	5	10	6	12	0.2	2	2	0.6
Lamp resistance, ohms	5				1.5	0.4		
Trigger voltage, kV	1					0.25		
Self-breakdown voltage, kV	2					1.2		

* Power supply in a circuit with forced switching in the discharge circuit in short cycles lasting 0.2 to 2 s.

** 15 s of continuous operation.

Table 8-6. Spherical Flashlamps

Parameter	ISSh-2	ISSh-4*	ISSh-4-1	ISSh-5* mode A	ISSh-5* mode B	ISSh-7 mode A	ISSh-7 mode B	ISSh-15* mode A	ISSh-15* mode B
Av. power, W	2	4	5	18	0.6	1.75	7.0	15	1
Max. flash freq., s^{-1}	100	10	2500	100	3	500	2000	500	0.1
Discharge energy, J	0.026	0.4	0.002	0.18	0.2	0.0035		0.03	10
Working voltage, kV	0.65	0.65	0.8	1.2	1.9	1		0.45	1
Peak trigger-pulse voltage, kV	5	3	5	5		1.5		6	
Max. time of continuous operation	–	–	–	–		20		–	
Operating life, million flashes	720	1	12.5	0.2		6	–	500	
Peak lum. intensity, kcd	10	40	2	50		7		4	300
Size of lum. vol., mm	Ø0.5x8	Ø0.5x8	Ø0.3x8	Ø1x4		Ø0.5x2		Ø1x2.5	Ø5x4
Size of lamp, mm	Ø22.5x36	Ø23x27x32	Ø22.5x36	Ø22.5x55		Ø22.5x58		Ø33x75	
Flash duration, μs	1	3	0.5	1	1	0.35		1.5	15
Integrated lum. intensity of flash, cd-s	0.01	0.14	0.001	0.045	0.05	0.005		0.006	5
Peak luminance, Gcd/m^2	2.5	3.3	0.85	12	10	7		1.5	50
Trigger voltage, kV	0.4	0.4	0.6	1.2		0.8		0.25	
Self-breakdown voltage, kV	0.9	0.9	2	2.5		2		1.2	

* Discharge through a ballast resistance.

Table 8-6, continued

Parameter	ISSh-12	ISSh-100-2 mode			ISSh-100-3M* mode		
		A	B	C	A	B	C
Av. power, W	12	100	50	5	100	150	5
Max. flash freq., s⁻¹	2	500	500	0.1	20	50	0.1
Discharge energy, J	6	0.22	0.11	50	5	3	50
Working voltage, kV	0.8		3		4.5	3.5	3
Peak trigger-pulse voltage, kV	7		6			6	
Max. time of continuous operation, s	–	30	–	–			
Operating life, million flashes	1	0.3	3	–	3.6	–	–
Peak luminous intensity, kcd	–	250	150	3000	1000	800	4000
Size of lum. volume, mm	Ø2x8	Ø0.7x3	Ø0.5x3	Ø5	Ø2x6	Ø2x6	Ø7
Size of lamp, mm	Ø30x40		Ø34x85			Ø35x97	
Flash duration, μs	50	1.3	1	15	2.5	2.2	15
Integrated lum. intensity of flash, kcd-s	1	0.2	0.1	50	2.5	2	60
Peak luminance, Gcd/m²	10		100			100	
Trigger voltage, kV	0.7		0.8			2.5	
Self-breakdown voltage, kV	1.5		3.5			6	

Table 8-6, continued

Parameter	ISSh-100-4	ISSh-100-5	ISSh-100-6	ISSh-300 mode		ISSh-400*	ISSh-500 mode	
				A	B		A	B
Av. power, W	55	62	20	300	10	375	500	10
Max. flash freq., s^{-1}	25	250	5	400	0.1	3000	100	0.75
Discharge energy, J	2.5	0.25	4	0.75	100	0.12	5	160
Working voltage, kV	6.5	2.6	4	6.3	6	5	9	7
Peak trigger-pulse voltage, kV	1	4	10	25		8	25	
Max. time of continuous operation, s	90	900	—	—		50	—	—
Operating life, million flashes	0.27	100	3	7.2	—	10	0.36	—
Peak luminous intensity, kcd	600	150	600	300	3000	100	1000	4000
Size of lum. volume, mm	Ø1.5x10	Ø0.6x3	Ø2x5	Ø1x6.5	Ø5x6.5	Ø0.5x3.0	Ø1.2x8	Ø5x8
Size of lamp, mm	Ø30x102	Ø36x81	Ø40x102	Ø82x150		Ø80x73	Ø50x110 x130	
Flash duration, µs	3	1.5	3	2	15	0.8	6	25
Integrated lum. intensity of flash, cd-s	1.8	0.25	1.8	0.6	50	0.08	6	100
Peak luminance, Gcd/m^2	50	70	50	50	100	60	100	100
Trigger voltage, kV	5	2.2	3	5.5		3	5	
Self-breakdown voltage, kV	10	3.5	6	10		7	15	

* Power supply with forced switching in the discharge circuit.

CHAPTER 9

CIRCUIT DIAGRAMS OF FLASHLAMPS AND
THE MAIN CIRCUIT ELEMENTS

9-1. The Functional Power-Supply Diagram of a Flashlamp

The radiation characteristics of gas-discharge flashlamps depend
largely on the conditions of power supply. The flash duration, peak
luminous intensity, flash repetition frequency, time dependence of
the luminance and spectral composition of the radiation during the
flash, and other quantities are determined by the design features of
the lamp and by the parameters of the power-supply devices: the capacitance of the storage capacitor and the voltage across it before
the flash, the inductance of the discharge circuit, the properties
of the trigger pulse, the method used to store energy in the capacitor, the presence of commutators in the charging and discharging
circuits, the effect of the previous discharge, and other factors.

Allowance must be made for the fact that many of the main parameters of a lamp (the trigger voltage and discharge extinction voltage, the nature of the time dependence of the lamp, the size and
location of the glow volume, etc.) vary as a function of power-supply
conditions, such as the flash frequency and duration, and the average
power, energy, and voltage of the trigger pulse.

As a first approximation, a flashlamp, viewed as a circuit element, may be considered a one-way control switch: in most cases
the trigger pulse initiates a current flow through the lamp. This
current is cut off when the energy in the storage capacitor has
nearly run out or when external devices act on the current, such as
a cutout in the discharge circuit or a quenching device which reduces
the voltage across the lamp to a value below the extinction voltage.

The most widely used and simplest method of providing power to
a flashlamp is to connect it to a storage capacitor (or to a group

319

of capacitors which form an artificial line, for example). When the
capacitor discharges, a current pulse capable of injecting large
amounts of energy into the lamp in a short time is generated. Direct
power supply to flashlamps from primary sources (a grid or chemical
current sources) also is possible in some cases, e.g., when flashes
of comparatively long duration are used. Flashlamps also can be
powered by inductive or electromechanical energy storage devices or
special pulsed MHD generators. However, the creation of appropriate
apparatus is technically complicated and is advisable only in exceptional cases.

The power-supply unit of a flashlamp is designed to shape current
pulses at specified times. Depending on the type of lamp and operating conditions, these pulses may have a duration ranging from fractions of a microsecond to tens of milliseconds or more, reaching a
maximum of $10-10^5$ A.

Despite the significant differences, nearly all power-supply
circuits for flashlamps may be reduced to a single functional diagram (Figure 9-1), which consists of the following basic elements
and units: FL - a flashlamp connected to a discharge circuit D with
an energy storage device E, which is a secondary supply for the lamp;
Ch - the charger; PE - the primary energy supply; TG -- the trigger-
pulse generator; and CSP - a unit containing control, synchronization,
and protective devices. The storage device is charged by the primary
source (a grid or a chemical source of DC current) through the
charger. The control, synchronization, and protective devices regulate the operation of the charger, activate the trigger-pulse generator, and control the operation of the discharge circuit ([9-1] and
West German patent No. 1,920,951).

Various circuit solutions can be used in the main units to vary
the flash parameters over a broad range and to change the technical
indices and economic indicators of the equipment: weight, cost, size,
efficiency, reliability, and so forth.

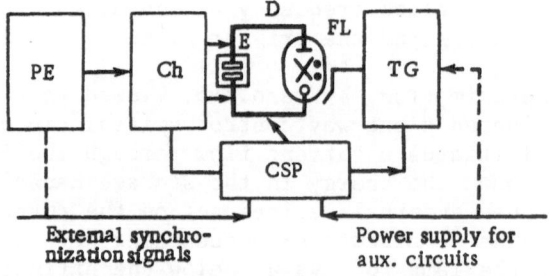

Fig. 9-1.   Block diagram of the power supply for a flashlamp,
             containing a storage device.

If the flashlamp can be powered without using an energy storage device, then the functional diagram is simplified: the storage device and charger are eliminated, but the tasks of controlling current switching in the discharge circuit and synchronizing the flashes with the voltage phase of the supplying network are additionally imposed on the control unit.

9-2. DC Power-Supply Circuits without Energy Storage Devices

A schematic diagram of the discharge circuit for direct connection of a tubular flashlamp to the primary power supply is shown in Figure 9-2a. The supply is characterized by an emf $U_{x.x}$, an external resistance R_s, and a self-inductance L_s, the connecting leads are characterized by a resistance R_1 (which includes the resistance of the fuses, the measuring instruments, etc.) and an inductance L_1, the switch K of the discharge circuit is characterized by a resistance R_c, and the lamp by a resistance R_L. Strictly speaking, the parameters of these elements may vary during a flash, but the resistance of the lamp is used for a rough evaluation of the circuit and the other parameters may be considered constant, having some average values for the flash duration T. The energy released in the lamp during a single pulse is

$$W = \frac{U_{x.x}^2 R_L}{(R_L + R_b)^2} T, \tag{9-1}$$

where R_b is the total ballast resistance, equal to $R_s + R_1 + R_c$.

Formula (9-1) is valid if $T \gg \tau = (L_s + L_1)/R_b$ and if T greatly exceeds the expansion time τ_{ex} of the plasma column in the lamp. The lamp releases 121 J of energy for $U_{x.x} = 220$ V, a duration T = 10 ms, $R_L = 1\ \Omega$, and $R_b = 1\ \Omega$, and the peak pulse current is equal to 110 A. (If T is not large enough compared with τ and τ_{ex}, then the computational estimate of W is made much more complicated.)

It follows from relation (9-1) that when this method of power supply is used in a flashlamp:

(a) in order to increase the flash energy, the supply voltage must be increased, high-power sources (i.e., sources with a small R_s) and lamps with low internal resistance must be used, and T must be increased.
(b) for a given R_b, the highest flash energy can be reached when $R_L = R_b$.

High-speed mechanical switches or blowout-type fuse links, for example, may be used as the switch K. Semiconductor switches (Table 9-1) are most convenient for this purpose. These can be used to

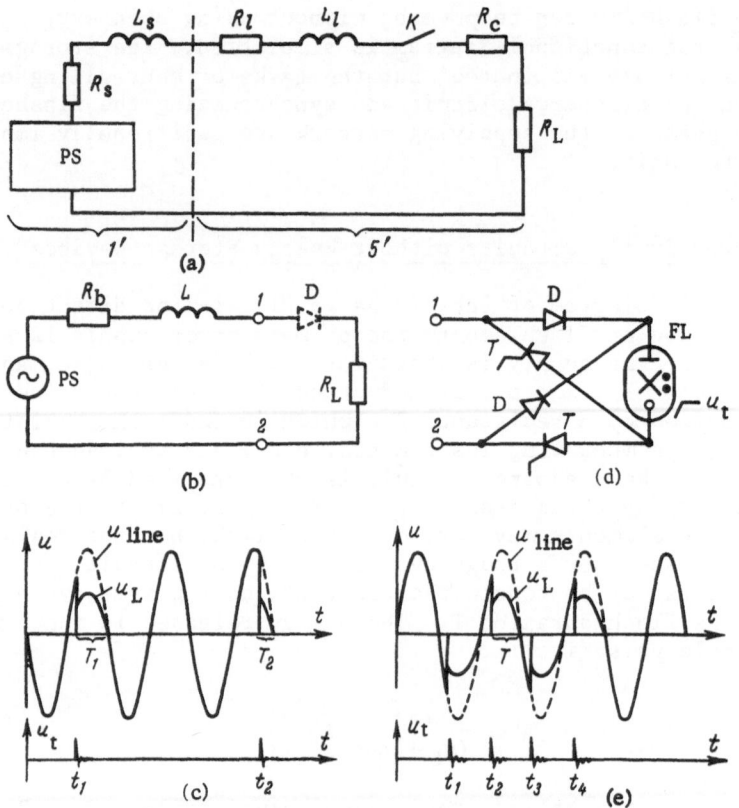

Fig. 9-2. Direct power supply to a flashlamp from a primary source.
a) Equivalent power-supply circuit with direct connection
of the lamp to a primary DC source; b) simplified equivalent power-supply circuit with direct connection of the
lamp to an AC source; c) variation of voltage across
flashlamp powered by an AC network (t_1 and t_2 are the
times when the trigger pulses are supplied, and T_1 and
T_2 are the corresponding flash durations); d) circuit
diagram of a flashlamp for connection across the diagonal
of a bridge rectifier; e) four successive flashes during
the "positive" and "negative" half-waves of the supply-
line voltage (t_1 to t_4 are the times when trigger pulses
are supplied).

design control circuits for pulsed light sources with very high
flash repetition frequencies. The possibility of using biooperational
thyristors series-connected to a flashlamp (to cut off the light
pulse after the required exposure is reached) in electronic photoflashes has been mentioned in the literature (British patent No.

Table 9-1. The Characteristics of Some Semiconductor Devices Which
Are Suitable for Current Regulation in Flashlamps

Type of device	Max. constant voltage, V	Max. pulsed voltage, V	Pulse current, A	Switch-on time, μs
KT704A transistor	500	1000	4	1-2
KT808A transistor	120	250	10	1-2
Mitsubishi bioperational thyristor [9-2]	300	-	300	5

1,290,313, and Belgian patents Nos. 724,245, 726,076, 727,159, 729,602, 735,353, and 739,170).

Figure 9-3 shows the power-supply circuit of a lamp using a low-voltage source with a transistor switch (East German patent No. 56,021). A DC source (35-100 V) is connected across high-voltage diodes D to the anode of flashlamp FL, which has a small spacing between the anode and cathode (about 1 mm) and a comparatively low trigger voltage. One or several parallel-connected transistors T are series connected to the lamp in circuit I. The transistors are kept open when the switch P_1 is closed. A discharge is triggered in the lamp when switch P_2 in auxiliary circuit II is closed. Circuit II contains a high-voltage source HVS (at a few kilovolts and a few milliamps). After the primary-discharge current is established in circuit I, switches P_1 and P_2 open after some time lag, and the current is cut off in the lamp. If a multivibrator which periodically opens and closes the transistor is connected to the base of transistor T, then the current through the lamp is modulated. At high frequencies the lamp is unable to deionize after a pulse, and thus does not lose its controllability under such operating conditions.

Power-supply circuits in which the lamps are supplied directly by DC sources make it possible to produce comparatively high-power, long-duration flashes with nearly constant radiation intensity during the pulse (such light sources are required for high-speed filming in many cases). For example (British patent No. 1,190,843), an XBO short-arc xenon lamp (2500 W) powered by a rectifier has been used in pulsed operation with a flash duration on the order of 0.1 s, a current of up to 400 A through the lamp, and an average power of about 30 kW during a pulse. The transformer of the rectifier unit served as the ballast in this circuit, and the current was switched by heavy-duty contactors series-connected to the transformer primary.

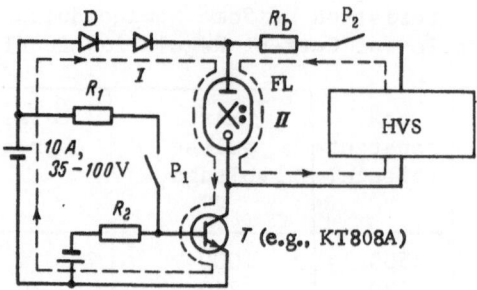

Fig. 9-3. Power-supply circuit of a spherical flashlamp FL, using
           a DC source with switching of circuit I by means of a
           transistor.

9-3. Power-Supply Circuits Using an AC Supply Line
      with No Storage Device

     When a flashlamp is powered by an AC line, the lamp usually
extinguishes after the instantaneous value of the supply-line voltage
falls below the discharge extinction voltage.* Hence, in this case
we can make do without a switch in the discharge circuit, or we may
limit ourselves to series connection of the diode to the lamp (Figure
9-2b) to prevent reverse triggering. In this case the longest duration of a single flash turns out to be somewhat shorter than the
duration of one half-wave of the supply voltage or is some small
fraction thereof (T_1 and T_2 in Figure 9-2c). If we wish to obtain
several flashes successively during the positive and negative half-
waves, then we should use a bridge rectification circuit with the
lamp connected across the diagonal of the bridge (Figure 9-2d). The
trigger-pulse generator should produce pulses (U_t in Figure 9-2e) at
times when the instantaneous value of the supply-line voltage exceeds the trigger voltage of the lamp. In this case, controlled
rectifiers such as thyristors should be used in two arms of the
bridge to prevent the lamp from going into a stationary burn.

     The duration of a single pulse in such a circuit at a line fre-
quency of 50 Hz can be regulated from about 1 to 8 ms, allowing for
the possible accuracy of synchronization, the trigger voltage of the
lamp, and the spread of its extinction voltage. The duration of a
series of flashes can be varied over a broad range, the limitation
being overheating of the lamp.

_____

*By contrast, large-diameter gas discharge lamps do not extinguish
when the voltage passes through zero. The power supply for high-
power no-ballast tubular lamps in particular is based on this
fact [0-6].

The energy W of a single flash of the lamp when power is supplied from an AC line with a peak voltage U_a can be estimated from
expression (9-2) if we assume the resistance R_L of the lamp constant
and assume that $L_1 = L_s = 0$:

$$W = \frac{U_a^2 R_L}{(R_L + R_b)^2} \int_{t_1}^{t_2} \sin^2(\omega t)dt = \frac{U_a^2 R_L}{(R_L + R_b)^2} [F(t_2) - F(t_1)], \qquad (9\text{-}2)$$

where t_1 and t_2 are the times when the discharge starts and ends;
R_b is the ballast resistance; ω is the circular frequency; and $F(t)$
$= \frac{t}{2} - \frac{1}{4\omega} \sin 2\omega t$. The plot of this function for a line frequency of
50 Hz is shown by the solid line in Figure 9-4. As we can see from
Figure 9-4 (curve 1), when $R_L = 1\ \Omega$ (e.g., the IFK-120 lamp) and R_b
$= 0$, the energy of a flash that starts at $t = 0$ and extinguishes
after a half-period is close to 500 J. It exceeds 120 J after as
little as 3.5 ms. Over a half-period the energy cannot exceed 120 J
when $R_b = 1\ \Omega$ (curve 2) or 40 J when $R_b = 3\ \Omega$ (curve 3).

In practice R_b is made up of the output resistance of the substation step-down transformer (usually about 0.1 Ω), the resistance of
the cables from the substation to the building, the resistance of
the indoor wiring, the resistance of the fuses and meter (usually
from 0.15 to 1.2 Ω), and an additional resistance which is series-
connected between the lamp and the illuminating device (e.g., 0.8 Ω).
Of these quantities, the most indefinite is the supply-line resistance, which may reach several ohms in extended rural lines. This
greatly complicates the use of devices using high-power flashlamps
which are powered directly from a supply line. (As we can see from
Figure 9-4, the flash energy of such lamps may vary by approximately
an order of magnitude.)

A much smaller spread of the flash energy as a function of the
supply-line resistance can be obtained by using low-power lamps with
a higher internal resistance. For example, as the supply-line resistance varies from 0 to 3 Ω, the spread of the flash energy for a
lamp with an effective resistance of 3 Ω (curves 4 to 6 in Figure
9-4, corresponding to $R_b = 0$; 1 and 3 Ω) is nearly a fourth that of
a lamp with a resistance of 1 Ω. A flash energy of 10 J is entirely
sufficient for some purposes, such as light signaling or decorative
effects, and it is reasonable to use such power-supply circuits.

In such circuits the flash energy can be regulated at a specific
supply-line resistance by introducing an additional active ballast,
by changing the phase of the trigger pulse, or by shunting the lamp
with a discharger or a pulsed thyristor.

A very simple example of a circuit for line power supply to a
tubular lamp is shown in Figure 9-5. This circuit can be used, for

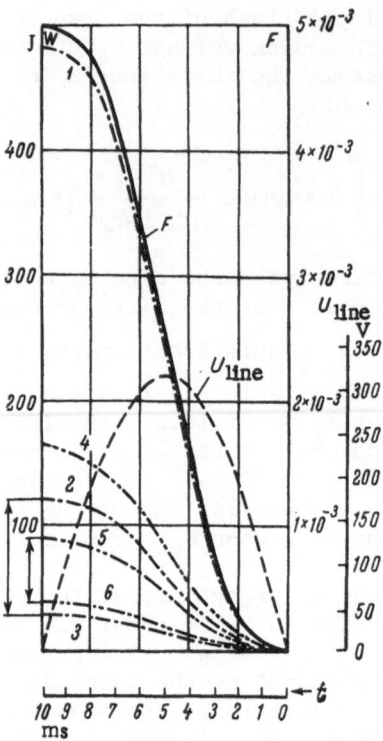

Fig. 9-4. Graph for estimating the flash energy of a flashlamp
           powered by an AC line as a function of the phase, the
           effective resistance of the lamp - 1, 2, 3) 1 Ω; 4, 5, 6)
           3 Ω - and the ballast resistance - 1, 4) 0 Ω; 2, 5) 1 Ω;
           3, 6) 3 Ω. The arrows indicate the energy spread of
           flashes 10 ms long as R_b fluctuates from 1 to 3 Ω for
           lamps with R_L = 1 and 3 Ω. Example: A lamp with a re-
           sistance R_L = 1 Ω fired 3 ms after the line voltage passed
           through zero (i.e., t_2 = 7 ms, U_{line} = 250 V, F_2 = 4.3 x
           10^{-3}), burned for 6 ms, and extinguished at t_1 = 1 ms
           (i.e., U_{line} = 90 V, F_1 = 0.1 x 10^{-3}). The flash energy
           for U_a = 310 V and R_b = 1 Ω is found from formula (9-2);
           $\frac{310^2}{(1+1)^2}$ (4.3 x 10^{-3} - 0.1 x 10^{-3}) = 99 J. This point lies
           on curve 2.

example, in signaling or warning devices or to produce decorative
effects. The flash energy can be regulated from a few joules to
several tens of joules, and the frequency from approximately 0.3 to
3 Hz. The lamp fires when the line voltage reaches the peak value.
The flash duration is about 4 ms. The windings of pulse transformer
PT are wound onto a form 8 mm in diameter and 12 mm tall. The high-
voltage winding (a wire with ϕ = 0.09 mm) is wound first (in layers),
followed by the primary (a wire with ϕ = 0.4 mm).

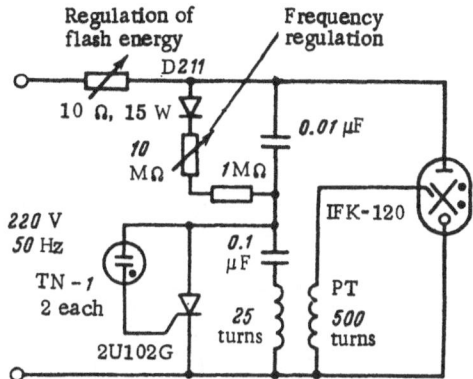

Fig. 9-5.   Example of a circuit for direct line power supply to an
            IFK-120.

The power-supply circuit of a tubular lamp connected to the
supply line through a step-up transformer in series with a thyristor
T is shown in Figure 9-6. The circuit (US patent No. 3,497,768) is
recommended for stationary photoillumination devices. If switch K
is closed during the "positive" half-wave of the voltage, then the
thyristor opens only after the polarity of the supply voltage changes.
When this occurs, the lamp fires and a keep-alive current is set up
across the lamp (a 10-µF capacitor discharges across a 330-Ω resistor). The keep-alive current supplies power to the lamp throughout the entire "negative" half-wave. After the second change in the
polarity of the line voltage, a current pulse about 8 ms long begins
to flow through the lamp from the supply line. Another flash is
possible only after switch K has been opened and reclosed.

9-4. General Characteristics of Circuits Using
      Capacitive Storage Devices

Flashlamp power-supply circuits using capacitive storage devices
are the most widely used type. They make it possible to vary the
parameters of the light pulses over a wide range with respect to
flash energy, flash duration, flash frequency, the steepness of the
pulse edge, and other characteristics.

There are many types of devices for the charging of capacitors,
ranging from a resistor in a DC charging circuit to highly complicated
converters containing nonlinear elements, transformers, intermediate
energy storage devices, two-way switches, and others.

Charging devices are characterized by the following parameters:

a)   the voltage $U_{x.x}$ of the primary power supply, which goes to the

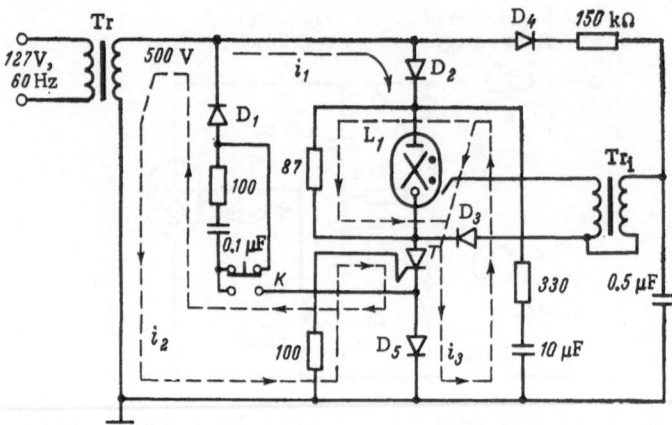

Fig. 9-6. Circuit for producing a single flash in a lamp directly
           powered by an AC supply line through a step-up trans-
           former. i_1) the operating current during the "positive"
           half-wave of the voltage; i_2) the current which shuts
           off the thyristor during the "negative" half-wave of the
           voltage; i_3) the keep-alive current through the lamp during
           the "negative" half-wave of the voltage.

      input of the charging device, and the type of current (direct or
      alternating);
b)   the maximum output voltage U_M of the charging device under open-
      circuit conditions, and the maximum voltage U_0 across the storage
      capacitor before a flash;
c)   the voltage conversion factor $K_c = U_M/U_{x.x}$ of the charging de-
      vice;
d)   the maximum input current I_M of the charging device, which de-
      termines the maximum output power $P_M = I_M U_{x.x}$ that can be taken
      from the primary power supply;
e)   the average output power of the charging device during a charging
      cycle T: $P_{av} = (U_{x.x}/T) \int_0^T i(t)dt$, where $i(t)$ is the output cur-
      rent of the primary power supply;
f)   the energy accumulated during the charging cycle in the storage
      capacitor of capacitance C: $W = C(U_0^2 - U_{res}^2)/2$, where U_{res} is
      the residual voltage across the capacitor after a flash;
g)   the average output power P_{av-out} of the charging device, which
      is equal to the product of the energy stored during a single
      charging cycle by the frequency f of the cycles: $P_{av-out} = Wf$;
h)   the utilization factor of the power supply: the ratio of the
      average input power into the storage device to the maximum out-
      put power taken from the power supply during periodic operation
      of the charging device, $K_u = Wf/P_M$;

i)   the efficiency of the charging device, defined as the ratio of
     the average input power into the storage capacitors to the aver-
     age input power of the charging device during a charging cycle,
     $\eta = Wf/P_{av}$;
j)   the power factor K_M;
k)   the maximum output current of the charging device;
l)   the weight and volume of the charging device;
m)   reliability;
n)   cost.

      The power-supply utilization factor has special importance when
the storage device is charged by a source having limited power. For
example, when the circuit used goes through a resistor, where $U_0 = 0.7U_M$, the net output power of the charging device is just one-fifth
the maximum power drawn from the power supply ($K_u = 0.2$). If all
the power drawn from the power supply went into the storage device
($K_u = 1$), then this would mean that there are no energy losses in
the charging device (100% efficiency), there are no reactive currents
in the input circuit of the charging device ($K_M = 1$), and the power
consumption at the input of the charging device during the charging
cycle, which can be characterized by some coefficient K_p, is uniform
($K_p = 1$), i.e., $K_u = \eta \times K_M \times K_p$ [9-2a]. The requirement that the
power supplied be utilized as completely as possible is one specific
feature of the power-supply circuits of flashlamps. It is especially
difficult to design circuits for charging devices whose input power
(and hence output power) is approximately constant throughout the
entire charging cycle. In the better types of charging devices the
factor K_u may reach a value of 0.7 under the most favorable conditions. However, if self-contained generators or batteries which
allow the output power to be exceeded repeatedly for a short duration
at a specified average power are used as the power supply, for example, then the factor K_u is defined as the ratio of the stored
energy to the average power over a cycle [9-11, 9-12, and 9-16] and
is expressed as the product of the efficiency by K_M. In this case
the value of K_u may reach 0.9 for better charging devices.

      Generators sometimes are used as charging devices powered by
DC supplies. During one part of the period of natural oscillations,
these generators draw current from the power supply, but during the
other part they partially return current to the source. The total
power in the input circuit, defined as the product of the power-
supply voltage by the effective value of the current in the circuit,
substantially exceeds the actual active power drawn from the power
supply. Such operation of a generator can occur after the process
of charging the capacitor to the specified voltage has ended. The
energy stored in the generator of the converter during its natural
oscillation period is dumped back to the power supply through a
special limiter winding in the transformer [9-3]. As examples,
Figure 9-7 presents oscillograms of the current in the input circuit
of a one-cycle thyristor converter [9-4 to 9-5] and an input filter

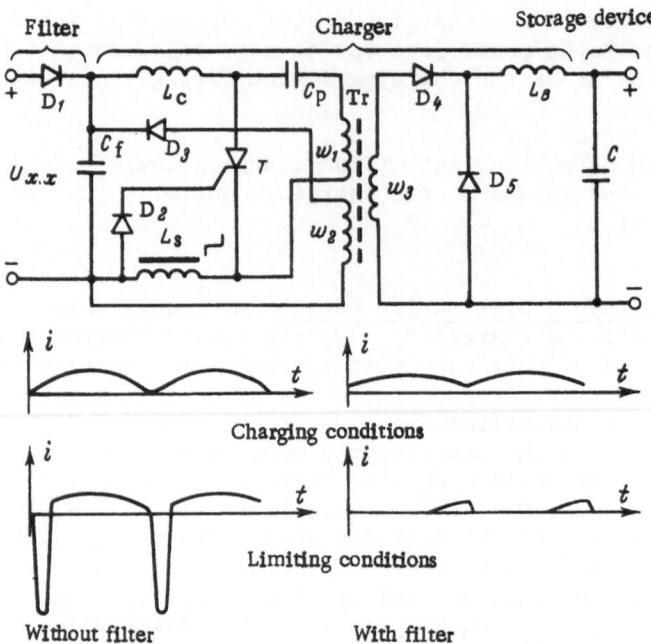

Fig. 9-7. One-cycle thyristor generator ($U_{x.x} = 27$ V, input current up to 70 A, output voltage 1.7 kV, frequency 5 kHz) with an input filter, and oscillograms of the current in the input circuit with (bottom) and without (top) a filter.

circuit which eliminates the flow of negative current pulses to the power supply. During limiting operation, the use of such a filter makes it possible to reduce the current drawn from the power supply to a value determined solely by the generator's own consumption under open-circuit conditions. The operation of the filter is based on the valve-type nature of the input element of the lumped circuit. Capacitor C_p is charged resonantly through D_w, L_c, w_1, and saturable reactor L_s. The core of reactor L_s undergoes magnetic polarity reversal at the start of the back-current half-wave in the circuit consisting of L_c (36 µH), C_f, L_s, and w_1 (3 turns), and a pulse is generated in the reactor which switches on thyristor T (a TCh-100) through diode D_2. When this occurs, there is a pulsed discharge of capacitor C_p (100 µF) through the thyristor to turn w_1, and a pulse current is generated in winding w_3 (96 turns). This pulse is fed to capacitor C through the transformer, diode D_4 (4 each, 2D206V), and reactor L_B (2.5 µH). After the end of the pulse, reactor L_B discharges through diode D_5 onto capacitor C. Such a one-cycle thyristor-type generator may have $K_c \approx 100$. The output voltage of the generator under open-circuit conditions is determined not only by the transformation ratio of the step-up transformer, but also by

the Q factor of the circuit formed by elements L_c and C_p. Winding
w_2 (1 turn) and diode D_3 are used to eliminate resonant step-up of
the circuit L_c-C_p above the permissible voltage (e.g., 120 V). At
this voltage the winding and diode sharply lower the Q of the generator circuit by dumping the energy stored in reactive elements into
the capacitor of filter C_f.

The significance of the <u>maximum output current</u> of the charging
device is determined by the fact that immediately after a flash,
when the gas-filled gap of the lamp is still filled with highly
ionized gas and has a low resistance, the output of the charging
device turns out to be shunted by this resistance. As a result, a
stationary arc discharge may occur in the lamp, and the lamp no longer
is controlled. In order to prevent the lamp from entering a continuous burn, some switching element must be included in the discharge
circuit to shut off the lamp after a flash from the storage capacitor,
or the output current of the charging device must be limited in such
a way that the storage capacitor is charged (the voltage across the
electrodes increases) more slowly than the electric breakdown
strength of the lamp recovers during the post-discharge period. The
rate of breakdown strength recovery increases most slowly in the
first 10^{-5}-10^{-2} s, and it is in precisely this period that the current strength of the charging device must be limited, especially
when flashlamps are used under more intensive frequency-power conditions. Here the decision sometimes must be made not to charge
through a resistance or to use circuits with uniform power consumption
at the input. Instead we must use either controlled switches, which
are employed to create a current "pause" in the charging circuit
(US patents Nos. 3,543,125, 3,780,344, and 3,828,222), or circuits
with a smooth current rise in the output circuit of the charging
device, corresponding to the graphs in Figure 9-8. The maximum output current of a charging device (curve II) is determined by the
function $i(t) = C du(t)/dt$, where $u(t)$ is the breakdown strength recovery function of the lamp (curve I). During such charging, the
voltage across the capacitor does not go beyond the breakdown
strength recovery curve, and at the same time the specified U_0 is
reached as fast as possible. Obviously, the storage capacitor can
be charged with a current pulse of any shape and amplitude, provided
that, by some arbitrary time T, the area that lies under the charging
current curve (the hatched area under curve III) is less than or
equal to the area under the curve of the maximum output current.
Here the peak of the charging-current pulse is not limited and may
be significantly higher or lower than the peak of the maximum output current pulse. (For example, if the breakdown strength recovery
process has ended, then charging can be accomplished with a current
pulse that has a duration of a few microseconds and a tremendous
amplitude.) However, in the general case, the use of high-strength
current pulses leads to decreased efficiency of the charging device
and to a worsening of the uniformity of input-power consumption,
while the intervals between flashes are lengthened when a low current

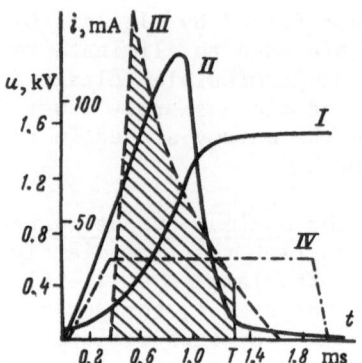

Fig. 9-8. Possible curves of the output current of a charging de-
           vice during operation of an ISSh-15 lamp when f = 500 Hz,
           W = 0.03 J, C = 0.05 µF, U_0 = 1100 V. I) Breakdown
           strength recovery curve; II) the corresponding curve of
           the maximum output current. i = C du(t)/dt; III) an ex-
           ample of a possible charging-current curve; IV) charging
           of the capacitor with the output direct current.

is used for charging. In the initial period of charging, the current
must be limited with allowance for the curve of the maximum output
current when the capacitor is charged with a current pulse of constant strength, under conditions where the energy losses in the output circuit of the charging device are minimal (curve IV in Figure
9-8).

    The efficiency of the charging device can be expressed as the
ratio of the energy W stored in the storage capacitor during a
charging cycle to the total energy that enters the charging device
during the same period of time. If we assume that the charging device contains no elements capable of storing any significant amounts
of energy, then we also may assume that the total energy that enters
the device during the charging cycle is equal to the energy stored
in the capacitor and the energy losses W_c in the elements of the
charging device:

$$\eta = \frac{W}{W + W_c}.$$ (9-3)

    Thus, the problem of increasing the efficiency of the charging
device essentially reduces to decreasing the energy losses in its
elements.

    The efficiency does not exceed 50% in circuits in which the
storage capacitor is charged through a resistance. These circuits
are uneconomical and require heat removal from the resistor. Therefore, preference is given to charging methods in which the charging

current is limited by reactive and switching elements having a low resistance (e.g., see the circuit in Figure 9-9). Such circuits may have an efficiency approaching 100%, provided that

$$W_C = \sum_n \int_0^T i_n^2(t) R_n \, dt + \sum_m W_m(T) \to 0, \qquad (9\text{-}4)$$

where i_n is the instantaneous value of the current in each of the n elements through which current flows from the power supply to the storage capacitor; R_n is the resistance of each element, with allowance for the skin effect, losses to electromagnetic radiation, losses in ferromagnetic materials and dielectrics, and losses in semiconductor elements, including energy losses due to back currents in semiconductor devices, etc.; and W_m is the energy consumed during a cycle by each of m auxiliary elements.

In other words, the number of elements through which current flows in the circuits between the primary power supply and the storage capacitor, the resistances of these elements, the number of auxiliary elements that draw energy during the charging cycle of the storage capacitor, and the power consumption of the auxiliary elements must be minimized to the extent possible. At the same time, the effective value of the current in each of the elements must be minimal at the required average value of the charger output current, i.e., the current pulses passing through the charging circuits (pulses which often have an intricate shape) should be as prolonged as possible, and their maximum excursions must be as small as possible, approaching the average value of the current in each circuit under consideration.

These recommendations make possible quantitative comparisons of different circuit and design solutions for charger elements in terms of efficiency.

For example, for a charging device consisting of a controlled switch T, a ballast reactor L, and a recharging diode D (Figure 9-9), the energy losses during a charging cycle consist of the energy that goes to power the rectangular pulse generator, losses in the open transistor and its switching losses, losses in the reactor, and losses in the diode. If we have oscillograms of the current in each element and know the resistance of each element during a current pulse, then the total energy losses can be estimated with sufficient accuracy. Formula (9-3) can be used to calculate the efficiency of the charger by referring the total energy losses to the energy stored in the capacitor.

To date insufficient work has been done on problems of minimizing the weight and volume of the charger while keeping the efficiency as high as possible. During the design of chargers using

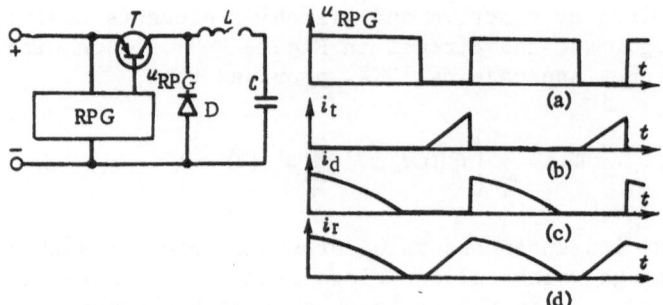

Fig. 9-9.   Circuit of a charger using a transistor and a ballast
            reactor, and graphs of the electrical processes in the
            circuit.   a) Output voltage of the rectangular pulse
            generator RPG; b) current in the transistor; c) current
            in the diode during discharge of the reactor onto the
            storage capacitor; d) current in the reactor.

reactive current-limiting elements, every effort should be made to
ensure that the ratio of the energy loss in an element of the charger
to the weight (or volume) of the element is approximately the same
for all elements.   For example, if up to 40% of all losses occur in
the transistor and reactor in a charger assembled according to the
diagram in Figure 9-9 when the semiconductor devices are operated
at average rated power, but the weight of the transistor accounts
for 3% of the total weight of the charger, and the weight of the
reactor for 80%, then it is advisable to use several parallel transistors in the circuit.   This reduces losses in the switching element
and increases its weight to, say, up to 10-15% of the total weight
of the charger by reducing the weight of the reactor and accordingly
increasing the fraction of losses in the reactor by 10%, for example.
The weight of the charger does not increase, but the efficiency is
improved.   Of course, during such optimization allowance should be
made for many parameters of the charging elements and the properties
of the circuit.

9-5.   A Survey of the Different Types of Chargers

        Charging devices perform basically two functions: they match
the voltage of the primary power supply to the voltage across the
storage capacitor, and they limit the charging current in order to
prevent overloading of the power supply and to protect the lamp from
going into a continuous burn.

        A diagram of the classification of chargers for supplying power
to flashlamps from storage capacitors is shown in Figure 9-10.

In general, the two types of primary power supplies (AC and DC) may be converted into each other by using an inverter or a rectifier [9-7]. The need to convert from alternating to direct voltage arises, for example, when a flashlamp is operated at a flash frequency that is not a multiple of the alternating current, and the reverse conversion is required when a step-up transformer is used (when only a steady low-voltage power supply is available), or when the charger used can operate only with an AC power supply. When these conversions are done, the alternating voltage obtained my be sinusoidal, but also may be trapezoidal or rectangular. The rectifiers of the power supply may use all sorts of rectification circuits: one- or two-period circuits, bridge circuits, three-phase circuits, voltage multipliers, and other circuits. Various semiconductor converters, vibropacks, and dynamoelectric converters are used as inverters.

Of particular importance are charging methods in which the primary-supply output voltage increases together with the voltage across the storage capacitor (such operation also is possible when chargers of groups 3, 4, or 5 are used).

Sometimes a generator (a primary power supply or a combination of such a supply with an appropriate charger from group 3, 4, or 5) having an emf that increases linearly from zero to U_M and an internal resistance R_{in} is used. In order to avoid a reverse discharge of the capacitor onto the generator, the generator is series-connected to a diode that has an internal resistance R_d (Figure 9-11a). In this case, given a lead resistance R_1, the charging current is equal to $i(t) = \dfrac{CU_M}{T}(1 - e^{-t/RC})$, where $R = R_{in} + R_d + R_1$, and T is the charging time. The energy losses over a charging cycle are equal to

$$W_c = R \int_0^T i^2(t)\, dt = \frac{C^2 U_M^2}{T^2} R\left(T - \frac{3RC}{2}\right).$$

The efficiency of such a process is equal to

$$\eta = \frac{1}{1 + \dfrac{2RC}{T} + \dfrac{3R^2 C^2}{T^2}} \approx \frac{1}{1 + \dfrac{2RC}{T}} \tag{9-5}$$

and approaches unity when T >> RC.

If the voltage increases smoothly according to any other curve, then an estimate of the efficiency gives slightly lower values when T >> RC than for a linear increase in voltage.

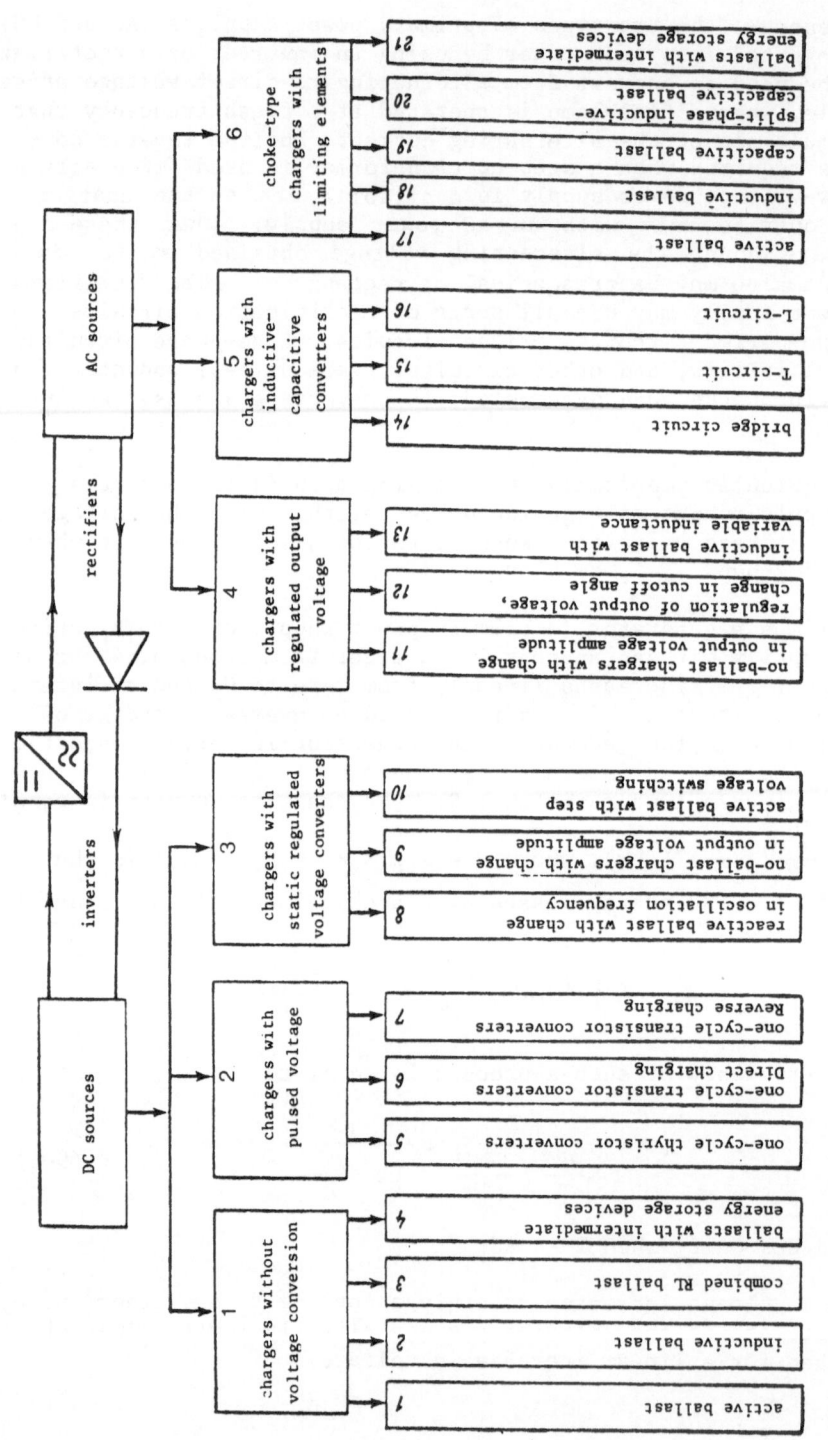

Fig. 9-10. Clasification of the chargers used to provide power to flashlamps from storage capacitors.

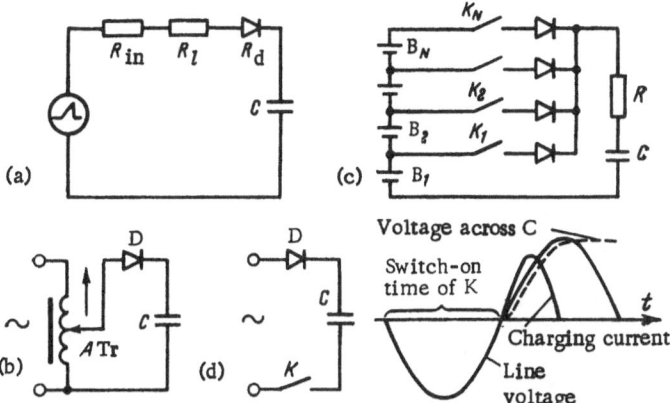

Fig. 9-11. Circuits for charging a storage capacitor from an in-
            creasing-voltage source. a) Linear increase in voltage;
            b) charging through an autotransformer with a variable
            transformation ratio; c) charging from a battery with
            series connection of elements; d) charging during one
            half-wave from a sine-wave alternating voltage source.

        This type of charging method can be used for any capacitance
of a capacitor operating in infrequent-discharge mode if a self-
contained generator with excitation that provides a practically constant output current at any value of the output voltage is used as
the power supply. A similar case obtains when an AC generator is
used with an output choke whose output current strength is limited
by the inductive resistance of the generator phases [9-8]. A generator of a linearly increasing voltage can be simulated by using an
autotransformer with a smoothly varying transformation ratio (Figure
9-11b), or a charging circuit containing n series-connected DC power
supplies which are successively connected to the storage capacitor
through their respective switches (Figure 9-11c). The efficiency
of such a circuit may reach $\eta = n/(n + 1)$ in the limit if the interval between the closure times of successive switches is much larger
than RC [7-45 and US patent No. 3,917,937].

        Finally, an analogous process is the no-ballast charging of a
low-capacitance capacitor during a single half-wave when it is connected to an AC supply through a switch and a diode (Figure 9-11d;
to avoid strong current surges, the switch must be closed with reverse voltage across the diode). If R is small, the efficiency here
also approaches unity, and the final voltage across the capacitor
is equal to the amplitude of the alternating voltage, whereas both
quantities may be substantially smaller if R is appreciable (Figure
9-12).

Circuits for no-ballast charging of storage capacitors should not be confused with charging circuits that do not use an external limiting resistance. A primary power supply such as a high-voltage battery with a high internal resistance (e.g., the battery for electronic photoflashes, type 330EVMTsG-1000, with an emf of 330 V and an internal resistance of 1-5 kΩ) can charge a capacitor without any charging device when directly connected (here the current limiter is the internal resistance of the battery).

It is advisable to consider the other types of charging in accordance with the classification given in Figure 9-10 (for convenience, the sequence shown in the diagram sometimes is violated).

(a) Charging a capacitor from a DC power supply with an emf $U_{x.x}$ and a ballast resistance R (group 1-1). The charging current is defined by the function

$$i(t) = \frac{U_{x.x}}{R}\left(1 - e^{-t/RC}\right),$$

and the power P supplied to the capacitor and the average power P_{av} over the charging time t are given by the expressions

$$P = \frac{U_{x.x}^2}{R}\,e^{-t/RC}\left(1 - e^{-t/RC}\right) \quad \text{and} \quad P_{av} = \frac{CU_{x.x}^2}{2t}\left(1 - e^{-t/RC}\right)^2.$$

P and P_{av} pass through maxima at the times 0.69RC and 1.24RC, respectively, when they take on values of $U_{x.x}^2/4R$ and $U_{x.x}^2/4.96R$. When 1.24RC is reached, the voltage across the capacitor is equal to 0.71U, and the energy stored in the capacitor is equal to half the total energy, i.e., $CU_{x.x}^2/4$. If charging is repeated periodically under conditions of maximum average output power, then the utilization factor K_u of the power supply is equal to 0.206 and the efficiency (as calculated from the formula $\eta = (1 - e^{-t/RC})/2$ [9-9]), is equal to 0.355. The efficiency can be increased to nearly 0.5 by charging the capacitor to a high voltage, such as $0.96U_{x.x}$, but K_u would be reduced by a factor of nearly 2.5 if this were done.

If the charging circuit uses a variable resistance which ranges from R_{max} to zero in such a way that the current always remains constant at $U_{x.x}/R_{max}$, then the characteristics of the charger are improved somewhat. K_u increases to 0.5, the power supply is loaded to full power, the charging time of the capacitor decreases to a minimum $t_{ch} = R_{max}C$, and the voltage across the capacitor by the end of t_{ch} reaches the value of the power-supply voltage. Such a charging resistance may consist, for example, of a resistor R_{max} in parallel with which a transistor is connected as a ballast that can be varied from infinity to zero.

Incomplete discharging of the capacitor (extinction of the lamp when there is a significant residual voltage across the storage

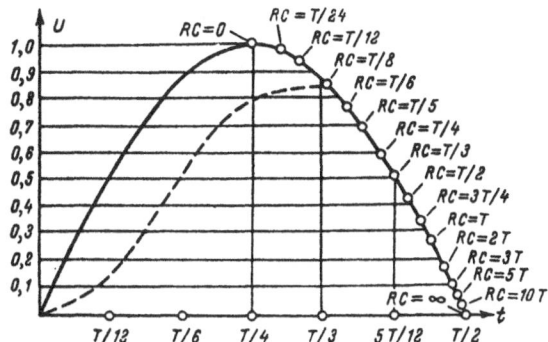

Fig. 9-12. Final values (the circles; the values are in relative
            units) of the voltage across a capacitor charged during
            one half-wave of an alternating current at different
            values of the time constant of the charging circuit.
            The dashed line represents the increase in voltage across
            the capacitor when RC = T/8.

capacitor) also leads to increased efficiency, since the energy
losses in the ballast resistance are particularly large during the
initial stage of charging, when the voltage across the capacitor is
low.

   (b) Charging a capacitor from a DC power supply through a
linear choke containing a diode (group 1-2). This process involves
doubling the voltage across the capacitor. A schematic diagram of
such a circuit is shown in Figure 7-8b. Charging is done with a
current pulse that has the shape of a half sine-wave when the period
between flashes is much longer than the half-period of the oscillations of a circuit consisting of a choke L and a storage capacitor
C. Such charging is characterized by a comparatively high K_u (about
0.64). The efficiency of such a circuit may approach unity if the
resistance of the choke and diode are small. However, in practice
these losses must be taken into account, and the efficiency lies in
the range 0.5-1. The efficiency can be determined from the expression $\eta = (1 - e^{-RT/2L})/2$ [9-9].

   For aperiodic charging, the efficiency of a charging circuit
formed by an inductive reactance and a resistance (group 1-3) should
be less than 0.5, and the presence of the choke only slightly slows
the increase in the voltage across the storage capacitor in the
initial stage of charging and speeds it up at the end of the cycle,
compared with charging through a constant resistance.

   The behavior of the voltage across the capacitor in this circuit is given by the expression

$$u(t) = U_{x.x}\left[1 - \frac{\omega_0}{n}\exp(-\alpha t)\sin(nt + \beta)\right]$$

where $\alpha = R/2L;\ n = \sqrt{\omega_0^2 - \alpha^2};\ \omega_0 = \sqrt{1/LC};\ \beta = \operatorname{arctg} n/\alpha$; the current in the charging circuit is

$$i(t) = (U_{x.x}/Ln) \exp(-\alpha t) \sin(nt),$$

and the time required for charging to maximum voltage is

$$T = \frac{\pi}{\sqrt{1/LC - R/4L^2}}.$$

Sample graphs constructed from these formulas are shown in Figure 9-13, together with plots of the voltage during charging through a constant resistance and through an inductance.

The use of nonlinear chokes in charging circuits makes it possible to slow the rate of increase of the voltage across the capacitor still further during the initial stage of charging. The shortcomings of a charging circuit working through a choke include the considerable size of the ballast elements when the charging cycle is long.

(c) <u>Charging a capacitor from a DC power supply through a circuit with intermediate inductive and capacitive storage devices</u> (group 1-4). Such charging makes it possible to reduce the size of chargers by using reactive ballast elements and switches [9-9a]. Examples of such circuits are shown in Figures 9-9 and 9-14. In the second circuit, a control pulse is supplied to thyristor T after capacitor C_1 is charged to twice the power-supply voltage, and the small capacitor C_1 is discharged by a short current pulse onto primary storage capacitor C through choke L_2. When the thyristor switches off, the energy stored in choke L_2 is pumped across to capacitor C through the circuit D_2-L_2-C (current pulse i_{D_2} in Figure 9-14b). At the end of the series of charging pulses, the voltage across capacitor C becomes approximately equal to the power-supply voltage. The efficiency of this charger may be quite high, and is determined by the Q factor of the circuits formed by the chokes and capacitors, and by energy losses in the semiconductor devices. K_u also is quite high, since the value of the input current is nearly constant throughout the entire charging cycle.

(d) <u>Rectifying circuits of chargers containing limiting elements</u> (groups 6-17 to 6-21). These rectifying circuits are designed to charge storage capacitors from AC supplies. These devices are close to devices of the first four groups in terms of the processes that occur in them, but their design is complicated by the oscillating nature of the voltage acting on the charging circuit. The calculations are labor-intensive, and computer methods are used to make them. Results in graphic form may be used for rough estimates. In a circuit containing a resistor (Figure 9-15), the charging at the beginning (to $u/U_M = 0.6$) goes approximately one-fourth as fast as

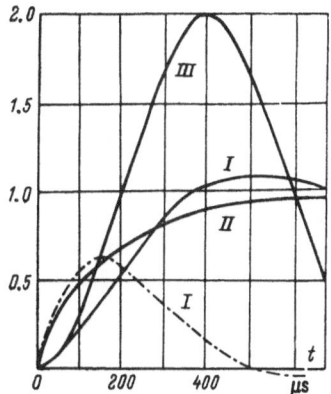

Fig. 9-13. Curves of the relative values of the voltage across the
            storage capacitor (0.2 μF) (solid lines) and of the
            current in the charging circuit (dot-dash line). I)
            Charging from a DC supply through a combined charging
            circuit (R = 800 Ω, L = 80 mH); II) charging through an
            800-Ω resistor alone; III) charging through an 80-mH
            choke alone.

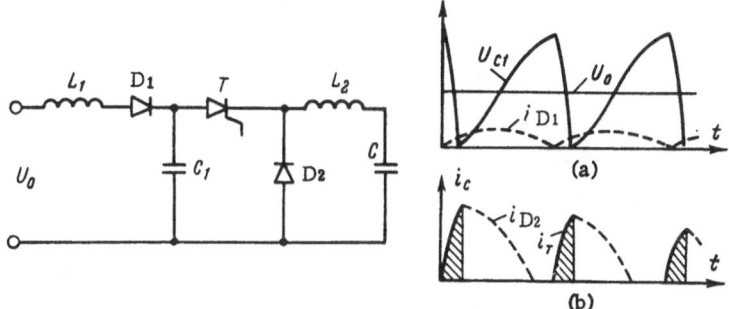

Fig. 9-14. Schematic diagram of a charger using reactive ballast
            elements which act as intermediate storage devices. The
            total energy entering the storage device is divided into
            portions which are transmitted successively to the
            storage device. a) Oscillogram of the voltage across
            storage device C_1 and of the current (dashed line) in
            the input circuit; b) oscillograms of the current in
            the output circuit. The current in (b) consists of a
            pulse from C_1 (hatched area) and a recharging pulse
            (dashed line).

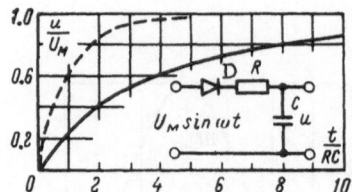

Fig. 9-15.   Graph of the increase in the voltage across a storage
             capacitor being charged by an AC power supply through
             a half-wave rectifier and a ballast resistor. The
             dashed line represents the charging curve when a DC
             power supply is used [9-10].

it would from a DC power supply. The use of a bridge-type rectification circuit doubles the charging rate. In this circuit the
nature of the charging curve is practically independent of the oscillatory frequency of the supply-line voltage if the period of the
oscillations is much shorter than the time constant RC of the charging
circuit. The efficiency of such a charger does not exceed 0.5 (in
practice, it lies in the range 0.3-0.4) if the storage capacitor is
charged to a voltage close to U_M. Distortion of the shape of the
current causes higher harmonics to appear in the charging circuit.
The K_M for this circuit is less than 0.1, and can be calculated
from the graph in Figure 9-15 (the same is true of Figures 9-16,
9-18, and 9-19) by using the formula $K_u = (u/U_M)^2/2\pi\tau$. The values
of u/U_M and of the reduced time $\tau = t/RC$ or $\tau = (t/T)(\omega_0^2/\omega^2)$ that
correspond to each point of the plot are inserted into the formula
[9-2a].

     (e) <u>Rectifier-type chargers with a ballast linear rectifier</u>
(group 6-18). Such chargers absorb a small fraction of the energy
that passes through them as the charging-current pulses are repeated.
A larger part of the energy stored in the valve during passage of
the current pulse is transferred to the storage capacitor. As a
result, given equal power consumption, charging is speeded up significantly in comparison with an active ballast (French patent No.
2,237,396). Figure 9-16 shows the corresponding plots of the dependence of the voltage on the time, referred to the charging-circuit
parameters when a bridge-type rectification circuit is used (T is
the oscillation period of the supply-line voltage, ω is the angular
frequency of the power supply, and $\omega_0 = 1/\sqrt{LC}$). Of course, the
charging time here is much longer than the oscillation period of
the supply voltage. (Calculated plots of the charging current, the
voltage across the choke, the power consumption, the utilization
factor K_u of the power supply, the efficiency, the power factor,
and others are given in [9-11 and 9-12].) Beginning at the minimum
value at the start of charging, the power factor of the charger
passes through a maximum value (when $u/U_M = 0.6-0.8$), decreasing

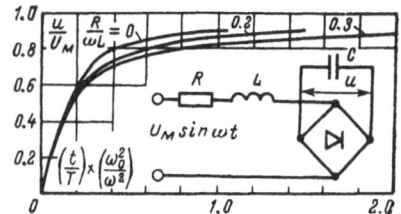

Fig. 9-16. Dependence of the relative voltage across a storage
capacitor on the reduced time for various R/ωL for a
circuit in which the capacitor is charged from an AC
line through a ballast choke [9-11].

toward the end of charging. The maximum value of the power factor,
0.75-0.82, depends on the ratio R/ωL. The limiting value of K_u for
this circuit is 0.17, and an increase in the ratio R/ωL decreases
K_u insignificantly. The efficiency of this circuit decreases with
increasing losses in the charging circuit, to 0.69 when R/ωL = 0.3.

When a choke is included after the bridge rectifier, the nature
of the operation of the charger and the methods used to design it
change, approaching conditions of charging through a choke from a
DC power supply (or more accurately, a ripple-voltage source).

Rectifier-type chargers with a ballast capacitor (group 6-19).
Such chargers are divided into two types: voltage-multiplying devices
(Figure 9-17) ([9-13 and 9-14], and US patents Nos. 3,600,996 and
3,780,344), and devices using a limiting capacitor. The last capacitor usually is used in cascade as a storage element, while all
previous capacitors are used as charging-current limiters. The
charging rate of the storage element is determined by the ratio of
the capacitance of the smallest limiting capacitor to that of the
storage device, by the voltage multiplication factor, and by the
supply-line frequency. In practice, the efficiency of such a charger
may reach 0.75, but the large number of capacitors in the circuit
leads to an increase in weight and size, especially at high power
levels. The current in the input circuit is of the pulsed, sign-
alternating type, and K_u is low.

A charger with only a limiting capacitor (and no choke) in the
charging circuit causes large distortions of the current shape.
These are detrimental to the performance of the power supply, transformer, and rectifiers. Therefore, we recommend that an inductive
element additionally be included in the circuit, in series with the
capacitor. When an inductive element is present, the shape of the
current in the charging circuit approaches a sine curve (Figure
9-18). The reactance X_C must be larger than the reactance X_L. The
limiting value $K_u \approx 0.3$ is reached when their ratio is optimal
$(X_C/X_L = 2.4)$ [9-11, 9-15, and 9-16]. If K_u is optimal, then the

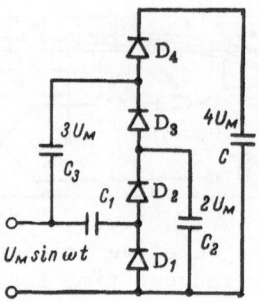

Fig. 9-17.   Example of the circuit of a voltage-quadrupling cascade
             generator.

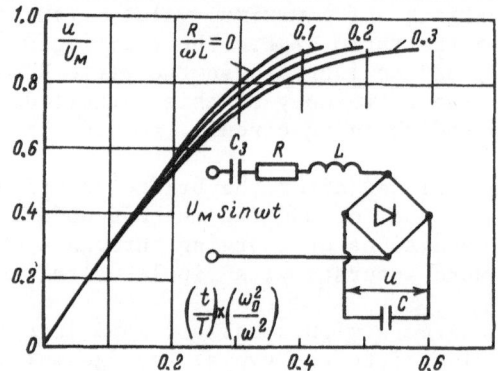

Fig. 9-18.   Charging circuit with a ballast capacitor and a choke
             (AC power supply). Graphs of the dependence of the
             relative voltage across the storage capacitor on the
             reduced time for $X_C/X_L = 2.4$ and for different values
             of $R/\omega L$ (where T is the period of the supply-line
             voltage, ω is the angular frequency, and $\omega_0 = \sqrt{1/LC}$)
             [9-11].

circuit makes it possible to charge the storage capacitor to $u/U_M =$
0.85, while K_u reaches just 0.17 for a circuit with an inductive
ballast. The charging current remains approximately constant through
nearly the entire charging cycle. The maximum power input into the
storage capacitor occurs at the middle of charging, when $u/U_M =$
0.72-0.55. The limiting efficiency of such a charger is equal to
unity, and when $R/\omega L = 0.3$ it is 0.75. The power ratio is 0.97-0.93
for $u/U_M = 0.8-0.85$. Having some of the best energy indices, such
a charger is characterized by low specific weight and volume.

     A combined charging circuit [9-2a and 9-17]. A combined charging
circuit has even better characteristics. It consists of parallel-
connected charging circuits containing an inductive ballast and a

capacitive ballast (group 6-20), and thus is a split-phase circuit
(Figure 9-19). The inductive and capacitive components of the reactive current are mutually compensated in this circuit throughout
nearly the entire charging period. The best compensation is reached
when X_C/X_L = 2-2.8 and L/L_c = 1.4 [9-11]. Under these conditions
the power ratio of the charger turns out to be close to unity throughout the entire charging cycle, i.e., the charging current mainly has
an active component. The utilization factor gradually decreases
with increasing R/ωL, and is equal to 0.69 when R/ωL = 0.2. The efficiency decreases from 1 to 0.8 as R/ωL increases from 0 to 0.3.
This circuit is the best type of charger when the weight and volume
are not limited. It allows the storage capacitor to be charged from
a limited-power AC line faster than with any other known charging
device. The use of a step-up transformer at the charger output makes
it possible to obtain practically any voltage across the storage capacitor. The absence of switching elements is a significant advantage
of the circuit, especially when the device is operated at high voltages. The charging current increases gradually in the initial
stage of charging (for the first few half-waves of the line voltage),
producing a sort of pause in the voltage rise across the lamp. This
is extremely favorable for extinction of the discharge. The average
current strength in the discharge circuit remains nearly constant
all the way to u/U_M = 0.7, decreasing smoothly after this value.
It should be noted that this circuit can operate reliably only if
the power-supply frequency is sufficiently stable.

Finally, rectifier-type chargers with ballasts that contain
intermediate storage elements (group 6-21). Such chargers have
switching elements in their charging circuits and are characterized
by an extremely high efficiency (which reaches 0.9) and by a quite

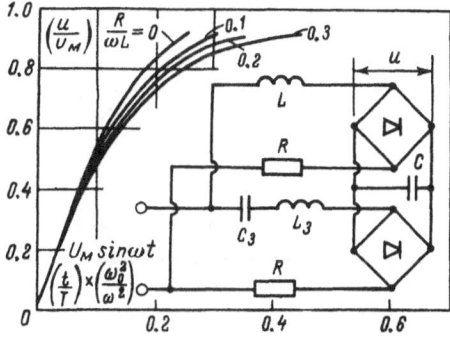

Fig. 9-19. Schematic diagram of a "split-phase" charger. Graphs
            of the dependence of the relative voltage across the
            storage capacitor on the reduced time for various R/ωL
            when X_{Cc}/X_{Lc} = 2.4 and L/L_c = 1.4 [9-11].

high K_u. The corresponding diagrams of one- and two-cycle charging
circuits are given in Figures 9-14 and 9-20. In Figure 9-20 an AC
power supply is connected to terminals A and B either directly or
through some inductance. (The circuit also can be powered by a DC
supply through an intermediate thyristor-type or transistor-type
two-cycle converter with a step-up transformer, as shown in the left
side of Figure 9-20.) The use of such circuits is limited to low
voltages (hundreds of volts to a few kilovolts) because of the difficulty of matching high-voltage switching thyristors. A characteristic feature of the circuits is constant power consumption during
the first half of the charging cycle, and a gradual decrease in the
output current.

     (e) <u>Chargers with inductive-capacitive voltage converters</u>
(groups 5-14 to 5-16). These chargers form a special group of devices with an output current that is constant during charging of the
capacitor [9-16].* Figure 9-21 shows the L-network, the bridge network, and the T-network of inductive-capacitive converters to which
storage capacitor C is connected through a diode or a bridge rectifier. The operation of these circuits, which sometimes are called
Bushero† circuits, is explained by Figure 9-21a. The input current
in this circuit is expressed as

$$\dot{I}_1 = \frac{\dot{U}_1}{Z_1 + \dfrac{Z_2 Z_L}{Z_2 + Z_L}},$$

the voltage across the load is expressed as

$$\dot{U}_2 = \frac{\dot{I}_1 Z_2 Z_L}{Z_2 + Z_L},$$

and the current in the load is expressed as

$$\dot{I}_L = \frac{\dot{U}_2}{Z_L} = \frac{\dot{U}_1}{Z_1 + (Z_L/Z_2)(Z_1 + Z_2)}.$$

---

*Universal power units of types BP-2000 and BP-5000, with ratings
 of 2 and 5 kV, respectively, and charging currents of 0.75-9 and 0.3-3.6
 A are such chargers. (The amperage is included in the name of the
 unit in two-digit form, e.g.: BP-5000-3.6.) These power units are
 produced commercially for the storage capacitors of laboratory
 lasers. They include modulator control systems which make it
 possible to regulate and stabilize the voltage and to produce single
 pulses or pulses repeated at a frequency of 0.05-20 Hz which are
 synchronized with an external contact such as an optical shutter.

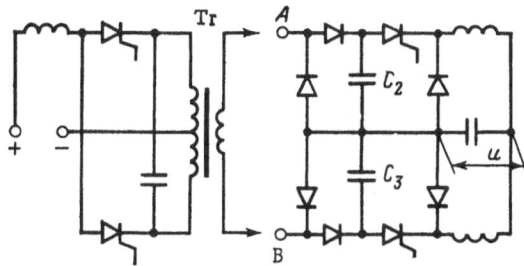

Fig. 9-20. Schematic diagram of a rectifier-type charger with intermediate capacitive storage elements C_2 and C_3. A two-cycle converter for providing power to the charger from a low-voltage DC source is shown at left.

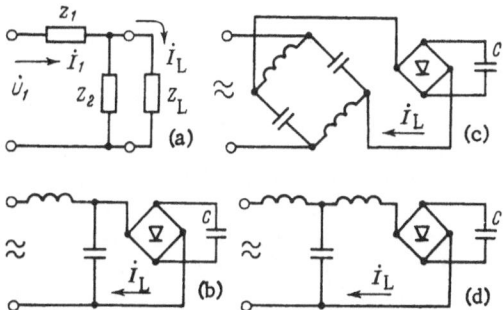

Fig. 9-21. Schematic diagrams of an inductive-capacitive four-pole (a), and the circuits of inductive-capacitive converters: an L-network (b), a bridge network (c), and a T-network (d).

If the sum $Z_1 + Z_2$ is equal to zero under any circumstances, then the current $\dot{I}_L$ turns out to be $\dot{U}_1/Z_1$, i.e., it is constant and independent of the load resistance. This condition is met in the circuit if we select as Z_1 and Z_2 a choke and a capacitor whose inductance and capacitance are related by the equation $\omega L = 1/\omega C$, where ω is the angular frequency of the supply voltage.

Resonant voltage step-up occurs in such circuits across reactive elements, whose current may exceed the output current of the charger by some large factor. According to formula (9-4), in this case the losses in the parasitic resistances of these elements should increase sharply. As a result, the efficiency is reduced sharply, generally turning out to be lower than in circuits using a reactive ballast. The load is connected through step-up transformers in order to reduce voltage step-up across the reactive elements. The output power of such chargers increases linearly with increasing voltage across the

storage capacitor. Therefore, if the oscillatory circuits have good
Q factors, the charger must be disconnected from the supply line
after the capacitor is charged to the specified voltage or the output circuit must be shunted with thyristors in order to avoid breakdowns due to overvoltages [9-9a].

A significant increase in efficiency (to 0.88 in an L-network
and to 0.92 in a T-network) and some increase in the power ratio can
be achieved in these circuits by tuning the oscillatory circuits to
their natural frequency, which is twice as high as the power-supply
frequency.

(f) <u>Chargers with regulated output voltage</u> (groups 4-11 to
4-13). Such chargers make it possible to build equipment for charging
capacitors without ballast devices or with ballasts of minimal size.
The rate of increase in the output voltage $du/dt = Ci(t)$ in these
chargers is determined by the function $i(t)$ during the charging process for a specified capacitance. In the circuit in Figure 9-11b
(group 4-11), a special pickup which monitors the charging current
must send a signal to a programmed drive which changes the transformation ratio of transformer ATr. In the circuit in Figure 9-22a
(group 4-12), this is accomplished by a smooth change in the connection phase of a symmetrical thyristor ([9-9a] and US patents Nos.
3,377,542 and 3,588,423), and in the circuit in Figure 9-22c (group
4-13) by varying the magnetizing current to change the inductance of
the choke in the charging circuit [9-18]. The circuits of groups
4-11 to 4-13 are characterized by high efficiency and a quite good
K_u and can be used in stationary units that are powered by a commercial AC line, especially when very high energies must be stored
over comparatively long periods of time. The shortcomings of some
of these circuits include the strong distortion of current shape in
the network. The desire to increase the efficiency and K_u of the
device and to prevent the lamp from entering a continuous-burn mode
usually leads to the selection of capacitor charging conditions intermediate between constant output current of the charger and constant
power. This is graphically evident from Figure 9-23. Here the
charger is represented as a four-pole (R_{in} and R_{out} are the resistances of its input and output circuits, and $R_{e.r}$ is the equivalent
resistance of the self-losses of the charger). If the power losses
of the charger are much smaller than the power transmitted, then
the plots of the input (p_1) and output (p_2) power should be approximately the same. At constant output current, the power of the
charger increases linearly (p_1 and p_2 in Figure 9-23a and 9-23b) because of the constant power-supply voltage $U_{x.x}$ and the linear increases in the output voltage u_o of the charger as the charging of
the capacitor progresses. At constant power, the output current decreases rapidly as the output voltage increases. In the first case
K_u does not exceed 0.5, since the maximum power consumption is twice
the average, but the energy losses $i_H^2 R_{out}$ in the output circuits are
minimal. In the second case K_u approaches unity, losses in the input

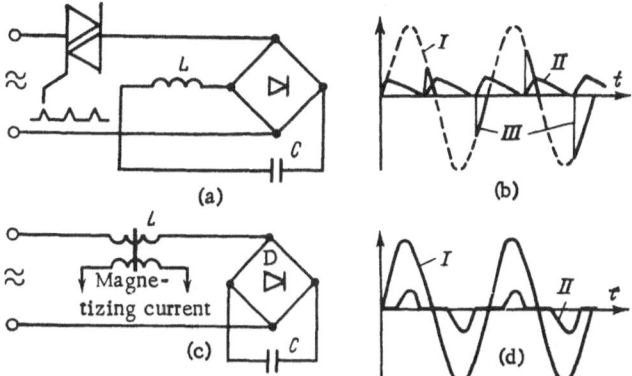

Fig. 9-22. Circuits of chargers using a thyristor-type current regulator (a) and a magnetized choke (c); b, d) the corresponding graphs of the voltage and current. I) Supply voltage; II) current through the choke; III) regulated voltage.

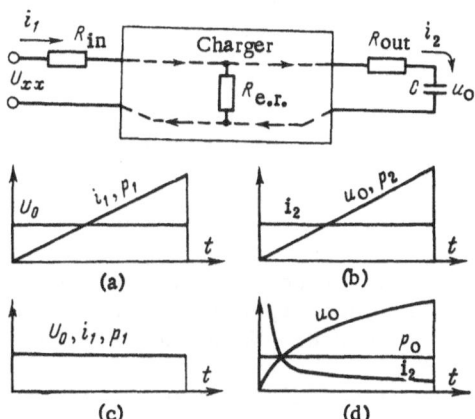

Fig. 9-23. Comparison of two modes of operation of the charger: at constant current (graphs a and b) and at constant power (graphs c and d).

circuits are minimal, but the output circuit operates under conditions close to a short circuit during the initial period of charging. This entails the danger of a continuous discharge in the lamp and considerable losses through R_{out}. During the practical implementation of a charger, allowance must be made for the equivalent internal resistance $R_{e.r.}$ of the losses. In many cases this resistance may lead to a significant increase in energy losses and reduce to naught the advantages of the most favorable charging conditions.

(g) <u>Chargers with adjustable static converters</u>. A smooth increase in output voltage is provided in such chargers, which are powered by a DC supply (groups 3-8 to 3-10 [9-4 and 9-19]), as in chargers of groups 4-11 to 4-13. These chargers are characterized by a high efficiency and K_u, but are so complex that it is advisable to use them only in especially high-power apparatus. In Figure 9-24a (group 3-8), a two-cycle converter with a sinusoidal or trapezoidal output voltage is connected to a storage element through a reactive ballast (a choke or a capacitor) and a rectifier, and feedback unit FB regulates the oscillation frequency of the converter so that the current through the pickup in the form of a resistance varies according to the requirement outlined in item "f". The frequency of the converter should decrease as charging proceeds when an inductive ballast is used, but should increase when a capacitive ballast is used (US patent No. 3,529,228).

The circuit in Figure 9-24b (group 3-9) uses two identical push-pull oscillators with a sinusoidal output voltage which goes to transformers Tr1 and Tr2. As the charging of capacitor C progresses, a special follow-up system smoothly changes the phase difference of their output voltages so that the oscillators operate in opposition at the start of charging and the output voltage is equal to zero; then the out-of-phase conditions decrease smoothly, and the output voltage increases. At the end of charging the oscillators work in phase, and the output voltage of the charger is equal to twice the voltage of each of the series-connected output windings of the transformers (the principle illustrated in Figure 9-11a is realized).

Finally, the circuit in Figure 9-24c (group 3-10) contains several (four, for example) identical push-pull oscillators operating in step with a sinusoidal or trapezoidal output voltage. The oscillators are loaded to storage capacitor C through the transformers and the small ballast resistor R. When the oscillators operate in phase, the output windings of the transformers are parallel-connected through the diodes, and the capacitor is charged through resistance R to the peak value of the oscillator voltage U. If the voltage phase of the first and second oscillators then changes by $180°$ relative to the phase of the third and fourth oscillators, for example, then the output voltage will double and the second period of the charging cycle, from U to 2U, will begin. The output voltage will be tripled when the phase of the first oscillator next changes by $180°$ relative to the phase of the second generator, and it will be quadrupled when the phase of the fourth oscillator changes relative to that of the third. The capacitor is charged in this way from a power supply having a stepwise-increasing voltage in accordance with Figure 9-11d.

(h) <u>Chargers of groups 2-5 to 2-7 with pulsed voltage converters</u>. Such chargers are used extensively in low-power equipment supplied by low-voltage DC sources and usually have simple circuitry.

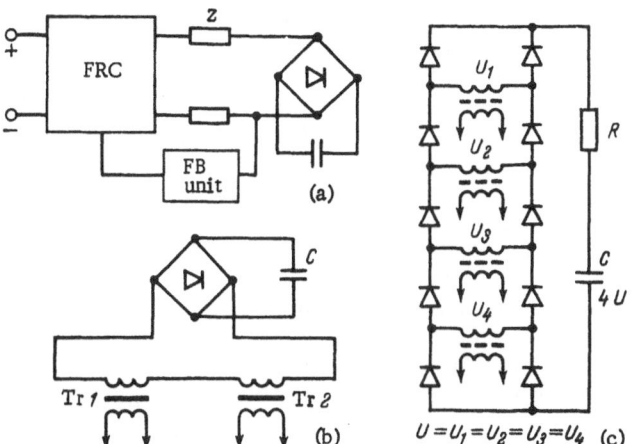

Fig. 9-24. Schematic diagrams of the use of converters to charge
            storage capacitors. a) With a frequency regulated during
            charging (FRC); b) with a smoothly variable oscillation
            phase in the two harmonic oscillators [9-3]; c) with
            four-stage switching of the output voltage of four os-
            cillators that have a rectangular output voltage.

However, their power is low when existing transistors and thyristors
are used: the power may be up to 1 kW for one-cycle thyristor converters (group 2-5 [9-4 to 9-6]; see Figure 9-7), up to hundreds of
watts for one-cycle transistor converters with reverse charging of
the storage element (group 2-7 [9-20 and 9-21] and US patents Nos.
3,435,320, 3,417,306, 3,515,973, 3,319,146, and 3,531,738; see
Figure 7-10), and a few watts for one-cycle transistor converters
with direct charging of the capacitor (group 2-6, see Figure 7-9).

9-6. Inductive Storage Devices

     Capacitors are convenient storage elements for providing power
to flashlamps. However, the weight, volume, and cost of a capacitive
storage element consisting of like capacitors increases in proportion
to the energy stored in them. Therefore, capacitor banks are extremely cumbersome and expensive at very high energies.

     The active part of a storage capacitor is its dielectric, which
contains an amount of energy $\varepsilon_a E^2/2$ (where ε_a is the absolute per-
mittivity) per unit volume at an electric field strength E. The
storage of energy by a magnetic field is possible in principle, like
the storage of energy by an electric field. An amount of energy
$\mu_a H^2/2$ (where μ_a is the absolute magnetic permeability) is contained
per unit active volume of the magnetic circuit of a magnetic storage

element, which usually is called an inductive energy storage element. The limiting energy concentration in the dielectric is absolutely bounded by the breakdown value of E. There is no such absolute constraint for H. This enables us to count on reaching much higher effective values of $\mu_a H^2$ than the attainable values of $\epsilon_a E^2$, especially in large designs. Indeed, let us take two geometrically similar chokes. All linear dimensions of one choke (D_1) are a factor n larger than those of the other (D_2), but the number of turns and the current density in the windings are equal. Let the chokes have the following relationships between parameters: the length of the winding wire $l_1 = n l_2$; wire cross section $S_1 = n^2 S_2$; resistance $R_1 = R_2$; winding inductance $L_1 = n L_2$; currents in winding $i_1 = n^2 i_2$; time constants ($\tau = L/R$) $\tau_1 = n \tau_2$; voltages of power supplies capable of producing the appropriate currents in the windings, $U_1 = n^2 U_2$; total weight $M_1 = n^3 M_2$; and energy stored ($i^2 L/2$) $W_1 = n^5 W_2$. Since the energy stored in a choke is proportional to the fifth power of its linear dimensions, the supply power to the fourth power, and the weight to the third power, the energy increases much more quickly with increasing size of the choke than the weight and (the size of) the power pack, and hence faster than the cost and volume of the unit as a whole. According to the estimates in [9-22], the characteristics of capacitive and inductive storage elements roughly level out at an energy of 0.25 MJ, and at 10 MJ an inductive storage element may turn out to be 5 times as advantageous as a capacitor bank. However, considerable difficulties must be overcome in order to implement devices using inductive storage elements.

A schematic diagram of the power supply for a flashlamp using an inductive storage element is shown in Figure 9-25. When switch K1 is closed, a current begins to flow through the choke, which has an inductance L:

$$i(t) = \frac{U}{R}\left[1 - \exp\left(-\frac{R}{L}t\right)\right],$$

where R is the resistance of the charging-circuit elements and U the power-supply voltage. In the limit the current reaches a value $I_M = U/R$, and the energy stored in the choke is $W = I_M^2 L/2 = L P_M/2R$, where $P_M = U_M I_M$ is the maximum power of the power supply. Switch K2 is closed with some advance just before switch K1 opens. As a result, a trigger pulse is generated when capacitor C_2 undergoes pulsed charging from capacitor C_1 (through the primary of pulse transformer Tr). The lamp fires and capacitor C_1 (of comparatively low capacitance) discharges through it, creating a plasma channel. When the resistance of the lamp drops to its minimum value, switch K1 opens and the choke current is switched from circuit I to circuit II, beginning to flow through the flashlamp. As this occurs, the maximum voltage $U_M = I_M R_L$ appears across the lamp and the switch, and the damping rate of the current pulse in circuit II is determined by the time constant:

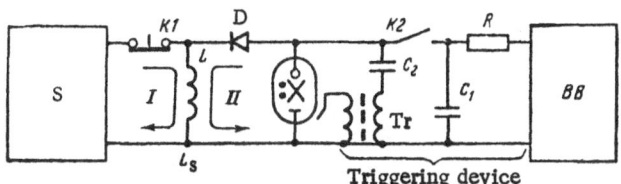

Fig. 9-25.   Schematic diagram of the power supply for a flashlamp
             using an inductive storage element. S - low-voltage DC
             source; K1 - high-current, high-voltage switch; K2 - a
             switch ahead of K1 in the trigger unit, which is powered
             by high-voltage rectifier BB.

$$\tau = \frac{L}{R_\text{L} + R_\text{ch} + R_\text{d}} ,$$

where R_ch and R_d are the resistances of the choke and diode, respectively. (In fact, this parameter varies slightly during a pulse,
since R_L is a function of the current flowing through the lamp.)

     The following are required in order to build actual devices
using inductive storage:

1. An extremely high-power low-voltage source of <u>direct current</u>:
    a rectifier, a storage battery, or a DC generator. Allowing for
    the weight and volume of the power supply reduces to some extent
    the gain due to the high specific energy capacity of an inductive
    storage element.
2. A switch K1, which must pass a high current and then, after the
    circuit is opened, must withstand a high voltage equal to the
    voltage across the discharging lamp. The performance and re-
    liability of the entire device using an inductive storage element
    hinge on the solution of the switching problem. The following
    may be used as the required switch: a high-speed air circuit
    breaker with an external air flow, which is shunted by a blow-
    out fuse link in the initial stage of switching [9-22]; explosive
    analog-type cut-outs [9-23]; vacuum arc-suppression chambers
    with a suppression circuit which creates (during opening of the
    contacts) a reverse current that makes it possible to break the
    briefly deenergized circuit [9-24 and 9-25]; superconductor-
    based breakers, which instantly go from the superconducting state
    to the ordinary state with significant electrical resistance
    [9-26 and 9-27]; and thyristor breakers with a suppression cir-
    cuit [9-28]. The search for the optimal type of switch continues
    [9-29].
3. An <u>inductive storage element (a reactor)</u> proper, with low re-
    sistance and sufficient mechanical strength. Aluminum and copper
    are the best materials for the storage element winding (excluding

the optimal solution: a superconductor). When these materials
are used, an energy-storage time constant on the order of a few
seconds can be ensured with energy storage of about 1 MJ. The
time constant of the coil increases with increasing storage
energy. The use of ferromagnetic cores in inductive storage
elements does not produce a significant gain in the weight or
volume of reactors. The use of a superconducting winding in an
inductive storage element [9-28] gives a qualitative jump in increased efficiency and an increased energy-storage time constant,
but requires the solution of a number of other problems inherent
in cryogenic technology. (These problems include high equipment
cost, the explosion hazard of the unit, the thermal insulation
of high-current leads, allowance for the critical current density
beyond which the superconductor enters the ordinary state, and
allowance for the effect of strong magnetic fields on the surrounding ferromagnetic structural elements, for the shielding
of these fields, and for the associated mechanical forces).

4. A discharge triggering device in a flashlamp or a group of lamps,
which produces a powerful predischarge with low resistance in
every lamp (the operation of switch K1 is facilitated when such
a discharge is present, since this reduces the voltage of extra
pulses in the circuit being switched).

Work toward creating equipment using inductive storage elements
to power flashlamps is in the experimental stage because of the difficulties in designing such elements. Therefore, it would be pre-
mature to evaluate the operating and technical characteristics of
such equipment in detail. However, the general criteria for evalu-
ating chargers with capacitive energy storage elements that were
presented in Section 9-4 also are applicable to systems using inductive storage. For example, the efficiency of the charger of such
a system may be 0.5 if the charging time of the storage device is
equal to its time constant L/R, and the efficiency should decrease
with increasing charging time. The utilization factor of the power
supply should be 0.3 at best.

The use of inductive storage with at least 10 MJ of stored
energy (see Figure 9-49, below) and of a pulse-type primary power
supply obviously is promising. The power supply may be, for example,
a mechanical storage device such as a homopolar generator, a pulsed
MHD generator, or an explosive-magnetic generator, which also can
power a pulsed light source directly by virtue of its short-duration
operation [9-30].

At present, inductive storage devices having an energy of up
to a few tens of joules are being used, for example, to pump active
media in lasers (West German patent No. 1,940,030). Bidirectional
thyristors are used as the current switch here.

9-7. The Discharge Circuit

In the power-supply circuits of flashlamps using a storage capacitor, the discharge circuit always contains at least three elements: the lamp, the capacitor, and connecting leads. Such a basic discharge circuit is characterized by the following main parameters: the voltage U_0 across the capacitor just before the flash; the capacitance C of the capacitor; the circuit inductance L_K, which consists of the inductances of the capacitor, the connecting leads, and the lamp; the circuit resistance R_K, which is determined by active energy losses in the capacitor upon its discharge and by the resistance of the leads, with allowance for the skin effect; and the resistance R_L of the flashlamp during a flash. The temporal and electrical parameters of the current pulse in the circuit determine the main parameters of the flash. The oscillatory or aperiodic nature of the discharge, the duration and strength of the current through the lamp, the time dependence of the power dissipated in the lamp, the energy dissipated in the lamp, and other characteristics are defined by the equation

$$i(R_\kappa + R_L) + L_\kappa \frac{di}{dt} + \frac{1}{C} \int i \, dt = U_0. \tag{9-6}$$

As we showed in Chapters 2 and 3, the electrical resistance of a tubular lamp after firing drops rapidly to some value which then is nearly constant. If plasma fills the entire inside cross section of the tube, then this value is given approximately by the formula

$$R_L = 4l/\pi\sigma d^2,$$

where l and d are the length and diameter of the working volume of the lamp, and σ is the conductivity of the plasma, which can be estimated from expression (2-1).

Hence, as the first approximation the current i(t) through the lamp and the power P and energy W dissipated in it can be found from the expressions:

$$i(t) = \frac{U_0}{R_L + R_\kappa} \exp[-t/(R_L + R_\kappa)C]; \tag{9-7}$$

$$P = i^2(t) R_L = \frac{U_0^2}{(R_L + R_\kappa)^2} R_L \exp[-2t/(R_L + R_\kappa)C]; \tag{9-8}$$

$$W = \int_0^\infty i^2(t) R_L \, dt = \frac{CU_0^2}{2} \frac{R_L}{R_L + R_\kappa}. \tag{9-9}$$

The expression for the power pulse length at the level of $1/e$ follows from relation (9-8) (this pulse length is approximately equal to the effective flash duration):

$$\tau = (R_\kappa + R_l)\,C/2. \qquad\qquad (9\text{-}10)$$

If follows from the formulas that an increase in the resistance R_K leads to a decrease in the amplitude of the current and power pulses through the lamp, to a prolongation of the flash, and to a decrease in the energy dissipated in the lamp that is proportional to R_L/R_K. An additional resistance rarely is introduced into the circuit for the purpose of regulating the flash energy. This is done only when other methods of control are unacceptable for some reason (US patents Nos. 3,783,338 and 3,496,411 and British patent No. 1,334,120). The introduction of an additional inductance L_{ch} changes the shape of the current pulse through the lamp, mainly in the first stage of the discharge (when the rate of change of the current is highest), when the inductance slows the current rise and stores some amount of energy. This results in the formation of a smoother bell-shaped current pulse through the lamp at a nearly un-changed integrated luminous intensity of the flash and ensures an increase in the operating life of the electrodes and the envelope. Increasing the inductance all the way to values at which the con-ditions for a critical discharge arise ($L_K = C(R_L + R_K)^2/4$) causes an insignificant prolongation of the flash.

A further increase in the inductance leads to oscillatory op-eration. If the amplitude of the voltage of reverse polarity does not exceed the extinction voltage of the lamp, then the lamp will die out after the first half-period. Some negative voltage is use-ful for delaying the next charging of the storage capacitor. An ap-preciable increase in the negative potential across the capacitor can be obtained, for example, by using a diode in the discharge circuit. However, such an increase would hardly be advisable, since the efficiency of some types of chargers would be lowered.

The flash can be lengthened without changing the direction of the current in the lamp by using the circuit shown in Figure 9-26 (US patent No. 3,465,203). Here choke L in the circuit formed by the lamp and capacitor C_1 stores energy at the start of the flash. After capacitor C_1 is completely discharged, the discharge current is switched into the circuit formed by choke L, diode D, and the lamp, and the energy stored by the choke gradually is delivered to the lamp. The flash duration also can be increased and the instan-taneous power of the flash reduced by lowering the supply voltage and increasing the capacitance of the storage capacitor, but in this case triggering of the discharge becomes complicated. The circuits shown in Figure 9-27 may be recommended to facilitate triggering. When switch K in the circuits in Figures 4-27a (sic) and 4-27b (sic) is closed (US patents Nos. 3,781,602, 2,724,792 (sic), and 3,600,996),

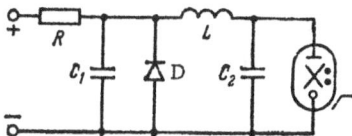

Fig. 9-26. Circuit for stretching the current pulse through a
            lamp by including an inductance L in the discharge
            circuit and shunting the storage capacitor with a diode.
            Low-capacitance capacitor C_2 is used to facilitate trig-
            gering of the lamp when the inductance of the discharge
            circuit is high.

the voltage on the lamp anode increases as a result of the addition
of the voltages across capacitors C_1 and C_2, and a trigger pulse is
generated at the same time by transformer Tr. The use of a diode in
the circuit in Figure 9-27b makes it possible sharply to reduce capacitance C_2 and the current pulse through K (see also US patent
No. 3,912,968). The circuits in Figures 9-27c and 9-27d (US patent
No. 3,600,996 and Japanese patent No. 50,151) contain an auxiliary
circuit with a small capacitance C_2 which powers the lamp in the
initial stage of the discharge. Most of the energy enters the lamp
from high-capacitance low-voltage capacitor C_1. It is convenient to
use these circuits to supply power to high-voltage lamps from circuits that contain low-voltage capacitors (e.g., electrolytic capacitors). If a widened square light pulse must be produced for
high-speed photography or laser pumping, several chokes are used.
These are connected between the individual storage capacitors in the
form of an artifical line with a characteristic impedance equal to
the resistance of the lamp ([2-4 and 9-31 to 9-34], British patent
No. 839,768, and East German patents Nos. 99,496 and 107,193). For
example, when R_L = 3 Ω, a square pulse 3 ms long is produced by
three sections (counting from the lamp): 0.9 mH – 80 µF – 1.22 mH –
100 µF – 1.58 mH – 200 µF.

    Switches sometimes are used in the discharge circuit of a tubu-
lar flashlamp in circuits that have elevated flash frequencies (see
Figure 7-11), and also when low-voltage lamps are powered by high-
voltage capacitor banks. (Some commutator dischargers that are
suitable for these purposes are characterized below, in Tables 9-7
to 9-9.)

    Switches sometimes are used to cut off the current through a
lamp before the capacitor has finished discharging ([9-2], US patents
Nos. 3,779,142 and 3,591,829, and British patent No. 1,290,313).
For example, to control the integrated luminous intensity in electronic photoflash circuits, the discharge in the lamp may be shunted
at some time by a low-resistance discharger or a thyristor which
passes a much larger current than the flashlamp does (US patents
Nos. 3,517,255, 3,727,100, 3,737,721, 3,758,822, 3,769,888, and
3,779,141).

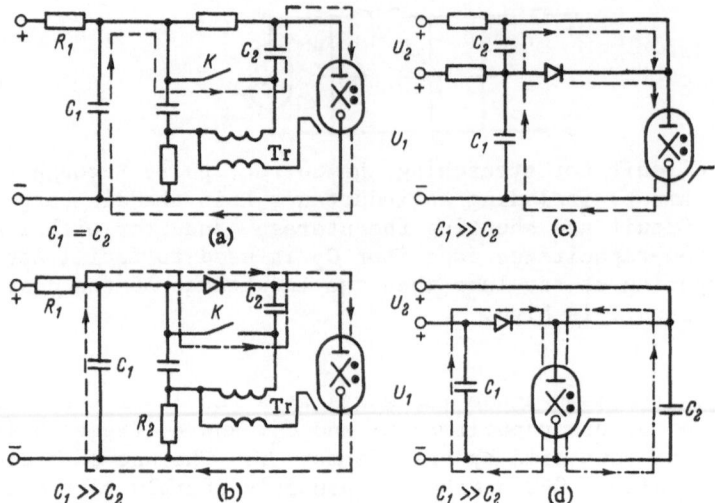

Fig. 9-27. Circuits which produce an increase in voltage across a
            lamp when low-voltage capacitors are used. a, b) Cir-
            cuits with voltage doubling when switch K is closed;
            c, d) circuits with series and parallel connection of
            a high-voltage capacitor. The dashed lines represent
            primary circuits, and the dot-dash lines indicate auxili-
            ary circuits.

     In some cases switches, particularly thyristors, may be used to
connect and disconnect the capacitors or chokes in the discharge
circuit of a lamp during a flash ([9-9a] and West German patent No.
2,045,319).

     In circuits that contain a switch, the lamp sometimes is con-
nected to the discharge circuit through an isolation step-up or
step-down transformer ([9-1], US patent No. 2,933,647). This makes
it possible to match the current pulse and the voltage across the
lamp to the current pulse through the switching element and the voltage of the capacitor in the discharge circuit. When a step-down
transformer is used to reduce the amplitude of the current pulse
through a thyratron, a high-voltage thyratron may be used with a
comparatively low-voltage lamp. However, the efficiency of the discharge circuit is reduced in this case. Furthermore, the weight,
size, and cost of the transformer may be quite considerable, especially if the flash energy is high (US patent No. 3,134,048).

     Nonlinear chokes with magnetic circuits which become saturated
when the discharge current of the circuit flows through them are
used in successive- and combined-triggering circuits (Figure 1-2d).
They have very high inductance when short trigger pulses pass through,
but saturate when the discharge current flows in the circuit. They
thus have practically no effect on the parameters of this pulse.

The electrical processes in discharge circuits containing spherical lamps also are described by equation (9-6). In contrast to tubular lamps, the R_L of spherical lamps varies sharply during the pulse, but its equivalent value decreases quickly with increasing flash energy and may reach extremely low values (thousandths of an ohm). Therefore, the primary factors here are the inductance of the circuit, the ballast resistance specially included in the circuit to damp electric oscillations, and the value of the voltage across the capacitor. A discharge in a circuit containing a spherical lamp usually is oscillatory ($R_L + R_K < 2\sqrt{L/C}$). Oscillations are not observed when L is very small (when ($R_L + R_K$) $> 2\sqrt{L/C}$), and the rate of energy input into the discharge is limited by the resistance of the discharge channel, which has not had time to expand [5-33]. The current curves for a varying circuit resistance do not differ markedly from the classical curves of an exponentially damped sine wave or a decreasing exponential that are known from electrical engineering. This makes it possible, for example, to use the rate of current droop or the damping decrement of the oscillations to determine some equivalent resistance R_L of the lamp for the flash duration under given power-supply conditions. Here the usual relations may be used to estimate the flash parameters of a spherical lamp:

$$\omega_0 = 1/\sqrt{LC}, \qquad \delta = (R_L + R_\kappa)/2L, \qquad (9-11)$$

where ω_0 is the angular frequency of the natural oscillations of the circuit, and δ is their damping factor.

The oscillation frequency of the circuit when active elements are present is

$$\omega_c = \sqrt{\omega_0^2 - \delta^2}, \qquad (9-12)$$

and the current through the lamp during oscillatory-discharge conditions is:

$$i(t) = -\frac{U_0}{L} e^{-\delta t} \frac{\sin(\omega_c t)}{\omega_c}. \qquad (9-13)$$

For the limiting case of an aperiodic discharge, when $\omega_c \to 0$, the current through the lamp is given by the expression:

$$i(t) = -\frac{U_0}{L} e^{-\delta t}. \qquad (9-14)$$

The flash power and energy released in the lamp are determined from formulas similar to (9-8) and (9-9). The duration of the current pulse in oscillatory operation is determined by the value of δ, and the number of current half-periods during a flash is determined by the value of ω_c.

Minimal flash duration at maximum power is encountered in the
limiting case of an aperiodic discharge: the length of the current
pulse is approximately equal to the half-period of the natural oscillations of the circuit, $T_{pulse} \approx \pi\sqrt{L_K C}$, and the average power
over the pulse is

$$P = \frac{CU_0^2}{2T_{pulse}} \frac{R_L}{R_L + R_K}.$$

A decrease in flash duration during oscillatory-discharge mode
can be achieved by damping the oscillations with a noninductive
series resistance somewhat smaller than $0.8R_{cr}$. (Here the integrated
luminous intensity may be more than halved, while the peak luminous
intensity may be reduced significantly.) The most effective method
of achieving the limiting characteristics of spherical lamps is to
lower the inductance of discharge-circuit elements, i.e., to use
low-inductance capacitors (special disk or cylindrical capacitors
having a very low self-inductance) and low-inductance connections,
and to raise the operating voltage of the lamp (US patent No.
3,914,648). Such connections between capacitors and flashlamps can
be provided by building the lamp into the capacitor or by using a
low-inductance cable, such as a rectangular cable (Figure 9-28).
In the first case the inductance of the circuit is determined almost
entirely by the inductance of the lamp (e.g., 90 nH for the ISSh-300
lamp). Using a 1-m coaxial cable increases the inductance by 120-
500 nH, and using the rectangular cable shown in Figure 9-28d increases it by 90 nH. The inductance can be reduced by using several
discharge circuits that are parallel-connected to and positioned
around a single lamp (Figure 9-28b) [2-45, 7-32, and 9-35]. In
some cases these methods of reducing the inductance turn out to be
inadequate, conventional flashlamps must be rejected, and they must
be replaced with spark gaps [5-32, 5-33, 9-35, and 9-36] or with
sliding-spark designs [9-36 to 9-38]. The absence of an envelope
makes it possible to reduce the area of the discharge circuit considerably and to design mechanical connections with the lowest possible inductance by using ceramic disk or cylindrical capacitors.

By contrast, if an additional inductance is introduced into a
discharge circuit containing a spherical lamp or a spark-type source
in order to widen the light pulse (US patent No. 3,749,975), then
the luminous flux of the flash can be modulated at twice the frequency of the circuit's natural oscillations.

Diodes, thyristors, and pulse transformers are rarely used in
discharge circuits containing spherical lamps because of their high
peak currents and the high rates of current variation. Perhaps the
only type of switch suitable for these purposes is a spark discharger
([2-45 and 7-32 to 7-34], and French patent No. 1,194,449). The
introduction of a discharger into a circuit containing a spherical

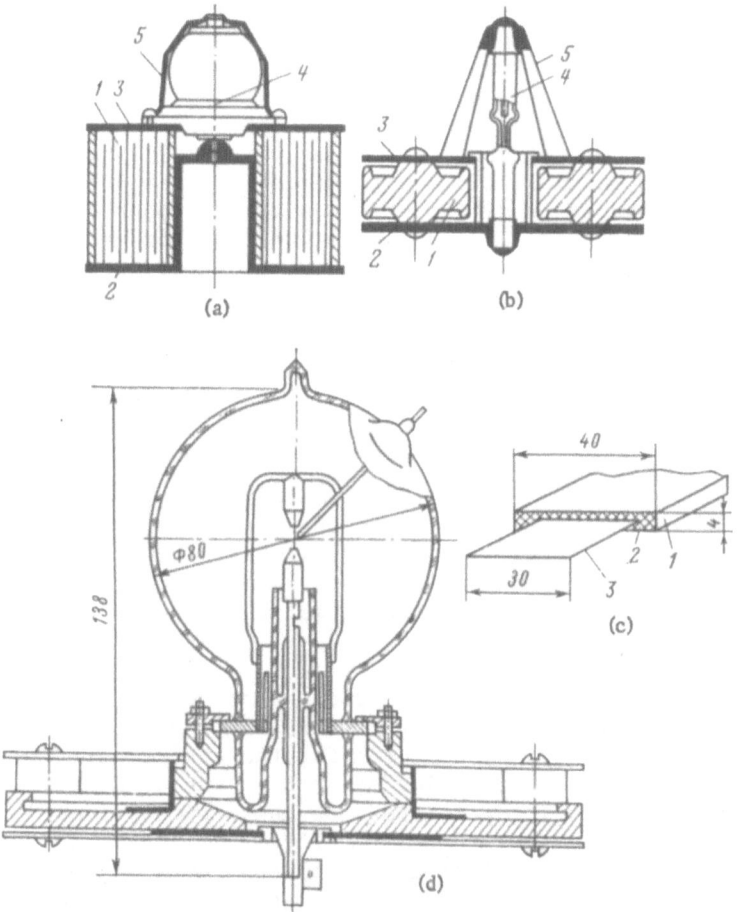

Fig. 9-28. Low-inductance connections between a flashlamp and a
            capacitor. a) A cylindrical capacitor is connected to
            an ISSh lamp by means of a current-carrying cap; b) a
            circuit assembled from several capacitors positioned
            concentrically around an ISP5 capillary lamp (1 -
            capacitor; 2 - lower bus; 3 - upper bus; 4 - lamp; 5 -
            low-inductance cap); c) a rectangular low-inductance
            cable for connecting the lamp to a disk-type cylindrical
            capacitor (1 - a box made of sheet brass 1 mm thick;
            2 - fluoro-plastic insulation; 3 - copper ribbon 0.3 mm
            thick); d) a low-inductance circuit for powering an
            ISSh-300 flashlamp with a ceramic disk capacitor.

lamp has the goal of controlling the time of pulse generation or of
improving the frequency response of the lamp (see Figure 7-11b).
This substantially reduces the amount of energy dissipated in it
(sometimes more than halving it).

Methods of series-connection of several lamps in a single discharge circuit in order to trigger them with the voltage redistri-
buted among lamps (or between the lamps and a controlled discharger)
are shown in Figure 9-29. When the lamps are parallel-connected, a
small choke must be series-connected to each lamp to provide conditions for uniform distribution of energy [9-25]. Sample circuits
for simultaneous initiation of a discharge in several lamps that are
independently powered by different circuits have been described in
West German patents Nos. 973,335 and 1,056,270, for example, and
some circuit concepts for triggering several lamps with a time difference have been described in [9-39 to 9-42]. Such circuits are
used to generate a series of light pulses during the photography of
high-speed processes, such as the flight of a bullet.

9-8.   The Main Elements of the Discharge Circuit

Many flashlamp circuits are assembled from elements that are
common in electronics, such as capacitors, resistors, transformers,
chokes, reactors, valves, rectifiers, switching devices, and cable
products. In this section we intend to examine only specific elements which are particularly characteristic of large units using
high-voltage flashlamps.

Capacitors. Capacitors often are the principal components of
large units containing pulsed light sources. The cost of capacitors
may amount to up to 90% of the total for the entire unit, and their
volume and weight up to 75%. The other features of a unit also are
determined largely by the parameters of the capacitors used.

The development of pulsed light sources required the invention
of capacitors especially designed for flashlamp power supplies.
Such capacitors must have the following properties:

(a) a high specific energy - the possibility of storing a large
    amount of energy per unit volume (or weight);
(b) a low internal inductance* and a relatively good Q factor;
(c) high dynamic stability of the internal connections;
(d) an adequate service life during operation with multiple dis-
    charges at low resistance;
(e) high reliability.

Requirements a, d, and e are mutually contradictory, since the
electric field strength in the dielectric of the capacitor must be

---

*This requirement is not so significant for units that have a large
 number n of parallel-connected capacitors, since the total inductance of the capacitor banks is lower by a factor n than the inductance of an individual capacitor.

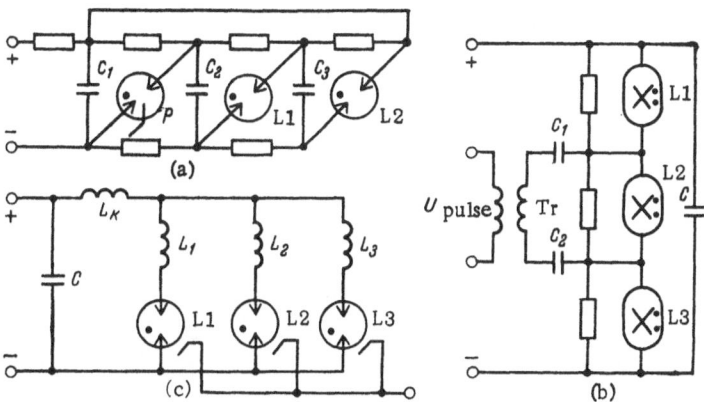

Fig. 9-29. Methods of connecting several lamps in a single discharge circuit. a) The Marks[†] shock circuit. Capacitors C_1, C_2, and C_3 are switched from parallel to
series connection upon breakdown of controlled discharger P (US patent No. 2,825,002); b) the triggering
of a discharge in lamp L2 due to a pulse from transformer Tr causes an increase in voltage and breakdown
in lamps L1 and L3 (West German patent No. 2,444,893);
c) the series connection of a choke to each of the lamps
promotes more uniform energy distribution among the
lamps, even when firing lags somewhat in one of the
lamps (L_1 = L_2 = L_3 >> L_K).

increased in order to increase the specific energy of the capacitor,
but this involves accelerated failure of the dielectric.

The list of capacitors designed to provide a power supply to
flashed light sources is expanding continually. The characteristics
of the main types of pulsed capacitors that are series-produced by
Soviet industry are given in Tables 9-2 and 9-3 (paper and plastic-
film capacitors) and in Table 9-4 (electrolytic capacitors).

Tables 9-5 and 9-6 present data on pulse-type oil-filled paper
and castor-oil-filled paper capacitors from the pilot plants of the
Khar'kov and Leningrad polytechnical institutes.

The dielectric between the capacitor plates usually is capacitor
paper or film-type insulating materials such as polystyrene, polyethylene, or lavsan* (Mylar® or Hostaphan®). The insulator is per-
meated with a liquid dielectric such as capacitor oil, castor oil,

_____

*Translator's note: lavsan is a Soviet polyester fiber. Western
 equivalents are Therylene, Daeron, and Tergal.

Table 9-2. Pulse-Type Capacitors Produced by the Electrical Components Industry

Type of capacitor	Rated voltage, kV	Rated capacitance, μF	Self-inductance, nH	Discharge current amplitude, kA	Discharge frequency, Hz	Overall size excl. isolator (size including isolators given in parentheses), mm	Weight, kg	Specific energy J/kg	Specific energy J/dm³	Operating life, hr* Warranted	Operating life, hr* Safe life	Tangent of loss angle at 50Hz	Temperature range, °C
IM-3-30	3	30	—	—	25	400×120×325(420)	30	4.5	9.1	10^3 hr	10^4 hr	0.0026	−35, +40
IM-3-100	3	100	430	2	1.7	400×110×370(463)	25	18	29	10^5 c	10^6 c	0.0045	−35, +50
IC-4-13	4	13	—	—	25	400×120×325(424)	28	3.7	7.03	10^3 hr	10^4 hr	0.0035	−50, +45
IMN-5-140	5	140	600	2.5	0.8	360×138×800(915)	57	31	51	10^3 hr	$3 \cdot 10^3$ hr	0.0045	−50, +50
IC-5-200	5	200	600	10	0.17	360×150×800(890)	63	39.7	79	10^4 c	$1{,}5 \cdot 10^5$ c	0.005	−10, +40
IMN-6-36×2	6	36×2	600	1	0.5	360×150×585(720)	48	26.6	47	10^4 c	$5 \cdot 10^4$ c	0.004	−10, +40
IC-6-5,5	6	5.5	—	—	25	400×120×325(440)	28	3.5	6.7	10^3 hr	10^4 hr	0.0035	−50, +45
IK-6-150	6	150	60	50	0.17	360×138×640	45	59	99	$2 \cdot 10^4$ c	10^5 c	—	—
IC-6-200	6	200	600	—	1.7	360×133×680	55	65	133	$2 \cdot 10^4$ c	$5 \cdot 10^5$ c	0.0035	—
IK-10-50	10	50	500	10	0.17	360×138×640(758)	50	50	91.3	10^4 c	10^5 c	0.006	+1, +40

IS-16-0.8	16	0.8	—	—	25	400×120×325(470)	28	3.6	6.9	10^3 hr	10^4 hr	0.0035	—50, +45
IS-20-0.5	20	0.5	—	—	25	400×120×325(470)	28	3.6	6.7	10^3 hr	10^4 hr	0.0035	—50, +45
IK-25-12	25	12	40	200	0.02	314×314×680(730)	110	34	54	10^3 c	10^4 c	0.01	+1, +40
IK-25-13	25	13	350	15	0.03	314×314×680(820)	120	34.4	60	10^3 c	$5·10^4$ c	0.01	+1, +40
IM-30-0.2	30	0.2	—	5	20	379×118×325	30	3	6.2	$7·10^5$ c	—	—	—
IK-40-5	40	5	40	200	0.02	314×314×680(730)	120	33.4	57	$3·10^3$ c	$3·10^4$ c	0.01	+1, +40
IMG-45-0.1	45	0.1	—	10	0.2	362×766×800(1200)	310	0.32	0.43	10^4 c	10^5 c	0.003	0, +35
IKG-50-1	50	1	500	20	1.0	394×314×680(907)	110	11.3	18.6	—	$5·10^7$ c	0.01	+1, +40
IK-50-3	50	3	40	200	0.02	314×314×680(730)	110	34	54	$3·10^3$ c	$3·10^4$ c	0.01	+1, +40
IK-50-3.3	50	3.3	400	20	0.17	314×314×680(730)	120	34.4	62	$3·10^3$ c	$3·10^4$ c	—	—
IKM-50-3	50	3	10	300	0.02	314×314×680(730)	110	34	54	$3·10^3$ c	—	—	—
IM-60-0.2	60	0.2	200	—	100	455×150×326	32	11.2	16.2	—	10^4 hr	—	—
IMN-100-0.1	100	0.1	200	10	0.05	455×150×326(345)	32	15.6	22.4	10^4 c	$3·10^4$ c	0.004	—10, +45
IK-100-0.25	100	0.25	150	80	0.06	455×150×326(345)	32	39	56	$2·10^3$ c	$2·10^4$ c	0.006	—
IK-100-0.4	100	0.4	150	50	0.06	455×150×326(345)	32	62.5	90	—	$2·10^4$ c	—	—

*c — cycles.

Table 9-3. Pulsed Capacitors Manufactured by the Electronics Industry

Type of capacitor	Rated voltage, kV	Rated capacitance, μF	Discharge frequency, Hz	Charging time, ms	Discharge time, μs	Time of continuous operation	Relaxation time	Type of discharge**	Max. discharge current, kA	Self-discharge time constant, MΩ–μF	Overall size excl. isolator (size isolator including isolators given in parenthses), mm	Weight, kg	Specific energy J/kg	Specific energy J/dm³	Operating life cycles*** Warranted	Operating life cycles*** Safe life	Tangent of loss angle at 50Hz	Temperature range, °C
K 75-40	0.75	40	4	3000*	60	Continuous		A	0.5	500	86×26×141(156)	0.6	18.7	35.5	10^4	10^5	0.008	−60, +70
K 75-40	0.75	60	4	3000*	60	Continuous		A	1.0	500	86×41×141(159)	0.95	17.8	34.0	10^4	10^5	0.008	−60, +70
K 75-40	0.75	80	4	3000*	60	Continuous		A	1.0	500	86×46×141(159)	1.0	22.5	40.4	10^4	10^5	0.008	−60, +70
K 75-40	1	100	4	3000*	60	Continuous		A	1.0	500	86×51×141(159)	1.1	25.6	45.6	10^4	10^5	0.008	−60, +70
K 75-40	1	20	4	3000*	60	Continuous		A	0.5	500	86×26×141(156)	0.6	16.7	31.8	10^4	10^5	0.008	−60, +70
K 75-40	1	40	4	3000*	60	Continuous		A	1.0	500	86×41×141(159)	0.95	21.0	40.3	10^4	10^5	0.008	−60, +70
K 75-40	1	60	3	3000*	60	Continuous		A	1.0	500	86×46×141(159)	1.0	30.0	53.8	10^4	10^5	0.008	−60, +70
K 75-40	1	80	3	3000*	60	Continuous		A	1.0	500	86×65×141(159)	1.4	28.6	50.0	10^4	10^5	0.008	−60, +70
K 75-40	1	100	3	500	60	Continuous		A	2.0	500	86×76×141(159)	1.65	30.4	47.0	10^4	10^5	0.008	−60, +60
K 42-13	1	20	5	—	—	2 min	10 min	—	0.45	1000	46×92(110)	0.310	32.3	60.6	$6\cdot10^4$		0.01	−60, +70
K 42-15	1	50	55	—	—	—	—	—	—	1000	289×52×138(150)	0.395	25.3	48.0	5000 hr		0.01	−60, +70
K 75-17	1	100	100	8	100	10 min	30 min	—	1.2	3000	85×50×140(188)	1.25	19.8	42.0	$3.6\cdot10^6$		0.008	−60, −30
K 75-18	1	100	3	190	10^1	2 hr	1 hr	PA	0.001	3000	86×71×111(161)	1.55	32.2	58.8	10^4		0.008	−60, +50
K 75-36	1.5	60	0.3	3000	60	1 hr	1 hr	A	2.0	2000	86×46×141(159)	1.0	30.0	53.7	$2\cdot10^5$		0.01	−60, +70
K 75-9	1.5	50	50	—	100	1 hr	1 hr	A	2.0	5000	122×122×220(285)	6.5	8.7	17.1	100 hr		0.01	−10, +40
K 421-1	1.6	20	3	—	100	1 hr	1 hr	A	1.0	5000	122×122×390(456)	11.5	9.8	19.4	100 hr		0.008	−10, +40
K 421-1	1.6	25	2.5	3000*	60	Continuous		A	0.5	1000	86×26×141(171)	0.65	39.4	81.4	10^4		0.008	−60, +70
K 75-40	1.6	40	2.5	3000*	60	Continuous		A	1.0	1000	86×46×141(171)	1.0	51.2	92.0	10^4	10^5	0.008	−60, +70
K 75-40	1.6	60	2.5	3000*	60	Continuous		A	2.0	1000	86×66×141(171)	1.4	54.8	96.0	10^4	10^5	0.008	−60, +70
K 75-40	1.6	80	2.5	3000*	60	Continuous		A	2.0	1000	86×81×141(171)	1.8	57.0	104.5	10^4	10^5	0.008	−60, +70
K 75-40	1.6	100	2.5	3000*	60	Continuous		A	2.0	1000	86×101×141(171)	2.2	58.0	104.5	10^4	10^5	0.008	−60, +70
K 75-40	2.0	40	2.5	3000*	60	Continuous		A	0.5	1000	86×26×141(171)	0.65	61.0	127.0	10^4	10^5	0.008	−60, +70
K 75-40	2.0	60	2.5	3000*	60	Continuous		A	2.0	1000	86×66×141(171)	1.4	80.0	144.0	10^4	10^5	0.008	−60, +70
K 75-40	2.0	80	2.5	3000*	60	Continuous		A	2.0	1000	86×91×141(171)	2.05	79.0	145.0	10^4	10^5	0.008	−60, +70
K 75-40	2.0	100	1.7	3000*	60	Continuous		A	2.0	1000	86×111×141(171)	2.35	85.5	148.0	10^4	10^5	0.008	−60, +70
K 421-1	2.0	10	10.0	—	10^1			—	—	1000	86×41×140(162)	1.9	22.2	40.6	250 hr		0.01	−60, +50
K 421-1	2.0	25	10.0	—	10^1			—	—	1000	86×91×140(162)	3.5	26.3	45.6	250 hr		0.01	−60, +50
K 75-11	2.0	50	0.5	1000	250	30 s	10 min	A	1.8	3000	86×176×140(162)	2.2	28.6	47.2	$5\cdot10^3$		0.008	−60, +40
K 75-19	2.0	100	10.0	80	600	Continuous		A	0.7	3000	260×85×230(284)	8.6	23.3	39.5	$3.6\cdot10^7$		0.008	−60, +60
K 75-27	2.0	100	0.1	8000	360	3 min	3 min	A	0.5	3000	126×86×141(159)	2.7	74.0	131.0	$25\cdot10^3$		0.008	−60, +50
K 75-34	2.0	200	0.1	8000	360	3 min	3 min	A	0.5	3000	177×106×178(202)	5.8	69.0	120.0	10^4		0.008	−60, +70
K 75-40	2.0	80	0.3	3000*	100	90 s		A	3.5	800	86×96×141(159)	2.2	76.0	137.5	500 hr		0.012	−60, +70
K 75-40	2.5	20	2	3000*	60	180 s		A	0.5	3000	86×31×141(181)	0.75	83.5	166	10^4	10^5	0.008	−60, +70
K 75-40	2.5	40	2	3000*	60	Continuous		A	1.0	3000	86×66×141(181)	1.4	89.3	156	10^4	10^5	0.008	−60, +70

Type																	
K 75-40	2.5	60	1.7	3000*	60	^	A	2.0	3000	86×86×141(181)	1.9	99	180	10^4	10^5	0.008	−60, +70
K 75-40	2.5	80	1.7	3000*	60	^	A	2.0	3000	86×111×141(181)	2.35	106	185	10^4	10^5	0.008	−60, +70
K 421-1	2.5	100	1.7	3000*	60	^	A	4.0	3000	86×141×141(181)	3.0	104	183	10^4	10^5	0.008	−60, +70
K 75-14	3	10	6	—	—	60 s / 1 min	—	—	1000	71×86×140(162)	1.55	29	52.6	250 hr	—	0.01	−10, +50
K 75-14	3	100	1	800	500	20 s	K	1.2	3000	140×85×375(429)	7.5	60	100	—	10^5	0.008	−60, +50
K 75-28	3	100	1	600	300	5 min	K	1.8	3000	140×85×230(258)	4.9	92	165	$5·10^4$	—	0.008	−60, +50
K 75-10	3	20	1.3	3000*	60	Continuous	A	1.0	3000	105×35×170(210)	1.25	72	144	10^4	10^5	0.008	−60, +70
K 75-10	3	40	1.3	3000	60	^	A	1.0	3000	105×65×170(210)	2.0	90	155	10^4	10^5	0.008	−60, +70
K 75-10	3	60	1.0	3000*	60	^	A	2.0	3000	105×90×170(210)	2.75	98	167	10^4	10^5	0.008	−60, +70
K 75-10	3	80	1.0	3000*	60	^	A	2.0	3000	105×120×170(210)	3.65	99	168	10^4	10^5	0.008	−60, +70
K 75-40	3	100	0.8	3000*	60	^	A	4.0	3000	105×150×170(210)	4.5	100	168	10^4	10^5	0.008	−60, +70
K 75-20	4	100	0.03	$3·10^5$	150	^	A	3.0	3000	140×110×454(508)	11.5	69.5	114	10^5	—	0.008	−60, +40
K 75-40	4	20	1.3	3000*	60	Continuous	A	1.0	3000	105×65×170(210)	2.0	80	138	10^4	10^5	0.008	−60, +70
K 75-40	4	40	1.0	3000*	60	^	A	2.0	3000	105×110×170(210)	3.3	97	163	10^4	10^5	0.008	−60, +70
K 411-7	4	60	0.8	3000*	60	^	A	4.0	3000	105×170×170(210)	5.0	96	158	10^4	10^5	0.008	−60, +70
K 75-40	5	100	0.0083	—	500	—	A	1.0	500	170×125×410(476)	15	83.2	143	$6·10^3$	—	0.012	−10, +40
K 75-40	5	20	1.0	3000*	60	Continuous	A	2.0	3000	105×90×170(210)	2.75	91	155	10^4	10^5	0.008	−60, +70
	5	40	0.8	3000*	60	^				105×170×170(210)	5.0	100	165	10^4	10^5	0.008	−60, +70
K 411-7	10	32	0.0083	—	500	—	PA	—	500	230×180×410(455)	30	53.4	94.5	$6·10^3$	—	0.012	−10, +40
K 75-22	40	0.1	5000	—	0.1−5	—	PA	180	—	150×130×210(350)	7.2	11.1	19.6	12 000 hr at +50°C	—	0.01	−60, +125
K 75-35	50	0.024	400	0.05	—	—	A, KP	2.5	1000	152×146×290(340)	12	2.5	4.7	1000 hr at 100 Hz	—	0.005	−10, +55
K 75-39	63	0.033	100	1.0	—	—	A, KP	2.0	1000	180×145×450(595)	18	3.6	7.3	2000 hr	—	0.0025	−60, +70
K 75-39	63	0.1	100	1.0	—	—	A, KP	3.0	1000	395×145×450(595)	48	4.1	8.7	2000 hr	—	0.0025	−60, +70

* For frequencies up to 0.1 Hz. For higher frequencies the charging time is at least 0.5 the discharge time.

** A – aperiodic, PA – partial aperiodic, K – ocillatory, KP – oscillatory with a 15–20% reverse.

Table 9-4. Pulsed Electrolytic Capacitors

Type of capacitor	Rated voltage, V	Rated capacitance, μF	Energy storage, J	Discharge repetition freq., Hz	Load resistance, ohms	Leakage current at 20 ± 5°C, mA	Overall size, mm		Weight, kg	Specific energy		Operating life, cycles		Range of operating temps, °C
							Diam.	Height, mm (numbers in parens, incl. lead-outs)		J/kg	J/dm³	Warranted	Safe life	
K50-13	350	250	15.3	0.1	—	1.5	30	56	0.070	218	395	10^5	$2 \cdot 10^5$	−10, +40
K50-17	300	400	18	0.1	0.45	1.0	28	60(65)	0.070	256	490	10^5	$2 \cdot 10^5$	−10, +50
K50-17	300	800	36	0.1	0.45	1.2	40	60(65)	0.140	256	475	10^5	$2 \cdot 10^5$	−10, +50
K50-17	300	1500	67.2	0.1	0.45	2.2	40	118(123)	0.270	250	455	10^5	$2 \cdot 10^5$	−10, +50
K50-17	400	200	16	0.1	0.45	1.0	28	48(53)	0.060	266	540	10^5	$2 \cdot 10^5$	−10, +50
K50-17	400	500	40	0.1	0.45	1.0	28	105(110)	0.120	335	620	10^5	$2 \cdot 10^5$	−10, +50
K50-17	400	1000	80	0.1	0.45	2.0	40	118(123)	0.270	295	540	10^5	$2 \cdot 10^5$	−10, +53
K50-17	500	200	25	0.1	0.45	1.0	28	85(90)	0.090	278	480	10^5	$2 \cdot 10^5$	−10, +50
K50-21	160	5000	64	3	—	2.0	55	140(152)	0.55	125	200	10^4	$2 \cdot 10^4$	−10, +50
K50-21	160	15 000	192	3	—	2.5	95	140(152)	1.65	125	200	10^4	$2 \cdot 10^4$	−10, +50
K50-21	250	1000	31	3	—	1.0	40	60	0.93	334	500	10^5	$2 \cdot 10^5$	−10, +50
K50-23	500	500	62.5	0.1	—	2.0	51×25*	130(142)	0.26	250	450	$5 \cdot 10^4$	10^5	−10, +40
K50-23	500	1000	125	0.1	—	3.0	55	133(145)	0.55	250	450	$5 \cdot 10^4$	10^5	−10, +40
K50-31	450	500	51	0.16***	—	2.5	65	110(130)	0.650	78.5	139	500 hr	1000 hr	−40, +60
K50-3F	300	500	22.5	0.1	1.0	1.0	32	107(127)	0.160	141	257	10^4	$2 \cdot 10^4$	−25, +40
K50-3F	300	1000	45	0.1	1.0	1.0	50	107(127)	0.4	141	257	10^4	$2 \cdot 10^4$	−25, +40
K50-3F	450	500	51	0.1	1.0	2.5	65	110(130)	0.65	141	257	10^4	$2 \cdot 10^4$	−25, +40
K50I-8	300	800	36	0.1	0.45	1.5	50×24*	130(142)	0.24	148	262	10^4	$2 \cdot 10^4$	−10, +40
K50I-8	400	500	40	0.1	0.45	2.5	50×24*	130(142)	0.24	167	289	10^4	$2 \cdot 10^4$	−10, +40
K50I-8	500	300	37.5	10**	24	1.5	50×24*	130(142)	0.24	157	270	$3.6 \cdot 10^6$	10^7	−10, +40

* Oval body.

** Discharge 30% with respect to voltage.

*** The time required for discharge of 50% of the energy is at least 25 ms.

Table 9-5. Pulsed Capacitors of the Khar'kov Polytechnical Institute

Type of capacitor	Rated voltage, kV	Rated capacitance, μF	Self-inductance, nH	Amplitude of discharge current, kA	Frequency of discharge circuit, kHz	Discharge repetition frequency, pulses/min	Overall size excl. lead-outs, mm (size including lead-outs is given in parentheses)	Weight, kg	Specific energy J/kg	Specific energy J/dm³	Operating life, cycles Warranted	Operating life, Safe life
MIOM-4	6	340	20	100	30	6	460×504×720(760)	300	20.3	36.8	10^5	$2 \cdot 10^5$
KIM-17A	10	200	25	100	30	6	356×600×955(1035)	275	36.4	49.0	10^4	$3 \cdot 10^4$
KIM-26	10	200	50			2	414×604×980(1033)	340	29.4	40.0	10^4	$3 \cdot 10^4$
KIM-17B	20	50	30	100	30	6	356×600×955(1035)	270	37	49.0	10^4	$3 \cdot 10^4$
KIM-24	10/20	156/39	40(at 10 kV)	10		0.2	534×322×705(745)	230	34	65.5	10^4	$2 \cdot 10^4$
KIM-25	25	32	30			2	414×604×980(1033)	340	29.4	40.0	10^4	$3 \cdot 10^4$
KIM-21	25	32	30	100	50	2	310×526×960(1000)	300	33.3	64.3	10^4	$3 \cdot 10^4$
KIM-19	40	6.25	30	60	100	1	282×526×685(725)	200	25	50.0	10^4	$3 \cdot 10^4$
KIM-20	40	1	20	20	50	1	295×255×450(555)	60	13.2	23.7	10^4	$3 \cdot 10^4$
KIM-14K	50	8	40	40	100	0.2	322×566×1000(1040)	270	37	55.8	10^4	$3 \cdot 10^4$
KIM-18	50	0.5	40	20	50	1	295×255×490(540)	60	10.4	16.9	10^5	10^6
KIM-23	50	8	20	100	50	1	322×566×700(740)	200	50	78.7	$8 \cdot 10^3$	10^4
KIM-22	100	0.5	10	170	100	2	274×475×660(700)	80	31.2	29.0	$3 \cdot 10^3$	$5 \cdot 10^3$
KIM-16	125	1.28	250	50			545×965×324(413)	300	33.3	59.0	10^4	$3 \cdot 10^4$

Table 9-6. Pulsed Capacitors Produced by the Leningrad Polytechnical Institute

Type of capacitor	Rated voltage, kV	Rated capacitance, μF	Self-inductance, nH	Q factor at F = 1, MHz	Amplitude of discharge current, kA	Overall size (excluding insulators), mm	Weight, kg	Specific energy		Warranted operating life, cycles
								J/kg	J/dm³	
KBLE14-78	14	78	40	10	4	315×230×310	38	200	350	10^3
KMK20-0,5	20	0.5	10	10	100	75×218×240	6	16.6	25	10^4
KMK20-25	20	25	10	10	500	765×175×630	140	35.6	60	10^4
KMK25-0,5	25	0.5	15	5	100	73×222×254	6	26	37.7	10^4
KMK25-5	25	5	15	10	400	445×225×640	67	23.3	44.4	10^4
KMK30-10	30	10	9	10	400	695×160×640	95	47	63.4	10^4
KMK35-5	35	5	15	5*	400	520×175×640	103	30	52.3	10^4
KMK40-0,7	40	0.7	18	10	150	205×220×265	24	23.3	48.5	10^4
KMVD50-0.1	50	0.1	16	100	100	365×144×335	22	5.7	7.1	10^5
KMK50-0.1	50	0.1	20	10	80	88×225×235	6.7	18	26	10^4
KMK50-0.7	50	0.7	30	5	150	275×205×370	36	24.2	42	10^4
KMK50-4	50	4	15	5*	500	525×220×675	98	37.2	65	10^4
KMK50-8	50	8	16	10	400	440×520×645	196	51	68.5	10^4
KMKI 60-2	60	2	20	10	300	460×155×325	33	110	160	10^3
KMK100-1,3	100	1.3	200	10	150	510×300×750	202	32.5	56.6	10^1
KK125-0.8	125	0.8	350	10	150	530×285×770	220	28.4	54	10^4

*Values of Q at resonant frequency of Capacitor.

trichlorodiphenyl, or pentachlorodiphenyl. Combined film-paper insulation consisting of alternating layers of capacitor paper and
film also is used. In this case the permissible electric field
strength may be increased to 150-200 kV/mm, as against the usual
value of 60-80 kV/mm (or up to 120 kV/mm under certain conditions)
for pulse-type paper capacitors.

It should be kept in mind that comparisons in terms of specific
energy per unit volume or weight are valid only for capacitors operating under identical conditions and having the same operating
life. For example, Mylar-film-insulated pulse-type capacitors produced by the Maxwell Company (United States), which are designed for
voltages of 2 to 12 kV, have an exceptionally high specific energy
(up to 480 J/dm^3, 277 J/kg), but their service life is just 5 x 10^3
to 5 x 10^2 pulses, depending on operating conditions [9-43]. (The
self-inductance of these capacitors does not exceed 5-10 nH.)

Electrolytic capacitors, which are lighter and more compact,
often are used to power low-voltage flashlamps (up to 500 V). A
peculiarity of electrolytic capacitors is the sharp increase in the
leakage current ("de-forming") after they remain unused for a prolonged period (over 2 weeks). In order to restore the normal leakage current following a break in operation, an electrolytic capacitor
must be re-formed, i.e., kept at rated voltage (or at a voltage
5-10% higher than the rated voltage) for 20-30 min.

A second peculiarity of electrolytic capacitors is their large
dielectric losses. Therefore, only capacitors that are especially
adapted for operation under forced air cooling may be used for repeating-flash operation (with interflash intervals of 1-2 s). Dielectric losses increase sharply at temperatures below -20°C, but
after a small number of flashes the capacitors themselves heat up
as a result of the losses, which consequently reach their normal
level.

It should be taken into consideration during the development
and operation of units containing capacitive storage devices that
the parameters of capacitors may differ significantly from their
rated values if they are used under conditions other than those en-
visaged in the specifications. For example, the dependence of operating life on operating conditions may be expressed by the following empirical formula for oil-filled paper capacitors [9-44 and
9-45]:

$$M = AE^{-m} f^b F^{-l} \Delta^{n/2} 10^{\beta T}, \qquad (9-15)$$

where M is the operating life (number of pulses); E is the working
electric field strength (kV/mm); f is the pulse repetition frequency
(s^{-1}); F is the current frequency in the discharge circuit (Hz); Δ
is the decrement of the oscillation of E (the ratio of two successive

amplitudes of E of the same sign), which characterizes the damping
rate of the oscillations*; T is the temperature inside the capacitor
(°K); and A, m, b, l, n, and β are factors that depend on the operating conditions of the capacitor, the processes used to make the
capacitor, and other factors. The following values may be assumed
for Soviet-made capacitors: A = 2.72 x 10^{21}, m = 8-10, b = 0.21,
l = 0.237, n = 2.33, β = 2.73 x 10^{-3}. The corresponding nomogram is
shown in Figure 9-30.

The dependences of the most widely used types of electrolytic
capacitors on the discharge resistance are presented in Figure 9-31
[9-46].

Dischargers. Spark dischargers are used extensively as switching elements in the power-supply circuits of flashlamps. Spark dischargers have a number of advantages over other devices with a similar function: good frequency response, high switch-on accuracy, low
resistance and inductance, and continuous operational readiness due
to the absence of electrode heaters. The advantages of spark dischargers also include their broad range of switchable voltages
(from hundreds of volts to hundreds of kilovolts) and currents
(kiloamps to megamps).

Dischargers with a higher flash frequency were mentioned in
Section 7-2 in our description of the corresponding circuits. Here
we should additionally consider high-current high-voltage dischargers
with a comparatively low frequency response. These may be conditionally divided into vacuum, air, high-pressure, and solid-dielectric
dischargers.

In vacuum dischargers the insulating medium between the primary
electrodes is a vacuum (10^{-5}-10 Pa). Such dischargers make it possible to switch currents up to a few megamps at voltages ranging
from 100-300 V to 150 kV. They are characterized by low self-inductance (a few nanohenrys) and a low voltage drop, are noiseless
in operation, and have a short lag time relative to the control
pulse. This permits the parallel connection of a large number of
dischargers. The need for a complex vacuum system is a significant
drawback of unsealed dischargers of this kind. Sealed dischargers
lack this shortcoming, but they can only switch comparatively low
currents and energies (Tables 9-7 and 9-8). There is no series production of unsealed dischargers, and they are being manufactured
only through local facilities.

---

*The damping rate of the oscillations also can be expressed as the
 degree of damping Δ', which is the inverse ratio of two adjacent
 amplitudes of opposite sign.

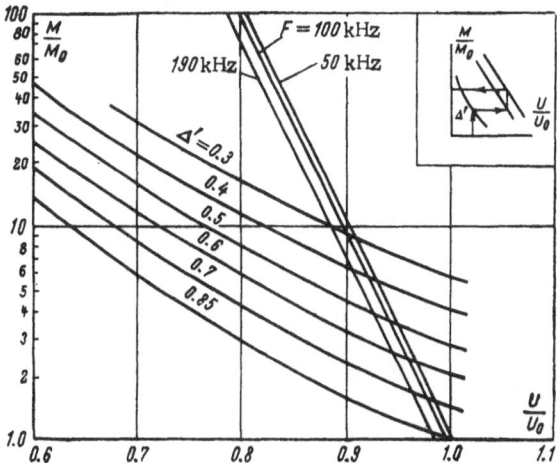

Fig. 9-30. Nomogram for determining operating life as a function of the operating conditions of the capacitor.

At present vacuum dischargers normally are used in units having an operating voltage of up to 30 kV. Reference [9-48] describes low-inductance vacuum dischargers designed for a voltage of up to 50 kV. These can be controlled in time to a high accuracy over a broad range of voltages (1-50 kV) and are suitable for repeated switching of pulsed currents of different shape (having an amplitude of 10^4-10^6 A and a duration of 10^{-3}-10^{-5} s) at energies over 50 kJ. The layout of the discharger is shown in Figure 9-32. The discharger in Figure 9-32a has flat electrodes, while the discharger in Figure 9-32b has coaxial electrodes (to reduce corrosion). The electrode spacing is 32 mm. The discharge is initiated near the insulating walls of the chamber. In the discharger in Figure 9-32b, electromagnetic forces cause the plasma channel to move through along the coaxial electrodes to their end faces. In order to reduce the lag time and ensure a uniform distribution of current, the discharge is initiated by means of three trigger electrodes positioned on the lower high-current electrode around the walls of the chamber. The trigger pulse is 25 kV with a 40-ns edge. The initial pressure in the discharger is 10^{-3}-0.1 Pa, and the inductance with coaxial connection is 6 nH.

KDV arc-suppressing chambers are distinctive mechanically controlled vacuum dischargers (see Table 9-9). Their distinguishing feature is the ability not only to close a circuit, but also to break currents of considerable strength.

Air dischargers, which operate at atmospheric pressure, are characterized by simplicity of operation (no vacuum devices or compressors are required), a trigger time with a small spread (a few

Table 9-7. Controlled Sealed Dischargers Produced by Soviet Industry [9-47 and 9-47a]

Type of discharger	Working voltage, kV	Amplitude of current pulse, kA	Current pulse length, μs	Switched energy, J	Average switched power, kW	Pulse repetition frequency, Hz	Control pulse Amplitude, kV	Control pulse Energy, J	Warranted life, pulses	Overall dimensions, mm	Weight, g
R-21	1.1—2.0	0.5	100	4.0	180	1	15	0.04	$1.5\cdot10^7$	Ø25×105	125
R-24	2—6	0.3	100	18	200	0.17	15	0.04	$2\cdot10^6$	Ø25×112	140
R-30	4.7—5.3	—	—	0.125	—	3000	—	0.04	$3.5\cdot10^6$	Ø61×62	500
R-37	0.8—2.0	0.25	10	2.0	20	0.17	—	0.07	$2\cdot10^6$	Ø20×62	40
R-41	16—27	0.2	70	1600	—	0.03	25	—	$3\cdot10^3$	Ø75×170	600
RT-39	45—80	1.8	2—3	300	—	0.03	45—55	—	$3\cdot10^3$	Ø75×243	800
R-47*	6—16	6 / 10	—	25 / 160	—	50 / 10	17	0.07	10^5 / 10^3	Ø24×105	130
R-52	10—25	1.1 / 1.3	480	800 / 1.7	—	—	28—50	1—5	$2\cdot10^8$ / $5\cdot10^5$	Ø75.5×170	600
RT-53*	0.6—1.6	10	—	60	—	10	4	—	300	Ø45×36	35
"Doktrin-2"	5—10	4	—	23	—	0.1 / 50	8	0.1	10^6	Ø75×45	80
VIR-19	1.5	1.0	15	180	—	50	1.0	10^{-4}	10^5	Ø18×25	—
VIR-21	1.7	1.0	15	360	—	100	4	10^{-4}	$5\cdot10^6$	Ø30×70	—
SKECh-30	1.5	1.0	1	—	10	30	2	10^{-4}	10^7	Ø30×80	—

* Mode I is given at top and mode II at bottom.

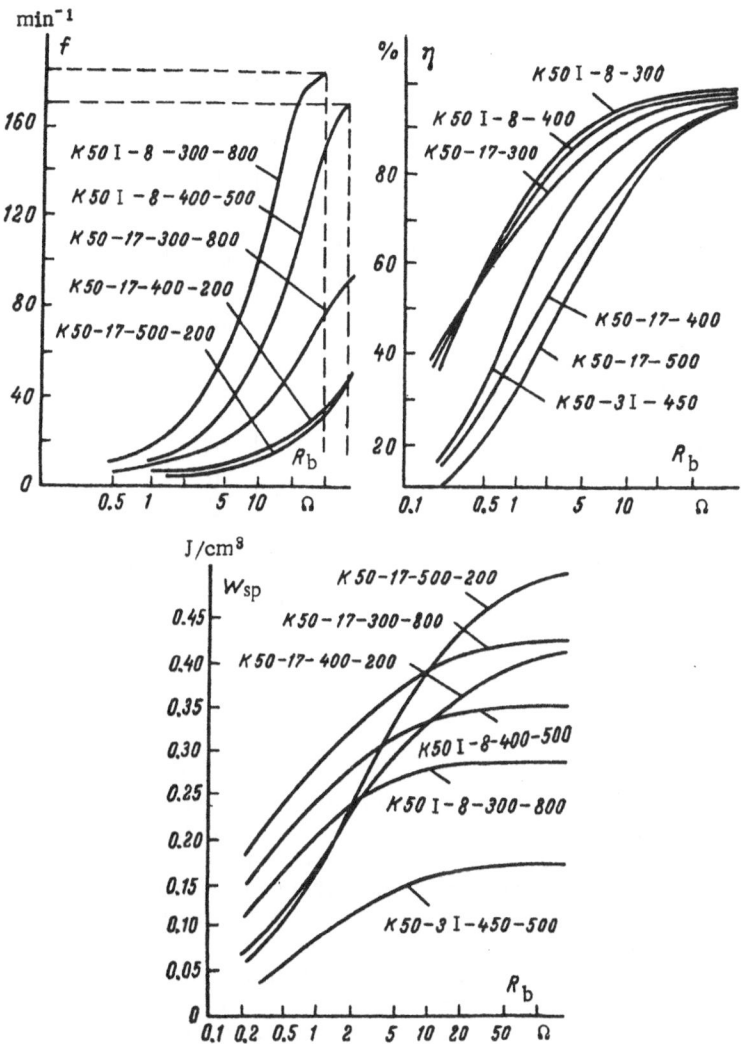

Fig. 9-31. Dependences of the permissible frequency f of discharge
cycles, efficiency η, and specific energy output W_{sp} on
the discharge resistance R_{dis} for electrolytic capacitors.
If n capacitors are parallel-connected, then $R_{eq} = nR_{dis}$
is inserted instead of R_{dis}.

Table 9.8. Characteristics of Metal-Ceramic and Metal-Glass Trigatron-Fired Discharges (for sparse flashes) Manufactured by the EGG Co. (Edgerton, Germshausen and Greer)

Type of discharge	Operating voltage range, kV	Static break-down voltage, kV	Maximum pulse current, kA	Current pulse duration, sec	Switched energy, kJ	Minimum firing-pulse voltage, kV	Firing delay time
GP-11B	1.3—3.4	4.2	5	20	25	5.5	0.10
GP-12B	10—24	30	100	10	2500	20	0.03
GP-14B	12—36	42	100	10	2500	20	0.03
GP-15B	25—69	86	100	10	4000	20	0.03
GP-16B	0.7—2.1	2.6	5	20	25	5	0.10
GP-17B	4.4—10	12.5	5	20	25	7	0.10
GP-20A	3.5—11	14	15	20	200	10	0.10
GP-22B	6—15	19	100	10	2500	20	0.03
GP-30B	2—6	7.5	100	10	2500	20	0.03
GP-31A	2—6	7.5	15	20	200	10	0.10
GP-32B	20—48	60	100	10	4000	20	0.03
GP-39	8—20	25	5	20	25	7	0.10
GP-41B	12—36	42	100	10	4000	20	0.03

tens of nanoseconds), suitability for program control, and high values of the switchable current. Their main shortcoming is the noise effect upon triggering. The simplest type of air discharger of the trigatron type, which is used extensively in various circuits and devices, has a circuit diagram and mechanism of operation which are close to those of analogous flashlamps (see Figure 7-2a). Given sufficient size, they can be used to switch current pulses up to a few tens of kiloamps having a duration up to tens of microseconds at voltages up to tens of kilovolts. The main shortcoming of this type of discharger is the significant inductance of the coupling to the discharge circuit (a few microhenrys).

The RVU-7M high-voltage low-inductance air discharger, produced by the pilot plant of the Khar'kov Polytechnical Institute, is a "cascade discharger," whose operating principle is made clear by Figure 9-33a. Principal electrodes 1 and 2 and trigger electrode E_3 form two series-connected spark gaps S_1 and S_2 with a uniform

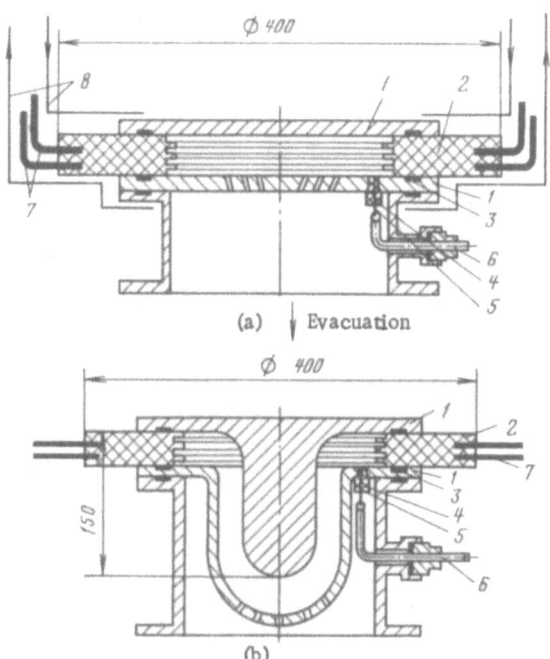

Fig. 9-32.   50-kV vacuum dischargers.   a) Discharger with flat elec-
             trodes; b) discharger with coaxial electrodes; 1) primary
             electrode (steel); 2) insulating body (organic glass);
             3) vacuum rubber; 4) bushing (fluoroplastic); 5) trigger
             electrode (stainless steel); 6) trigger cable; 7) insu-
             lating ring; 8) current-supply cylinder.

voltage drop. When a high-voltage trigger pulse is supplied to
electrode E_3, the field is amplified in one of the gaps and the gap
breaks down. As a result, the total voltage is applied to the
second gap, which then breaks down as well (there may also be more
than two gaps). Since breakdown occurs at voltages which greatly
exceed the static discharge voltage, the firing lag time of the discharger may be very small. The design of the RVU-7M discharger is
shown in Figure 9-33b, and its circuit diagram is presented in
Figure 9-33c. Primary electrodes 1 and 2, which have toroidal
stainless-steel cover plates 1' and 2', are insulated from each
other by an organic glass plate 8. Trigger electrode E_3 passes
through the opening in the center of the plate. This electrode consists of an inside metal cylinder 3 and two outer metal cylinders
4 and 5, which are mounted on it by means of a metal bushing 6 and
a fluoroplastic bushing 7. The plastic body 9 of the discharger
is detachable, consisting of two parts. The trigger pulse is supplied to cylinder 3 via cable 10. The discharger can switch current

Table 9-9.   Arc-Suppressing Chambers [9-49]

Parameter	KDV-26⁻	KDV-21	KDV-19	KDV-16
Rated voltage, kV	15	20	35	35
Test voltage (1 min), kV	56	66	90	95
Rated current, A	300	300	300	200
Reswitchable current, A	900	900	900	-
Max. disconnectable current, A	2000	2000	1600	4500
Voltage at cutoff, kV	15	20	35	35
Switchable power, MV-A	30	40	56	157.5
Service life, switchings	$3 \cdot 10^4$	$4 \cdot 10^4$	$3 \cdot 10^4$	-
Capacitance, pF	8.0	6.8	20	6.8
Inductance, mH	0.3	0.2	0.5	0.5

pulses having an amplitude of up to 150 kA in the voltage range 15-50 kV for pulse lengths up to tens of microseconds. The discharge has a self-inductance of 40 nH and is used to switch discharge circuits in high-power units [9-50].

The advantages of dischargers that operate at high pressure are a trigger time with a small spread, low inductance, an electric breakdown strength recovery time that is short compared with vacuum dischargers, a broad range of controllable voltage without adjusting the length of the spark gap (0.15-1.0 the rated voltage), and a low noise level in operation. Figure 9-34a shows the design of a high-pressure discharger developed at the Moscow Power Engineering Institute [9-51]. The discharger employs a ring-type system of electrodes which ensures more even wear than in other designs, and makes the working gap independent of the degree of tension on the detachable joints. Electrodynamic and gas-dynamic forces cause the main discharge channel to blow out into the damping volume formed by the cover of the discharger. Therefore, the main discharge channel has no adverse effect on the trigger electrode. The operating voltage of the discharger is 5 kV at an internal air pressure of 0.5 MPa. The switchable current is 160 kA and the self-inductance is 15 nH.

The layout of a discharger containing a solid dielectric is shown in Figure 9-34b [9-52]. Ring 3, together with detonator 4, is placed in the recess of electrode 1. The detonator explodes when the initiating pulse is supplied. The resultant metal jet breaks down insulation 2 and closes upper and lower electrodes 1. Organic glass, polyethylene, lavsan, a fluoroplastic, or other materials may be used as the dielectric, and exploding wires, metal rivets, and electrodynamic hammers may be used to break down the dielectric. Solid-dielectric dischargers have extremely low values

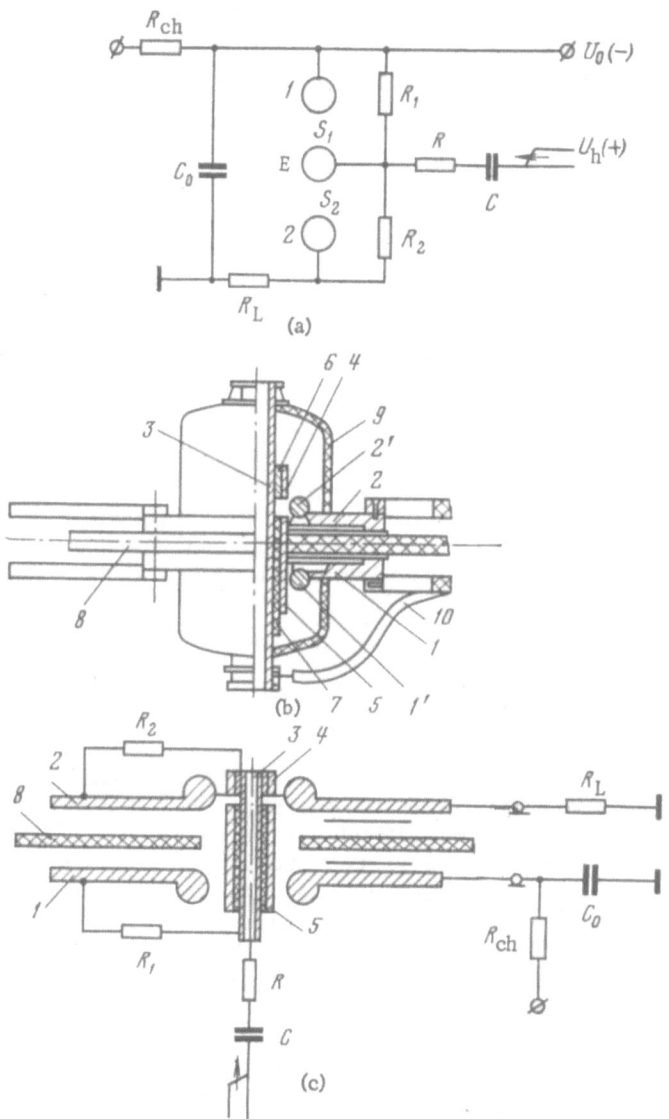

Fig. 9-33. The RVU-7M discharger. a) Schematic diagram; b) design of the discharger; c) circuit diagram of the discharger. 1,2) Primary electrodes; R_1, R_2) voltage dividers; R_L) load; R_{ch}) charging resistor; C_0) capacitor bank; C and R) coupling capacitor and resistor; U_h) high-voltage trigger pulse; U_0) working voltage of bank.

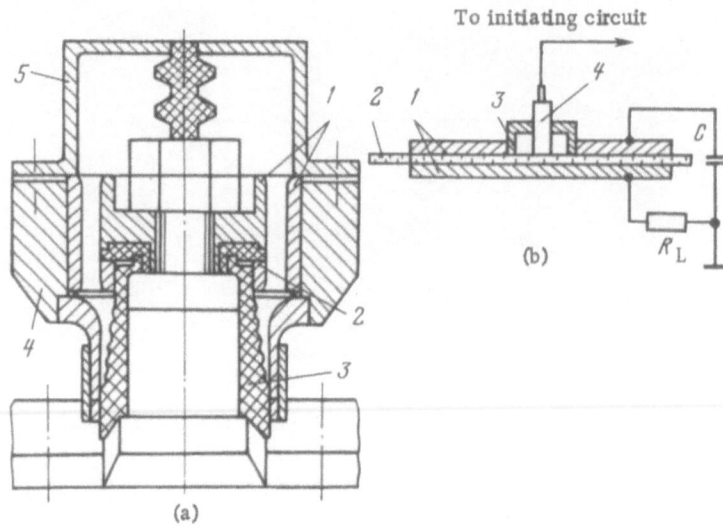

Fig. 9-34. Dischargers. a) The high-pressure discharger of Moscow
Power Engineering Institute. 1) Primary electrodes;
2) trigger electrode; 3) insulation; 4) body; 5) cover.
b) Discharger containing a solid dielectric. 1) Primary
electrodes; 2) dielectric; 3) metal ring; 4) detonator.

of the spark-gap inductance and an extremely compact design, though
they are not widely used.

In recent years work has been done in the USSR and abroad to
develop high-power mercury ignitron dischargers. Their advantages
are the lack of damage to the cathode, a broad range of operating
voltages, a small voltage drop in the arc, and noiseless operation.
The series-produced IRT-1 ignitron is designed for a current of 100
kA, a voltage of 100 V to 10 kV, and a pulse repetition frequency of
10 pulses/min. It has a self-inductance of 40 nH and an operating
life of 10^5 connections. The short triggering time of this ignitron
(less than 0.5 μs with an instability not exceeding 0.05 μs) makes
possible the parallel operation of many dischargers. Ignitrons designed to handle 5 and 25 kA at a voltage of 50 kV with a pulse repetition frequency of 2 and 50 Hz also have been developed [9-53].

Cables. Special cables are used to transfer energy from the
storage element to the load and to complete other high-voltage circuits. The number of cable sections in large units having a large
number of flashlamps may reach several thousand, with a section
length of up to a few tens of meters. Therefore, the reliability of
the unit as a whole depends largely on the reliability of the cable.
In some cases the inductance of the cables may amount to up to 50%
of the inductance of the entire discharge circuit, and the output
parameters of the unit often depend on the cable inductance. Accord-

ingly, much attention is paid to the selection of the cable during
the design of such units for supplying power to pulsed light sources.

One of the main characteristics of a cable is its inductance.
For a single-conductor cable the inductance (in henrys) is equal to
[9-54]

$$L = 2l \ln \frac{D}{d_c} 10^{-7},$$ (9-16)

where l is the cable length (m), and D and d_c are the diameters of
the insulation and the cable conductor (m), respectively.

The most widely used cables have stabilized-polyethylene insulation. Typical designs of high-voltage pulsed cables are shown
in Figure 9-35, and the basic data on some commercially produced
brands are given in Table 9-10. In addition to layers of insulation,
cables often have semiconducting coatings (usually made of semiconductive polyethylene) which improve the ionization characteristics
and equalize the electric field.

When cables are installed, the selection of the type of cable
termination and its quality are of great importance. Operating experience demonstrates that cable failures basically result either
from the breakdown of the braid onto the cable core in the terminal
zone, or from overlapping of the insulation or the semiconductive
layer on the surface (if the layer is left on the surface of the
cable). Figure 9-36a enables us to determine the length of the cable
termination for KPV-1/50 and RK 75-9-13 cables having insulation 4 mm
thick [9-52]. The operating voltages usually are selected to be
30-50% below the discharge voltages for given lengths of the terminations. It should be kept in mind that an increase in the length of
the termination leads to an increase in its inductance. Furthermore,
increasing the length of a termination beyond some specific size
leads to an insignificant increase in the discharge voltage, since
this voltage depends on the thickness of the insulation, the properties of the semiconductive coating, and the nature of the surrounding
medium. Whenever a high-voltage cable is operated only under a
pulsed voltage with a pulse length of a few tens of microseconds,
the semiconductive layer may be left on the surface of the insulation
at the cable termination. This results in a more uniform distribution of the voltage over the length of the termination. The semiconductive coating must be removed if there is to be prolonged exposure to the voltage, since otherwise it is destroyed by the heat
and damages the underlying insulation. Examples of cable terminations
are shown in Figures 9-36b and 9-36c.

Table 9-10. The Product Variety of High-Voltage Coaxial Lines

Make of cable*	Voltage, kV				Inductance, nH/m	Capacitance, pF/m	Characteristic impedance, ohms	Outside diameter, mm	Weight, kg/km	Safe life	Remarks
	Working (pulsed)	At start of internal discharges in insulation	Test voltage at 50 Hz**	Constant test voltage**							
RK-50-0,6-22	—	0,3	$\frac{0,6}{1}$	—	240	96	50	1,2	3,5	3000 hr	
RK-50-1-22	—	0,5	$\frac{1,0}{1}$	—	240	95	50	1,7	7,0	1000 hr	
RK-50-1.5-22	—	0,9	$\frac{1,8}{1}$	—	240	97	50	2,0	17,7	10 000 hr	
RK-50-1-23	—	0,6	$\frac{1,2}{1}$	—	230	93	50	1,5	11,2	10 000 hr	
RK-50-2-11	—	1,5	$\frac{3,0}{1}$	—	250	100	50	4,0	21,4	5000 hr	
RK-50-2-15	—	1,5	$\frac{3,0}{1}$	—	290	115	50	4,4	37,4	5000 hr	
RK-50-3-23	—	2,0	$\frac{4,0}{1}$	—	240	95	50	4,4	47	10 000 hr	

RK-50-4-11	—	3,0	$\frac{6,0}{1}$	—	250	100	50	9,6	123	10 000 hr
RK-50-7-23	—	5,0	$\frac{9,0}{1}$	—	250	100	50	12,2	331	10 000 hr
RK-50-9-12	—	5,0	$\frac{10}{1}$	—	275	110	50	12,2	229	10 000 hr
RK-50-11-11	—	5,5	$\frac{14}{1}$	—	290	115	50	14	300	10 000 hr
RK-50-24-17	—	13	$\frac{25}{1}$	—	250	100	50	27,8	1096	10 000 hr
KVNTE	7,5	4,0	$\frac{6}{1}$	$\frac{50}{3}$	180—95	127—240	—	6,9—9,2	80,5—168	1 000 hr
KVNTE	15	7,5	$\frac{12}{1}$	$\frac{100}{3}$	29—170	80—134	—	9,5—11,8	164—271	1 000 hr
KVNTE	25	12	$\frac{20}{1}$	$\frac{200}{3}$	380—240	60—95	—	13,1—15,4	319—456	1 000 hr
KVN-10/75	10	—	6	50	365	65	75	8,3	93,5	1 000 hr
KVN-20/75	20	—	12	100	365	65	75	12,3	186	1 000 hr
KVN-20/50	20	—	12	100	250	100	50	11,3	178	1 000 hr
KVN-35/100	35	—	20	200	500	50	100	24,3	553	1 000 hr
KVNS-20/75	20	—	12	100	365	65	75	12,3	186	10 000 hr
KVNS-20/50	20	—	12	100	250	100	50	11,3	178	10 000 hr

* Series RK-75 also is produced. It is similar to series RK-50, but has a 75-Ω resistance.

** The top figure represents the voltage (kV), and the bottom figure the time (min).

Table 9-10, continued

Make of cable*	Voltage, kV Working (pulsed)	At start of internal discharges in insulation	Test voltage at 50Hz**	Constant test voltage**	Inductance, nH/m	Capacitance, pF/m	Characteristic impedance, ohms	Outside diameter, mm	Weight, kg/km	Safe life	Remarks
KVN-40	40	—	—	120/1	250	100	50	18,6	447	10 000 hr	
KPV-1/20	20	—	—	100/10	260	96	50	23,5	600	10 000 hr	
KPVM-1/30	30	—	—	60/5	100	40	20	32	1200	$5 \cdot 10^4$ pulses	
MKPVM-1/30	30	—	—	50/5	110	54,5	25	16	600	10^4 pulses	
FKP	50	—	—	100/1	60	24	12	16,2	434	500 pulses	10^4 pulses at 30 kV
AKPVM-1/50	50	—	—	75/5	110	54,5	25	20	500	$3 \cdot 10^4$ pulses	
KPVMG-1/50	50	—	—	140/5	180	90	40	36,5	1400	$1,5 \cdot 10^7$ pulses	
KPV-1/50	50	—	—	75/10	220	110	50	18,6	450	10^4 pulses	10^3 pulses at 60 kV

KPV-1/60	60	—	—	100/10	—	—	—	21,6	600	10^4 pulses	10^3 pulses at 70 kV
KVP-1/75	75	—	—	200/1	160	—	31	45	2500	10^6 pulses	
KVIM	60	—	—	140/1	120	—	22	18	473	10^4 pulses	500 pulses at 80 kV; 10^4 hr for DC voltage
KVI-120	120	—	100/1	120/1	250	100	50	15	335	10^4 pulses	
KVIO-150	150	—	50/1	—	100	40	20	19,6	806	1 pulse	
IK-2	25	—	9	35/1	250	100	50	20	400	10^3 hr	
IK-4	80	29	50/1	—	500	200	100	35,8	1297	10^3 hr	
IKSh-16	25	9	—	36/1	80	32	16	29,5	1620	$5\cdot10^3$ hr	
IKSh-24	25	9	—	36/1	120	48	24	28	1228	$5\cdot10^3$ hr	
IKSh-30	35	12,5	—	50/1	150	60	30	26	1095	$5\cdot10^3$ hr	

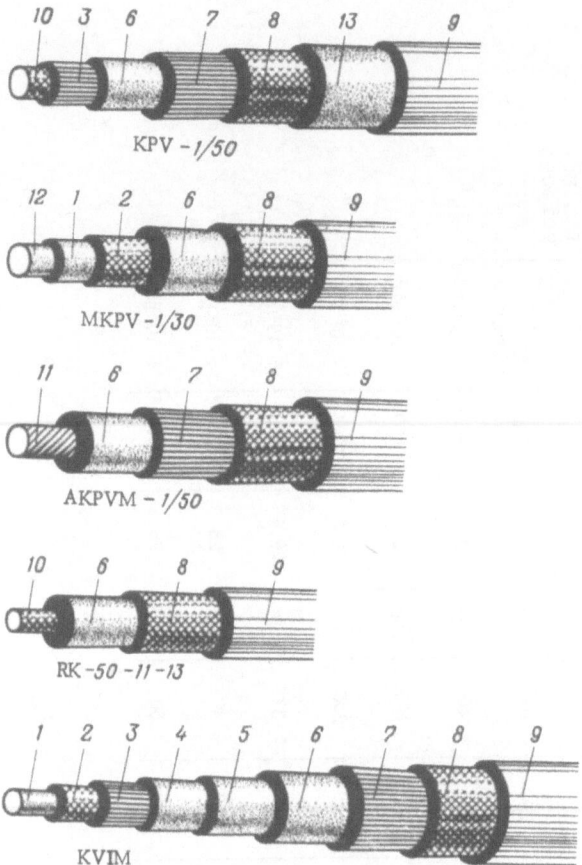

Fig. 9-35. Designs of high-voltage pulsed cables. 1) Polyethylene
            core; 2) copper-wire braiding; 3) semiconductive poly-
            ethylene; 4) polyethylene insulation; 5) winding of a
            polytetrafluorethylene (PTFE) film with a liquid organo-
            silicon grease; 6) polyethylene insulation; 7) semicon-
            ductive polyethylene; 8) copper-wire shielding (reverse
            conductor); 9) light-stabilized polyethylene sheath;
            10) copper-wire conductors; 11) aluminum core; 12) in-
            dustrial core; 13) layer of rubberized or plasticized
            ribbon.

9-9. Triggering Devices

A discharge is triggered in a flashlamp if the potential between the anode and the cathode exceeds the self-breakdown potential
U_{self}. At voltages below U_{self} but higher than U_b (see p.22), the
lamp can be fired by supplying a trigger pulse which initiates an

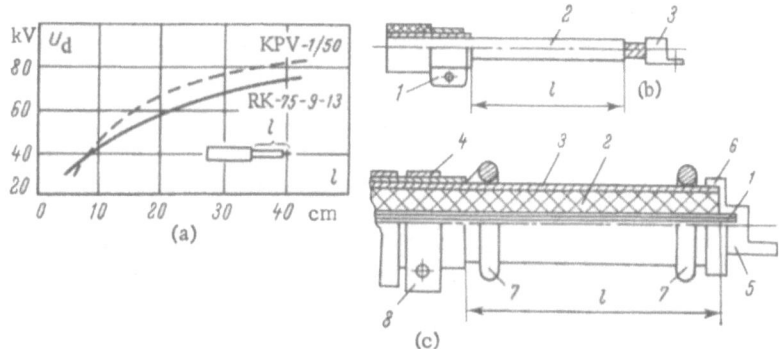

Fig. 9-36. Cable terminations. a) Dependence of discharge voltages
on the length of cable terminations (F = 30 kHz, Δ = 1.1);
b) termination of the KPV-1/50 cable. 1) Current-collecting clamp; 2) dielectric of termination; 3) cable
lug pressed onto the core. c) Termination of a cable
with a semiconductive layer. 1) Conductor; 2) primary
insulation; 3) semiconductive layer; 4) current-carrying
sheath; 5) cable lug; 6) current-carrying bridle connecting cable lug 5 to semiconductive layer 3; 7) collecting
rings soldered to 6 and 4; 8) current-collecting clamp.

auxiliary discharge throughout the entire discharge gap or in some
portion thereof. The lowest value of the voltage (U_t) at which a
lamp regularly fires when supplied a trigger pulse having specific
parameters is strongly dependent on the parameters of the pulse and
on the method used to apply the pulse to the lamp.

The circuit in which triggering of a discharge results from
self-breakdown may consist of just a charger, a storage capacitor,
and the lamp. The energy and frequency of the flashes, which occur
at indefinite times, are unstable in such a circuit because of the
instability of U_{self}.

A controlled switch such as a thyratron, a discharger, an ignitron, or a thyristor is series-connected to the lamp in the discharge circuit in order to stabilize the flash energy and fix precisely the time of occurrence of the discharge (US patent No.
3,749,975 and French patent No. 1,194,449). Such a circuit sometimes is made as a ringing voltage-multiplying circuit (see Figure
9-29a).

Two-electrode spherical lamps in switching circuits can be used
to achieve better spatiotemporal stability of the discharge channel
and a longer operating life for lamps than when three-electrode
lamps with an internal trigger electrode are used, as such an electrode is destroyed quickly by exposure to the primary discharge.
When tubular lamps are used, a continuous low-power ("keep-alive")

arc sometimes is employed in switching circuits (Figure 9-37). Such an arc makes it possible sharply to reduce the working voltage of the primary discharge circuit and to decrease the trigger lag. Good stability of the "keep-alive" arc is achieved by using a quite high-voltage auxiliary power supply to feed the arc. Such stability also is achieved by using inductances or current stabilizers in the auxiliary power-supply circuit to prevent arc suppression after a discharge (the stability increases with increasing current). Devices containing pulse transformers, which automatically generate a trigger pulse when the circuit is closed and the arc is arrested accidentally, may be employed for arc ignition.

Trigger circuits using pulse transformers are the most widely used. Some of these circuits were presented in Figure 1-2. In the circuit in Figure 1-2c, the voltage pulse across the transformer secondary is added to the voltage across the capacitor, and the lamp is broken down by the sum of the voltages. The secondary inductance L widens the current pulse, and the secondary resistance R lowers the efficiency. In order to reduce L and R, it is recommended to base the transformer on a magnetic circuit (with no gap) made of a material having a rectangular hysteresis loop and high magnetic permeability (East German patent No. 105,703). If the magnetic circuit has a sufficiently large cross section, a pulse having the length and amplitude required for triggering can be obtained with a minimum number of turns in the secondary, which is made of a large-cross-section wire. (The inductance of the secondary is high only during the generation of the trigger pulse, and is insignificant during the discharge-current pulse.) The weight and size of such a transformer usually are large.

If the transformer is connected in parallel with the lamp (Figures 1-2d and 1-2e), then the current pulse from the discharge circuit does not pass through the transformer secondary (low-capacitance capacitor C_1 prevents shunting of the lamp by the secondary). The transformer may be made without a magnetic circuit, from a wire of small cross section, and may have low weight and small dimensions. In the circuit in Figure 1-2d a choke with a saturable magnetic circuit is included in the discharge circuit. This eliminates shunting of the trigger pulse by the storage capacitor. In this case the duration and amplitude of the voltage pulse applied to the anode of the lamp are determined by the volt-second characteristic of the choke. If a discharger which breaks down when the voltage approaches the peak value is included in the circuit of the transformer secondary, then the steepness of the leading edge of the pulse increases, the pulse "peaks", and we can get along with a choke that has a lower volt-second characteristic.

We can make do without a choke and discharger by using the circuit in Figure 1-2e (Swiss patent No. 513,564). In this circuit the trigger pulse can have a comparatively gentle (leading) edge and

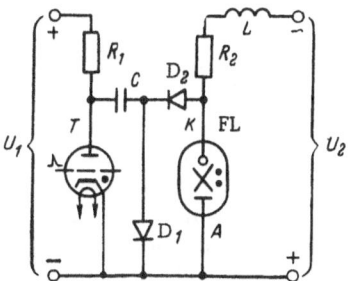

Fig. 9-37. The power-supply circuit of a flashlamp with a "keep-
            alive" arc and a thyratron as the switch in the dis-
            charge circuit. L, R_2, FL, U_2 - the circuit of the
            "keep-alive" arc. R_1, C, D_1, U_1 - the charging circuit.
            T, FL, D_2, C - the discharge circuit.

practically any shape, but a pulsed diode is needed with a permissible reverse voltage higher than the self-breakdown potential of
the lamp. When the current pulse flows through the diode, it must
have a resistance lower than that of the lamp during the discharge.
(When short-duration reverse-voltage pulses are applied, some types
of pulsed junction-type diffused silicon diodes [9-55] are capable
of withstanding voltages severalfold higher than under static conditions, and their transient resistance decreases sharply when high
pulsed currents pass through.)

     When the trigger pulse is supplied to one of the primary elec-
trodes of a lamp, triggering is facilitated if the external electrode
of the lamp is connected to one of the primary electrodes, either
directly or through a small coupling capacitor. In the triggering
methods described above, the voltage pulse is applied to one of the
primary electrodes. For example, a positive pulse is applied to the
anode and a negative pulse to the cathode. These methods can be used
to lower the trigger voltage quite sharply. However, the methods
most widely used in practice do not require additional separation
elements such as chokes and dischargers. These triggering methods
supply a trigger pulse to the external trigger electrode of a tubular
lamp (see Figure 1-2a) or to the internal trigger electrode of a
spherical lamp (see Figure 1-2b). In the latter case the power of
the firing device does not necessarily have to be increased in order
to increase the current strength of the auxiliary discharge, as this
can be accomplished by predischarging the coupling capacitor in the
trigger-electrode circuit [9-56].

     When a pulse is supplied to the auxiliary electrode, the re-
liability of firing of the lamp is improved if the voltage across
the primary discharge gap increases simultaneously with the pulse.
This can be done, for example, by using an auxiliary pulse from the
firing transformer tertiary (a combined trigger circuit, as in

Figure 1-2f [9-57]) or from a second transformer (US patent No.
3,859,562), by switching the parallel connection of the capacitors
to series connection (see Figures 9-27a and 9-27b), or by using an
auxiliary high-voltage circuit (see Figures 9-27c and 9-27d) to form
a high-power spark in the primary gap. It also becomes easier to
trigger lamps (and the spread of the triggering lag is reduced) if
the primary discharge gap is irradiated when the trigger pulse is
supplied (e.g., irradiation may be done with an auxiliary sliding
discharge over the surface of the ceramic insulator inside the lamp).
Triggering also can be facilitated if a high-strength pulsed electric
field is generated near one of the primary electrodes simultaneously
with the trigger pulse (US patent No. 3,775,641).

     The voltage pulses produced by the trigger-pulse generator (TPG)
under open-circuit conditions are characterized in terms of their
pulse repetition frequency, amplitude and shape (steepness of the
leading edge), the number and duration of half-waves (total pulse
length), polarity, and harmonic components. When the trigger-pulse
generator is connected to the lamp, the pulse parameters may change
significantly, since the equivalent circuit of the lamp using external triggering represents some combination of capacitances and
resistances, such as the one shown in Figure 1-3. The voltage and
length of the pulse decrease, and the rate of pulse damping increases
when the spark gap breaks down. Near short-circuiting of the transformer occurs in the circuits in Figures 1-2b, 1-2c, 1-2d, and 1-2e.
In this case the intensity of the auxiliary discharge is characterized
by the current strength of the short-circuited generator.*

     To date, quantitative methods of estimating the energy input into
the lamp by the trigger pulse have not been developed at all, and
the accepted practice is to talk of the energy switched in the primary
circuit of the trigger-pulse generator: for low-power lamps it
usually is a fraction of a millijoule, for medium-power lamps tens
of millijoules, and for high-power lamps up to a few joules.

     Pulse transformers without a magnetic circuit or with a core in
the form of a rod inserted inside the coil are used to trigger low-
power lamps and in circuits with a high flash repetition frequency.
A schematic diagram of such a transformer is presented below in
Figure 9-41b. The nature of the electrical processes with a transformer operating in the trigger circuit differs somewhat here from
the processes that occur in ordinary pulse transformers (e.g., in
blocking oscillators), where the main requirement is that the voltage
pulses be transmitted without distortion of shape. These processes
therefore should be treated in greater detail.

---

*The resistance of the auxiliary circuit is on the order of magnitude
 of a few units to hundreds of ohms. For a low-power trigger-pulse
 generator, this is practically equivalent to short-circuit conditions.

A simplified equivalent circuit of a trigger-pulse generator
under open-circuit conditions is presented in Figure 9-38. Together
with capacitor C_1, the connectors, and switch K, the pulse transformer forms a system of two coupled oscillatory circuits in which
free oscillations are excited by the energy of precharged capacitor
C_1 [9-58]. The first circuit is formed by capacitor C_1 and the inductances of transformer primary L_1 and connectors L_K. It includes
the loss resistance R_1, which determines the damping rate of oscillations in this circuit. The transformation ratio of the transformer
usually is high (the number of turns in the primary is small compared
with the number in the secondary). Therefore, the self-capacitance
of the primary may be ignored ($C_p << C_1$). Since the intent during
the design of equipment is to make the inductance of the connectors
and elements of the first circuit lower than that of the transformer
primary ($L_K < L_1$), the inductance L_K also may be omitted. The second
circuit is formed by the inductance of transformer secondary L_2 and
capacitance C_2, which consists of the self-inductance of the transformer secondary, the capacitance of the connectors (from the transformer to the lamp), and the capacitance of the lamp. Let the equivalent resistance of the losses in the second circuit be equal to R_2,
and the mutual inductance of the transformer primary and secondary
be M.

Here we introduce the notations: $\alpha_1 = R_1/2L_1$; $\alpha_2 = R_2/2L_2$;
$\omega_1^2 = 1/L_1C_1$; $\omega_2^2 = 1/L_2C_2$; $k = M/\sqrt{L_1L_2}$; $\sigma = 1 - k^2$; and $a = \omega_1/\omega_2$.
Here α_1 and α_2 are the damping factors of the first and second cir-
cuits; ω_1 and ω_2 are their natural frequencies of oscillation; a is
the ratio of these frequencies; k is the coupling factor of the circuits; and σ is the dispersion coefficient of the field lines. When
K is closed, the current i_1 that appears in the first circuit is
transformed into current i_2 in the second. The voltages u_1 and u_2
across capacitors C_1 and C_2 (the input and output voltages of the
trigger-pulse generator) are determined by the equations

$$u_1 = R_1i_1 + L_1\frac{di_1}{dt} + M\frac{di_2}{dt}, \quad u_2 = R_2i_2 + L_2\frac{di_2}{dt} + M\frac{di_1}{dt},$$

and the currents are given by the equations

$$i_1 = -C_1\,du_1/dt, \quad i_2 = -C_2\,du_2/dt.$$

By introducing simplifications that do not affect the qualitative nature of the results, we can solve this system of equations for
the output voltage $u_2(t)$ [9-59 and 9-60]:

$$u_2(t) = -U_0A\left(e^{-b_1t}\cos\Omega_1t - e^{-b_2t}\cos\Omega_2t\right). \tag{9-17}$$

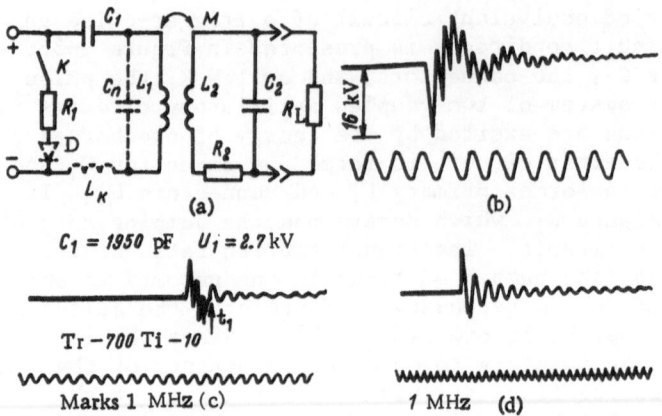

$C_1 = 1950$ pF $U_1 = 2.7$ kV

Tr -700 Ti -10

Marks 1 MHz (c)   1 MHz   (d)

Fig. 9-38. Simplified equivalent circuit of a trigger-pulse generator containing a transformer with no magnetic circuit
(a), and sample oscillograms of these pulses for switches
with bilateral (b) and unilateral conductivity with long
(c) and short (d) pulses in the primary circuit. The
oscillations put out by a calibration oscillator at a
frequency of 1 MHz are shown at bottom.

Thus, the transformer output voltage is proportional to the
initial voltage U_0 across capacitor C_1. It is determined by some
factor A and is the sum of two damped cosine waves having frequencies
Ω_1 and Ω_2. The analytic expressions for the damping factors b_1 and
b_2 in terms of the factors α_1, α_2, k, σ, a, and the ratio of the
frequencies Ω_1 and Ω_2 are quite complicated. Under no-load conditions
the damping in the circuits is small, and about 10 superposed oscillations having frequencies Ω_1 and Ω_2 can be observed. The resonant
frequencies Ω_1 and Ω_2 of the oscillations are called the coupling
frequencies. (The frequency Ω_2 always is higher than the higher
of the natural frequencies ω_1 and ω_2 and is called the fast coupling
frequency, while Ω_1 always is lower than the lower natural frequency
and is called the slow coupling frequency.) These frequencies are
given by the expression

$$\Omega_{1,2} = \sqrt{\frac{\omega_1^2 + \omega_2^2 \pm \sqrt{(\omega_1^2 + \omega_2^2)^2 - 4\sigma\omega_1^2\omega_2^2}}{2\sigma}} \qquad (9\text{-}18)$$

The quantities L_1, L_2, M, and C_2 can be calculated for a pulse
transformer by using known formulas [9-54] or can be measured, and
C_1 can be specified on the basis of the energy stored in the first
circuit. Therefore, determining the fast and slow coupling frequencies poses no difficulties. (Even if the natural frequencies ω_1 and
ω_2 are equal by chance, oscillations having the two frequencies Ω_1
and Ω_2 are present simultaneously in a system of coupled circuits

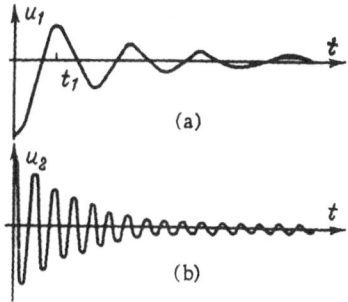

Fig. 9-39. Damped cosine waves of slow (a) and fast (b) oscillations
            in a system of coupled circuits, as constructed from
            calculated data.

when there is strong coupling between the circuits, i.e., when $\sigma \neq$
0 and k = 0.8-0.5.) The capacitance C_1 of the first circuit usually
is quite high and $\omega_1 \ll \omega_2$. Therefore, the inequality $\Omega_1 < \Omega_2$ is
quite pronounced in a system of coupled circuits. Hence, a system
of coupled circuits in a trigger-pulse generator is not characterized
by beating (when $\Omega_1 \approx \Omega$), and fast oscillations always are superposed
on slow ones (Figure 9-39).

     The amplitude of the output voltage depends on the factor A in
formula (10-17). If the natural frequencies ω_1 and ω_2 of the circuits are equal and the resistances R_1 and R_2 may be ignored, then
the expression for A has a quite simple form [9-59]:

$$A = \frac{1}{2} \sqrt{\frac{C_1}{C_2}} = \frac{1}{2} \sqrt{\frac{L_2}{L_1}} \approx \frac{n}{2} , \qquad (9-19)$$

where n is the transformation ratio.

     Since $R_1 \to 0$ and $R_2 \to 0$ in this case, $b_1 \to 0$ and $b_2 \to 0$ and the
maximum value of the parenthetic expression in formula (9-17) approaches 2, and the amplitude of output voltage U_2 is defined as

$$U_{2M} \approx U_0 n, \qquad (9-20)$$

i.e., in this case the maximum output voltage may not exceed the
product of the transformation ratio by the initial voltage across
storage capacitor C_1. The actual value of the voltage across the
primary winding is lower than U_0 because of the presence of the self-
inductance L_K of the primary circuit. The resistance R_1 and R_2 also
reduce somewhat the actual output voltage of the generator.

     The condition that the natural frequencies of the first and
second circuits be equal is not optimal from the standpoint of ob-
taining an output voltage having maximum amplitude. The output

voltage of the transformer can be practically doubled by significantly lowering the frequency ω_1 through an increase in capacitance
C_1. For the case $\omega_1 \neq \omega_2$ the factor A is defined to sufficient accuracy by the expression [9-60]

$$A = \frac{k\sqrt{L_2/L_1}}{\left(1 + \omega_1^2/\omega_2^2\right)^2 - 4\left(1 - k^2\right)\omega_1^2/\omega_2^2}. \qquad (9-21)$$

Assuming that when $\omega_1 \ll \omega_2$ for a transformer with strong
coupling $(k \to 1)$, we have $\sqrt{L_2/L_1} \simeq n$ and we find $A \simeq n$. Hence, when
the maximum value of the parenthetic quantity in relation (9-17) is
2,

$$U_{2M} \approx 2nU_0. \qquad (9-22)$$

Thus, the maximum instantaneous value of the output voltage of
the trigger-pulse generator cannot be over twice the transformation
ratio multiplied by the initial voltage across capacitor C_1. If we
reduce the influence of adverse factors (by lowering the values of
L_K, C_2, R_1, and R_2) and increase the influence of favorable factors
(by increasing the factor M and the frequency ratio ω_1/ω_2), we can
approximate the maximum value of the output voltage, as defined by
formula (9-22).

The significant increase in the transformation ratio due to the
increased number of secondary turns leads to an increase in L_2, C_2,
and ω_2, but this causes a lowering of the fast coupling frequency
and slows the increase in the factor A by reducing ω_1/ω_2. Increasing
the transformation ratio by reducing the number of primary turns
lowers the ratio L_1/L_2 and leads to an increase in the amplitude of
the current pulse i_1. This imposes constraints on the selection of
the switch in the first circuit.

Connecting some load resistance R_L to the second circuit
(Figure 9-38) leads to a speedup of the damping of the fast and slow
coupling frequencies. But if this resistance is so small that the
second circuit becomes aperiodic, then the system of coupled circuits
is broken, oscillations are preserved only in the first circuit, and
the oscillation frequency of the first circuit increases ($\omega_1' = \omega_1/\sqrt{1 - k^2}$). The equivalent resistance introduced into the first
circuit is expressed as

$$R_{1e} = R_1 + \left(M/L_2\right)^2 R_2',$$

where

$$R_2' = R_2 + L_2/C_2 R_L.$$

If there is a short circuit at the output of the secondary winding (which occurs when the control electrode-cathode gap breaks down in a spherical lamp), the system of coupled circuits also is disrupted and the last relations obtain. (Here R_2 is determined by the resistance of the auxiliary spark and the resistance of the transformer secondary.) The current strength at the output of the trigger-pulse generator under short-circuit conditions is directly proportional to the mutual inductance of the windings and to the initial voltage across capacitor C_1. It is inversely proportional to the inductance of the first and second circuits and to the resistances in these circuits [9-61].

To illustrate this, let us present an example of a circuit and the parameters of a trigger-pulse generator for the ignition of flashlamps.

The primary of a pulse transformer is wound into the threaded groove of an organic glass cylinder 48 mm in diameter with a 1-mm spacing between turns. PELSShO wire 0.23 mm thick is used, and the primary contains 40 turns. The cylinder is inserted into an organic glass vessel with an outside diameter of 52 mm and an inside diameter of 48 mm on whose outside surface 20 circular grooves 1 mm wide and 1 mm deep are cut at 1-mm spacings. The secondary, made of 120 turns of 0.18-mm PELSShO wire, is wound on, six turns per groove. The transformer is immersed in a can containing transformer oil. It has the values L_1 = 57 μH, L_2 = 566 μH, M = 167 μH, and C_2 = 55 pF. The external appearance of the transformer is shown in Figure 9-40.

During single-flash operation, the primary circuit of a trigger-pulse generator containing such a transformer can be closed by a switch or an air gap. If a mica or ceramic capacitor having a capacitance of a few thousand picofarads is discharged through the transformer primary via a TGI1-325/16 trigatron at an initial voltage of 5 kV, the trigger-pulse generator will produce trigger pulses at repetition frequencies up to 10-15 kHz. These are sufficient to fire practically any low- or medium-power spherical and tubular lamps. Assuming C_1 = 1300 pF and an inductance L_K = 9 μH for the other elements of the first circuit ($L_1 + L_K$ = 66 μH), under no-load conditions we find ω_1 = 3.4 x 10^6 rad/s; ω_2 = 5.7 x 10^6 rad/s; σ = 1 - k^2 = 0.25; Ω_1 = 2.8 x 10^6 rad/s; and Ω_2 = 12.9 x 10^6 rad/s. An oscillogram of the transformer output voltage for the conditions indicated when the primary circuit is closed by an air discharge is shown in Figure 9-38b. The highest instantaneous value of the voltage pulse is reached when the first negative maximum of the slow-frequency oscillations is added to the first negative maximum of the fast-frequency oscillations. Assuming that the parenthetic expression from relation (9-17) is equal to 2 for this time, we find U_{2M} = 2U_0A, where A from relation (9-21) is equal to 2.1. As might have been expected, A < n (the transformation ratio is equal to 3). If U_0 = 5 kV, we find U_{2M} = 21 kV. Allowing for the effect of the inductance L_K (which is

Fig. 9-40.   Transformer containing no magnetic circuit.   a) External
            appearance; b) cap with windings, removed from the can.

15% of the total inductance of the circuit) and the loss resistances
R_1 and R_2 (R_1 includes the resistance of the air discharger), we
find that the voltage applied to the transformer primary should be
approximately 25% below the voltage across capacitor C_1 just before
switching occurs. Hence the amplitude of the transformer output
voltage should be 25% below the estimate given, and should come to
15 kV. This is in agreement with the oscillogram in Figure 9-38b.

     This type of oscillogram of the output voltage can be obtained
only when a switch with bilateral conductivity (a discharger) is
used in the first circuit. If a thyratron or thyristor is used in
the first circuit, only one half-wave of current flows. Then the
current in the first circuit is cut off, the system of coupled oscillatory circuits is broken, and only natural oscillations having a
frequency ω_2 are preserved in the second circuit. Such operating
conditions for a generator using the equivalent circuit in Figure
9-38 correspond to the inclusion of diode D in the first circuit.
Oscillations at the fast and slow frequencies exist only during the
first half-period of oscillations at the slow coupling frequency (up
to time t_1 in Figure 9-39a), for which the pulse shape can be calculated by the method indicated above. An oscillogram having such
an output-voltage pulse is shown in Figure 9-38c for the case where
the transformer described is used, a hydrogen thyratron is used as
switch K, and C_1 = 1950 pF. Here Ω_1 = 2.5 x 10^6 rad/s and Ω_2 = 12.5
x 10^6 rad/s. The second part of the pulse (after time t_1) has
natural oscillations with an angular frequency of about 6.3 x 10^6
rad/s. If the capacitance C_1 is low, then the current pulse in the
first circuit is very short and the oscillogram of the output voltage
shows only the natural oscillations of the second circuit with a
short blip at the start during transmission of the pulse in the
first circuit (Figure 9-38d).

During the design of transformers for trigger-pulse generators, an attempt should be made to achieve maximal inductances L_1 and L_2 with a minimum number of turns and the largest possible M. Since interlinkage between the outermost turns of the coils increases with decreasing height of the coil and increasing diameter, windings wound onto larger-diameter, shorter coils have a higher inductance, given the same number of turns. The quantity M increases as the gap between the turns of the transformer primary and secondary decreases, i.e., when the turns of one winding are in direct proximity to the turns of the other. (The goal is to make the area of a turn of the primary as close as possible to the area of a turn of the secondary. It also is useful to tap windings, with alternate emplacement of sections of the primary and secondary, and to use materials of minimal thickness having the highest electric strength for inter-winding insulation.)

A slip-over current transformer is a special type of transformer for a trigger-pulse generator. A coaxial cable is wound into a coil, and its outer braid is first cut into several approximately equal pieces.

Several parallel-connected pieces of braid form the transformer primary, and the central core of the cable forms the secondary. The insulation between the windings of such a transformer is provided by the cable insulation. The transformation ratio is equal to the number of pieces of which the primary consists, and the area of a turn of the primary is equal to the area of a turn of the secondary. Such transformers can be used to trigger high-power and sometimes medium-power lamps. Their shortcomings are their considerable inter-winding capacitance and comparatively high weight and large size.

It is desirable to reduce the number of secondary turns in order to lower the self-capacitance C_2 of the second circuit. However, if the transformation ratio is preserved, the reduction of the number of secondary turns leads to a reduction of the number of primary turns, to a lowering of the primary inductance, and hence to a short-ening of the current-pulse length and an increase in the amplitude of the current pulse through the switch. This is not always permis-sible. The introduction of a magnetic circuit containing an air gap into the transformer makes it possible greatly to increase the inductance of the windings and the coupling factor while keeping the number of turns small. This solution eases the switching con-ditions in the first circuit, but often leads to a reduction of the high-frequency components of the trigger pulse and increases the transformer's weight and the energy losses. The electrical processes when the transformer operates with an unclosed magnetic circuit are similar to the processes in a transformer with no magnetic circuit [9-57]. (Additionally, allowance must be made for the nonlinearity of the characteristics of ferromagnetic materials in the region of strong magnetic fields, and for additional energy losses in the magnetic circuit [9-62].)

The magnetic circuit may become saturated when pulse transformers having a closed magnetic circuit (e.g., in the form of a ferrite ring) are used [9-62a]. As a result, the amplitude of the current pulse increases sharply in the first circuit. It is convenient to use such transformers in series-triggering circuits (see Figure 1-2f), in which the high-power current pulse from the discharger circuit of the lamp causes magnetic polarity reversal on passing through one of the transformer windings.

Pulse transformers based on closed magnetic circuits containing a rectangular hysteresis loop generate output-voltage pulses only during magnetic polarity reversal. If the inductance of the primary of such a transformer is much higher under conditions of magnetic polarity reversal than the inductance L_K of elements of the first circuit, then a packet of short, steep pulses having a polarity that changes when the direction of the current in the first circuit changes is generated in the second circuit when there are oscillatory discharge conditions in the first circuit [9-63]. This packet contains an intense spectrum of hf components and can be used effectively to trigger tubular lamps that have an external trigger electrode. Here there is strong magnetic coupling between the first and second circuits only upon magnetic polarity reversal. Therefore, the concepts described above regarding a system of coupled circuits cannot be extended to this case. The pulse frequency in the packet is twice the natural frequency of oscillations of the first circuit, the pulse length depends on the voltage across the capacitor of the first circuit and on the volt-second characteristic of the magnetic circuit, and the form and amplitude of the pulses generated depend on the parameters of the secondary winding. If the conditions of the first circuit are selected so that the closed magnetic circuit does not saturate during oscillations and the natural frequency of the second circuit is much higher than the frequency of the first circuit, then damped oscillations appear in the first circuit and are transformed in the second. The frequency of these oscillations decreases smoothly with decreasing current amplitude of each successive half-wave in the first circuit. High-frequency oscillations appear in the second circuit ($\omega_2 = \sqrt{1/L_2 C_2}$) only at the start of a pulse and usually damp very quickly, since the losses in most types of magnetic circuits rapidly increase with frequency. Efforts are made to suppress these hf oscillations in the ordinary pulse transformers which are used, for example, in radar, in order to reduce distortions of the pulse being transformed. The hf component is useful in the trigger circuits of flashlamps and can be intensified, for example, by introducing an additional capacitor in parallel to C_2 in the second circuit. When this component is present, the amplitude of the output pulse may be nearly double the amplitude calculated as the product of the voltage across the circuit by the transformation ratio. Transformers of this type can be used to fire high-power tubular lamps or even for multiple firing of lamps [9-64]. The method used in the design of such transformers are examined in [9-65 and 9-66].

Coils (induction coils) sometimes are used instead of pulse transformers in trigger-pulse generators. In contrast to a pulse transformer, a coil stores electric energy when a current pulse flows through it. This energy then is converted into a high-voltage pulse. When switch K is closed, a current appears in the circuit of the primary winding of the coil (Figure 9-41a). This current may reach $I_M = U_1/R$, where U_1 is the power-supply voltage and R is the resistance of all elements in the circuit of the coil primary circuit. The energy stored in the coil is $W = I_M^2 L_1/2$, where L_1 is the inductance of the coil primary. When the switch is opened, a voltage pulse is generated across the windings. The magnitude of the pulse is proportional to the number of turns in the windings. If the ratio of the turns is such that $\omega_1 \ll \omega_2$, then the amplitude of the coil output voltage is determined mainly by the energy stored in the coil and by the capacitance C of the secondary circuit, which consists of the self-capacitance of the winding, the capacitance of the wiring, and the load capacitance ($U = \sqrt{2W/C}$). This voltage is lower under actual conditions because of energy losses in the switch, the core of the coil, and other circuit elements, because of the presence of magnetic leakage fluxes, and because of the self-capacitance of the primary winding and other factors. Such a firing system, which is similar to the ignition system of internal combustion engines, can be powered by a comparatively low-voltage DC source and does not require the use of a storage capacitor. A mechanical breaker or a transistor may be used as the switch. The use of such a system entails some inconvenience because the trigger pulse is generated when the switch is opened, and an advance preliminary command for closure of the switch is required. The trigger circuits of lamps containing coils are used successfully in some photographic systems to synchronize the operation of an electromechanical shutter with a photoflash unit (US patent No. 3,690,237).

High-voltage pulses for the firing of lamps also can be produced without a step-up pulse transformer or a coil. Figure 9-42a shows a circuit in which capacitor C_1 is charged to high voltage by any low-power source - e.g., by using a voltage multiplier circuit. When switch K is closed, capacitor C_2 ($C_2 \ll C_1$) is charged through choke L. The voltage u across capacitor C_2 (Figure 9-42b) is equal to the sum of the voltage U_1 and the voltage of damped harmonic oscillations

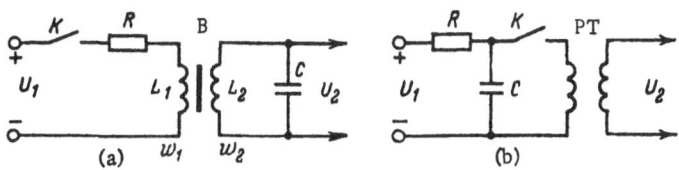

Fig. 9-41. Schematic diagrams of trigger-pulse generators. a) With a coil; b) with a pulse transformer.

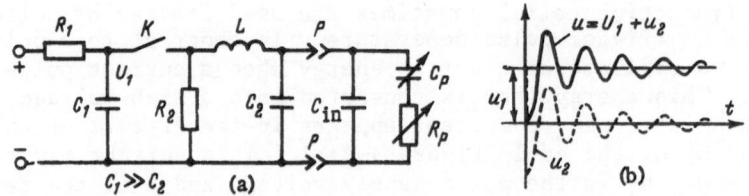

Fig. 9-42. High-voltage pulse generator without a transformer.
            a) Schematic diagram (the equivalent circuit of a
            tubular lamp with an external electrode for the trigger
            pulse is shown to the right of connector P-P); b) oscil-
            logram of the output voltage of such a generator.

which have a period determined mainly by the capacitance of C_2 and
the inductance of L. When switch K is opened, the charge flows from
capacitor C_2 through resistor R_2. The generator then is ready to
generate the next pulse. The equivalent circuit of a tubular lamp
with respect to the trigger pulse is shown to the right of the con-
nector in Figure 9-42a. When a lamp is connected to such a trigger-
pulse generator, the damping rate of the harmonic component and the
oscillation period increase somewhat. Several pulsed-voltage sources
can be obtained by connecting several circuits of L, C_2, and R_2 to
switch K. These sources can be used for simultaneous multiple firing
of several tubular lamps. The use of this scheme is limited by the
need to use a high-voltage source and a controlled high-voltage
switch (a discharger). Furthermore, a high-voltage coupling capacitor
connected in parallel with a high-resistance resistor also is needed
in order to use such a trigger-pulse generator to fire lamps that
have an internal control electrode.

     One possible way of producing high-voltage pulses to trigger
flashlamps is based on the use of piezoelectric devices which generate electrical pulses upon a sharp deformation or impact, for example
[9-67], US patent No. 3,880,572, and West German application No.
2,516,414. When struck a light longitudinal blow, one piezoelectric
stack made of lead zirconate-titanate (TsTS-19) 30 mm tall with a
cross section of 10 x 10 mm produces a voltage pulse having an
amplitude of a few kilovolts. This is sufficient to fire such lamps
as the IFK-120, the ISSh-7, and the ISSh-2.

     When a voltage pulse on the order of a kilovolt is required to
trigger flashlamps (e.g., low-power spherical flashlamps), it can
be easily obtained from an ordinary transistor oscillator, such as
a blocking oscillator, during forward or back current.

9-10. Control, Synchronization, and Protective Devices (CSP)

     The CSP unit may include devices for switching equipment on and
off; devices that regulate the discharge energy; devices for indi-

cating the operating conditions; a master generator, which determines the flash frequency and synchronizes flashes with an external event; devices for overload protection; devices for protection from miscellaneous damage; and devices for switching on standby elements.

Let us note some specific features of the devices indicated above.

In addition to ordinary on-off switches, devices for switching equipment on and off may contain various sensors. For example, these may be the sensors that often are used in light beacons and light warning devices, and electronic photorelays which switch on devices and change their operating conditions (US patent No. 3,846,750), depending on the level of natural illumination, the light from the headlights of a motor vehicle approaching a danger zone, and other factors [9-68 and 9-69]. It also is possible to use acoustic, thermal, and other sensors which actuate and trigger a pulsed light-signaling device (e.g., when some hazard occurs in production) [9-67 and 9-69].

Devices which regulate the discharge energy are used, for example, in automatic electronic photoilluminators. When the required level of illumination is reached on the object being photographed, an electronic integrator produces an output pulse which fires a specially developed spark discharger (US patents Nos. 3,769,888, 3,758,822, 3,779,142, 3,779,141, 3,797,721, 3,727,100, 3,842,428, 3,875,471, and 3,914,647; Belgian patents Nos. 724,245, 735,353, and 739,170; Australian patent No. 460,519; etc.) or which switches on a thyristor (US patents Nos. 3,591,829 and 3,980,924) or a photothyristor (US patent No. 3,517,255) which shunts the flashlamp (Figure 9-43a). It is possible to regulate the flash energy by means of a pulsed power transistor or thyristor that is series-connected to the flashlamp in the discharge circuit and that performs the function of an adjustable series resistor (US patents Nos. 3,671,649, 3,878,429, 3,878,433, and 3,946,269). In some electronic photoflashes the light pulse is cut off when the discharge circuit is broken (West German patent No. 2,045,319). Bioperational thyristors or a thyristor switch is used as the circuit breaker [9-2] (US patents Nos. 3,591,829, 3,980,536, and 3,896,333; British patent No. 1,290,313; West German patent No. 1,904,901; and the Belgian patents mentioned above, as well as Belgian patent No. 755,196). When the circuit is broken, the remaining energy in the capacitor is not expended. This makes it possible to shorten the duration of the next charging cycle and to save on energy in the primary power supply.

Tubular dischargers which shunt a flashlamp after the required integrated luminous intensity is produced (US patent No. 3,753,039) also are used in airport "running" flashlamp systems on landing strips to control the flash energy as a function of atmospheric visibility. A circuit has been described (US patent No. 3,946,411)

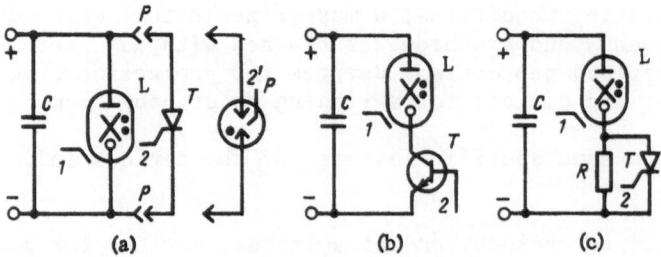

(a)   (b)   (c)

Fig. 9-43. Methods of automatically controlling the flash energy.
            The trigger pulse is supplied through channel 1, and the
            control pulse for the control device is supplied through
            channel 2. a) Shunting of the lamp by a thyristor or a
            discharger; b) series connection of a transistor acting
            as an adjustable resistance to the lamp; c) series con-
            nection of a resistor shunted by a thyristor to a lamp.

in which the integrated luminous intensity of the flash is regulated
when a thyristor connected in parallel with a resistor is included
in the discharger circuit (Figure 9-43c). The sooner the thyristor
is switched on after the start of a flash, the smaller the fraction
of energy expended across the resistor and the higher the integrated
luminous intensity. Extensive use also is made of very simple
methods of energy control. For example, these include varying the
voltage across the storage capacitor before the flash (US patent
No. 3,588,523), varying the capacitance (switching) of the storage
capacitors (US patents Nos. 3,751,714 and 3,792,309 and West German
patent No. 102,546), and introducing an additional active ballast
in the discharge circuit (British patent No. 1,334,120).

     In addition to neon warning lamps or light-emitting diodes (US
patents Nos. 3,532,961, 3,569,779, 3,586,906, 3,764,849, and
3,890,538, British patent No. 1,334,120, and West German patent No.
1,940,383) and measuring devices, devices for indicating the operating conditions of equipment sometimes contain very simple calculating circuits, such as a special calculator connected to the exposure meter of an electronic photoilluminator (US patent No.
3,723,810). This calculator shows whether the voltage across the
storage capacitor is sufficient to produce the required integrated
luminous intensity or whether the photoexposure is adequate (US
patent No. 3,993,928). One photoflash design (US patent No.
3,738,239) uses an electroluminescent panel for indication. The
panel both provides illumination for the calculator and is a load
for a one-cycle converter after the charging of the storage capacitor
is completed. Some multiple-flashlamp illuminators [9-70] have indicators for each flashlamp. These burn for some period of time
after a flash and make it possible to determine whether all flashlamps have been triggered normally. An indicator circuit has been

described which points out a flashlamp malfunction or failure and
missed flashes (US patent No. 3,626,401).

Protective devices shut off the entire unit containing flashlamps or individual parts of the unit in case of damage. For example,
if the lamp shown in the circuit in Figure 7-8b enters the continuous-
burn mode, the circuit must be briefly disconnected from the power
supply by means of an electromagnetic relay which breaks the current
of the stationary discharge (to prevent overloading of the circuit
elements with respect to average current).

A backup flashlamp can be connected in place of an inoperative
one by using electromagnetic relays or special thyristors. However,
flashlamps also allow "hot" backup, in which two lamps are connected
in parallel in a single discharge circuit. The operating voltage
and the trigger pulse are applied to both lamps simultaneously.
Either one lamp or both light up. The integrated luminous intensity
of the flash turns out to be about the same in the second case as in
the first, since the energy supplied to the lamp is no different.
The peak luminous intensity is increased and the flash duration is
shortened because of the lower impedance of the two lamps operating
in parallel. More complex units utilize electronic logic elements
which monitor each flash and initiate a discharge in a backup lamp
when a regular flash is missed (British patent No. 1,360,906).

Master pulse generators (MPG) may be self-excited (the frequency
of the output signals is determined by the natural frequency of the
oscillator) or may be synchronizable with an external trigger signal
(the time of an output pulse is synchronized with an external event,
such as the opening of a camera shutter, the phase of the line voltage, etc.). In many cases master pulse generators have several output channels via which signals are supplied to the triggering unit
and to the charger. The pulses in these channels may have different
amplitude, shape, and duration and may be mutually phase-shifted.
For example, a master pulse generator with a two-channel output is
needed to control a circuit with a "flip-flop" arm and a circuit
with an intermediate storage capacitor (see Figures 7-8e and 7-8f).
The first channel carries pulses which switch on the thyratron of
the "flip-flop" arm or the switch for recharging capacitor C_1 onto
capacitor C, while the second channel carries the signal to the triggering unit, which is shifted relative to the pulse in the first
channel by an amount of time sufficient for recharging processes to
go to completion in the circuits. A two-channel master pulse generator which controls the first two triggering units is needed to control a two-cycle circuit (Figure 7-8c). The pulses in the output
channels must alternate with a 180° phase difference. For the circuit in Figure 7-8d, the master pulse generator produces signals for
actuating the triggering unit in the first channel, and in the second
channel it produces rectangular signals of a specific length to open
switch K in the charging circuit (these signals must be generated

synchronously with and just before the pulses in the first channel
(US patent No. 3,543,125)).

Blocking oscillators operating at frequencies ranging from a
few hertz to tens of kilohertz, various relaxation LC oscillators
having a unijunction transistor, a thyratron, a stabilitron, a neon
lamp, or a discharger as the threshold element may be used as master
pulse generators. In two-channel master pulse generators it is convenient to use multivibrators with pulse shapers from the front of
each of the arms. Sine-wave generators using subsequent shaping of
control pulses by means of amplitude limiters and differentiating
circuits also may be employed as master pulse generators in cases
where elevated demands are placed on frequency stability.

When there is external triggering, the same generators or oscillators may be used on standby. The external trigger pulse may cause
one or a series of control pulses to be generated (US patent No.
3,767,969). The methods used to supply synchronizing pulses may
vary: mechanical closure of the control circuit of the master pulse
generators of a photoilluminator; the generation of a pulsed magnetic
field which acts on a pickup coil - e.g., in some types of high-speed
movie cameras or in photoilluminators for underwater photography (US
patent No. 3,711,741); the use of devices synchronized by radio signals - e.g., to control flashlamps mounted on earth satellites
[9-19]; synchronization by using a photocell that reacts to a light
pulse - e.g., to switch on auxiliary photoflashes that are controlled
from the main photoflash on a camera (US patents Nos. 3,751,714,
3,930,184, and 3,944,877); and other methods.

The repetition frequency of the control pulses may be constant,
but it also may vary. A master pulse generator may contain a pro-
gramming system which varies the flash energy and frequency of an
aerial photoilluminator (depending on the altitude and speed of the
aircraft), for example, or which produces interval-coded packets of
flashes for an electronic lock controlled by a photocell (British
patent No. 1,135,422). Master pulse generators having two fixed
flash frequencies may be used for flashing aircraft navigation lights.
One frequency (e.g., 1 Hz) is for flight between airports, while the
second (e.g., 2 Hz) is for the movement near an airport of an air-
plane that has received landing permission.

Devices for overload protection of the storage capacitor usually
contain a threshold element. After the charge on the capacitor
reaches a specified voltage, the threshold element generates a signal which acts on the control units (West German patent No.
1,920,951). When this occurs, either the charger is disconnected
from the power supply (US patents No. 3,772,564, 3,417,306,
3,751,714, and 3,819,983) or from the storage element, or there is
a change in operating conditions so that the voltage rise stops or
slows sharply [9-20 and 9-21], and (US patent Nos. 3,319,146 and

3,515,973). Such devices are used extensively in photoilluminators, in which the storage capacitor is charged through a semiconductor converter from a low-voltage DC source. The feedback system, which cuts off generation by the converter in the charging circuit, is actuated when a neon indicator lamp connected to the storage capacitor lights up (US patents Nos. 3,532,961, 3,569,779, 3,586,906, and 4,001,640), when a stabilitron connected to the capacitor opens (French patent No. 2,155,000), or when the current in the charging circuit decreases to some value (US patent No. 3,764,849). Reference-voltage follow-up systems which regulate the output voltage of the charger are used in cases where the voltage across the storage capacitor must be kept in a narrow range (US patent No. 3,624,446 and West German patents Nos. 2,032,731 and 2,446,960).

One possible method of stabilizing the voltage across a storage capacitor (e.g., in light-signaling devices, in which considerable variations of the flash repetition frequency can be tolerated) is to supply trigger pulses at times when the voltage across the capacitor reaches a specified value. When a scheduled flash is missed in such circuits, the pulse generator automatically is converted to the mode of frequent trigger-pulse repetition. This loads the power supply, which is common to the pulse generator and the charger, to such an extent that the voltage rise across the storage capacitor stops.

We should pay particular attention to special measures for protecting against overcurrents and overvoltages which may develop during the normal and emergency operation of units that contain high-energy capacitive storage elements. Such elements are extremely complicated, costly devices containing a multiplicity of capacitors, switches, cables, collecting bars, and other components. In such apparatus, overvoltages and overcurrents may result not only from breakdowns of insulation, but also from breakages of current-carrying cables, the overheating of fuse links, or the nonsimultaneous triggering of parallel-connected dischargers. Significant overvoltages also may occur under operating conditions when a storage capacitor discharges onto a pulsed light source having sharply varying conductivity (e.g., a source using an exploding wire).

We also should allow for the possible occurrence of considerable overvoltages in low-voltage power and measuring circuits. Potentials of up to a few tens of kilovolts may occur in such circuits as a result of transients during a high-voltage pulsed discharge. These potentials are dangerous to electrical insulation and maintenance personnel and create interference during measurements. Therefore, such equipment requires special protective devices: the use of isolation transformers to power measuring and electrical equipment, the grounding of all high-voltage sections at one point, the use of a grounding circuit of the shortest possible length (in order to reduce the inductance), and others [9-71].

Overcurrents and overvoltages can be eliminated by proper selection of the circuit and by building special monitoring and protective
devices. For example, it should be kept in mind that the series connection of capacitors leads to overvoltages in many cases [9-72 to
9-74]. In determining the number of capacitors that are connected
in parallel with a single discharger, allowance should be made for
the maximum permissible energy for the given type of capacitor (in
terms of the hazard of explosion of a capacitor that is breaking
down because of the discharge of other capacitors that are connected
to it) [9-75]. It is advisable to sectionalize the capacitor bank
in units that have no switches in the discharge circuit, so that
each section has its own load element. In many cases, monitoring
the condition of capacitors during operation makes it possible to
predict breakdown, to replace a defective capacitor, and thus to
avoid accidents and associated consequences. This possibility is
based on the fact that breakdown usually is preceded by a "prebreak-
down" state in the form of partial discharges at points where internal defects have appeared (increased gas liberation in the capacitor is a consequence of partial discharges). Some types of
foreign capacitors are fitted with appropriate signaling devices
[9-76]. A bellows 2 (Figure 9-44) is placed inside the capacitor
body. At a certain pressure, the bellows, working through pin 1,
which is connected to it, closes the contacts in the signal circuit.
(At critical pressures, the bellows acts as a valve. Spring-type
separators are provided for this purpose between the capacitor body
and the body of the bellows.)

Partial discharges also can be used for electroacoustical methods
of monitoring [9-77 to 9-80]. (Devices using contact pickups mounted
on the walls of the container can be used for large capacitors,
while devices containing microphones mounted on a group of capacitors
can be used for small capacitors. The pickup signals are fed through
amplifiers to signal lamps, telephones, loudspeakers, or oscillographs.)

Current-limiting resistors, which shunt dischargers and fuse
links, may be used to protect capacitors from the accidental currents
which occur when insulation breaks down and which can lead to an explosion, to ignition, or to short-circuiting of the charger when
there is a large number of parallel-connected capacitors.

Protective resistors included in the circuit of each capacitor
or group of capacitors limit the accidental current, but lower the
efficiency of the unit. They are unacceptable when the highest possible current must be obtained from the storage device.

Bridging gaps close the circuit which bypass the fault and col-
lect part of the current to ease the emergency conditions. The
design of a protective discharger is shown in Figure 9-45a. In it
one of the electrodes is connected to spring 1, which is compressed

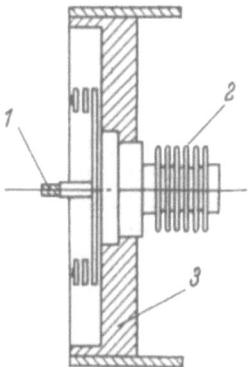

Fig. 9-44.   Pressure-change signaling device.   1) Pin; 2) bellows;
             3) cap of capacitor body.

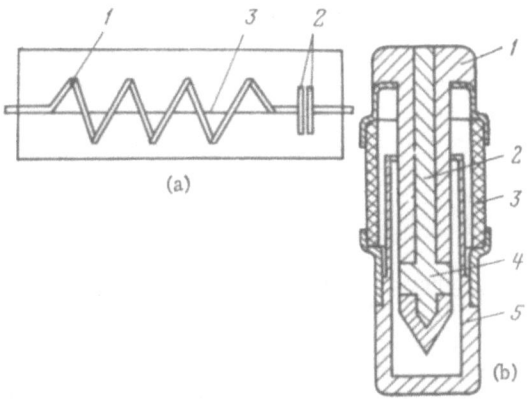

Fig. 9-45.   Protective dischargers.   a) With a self-shorting dis-
             charge gap.   1) Spring; 2) electrodes; 3) fuse link.
             b) With a fusible filler.   1, 5) Electrodes; 2) fusible
             filler; 3) ceramic insulator; 4) openings in electrode 1.

by means of wire 3. In case of an overvoltage, the gap between the
electrodes breaks down, the wire holding down the spring melts, the
spring opens up, and the electrodes are closed, providing a short-
circuit to ground [9-81]. In the discharger shown in Figure 9-45b,
electrodes 1 and 5 are separated by insulator 3. The hollow space
inside electrode 1 is filled by metal 2, which has a low melting
point and high thermal conductivity (e.g., aluminum or copper).
Filler 2 melts under excessive overloads, flows through openings 4,
and shorts the gap between electrodes 1 and 5. The discharger is
small: its diameter is 8.3 mm and its length is 70 mm.

    The voltages across the buses of faulted and unfaulted units
do not vary identically under emergency conditions in high-power

equipment using capacitive storage devices. This (fact) can be used
to design a protective circuit which compares the voltages across
the buses of the units and connects the bypass circuit when an emergency situation arises. The schematic diagram of such a protective
circuit using bypass dischargers (a "crowbar") is shown in Figure
9-46. The voltages from two adjacent units are fed to an electronic
comparison circuit through coupling capacitors C_p and voltage dividers
R and r. When an emergency situation develops, a potential difference
$\Delta U = U_{C1} - U_{C2}$ appears. At some value of this difference, the comparison circuit generates a pulse for the controller to be triggered
by bypass discharger S_{cr}.

It is quite simple to provide protection by using safety cutouts
or fuses. This method make it possible to prevent the explosion of
a capacitor, to obtain maximum currents, and to preserve the operational status of the storage device after breakdown of the insulation and actuation of the protective devices. (A properly selected
fuse cuts off the accidental current at its leading edge, practically
at the rated voltage.) A sample fuse design is shown in Figure 9-47
[9-82 to 9-87]. The rough minimum permissible diameter of the fuse
link can be determined from the formula (d is given in meters)
[9-83]

$$d = \sqrt{\frac{4i_p^2 \tau\beta\rho_0\left[1 + (\omega\tau)^2\right]}{\pi^2 c\delta\,(\omega\tau)^2\ln\left[1 + \beta\,(\theta - \theta_0)\right]}} \,, \qquad (9\text{-}23)$$

where i_p is the peak current in the link (A); τ is the time constant
of the discharger (s); β is the temperature coefficient of resistance
of the fuse link (°C^{-1}); ρ_0 is the resistivity of the fuse link (Ω-m); ω
is the angular frequency (rad/s); c is the specific heat of the fuse-
link material (J/(kg-°C)); δ is the density of the material (kg/m^3);

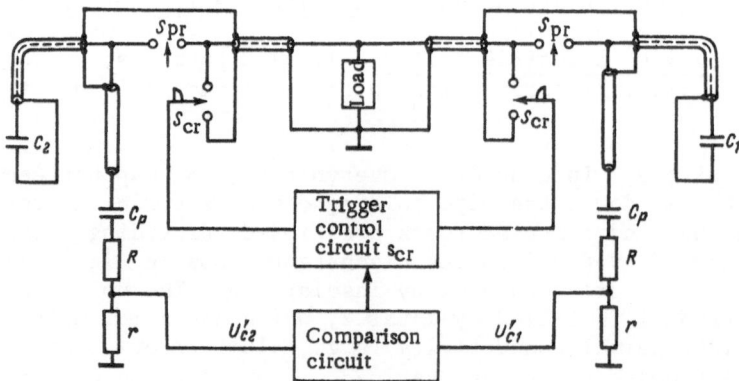

Fig. 9-46.   Protective circuit for capacitors. C_1 and C_2 are blocks
            of working capacitors; C_p are coupling capacitors; S_{pr}
            are the primary (startup) dischargers; and S_{cr} are
            bridging dischargers.

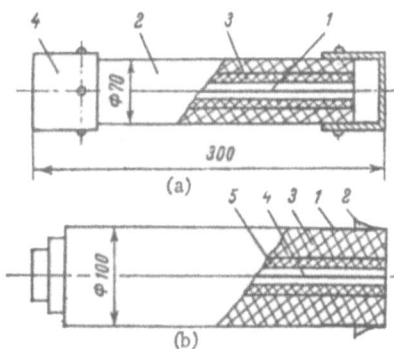

Fig. 9-47.   Quick-break high-voltage fuses.   a) 5-kV fuse: 1) fuse
            link; 2) body; 3) wax; 4) outlet chamber.   b) 50-kV fuse:
            1) insulating rubber; 2) reverse conductor; 3) glass-
            textolite tube; 4) fuse link; 5) wax.

θ is the permissible temperature to which the link can be heated
($^{\circ}$C); and θ_0 is the initial temperature of the link ($^{\circ}$C).

    The selection of a fuse link with respect to accidental current
usually is done experimentally for each specific circuit.   In some
cases standard high-voltage cutouts can be used to protect a group
of capacitors [9-84].   High-capacitance pulsed capacitors designed
for operating voltages up to 10 kV usually contain internal sectional
fuses.   In this case the breakdown of one or even several sections
of a capacitor should not lead to failure of the capacitor, since
the fuses disconnect the faulted sections.   However, as experience
shows, internal fuses are not always effective enough in units that
have a large number of capacitors in parallel.   Therefore, external
(individual or group-type) safety cutouts or other protective means
must be used.

    The explosion of a capacitor also can be prevented by using an
explosion-proof capacitor body.   One possible design is shown in
Figure 9-48.   The cap of the body deforms when there is strong internal gas liberation due to breakdown of the capacitor.   Strip 6
pushes against the edges of the cap.   As a result, flexible conductors
4 break and the capacitor sections are disconnected from the power
supply.

    Special attention should be paid to protection problems in
large units that have unsectionalized energy storage devices (e.g.,
an inductive storage element).   In this case, when many flashlamps
(or groups of lamps) are connected in parallel, the failure of some
may lead to a significant increase in the load on operating lamps
or to their failure.   One possible protective circuit is shown in
Figure 9-49 [9-88].   When switch 6 is closed, inductive storage
device 1 is charged by current source 5.   At the end of charging,
the switch disconnects the storage device from the source and at the

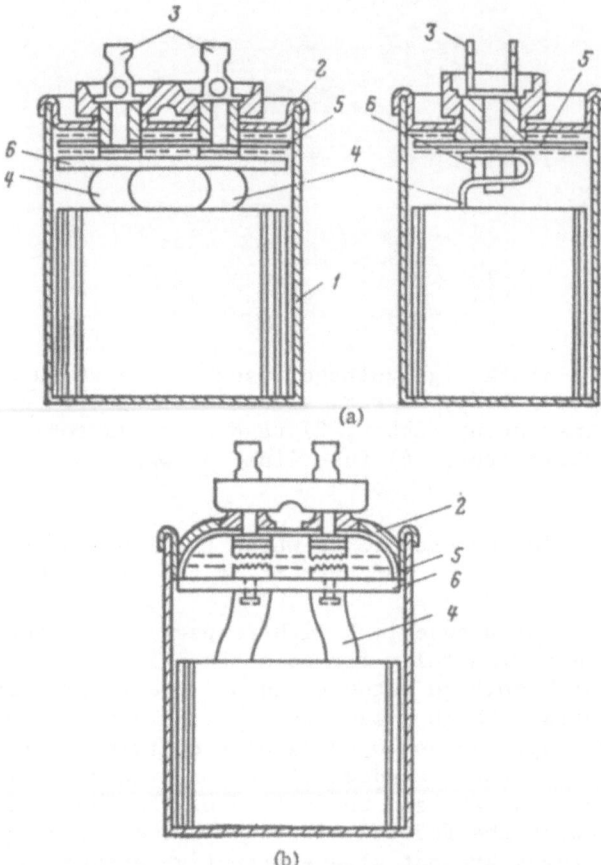

Fig. 9-48. Explosion-proof capacitor (US patent No. 3,248,617).
            a) In normal state; b) when pressure increases above
            the permissible pressure because of a breakdown between
            plates. 1) Body; 2) cover; 3) lead-outs; 4) flexible
            conductors; 5) flexible insulating separator; 6) narrow
            rigid strip.

same time trips time relay 7 and triggering unit 11 of the flashlamps
(the load). The inductive storage device begins to discharge onto
the load. The time relay actuates the control-pulse generator (CPG)
8 after some period of time that is necessary for a discharge to
develop in flashlamps (usually ranging from a few tens to a few hundred microseconds). The control pulses go to all current-recording
cells 10. The same cells also receive pulses from current sensors
9, which register the current in the branches of load 2. If some
cell 10 has not received a pulse from the appropriate current sensor
9, then a control pulse is sent from the control-pulse generator to
the output of cell 10 and then to discharger triggering unit 4. The
triggering unit actuates discharger 3, which goes into operation and
bypasses load 2.

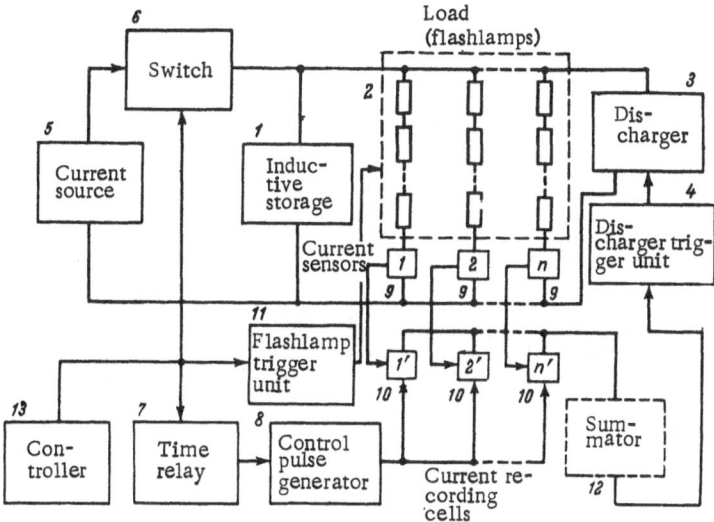

Fig. 9-49. Block diagram of the protective system of flashlamps.

If some number of failures is tolerated out of the total number of load branches, then a summator 12 is included in the circuit. The summator stores information on the number of failures of parallel branches of the load and switches on discharger triggering unit 4 if the number of failures exceeds a specified level. The summator can be based on the principle of simultaneous summing or on time-sequenced polling of the state of cells 10 by using a scaling circuit.

The time spread of the actuation of dischargers must not exceed some specific value in some cases involving the use of units that have a large number of dischargers operating in parallel. Figure 9-50 shows a circuit for monitoring the triggering lag of dischargers in the Scyllac unit (United States) [9-89]. The contacts of relay K1 are closed in the starting position, and K_2 and K_3 are open. A signal u_R, which trips the charging-pulse generator, is generated when the gap firing signal is received by the trigger registration sensor. The amplified charging pulse u_C goes from the amplifier through resistor R_C, diode D3, and contacts K1 and charges measuring capacitor C_T. The time constant of the charging circuit $R_C C_T$ is selected so that it is large compared with the pulse length u_C. Therefore, the voltage across C_T increases almost linearly in time.

A signal u_L is generated in the sensor mounted on each discharger when the gap fires. The positive component of this signal, which exceeds the threshold value of −4 V, is rectified by means of circuit D1-C3, and signal u_s is formed as a result. This signal closes diode D3 and the charging of C_T comes to a halt. The voltage across C_T turns out to be proportional to the trigger lag of the discharger.

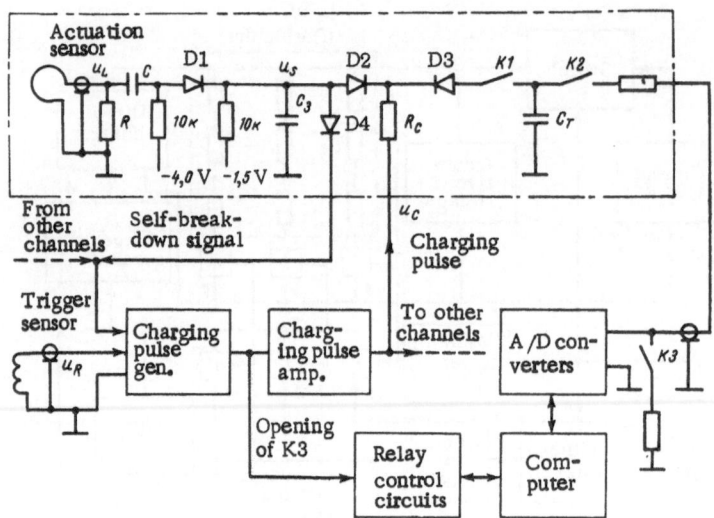

Fig. 9-50.   Discharger-triggering monitoring circuit.

In order to prevent the discharging of capacitor C_T through D3, all contacts K1 (every measuring channel contains contacts K1 and K2) are broken immediately after the capacitor bank fires (after approximately 1 ms). Then the contacts of each relay K2 successively close, and the voltage from the corresponding capacitor is read in order by means of A/D converter ADC. After all signals are read, the contacts of relays K2 and K3 close and all measuring capacitors C_T discharge. Then the contacts of relays K2 and K3 reopen, while those of K1 close, and the circuit is ready to start the next measuring cycle.

If just one of the dischargers fires spontaneously before the trigger pulse is supplied, then a signal u_L appears in the corresponding lag-measuring circuit in the absence of a signal u_R. The signal u_s formed in this circuit passes through diode D4 and is used to actuate the charging-pulse generator. Capacitors C_T are charged to the limiting voltage in circuits for measuring the firing lag of all dischargers except the one that triggered spontaneously. (The capacitor in this channel will not be charged at all, since throughout the entire charging pulse u_C the circuit contains the signal u_s, which cuts off the charging circuit.) Then the voltages across capacitors C_T are read in the usual fashion, and as a result the gap that fired spontaneously is readily determined. If it is unacceptable for one or several gaps to fire, then a premature-triggering signal can be used to fire all other dischargers. This circuit can be used for a large number of channels (tens of hundreds).

Figure 9-51 shows the circuit of one channel of the monitoring system for Burevestnik ignitron dischargers operating in parallel [9-90]. The trigger pulse initiates reset-pulse shaper S1. The

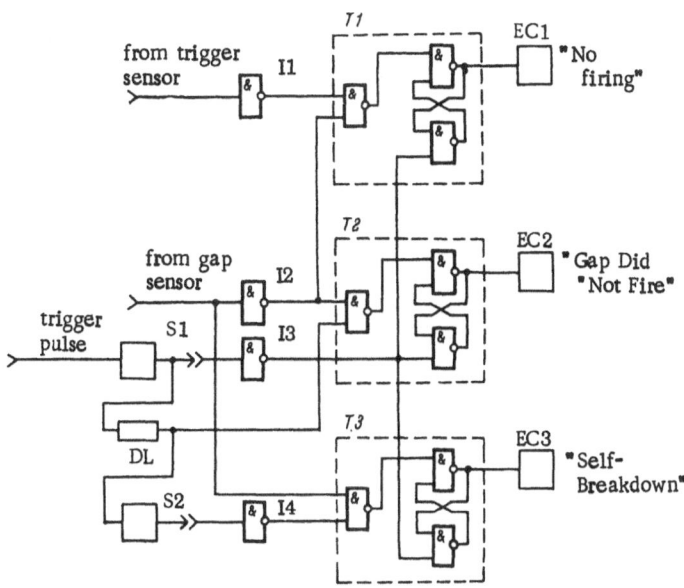

Fig. 9-51. Block diagram of the monitoring channel. S1, S2 -
shapers; I1 to I4 - inverters; T1 to T3 - triggers
with logic modules; EC1 to EC3 - signal units.

reset pulse is supplied to inverter I3 and at the same time to delay
line DL. The inverted reset pulse, taken from inverter I3, sets
triggers T1 to T3 in their initial position. The pulse from delay
line DL, delayed by 4 µs, is used as reference gate 1, and also triggers the shaper of gate 2 (S2). When the gap fires, a signal from
the corresponding sensor goes to the input of inverter I2. The inverted signal goes from I2 to the summing input of trigger T2 and
is added to reference gate 1. Gate 1 has a length shorter than that
of the inverted pulse from the sensor. Therefore, the trigger remains in its original state and the signal lamp does not light. In
the absence of a signal from the ignitron sensor, only reference
gate 1 acts on the trigger input, resetting the trigger to a new
state, and the "Gap Did Not Fire" lamp lights up. The "No Firing"
circuit works analogously.

Gate 2, which is 500 µs long, closes trigger T3 for the duration
of the discharging of the capacitor bank. If a gap fires prematurely
(spontaneously), then trigger T3 is reset by the signal from the
ignitron sensor and the signal lamp "Self-Breakdown" lights. The
signal also can be used in the protective circuit of the charger.

Integrating Rogowsky loops are used as the sensors in this
system. The use of integrated circuits enables the system to operate
at the level of 120 ns.

CHAPTER 10

THE INTERACTION OF RADIATION PULSES WITH DETECTORS

10-1. Pulsed-Radiation Detectors

The width of the spectral interval that is characteristic of
flashlamps (160-4500 nm) and the range of their flash durations (from
fractions of a microsecond to tens of milliseconds) make it possible
to use them in conjunction with all sorts of radiation detectors:
the eye and other biological objects; photographic materials and other
photochemical media; vacuum photoelectric devices, such as photocells,
photomultipliers, and image converter tubes; semiconductor photoelectric
devices, such as photovoltaic cells, photodiodes, phototransistors, and
photoresistors; thermal detectors, such as thermoelements, bolometers,
thermoresistors, and calorimeters; and photoluminescent and laser media
(solid, liquid, and gaseous). These detectors may be subdivided in the
usual way into surface and body-type detectors and into spectrally
selective and nonselective detectors. When such detectors are exposed
to pulsed radiation, the dynamic properties of the detectors become
especially important. These properties may be described in terms of
the corresponding characteristics: transient response, pulse characteristics, and frequency response. An extensive literature is devoted
to the detailed analysis of these characteristics [10-1 to 10-13].
The two extremes of dynamic characteristics which can be approached
by the specifications for pulsed radiation detectors are the ideal
no-lag detector and the ideal slow-responding detector. In the first
idealized case the shape of the detector's reaction pulse is similar
to the shape of the acting radiation pulse. Since the equal peak
reaction recorded by the fast-responding device can be produced by a
very short flash having low radiant energy and a prolonged high-energy
flash, the corresponding source must have the shortest possible flash.
In the other idealized case the effect due to the flash is determined
solely by the energy of the radiation incident on the detector,

regardless of the flash duration. Here the quest for economy, compactness, and long operating life of the light source requires that the
discharge conditions be eased, i.e., that the flash be broadened.
Under certain conditions, real detectors, combined with recording
devices, approach one of these extreme idealized cases, while under
other conditions they have a mixed reaction to flashes: their signal
may depend directly on two parameters and may differ, for example, at
identical energies but for different flash durations. At the same time,
they may not be determined by the peak radiant flux alone.

    The foregoing can be explained by examining the sample equal-
signal graphs in Figures 10-1 to 10-4. It is convenient to use such
graphs to characterize objects which interact with pulsed light sources.
The values of the flash durations are plotted along the abscissa of
the equal-signal graph, while the exposure required for a given flash
duration in order to obtain a specified signal (or effect) is plotted
along the ordinate. The first of these idealized cases is characterized by a straight line segment having a 45° slope. Given a linear
scale along the axes, the extension of this segment passes near the
origin (Figures 10-1 and 10-3). The exposure on this segment increases linearly with increasing flash duration, i.e., the signal
does not change if the peak illuminance is constant. The second
idealized case pertains to the horizontal segment. The signal is
constant on this segment, regardless of the flash duration, as long
as the exposure is constant. All other segments of this characteristic of an object receiving radiation correspond to intermediate
cases. If the equal-signal graph of an object has an ascending

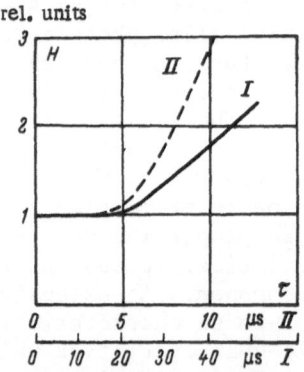

Fig. 10-1. Equal-signal curves of photodiodes. The solid line, scale
            I along the abscissa, is for a germanium photodiode (a
            supply voltage of 16 V and a 10-kΩ load resistor). The
            dashed line, scale II, is for a silicon photodiode (15 V,
            500 kΩ). The exposure H necessary to obtain a specified
            peak voltage across the load at flash durations τ is
            plotted along the ordinate in relative units.

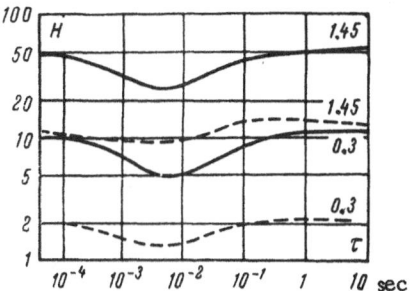

Fig. 10-2.   Sample type-VI equal density lines constructed for two
wavelengths (the solid lines represent 577 nm, and the
dashed lines 400 nm) for two optical densities (indicated
in the figure).   The exposure H is given in arbitrary
units [10-18].

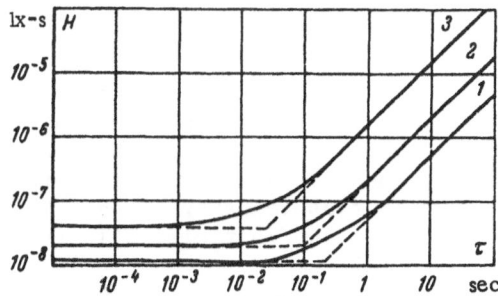

Fig. 10-3.   Equal-signal graphs of the dependence of the exposure at
the pupil of the eye on the flash duration for threshold
stimulation (1), for stimulation corresponding to the
"useful" threshold (2), and for stimulation 10 times
stronger than the illumination from a constant source (3).

branch, then pulsed light sources have an advantage over continuous
sources in all optical problems in which the use of short-duration
illumination does not entail a fundamental limitation of the radiation-
detection time (e.g., during the photography of a moving object, in
location devices, etc.).   The gain in exposure and hence in the power
consumption then can compensate for the reduced efficiency of the
power supply and for its greater complexity, which are due to the
pulsed mode of operation.

A radiation detector has specific properties in some cases:
for example, when flashlamps are used in lasers, photochemistry,
biology, and printing.   The investigation of a family of equal-signal

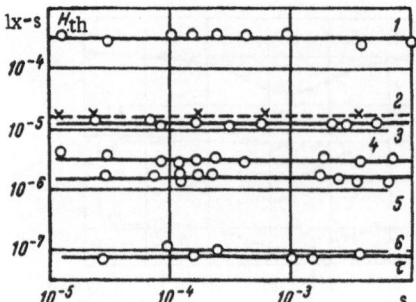

Fig. 10-4. Equal-signal graphs of the dependences of the threshold
           point brilliance H_{th} (the exposure at the pupil) on the
           flash duration of flashlamps [10-21] for different
           brightness of the adaptation background. 1) 300 cd/m^2;
           2) 13 cd/m^2; 3) 1.3 cd/m^2; 4) 0.13 cd/m^2; 5) 2.5 x 10^{-3}
           cd/m^2; 6) 0 cd/m^2. The solid lines are for a source with
           an angular dimension of 5.3', and the dashed line is for
           9.5'. The absolute values of the brightness are rough
           values.

graphs (which in general are similar to the ones presented in the
figures) that correspond to different operating conditions of such
a detector obviously is a special subject which cannot be covered
in detail here. However, the same standard detectors are employed
in many other cases in which flashlamps are used. Here it would be
useful to give a brief idea of the characteristic form of the equal-
signal graphs for these detectors.

Electronic photocells using the external photoelectric effect
and photoelectric multipliers come closest to idealized detectors of
the first type. Under certain conditions, their family of equal-
signal graphs is a bundle of straight lines which radiate from the
origin with slopes proportional to the peak signal. The slopes may
be proportional, for example, to the peak voltage across the load
resistance or to the maximum deflection of the beam of an electronic
oscillograph. The specific conditions are [10-1]:

a) The entire circuit through which the signal passes (the primary
   circuit of the detector, the amplifier, and a recording device
   such as an oscillograph) must be of a high enough frequency to
   pass the signal without distortion. Attention should be paid to
   ensuring sufficiently low resistance of the photocathode and its
   base and to ensuring that the photocathode is not fatigued. The
   use of photocathodes on a metal base make possible the undistorted
   conversion of the shortest illuminance pulses into electrical
   signals.
b) The peak illuminance of the photocathode must not exceed some limit,
   so that the space charge of the electrons ejected from the photocathode does not affect the photocurrent. The higher the power-

supply voltage, the later the electron current from the cathode due to the space charge becomes saturated in all electric vacuum devices. However, the supply voltage also must not be too high, so that it does not cause distortions due to secondary electron emission unforeseen in the design of the device or to other types of amplification of the photocurrent.

A number of devices, including instruments for measuring the peak luminous intensity and peak luminance of flashlamps, are based on the use of electronic photocells and photomultipliers as idealized detectors of the first type [10-1, 10-3 to 10-6, 10-14, and 10-15]. At the same time, electronic photoelectric devices can be used to create idealized detectors of the second type. To do this, a capacitor is connected in series with a vacuum photocell or a photomultiplier in the circuit. The capacitor is charged by the photocurrent throughout the entire flash and thus acts as an integrating element. The equal-signal graph of the voltage across the capacitor, which is proportional to the exposure of the photocathode, is a horizontal line within the linearity range of the lux-ampere characteristic of the photocell. A number of devices for measuring the integrated luminous intensity of pulsed light sources are based on this principle [10-1, 10-3 to 10-6 and 10-14 to 10-16]. The characteristic lag properties of modern electronic photocells and photomultipliers designed to register and measure pulsed radiation begin to be exhibited only in the nanosecond range of pulse lengths [10-1 to 10-6 and 10-17], i.e., outside the basic range of flashlamp flash durations.

Given sufficiently long flashes, semiconductor detectors (photodiodes) behave like idealized detectors of the first type: the peak signal generated is proportional to the peak illuminance and is independent of the flash duration. By contrast, for shorter flashes semiconductor detectors behave like idealized detectors of the second type: the signal generated is determined by the exposure of the photodiode over the entire flash, regardless whether the exposure came from a short, intense flash or from a weaker but longer flash. The time constant of conventional industrially produced germanium and silicon photodiodes, which characterizes their lag properties, usually is on the order of $10^{-5}-10^{-7}$ s [10-2 and 10-17]. The time constant decreases to 10^{-8} s as the light-sensitive area and the depth of the p-n junction of the photodiode decrease. Thus, the equal-signal curves of different types of photodiodes loaded with different resistances may differ considerably. Sample equal-signal graphs are shown in Figure 10-1 for the two types of photodiodes, the linearity of their light characteristics up to high levels of illuminance, the continuing increase in their operating speed, and the expansion of product variety, we may assume that they are promising detectors for use with flashlamps. They can replace electronic photocells and photomultipliers in many recording and measuring instruments.

The equal-signal graphs of a <u>photographic emulsion</u> are called
"lines of equal density." The simplest law of interchangeability was
the original Bunsen-Roscoe law. According to this law, the darkening
of a photographic emulsion is determined by the exposure H = Eτ, and
a decrease in the illuminance E can be compensated for by increasing
the exposure time τ. (This law essentially corresponded to the hypothesis that an emulsion is an idealized detector of the second type.)
The law later was refined by introducing the "Schwartzschild correction," and still later by introducing the concept of a complete family
of lines of equal density which characterizes a given photographic
emulsion [10-10 to 10-12]. Since some photographic emulsions have
lines of equal density with horizontal segments that fall in a rather
wide range of exposure times, "photographic photometry" becomes possible. Such photometry also can be used on pulsed light sources
[10-7]. However, the complexity of the measurement process, the
nonuniformity of a photographic emulsion, the deviations of lines of
equal density from a horizontal path (the deviations vary in different
spectral regions and for different optical densities), and other factors usually make this method inconvenient and insufficiently accurate
[10-18].

The Schwartzschild correction involves the use of the following
equation instead of the constant product Eτ implied by the Bunsen-
Roscoe law:

$$E\tau^{1+p} = \text{const.} \tag{10-1}$$

Here p is a quantity called the Schwartzschild constant, which
is small compared with unity. On the horizontal segment of the line
of equal density p = 0, on the descending branch p > 0, and on the
ascending branch p < 0. We may estimate quite roughly for most
emulsions that p = 0 in the interval $10^{-3} < \tau < 10^{-2}$ s; p = 0.1-0.2
for $\tau < 10^{-4}$ s; and p = -0.1 for $\tau > 1$ s. However, a more exact
approach requires that a separate value of the Schwartzschild constant
be adopted for each small interval of τ, each interval of density,
and each spectral interval. Thus, the Schwartzschild constant becomes meaningless. An example of lines of equal density obtained
for different spectral regions and different optical densities is
shown in Figure 10-2. It follows from this figure that a photographic
emulsion is not an idealized detector of either the first or second
types, but rather is a mixed-type detector. The optimal (from an
energy standpoint) flash duration for a pulsed source intended for
photographic illumination is a few milliseconds when the lines of
equal density have a minimum (p = 0). This minimum is more pronounced
for some emulsions, and less so for others. There are emulsions for
which the lines of equal density in the region up to 10^{-5} s do not
have an upswing on the left side (they are horizontal in the range of
τ from 10^{-5} to 10^{-2} s). When such emulsions are used, the flash
duration of a light source may be selected arbitrarily within this
range. Data have been published to the effect that the Bunsen-Roscoe
law usually is fulfilled for a photographic emulsion in the range of
τ from 10^{-7} to 10^{-5} s.

Thermoelements or thermopiles are spectrally nonselective radiation detectors which come closest to an idealized detector of the second type over a broad range of flash durations. The long thermal lag of thermopiles makes it possible to use them for reliable integration of the incident radiant flux. If a galvanometer operating under ballistic conditions is used, a thermopile makes it possible to obtain a signal that is strictly proportional to the radiant exposure on its working area. The same linearity obtains when there is considerable variation of the total duration of the light signal: e.g., during flashes of two flashlamps that occur either simultaneously or at 0.1-s intervals. Hence a thermopile can be used for photometric purposes [10-18]. Nonselective thermal detectors of other types, such as a conical calorimeter [5-37], also may be used for spectrally integrated radiant measurements of flashlamps.

The human eye is a mixed-type radiation detector. According to Blondel and Rey [10-10] and the latest research by many investigators [10-9], the equal-signal graphs for the eye (Figure 10-3) can be expressed by the empirical relation

$$H = (a + \tau) E_{eff}. \qquad (10-2)$$

Here $H = \int E\, dt$ is the exposure on the pupil (the "point brilliance") required to produce a specific visual stimulus (e.g., the threshold stimulus or a stimulus which exceeds it by a factor n); E_{eff} is a constant equal to the illuminance from a constant light source which produces the same stimulation (in the case of the threshold point brilliance E_{eff}, this is the threshold illuminance as perceived from a constant light source); τ is the flash duration; and a is a constant that depends on the level of stimulation. For threshold stimulation corresponding to an illuminance of about 5×10^{-8} lx from a constant source, $a = 0.21$ s. The value decreases for stimuli above the threshold, coming to approximately 0.1 s at the "useful threshold," which corresponds to a constant illuminance of 10^{-7} lx, and to 0.02 s for stimulation corresponding to a constant illuminance of 10^{-6} lx. The eye may be considered an idealized detector of the first type for τ longer than 1 s for threshold point brilliance, and longer than 0.1 s for point brilliance substantially above the threshold. The eye may be considered an idealized detector of the second type for τ below 2×10^{-2} and 10^{-3} s, respectively. In the latter region visual photometric measurement of pulsed light sources (by the usual methods of field comparison) may be used if the flash repetition frequency is high enough that the eye does not perceive the discontinuity of the light (over 50–80 Hz).

Up to some point in time the literature contained contradictory data on whether the eye corresponds to an idealized detector of the second type in the region of extremely short flashes (shorter than 10^{-3} s). A strictly horizontal position of the equal-signal plots was found in [10-9 and 10-20 to 10-23] for different sizes of the

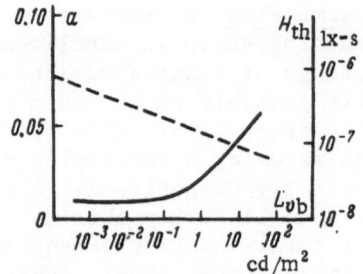

Fig. 10-5. Dependence of the threshold brilliance H_{th} for short
            flashes (solid line) and of the quantity a (dashed line)
            on the brightness L_{vb} of the visual adaptation background
            [10-9].

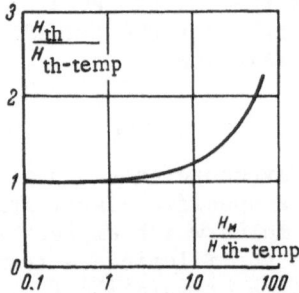

Fig. 10-6. Dependence of the threshold point brilliance on the
            exposure H_M produced on the pupil by an interfering
            pulsed light source with an angular width of 9.5',
            positioned at an angle of 9° to a weak pulsed light source;
            $H_{th-temp}$ is the threshold point brilliance when $H_M = 0$
            [10-21].

glow spot, different luminances of the visual adaptation background
and different levels of stimulation above the threshold. In particular, this conclusion was drawn in [10-22] by constructing threshold plots for flashlamp flashes (Figure 10-4). Figures 10-5 and 10-6
show the dependences of the threshold point brilliance on the brightness of the adaptation background and on the exposure $H_M = \int E_M \, dt$
which was produced on the pupil by a second (interfering) pulsed
light positioned at some angle relative to a weak source.

Allowing for the proportionality between the illuminance and
the luminous intensity of the source, we can transform relation
(10-2) into the formula

$$I_{eff} = \frac{\theta}{a + \tau},$$

(10-3)

where I_{eff} is the luminous intensity of a constant source. This is equal to a given pulsed source having an integrated luminous intensity θ with respect to the efficiency of the source.

When $\tau \ll a$ (which practically always is true for flashlamps), this expression assumes the form

$$\frac{I_{eff}}{f\theta} = \frac{1}{fa},$$

(10-4)

where f is the flash frequency.

If the luminous efficacies of the pulsed and continuous light sources are included (including the pulsed power supply), then the ratio at left is equal to the ratio of the electric power consumed by the pulsed light source to that consumed by a continuous light source of equal efficiency. Therefore, the quantity $1/fa$ represents the energy gain produced by a pulsed source because the equal-signal graph of the eye has an ascending branch. For example, for a signal frequency of 0.5 Hz (one flash every other second) and a stimulus equal to the useful threshold ($a = 0.1$ s), we find from relation (10-4) that the energy gain for a pulsed light source is a factor of 20. For flash frequencies above 10 Hz, a pulsed source no longer produces an energy gain compared with a continuous source. (A smooth transition to Talbot's law is observed. According to Talbot's law, the effective luminance of a pulsed light source is equal to its time-averaged luminance [10-9].)

It was shown in [10-24] that the constant a in equation (10-4) does not change when a pulsed light source acts on the eye through a turbid medium (a water mist) which strongly dissipates light. Thus, indications of the supposedly large gain produced by pulsed light sources when observed through a mist or fog were refuted.

In considering the interaction between a light source and the human eye, we should draw attention not only to the question whether a signal is adequate for perception, but also to whether the signal may cause excessive stimulation (a blinding effect). The study of the blinding effect of pulsed light sources [10-25 and 10-26] has showed that the blinding time (the time interval from the time of a flash until the observer begins to distinguish the test object) increases from 4 to 50 s as the exposure on the pupil increases from 5 to 150 lx-s when the brightness of the adaptation background is $L_{vb} = 2 \times 10^{-4}$ cd/m^2. The nature of the test object can alter the blinding time by a factor of 2-3. The blinding time is halved if the angular size of the blinding object is quadrupled. When $L_{vb} \geq 0.03$ cd/m^2, the blinding time is practically zero for all exposures up to 640 lx-s. Reducing L_{vb} below this level causes an increase in the blinding time that is inversely proportional to the logarithm of L_{vb} (Figure 10-7).

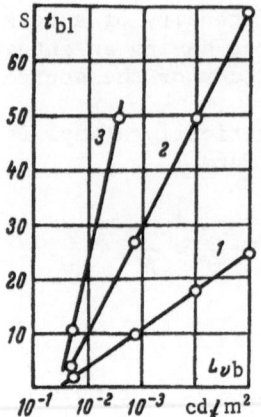

Fig. 10-7. Dependence of blinding time on brightness of adaptation
background [10-24]. The angular size of the source is
20'. 1) Exposure on the pupil is 40 lx-s; 2) 155 lx-s;
3) 640 lx-s.

Laser media are three-dimensional detectors that have complex
dynamic properties. The ultimate effect of the generation of narrowly
direction monochromatic radiation is intricately dependent on the
shape, length, and energy of the optical exciting pulse produced by
a flashlamp that has a reflecting system. The excitation of the medium
that is required for lasing (a population inversion of the working
levels of the atoms of the medium) occurs only when some threshold
value of the flashlamp's radiant energy is exceeded. Lasing generally
stops even before the end of the exciting pulse when the decreasing
radiant flux of the flashlamp that is absorbed by the medium proves
insufficient to maintain the population inversion. The radiant energy
emitted by the flashlamp after this time thus is expended uselessly.
Therefore, operating conditions which ensure steep leading and trailing
edges of the radiation pulse are more efficient from the standpoint
of energy. Furthermore, the efficiency of a laser depends on the
ratio of the exciting pulse length τ to the luminescence time constant
τ_f of the laser medium. Calculations show that increasing τ/τ_f from
the minimum value to 4 reduces the efficiency of the laser by a factor
of approximately 5 (Figure 10-8).

10-2. Methods of Measuring the Parameters and Characteristics of Pulsed Light Sources

In measuring the radiation characteristics of pulsed light
sources, one must take into account that the shape and length of
their radiation pulses are different for different parts of the
luminous body and are dependent on the direction of observation and
on the wavelength [5-1]. For example, the length and shape of the

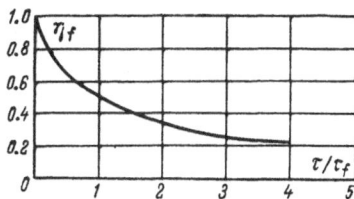

Fig. 10-8. Calculated dependence of the luminescence factor η_f of the efficiency of a laser on the ratio of the flashlamp's pulse length τ (at the 0.5 level) to the luminescence time constant τ_f of the laser medium [2-26].

luminance and luminous-intensity pulses of spherical flashlamps are not the same. The length and shape of pulses of a spectral line and of the adjacent radiation background of gas-discharge tubes also differ substantially.

Mainly photometric quantities are used in the photometry of continuous radiation sources. Since pulsed sources are used not only in interactions with the eye, pulsed-photometry practice more and more urgently requires the measurement of radiometric parameters in various precisely defined spectral intervals. To do this, extremely high-precision spectral absorbers and fast-responding nonselective radiation detectors are needed.

However, in technical manuals flashlamps commonly are characterized in terms of photometric rather than radiometric parameters. The reasons for this are as follow [10-27]:

1) there is a detailed international system of photometric measurements which is based on national photometric standards;
2) combined with standard spectral characteristics, photometric parameters characterize quite unambiguously the efficiency of widely used xenon flashlamps in the visible region and the adjacent spectral regions which are of the greatest relevance to practical applications;
3) the radiometry of pulsed sources of incoherent radiation is still in the development stage, and as yet does not have officially certified measuring facilities.

The standardized methods of measuring the luminous parameters of pulsed sources (GOST 22466.1-77) are based on the use of photoelectric detectors [10-27]. The spectral response of these detectors is adjusted to the relative spectral response of the eye by using a correcting absorber. (A detailed treatment of problems related to pulsed photometry based on the use of photoelectric detectors can be found in [10-1], a monograph which may be recommended as a primary source.)

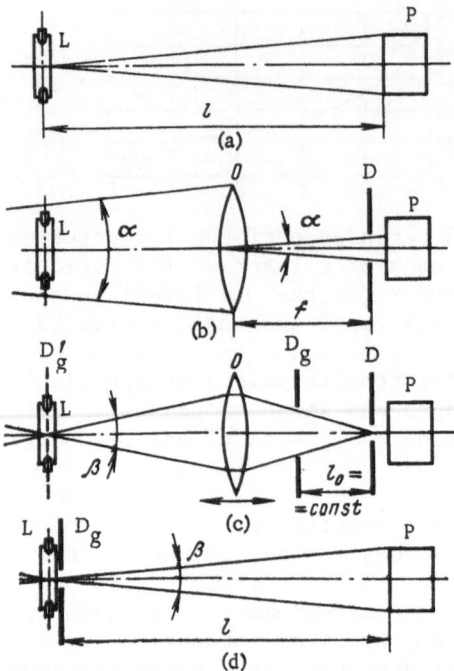

Fig. 10-9. The principal optical schemes for measuring illuminance
            and exposure (a), the luminous intensity and integrated
            luminous intensity with a telecentric system (b), the
            luminance and integrated luminance with a system containing
            a lens (c), and the luminance and integrated intensity
            with a system containing no lens (d). L - flashlamp;
            P - detector; O - lens having a focal length f; D -
            diaphragm in front of the detector; D_g - diaphragm which
            limits the size of the volume of the source that is
            measured photometrically; D_g' - image in the reverse path
            of rays from diaphragm D, which limits the size of the
            volume being measured photometrically; D_a - aperture
            diaphragm which is stationary with respect to the detector.

     The optical diagram of a photodetection device is determined
mainly by the spatial parameters of the measuring problem: the size
of the luminous body and the luminous quantity to be measured (luminous intensity, luminance, illuminance, etc.; see Table 5-1).

     The luminous intensity I(t) at different times and the integrated
luminous intensity θ are determined by two methods: by measuring
the illuminance E(t) and the exposure H (Figure 10-9a), or by direct
measurements by means of a telecentric optical system [3-80 and 10-1].
The quantities E(t) and H are measured in the first case; I(t) and

θ are determined indirectly with a photometer at a distance l from the center of the luminous volume. Then the relations for point light sources are used to make the calculation:

$$I(t) = E(t) \, l^2, \tag{10-5}$$

$$\Theta = H l^2. \tag{10-6}$$

The error given by these relations does not exceed 1% at distances exceeding the 10 largest dimensions [sic] of the luminous body of a source [10-28 and 10-28a].

The telecentric method of measuring I(t) and θ (Figure 10-9b) is based on the use of an aplanatic optical system. From the multiplicity of rays emanating from the test source, such a system singles out bundles of rays which are directed along the optical axis of the system and which are confined in some solid angle. This angle is equal to the ratio of the area of diaphragm D in the focal plane of the entrance objective to the square of the focal length f of the objective. In this case the source being measured photometrically should be placed in front of the objective in a cone having a vertex angle α = d/f, where d is the diameter of diaphragm D.

The essence of the method for measuring the luminance $L_V(t)$ and the integrated luminance pulse $\int L_V(t)dt$ according to the scheme in Figure 10-9c consists in measuring the illuminance E(t) and the exposure H of the optical image of the luminous body that is produced by lens O. The rays within an invariant solid angle limited by stationary aperture diaphragm D_a form the image, regardless of the position of the lens O, which moves during focusing. Therefore, the E(t) and H of diaphragm D are proportional to the $L_V(t)$ and $\int L_V(t)dt$, respectively, of the portion of the source being measured photometrically [10-1]. Such a system has a constant calibration characteristic. If there is no diaphragm D_a, then the photometer must be calibrated after each refocusing. The quantities $L_V(t)$ and $\int L_V(t)dt$ also can be determining without focusing optics by measuring E(t) and H at a distance l from the portion of the source being measured photometrically that is confined by the calibrated diaphragm D_g (Figure 10-9d). Diaphragm D_g should be positioned as close as possible to the luminous volume of the source in order to provide the sharpest possible definition of the area being measured.

When measuring the $L_V(t)$ and $\int L_V(t)dt$ of flashlamps, one should keep in mind the following circumstances, which are absent during the photometry of flat surface sources. The shape and size of the volume being probed depend not only on the diameter of the diaphragm (or of the image of the diaphragm formed by the lens in the return path of the rays) and the angular aperture β of the measuring system, but also only the location of the diaphragm (or the image of the diaphragm) with respect to the luminous volume of the source (Figure 10-10).

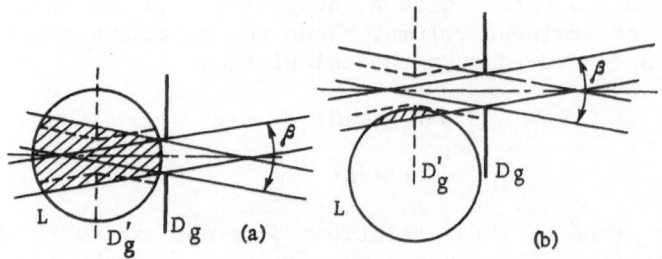

Fig. 10-10.   Measurement of the luminance of three-dimensional
             radiators. L - cross section of discharge column;
             D_g and D_g' - material diaphragm (Figure 10-9d) or the
             image of diaphragm D (Figure 10-9c) in the return path
             of the rays. The hatched areas and the dashed lines
             indicate the probed regions for two positions of the
             limiting diaphragm (or focal plane).

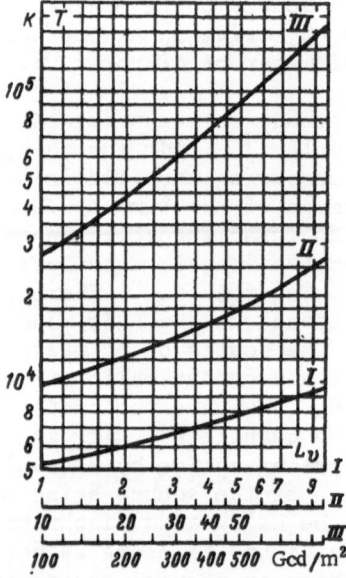

Fig. 10-11.   Relation between the temperature T and the luminance L_v
             of a blackbody.

If the focal plane is in front of or behind the radiator, there may
be an incorrect finding that radiation is present outside the luminous volume (Figure 10-10b). Not only the size of the limiting
diaphragm, but the angular aperture of the measuring system must be
quite small in order to obtain good spatial reoslution in the study
of the spatial dependences of flashlamp luminance.

The relation between the temperature and the blackbody luminance (Figure 10-11) can be used to determine the blackbody temperature of flashlamps that have a practically continuous emission spectrum in the visible region. If the optical density of the discharge column being probed with a photometer in a flashlamp is sufficiently high, then the blackbody temperature may be taken as the plasma temperature.

The luminous flux F(t) and the luminous energy Q can be determined by two methods: by using an integrating globe photometer, or by using the spatial distribution of the luminous intensity and the integrated luminous intensity. The essence of the first method lies in measuring the illuminance E(t) and the exposure H on the inside surface of the photometric globe. These quantities are proportional To F(t) and θ, respectively. The essence of the second method of determining Q, for example, lies in the numerical integration of the spatial distribution of the integrated luminous intensity θ: the indicatrices of the integrated luminous intensity, as measured with a distribution photometer. The distribution photometer is used to rotate the flashlamp or photodetector fixed angles about the center of the luminous volume. After each rotation the average integrated luminous intensity is determined in the given direction for several flashes. The complete indicatrices of the integrated luminous intensity can be determined for several tens or hundreds of successive flashes by virtue of the quite good reproducibility of the parameters of the radiation pulses (see Section 5-9). The indicatrices are used to calculate the equivalent solid angle Ω_{eq} (see Section 5-3), which usually is assumed constant for a given type of lamp and given operating conditions. The luminous energy Q is determined by multiplying the measured integrated luminous intensity θ by Ω_{eq}.

The signal produced by radiation detectors in the photometric devices described above are recorded by means of various electric-measuring and electronic instruments and devices. Their operating principles are examined in detail in [10-1]. Oscillographs (memory-type or with photographic attachments) usually are used to obtain the time dependences of the radiation parameters. The peak reactions of the detectors are measured with oscillographs, special electronic instruments, pulsed peak voltmeters, pulse analyzers, and pulse discriminators. Electronic instruments with input storage-capacitor banks or ballistic galvanometers are used to measure detector reactions integrated over a flash. The calibration characteristics of pulse photometers are determined by using photometric incandescent lamps or reference flashlamps [10-1 and 10-29].

Table 10-1 gives the characteristics of photometers for use with flashlamps. These photometers are made by Soviet industry and have been described in [10-1]. Additionally, extensive use is made of pulse photometers (pulse luxmeters and exposure meters) that consist of a photometric head and standard electric-measuring equipment

Table 10-1. Parameters of Soviet-Made Industrial Pulse Photometers

Model	Quantities measured	Range of measurements	Pulse length, s	Pulse rep. freq., Hz	Type of photo-detector	Size of luminous body, mm	Type of indicator	Basic error of measurements, %
SMI	θ_v	$10\text{--}5\cdot10^5$ cd-s	$10^{-5}\text{--}1$	0	F-13	$\leqslant200$	pointer	15
FIM	L_vb	$5.10^2\text{--}10^7$ cd	$10^{-7}\text{--}10^{-2}$	0	F-22	$\leqslant80$	digital indicator, and output to oscillograph	20
	θ_v L_v	$1\text{--}10^7$ cd-s $2\cdot10^5\text{--}10^{13}$ cd-m^{-2}						
	$\int L_v dt$	$6\cdot10^3\text{--}10^8$ cd-s-m^{-2}				$\geqslant0.2$		
	Pulse shape							
FIS	L_vb Spread of pulses Pulse shape	$5\cdot10^4\text{--}10^8$ cd	$10^{-7}\text{--}10^{-2}$	$0\text{--}3\cdot10^3$	F-22	$\leqslant80$	scale, digital indicator and output to oscillograph	20
ISI*	L_vb	$3\cdot10^5$ cd $7\cdot10^5$ cd	$3\cdot10^{-6}$ $3\cdot10^{-4}$	1/15	–	40 2.5	pulse counter	5
IMO-2**	H_e	$10\text{--}105$ J-m^{-2}	$10^{-6}\text{--}1$	0	conical calorim.	–	pointer	10

* A pulsed photometric source.
** With a calibrated diaphragm at the calorimeter entrance.

(oscillographs, peak voltmeters, and ballistic galvanometers). The
radiation detector (a photocell or a photodiode) and absorbers which
correct its spectral response are enclosed in the metal body of the
head. Removable light absorbers are mounted in front of the entrance
window of the head. The same design principle was used to construct
optometer-type portable multirange instruments containing a silicon
photodiode, such as the 40K OPTO-METER made by the American company
United Detector Technology. These instruments are designed to measure
not only illuminance and exposure, but also radiance and radiant exposure in the spectral interval 450-910 nm.

During measurements, the detector and all electric-measuring
circuits of pulse-photometry instruments must be carefully shielded
from electromagnetic interference due to the discharge through the
flashlamp. That the interference has no effect is confirmed by the
zero signal when the entrance window of the photometer is completely
closed by the opaque nonmetallic screen.

The spectral distributions of flashlamp radiation are measured
by methods analogous to the standard methods used for continuously
burning gas-discharge light sources (GOST 17203-71) [10-30]. The
differences are found only in the recording equipment, which includes
pulsed electric-measuring and electronic instruments. The spectral
distributions of flashlamp radiation are obtained by successive
measurement of the spectral density of the radiation at fixed wavelengths (or in fixed spectral intervals) for individual flashes or,
during repeating-flash operation, for packets of pulses.

The methods of quantitative photographic spectrophotometry are
used comparatively infrequently. (Photographic recording is used
more often to obtain time scans of spectra.)

As yet the methods for measuring the radiation parameters of
coaxial-type lamps [5-6 and 10-1] have not been developed adequately.
In addition to globe and cylindrical photometric probes, the efficiency of the operation of coaxial-type lamps containing laser media
can be estimated by using the luminescent radiation of these media
or the thermal expansion of liquid active media [5-94 and 10-31].

Thermal radiation detectors (thermopiles and calorimeters)
generally are used as radiometers for measuring the radiant parameters
of flashlamps. A conical calorimeter for measuring the narrowly
directional radiation of a pulsed laser can be converted to a high-
current instrument for measuring the radiant exposure from flashlamps
by mounting a calibrated diaphragm at its entrance [5-37]. Closed
calorimeters are used to measure the total radiant energy of flashlamps.

Photometric measurement of flashlamps operating in the frequent-
flash mode ($f \geq 50$ Hz) can be done visually by the method of comparison

surfaces for comparison with a continuously burning lamp (an incandescent lamp). However, the accuracy of the measurements turns out to be low because of the sharply different colors of the comparison surfaces.

The errors in measurements of the radiation parameters of flashlamps are strongly dependent on the procedure used to make the measurements and on the conditions of measurement. For example, luminous and radiant pulse measurements can be made with an error of about 10% under laboratory conditions if a system of measuring equipment and sample sources are available. These make it possible to calibrate the photometer or radiometer immediately before the measurements and to eliminate most of the systematic errors due to deviations of the spectral response from a given response (the standard response) and to differences between the spectral distribution of the radiation of the sample source and the source under study [10-1 and 10-29]. It is difficult to make measurements with a 5% error. The error of measurement lies in the range 10-20% in most of the ordinary cases encountered in practice. (A slight oversight in measuring techniques easily may lead to an error of over 50% [10-32].)

REFERENCES

O-1. I. S. Marshak. Impul'snye istochniki sveta (Pulsed Light
Sources). Moscow-Leningrad, Gosenergoizdat Publishing House,
336 pages (1963).

O-2. A. C. Malliaris et al. Inst. Environment Science 16th Ann. Techn.
Meeting and Equipment Expos. Proc., Boston, Mass.,1970.
Mt. Prospect, Ill., s.a., pp.1-13 (1970).

O-2a. I. S. Marshak. Sovremennye vysokointensivnye istochniki sveta
(Modern High-Intensity Light Sources). Itogi nauki i tekhniki.
All-Union Institute of Scientific and Technical Information
(VINITI), USSR Academy of Sciences (USSR AS). Moscow,
87 pages (1976).

O-3. W. E. Thouret et al. Journal of the Illuminating Engineering
Society, vol. 2, No. 1, pp. 8-18 (1972).

O-4. K. Taujimoto et al. Proc. 9th Intern. Symposium on Space Technology and Science. Tokyo, pp. 495-502 (1971).

O-5. H.-P. Popp, G. Weniger. Lichttechnik, vol. 25, No. 11,
pp. 527-530 (1973).

O-5a. V. G. Baryshnikov et al. In Impul'snaya fotometriya (Pulse
Photometry). Leningrad, Mashinostroenie Publishing House,
No. 4, pp. 151-155 (1975).

O-6. I. S. Marshak et al. Proc. of the 6th Intern. Congr. on High-
Speed Photography. Haarlem, Tjeenk Willink and Zoon,
pp. 134-142 (1963).

O-7. L. E. Belousova. Svetotekhnika, No. 12, pp. 14-15 (1974).

O-8. A. V. Guzhevskaya et al. In Impul'snaya fotometriya. Leningrad,
Mashinostroenie Publishing House, No. 4, pp. 155-157 (1975).

O-8a. S. G. Ashurkov et al. Svetotekhnika, No. 10, pp. 22-24 (1976).

O-9. H. D. Chevali and T. Cote. Lighting Des. and Applications,
vol. 3, No. 12, pp. 25-30 (1973).

O-10. L. M. Nuland and J. Schroder. Dutch patent 78d2 (C 06 1/10),
No. 137,539, application date Sept. 23, 1963, Publication
date Sept. 17 (1973).

O-11. J. D. Harper. Rev. of Sci. Instruments, vol. 39, No. 4,
pp. 509-510 (1968).

433

434   REFERENCES

0-12. J. L. Brewster et al. Proc. of the 19th Intern. Congr. on High-Speed Photography. New York, pp. 303-309 (1970).

0-13. W. H. Fox Talbot. British patent No. 13,664, June, 12 (1851).

0-14. A. Kornetzki et al. Z. für Technische Physik, vol. 14, pp. 274-280 (1933).

0-15. H. E. Edgerton, K. J. Germeshausen. Rev. of Sci. Instruments, vol. 3, pp. 532-542 (1932).

0-16. M. Laporte. Le lampes à éclairs lumière blanche et leurs applications. Paris, Gauthier-Villars, 90 pages (1949).

0-17. H. E. Edgerton et al. Photographic Journal, vol. 76, p. 198 (1936).

0-18. S. Ya. Bogdanov, K. S. Wul'fson. Doklady AN SSSR, vol. 30, No. 4, pp. 309-312 (1941).

0-19. I. S. Abramson, I. S. Marshak. Zhurnal tekhnicheskoi fiziki (ZhTF), vol. 12, No. 10, pp. 632-639 (1942).

0-20. E. G. Shaer. Primenenie fotografii v meditsine (The Use of Photography in Medicine). Moscow, Meditsina Publishing House, 298 pages (1974).

0-21. Electric Rev. (Great Britain), vol. 190, No. 26, p. 930 (1972).

0-22. Internationale Photo-Technik, No. 3, pp. 86-87 (1971).

0-23. H. E. Edgerton. Electronic flash, strobe. McGraw-Hill, New York, 362 pages (1970).

0-24. A. L. Mikaelyan, M. L. Ter-Mikaelyan, Yu. G. Turkov. Opticheskie kvantovye generatory na tverdom tele (Solid-State Lasers). Moscow, Sovetskoe Radio Publishing House, 384 pages (1967).

0-25. A. M. Bonch-Bruevich et al. Optiko-mekhanicheskaya promyshlennost', No. 12, pp. 49-59 (1973).

0-26. B. N. Stepanov, A. N. Rubinov. Uspekhi fizicheskikh nauk, vol. 95, No. 1, p. 45 (1968).

0-27. E. B. Gordon et al. Zhurnal eksperimental'noi i teoreticheskoi fiziki (JETP), vol. 63, No. 4, pp. 1159-1172 (1972).

0-28. Yu. G. Anikiev et al. Kvantovaya elektronika, vol. 2, No. 1, pp. 7-12 (1975).

0-29. G. D. Salamandra. Fotograficheskie metody issledovaniya bystroprotekayushchikh protsessov (Photographic Methods of Investigating High-Speed Processes). Moscow, Nauka Publishers, 200 pages (1974).

0-30. F. Van Vin.† Stroboskopiya (Stroboscopy). Translated from English. Moscow, Energiya Publishing House, 88 pages (1974).

0-31. Elektromeister+Dtsch. Elektrohandwerk, vol. 27/49, No. 2, pp. 56-58 (1974).

0-32. A. L. Vasserman et al. Vsesoyuznaya nauchnotekhnicheskaya konferentsiya "Sovremennoe sostoyanie i perspektivy razvitiya vysokoskorotsnoi fotografii i kinematografii i metrologii bystroprotekayushchikh protsessov". Tezisy dokladov. VDNKh-Vsesoyuznyi nauchno-issledovatel'skii institute optiko-fizicheskikh izmerenii (The All-Union Scientific and Technical Conference on the Current Status and Prospects for the Development of High-Speed Photography and Cinematography and the Metrology of High-Speed Processes. Abstracts of Papers. Exhibition of Achievments of the National Economy of the USSR — All-Union Scientific Research Institute of Opto-physical Measurements). Moscow, p. 111 (1975).

0-33.   The Perception and Application of Flashing Lights. Intern.
        Symposium, London, 1971. Printed by Billing and Sons Ltd.,
        Guilford and London. Published by A. Hilger, 429 pages.

0-34.   N. L. Ivanov and G. N. Senilov. Trudy Moskovskogo energetich-
        eskogo instituta, No. 147, pp. 151-158 (1972).

0-35.   T. Balmer. Design News, vol. 25, No. 2, pp. 78-79 (1970).

0-36.   E. G. Niemann and M. Kleinert. Applied Optics, vol. 7,
        No. 2, pp. 295-299 (1968).

0-37.   Pfail',[†] Porter. Pribory dlya nauchnykh issledovanii (Instru-
        ments for Scientific Research) (translated from English),
        No. 12, pp. 172-174 (1970).

0-38.   Taylor et al. Pribory dlya nauchnykh issledovanii, No. 12,
        pp. 63-66 (1972).

0-39.   I. V. Mes'kin. Fotoelektricheskie preobrazovateli uglovoi
        velichiny v tsifrovoi kod (Photoelectric Converters for
        Converting an Angular Quantity into Digital Code). Leningrad,
        Sudpromgiz Publishing House, 80 pages (1962).

0-40.   Fotoelektricheskie preobrazovateli informatsii (Photoelectric
        Data Converters). Edited by L. N. Presnukhin. Moscow,
        Mashinostroenie Publishing House, 376 pages (1974).

0-41.   J. Rüdiger. Poligraph-Jahrbuch, 1969, 7 Folge. Frankfurt/Main,
        pp. 121-140 (1968).

0-42.   Form und Technik, vol. 22, No. 3, pp. 141-142 (1971).

0-43.   R. W. Graham and R. Bridges. Journal of Photographic Sci.,
        vol. 18, No. 6, p. 244 (1970).

0-44.   P. W. van Maaren. Pract. Mettalogr., vol. 11, No. 9, pp.
        535-547 (1974).

0-45.   Fotolitografiya i optika (Photolithography and Optics).
        Edited by Ya. A. Fedorov and G. Pol'. Moscow, Sovetskoe
        Radio Publishing House, 392 pages (1974).

0-46.   A. Saito. J. Illuminating Engineering Inst. of Japan (Shomei
        Gakkai Zasshi), vol. 58, No. 7, pp. 314-324 (1974).

0-47.   N. Yamaki. Journal of Illuminating Engineering Inst. of Japan
        (Shomei Gakkai Zasshi), vol. 58, No. 4, pp. 130-133 (1974).

1-1.   Dzh. Mik[†] and Dzh. Kreggs[†]. Elektricheskii proboi v gazakh
        (Electric Breakdown in Gases). Translated from English.
        Moscow, Foreign Literature Publishing House, 528 pages (1960).

1-2.   E. Nasser. Fundamentals of Gaseous and Plasma Electronics.
        New York, Wiley-Interscience, 320 pages (1971).

1-3.   S. C. Brown. Proc. of the Inst. of Radio Engineering, vol.
        39, No. 12, pp. 1493-1570 (1951).

1-4.   I. S. Marshak. ZhTF, vol. 13, No. 1-2, pp. 59-84 (1943).

1-5.   A. von Hippel and J. Frank. Z. fur Physik, vol. 57, pp. 696-
        704 (1929).

1-6.   H. Raether. Z. für Physik, vol. 117, I, pp. 375-398; II, pp.
        524-542 (1941).

1-7.   K. Dehne et al. Dielectrics, vol. 1, No. 3, pp. 129-138 (1963).

1-8.   C. U. Däcke. Z. für angewandte Physik, vol. 19, pp. 453-460
        (1965).

1-9.   A. Z. Efendiev. Sbornik nauchnykh soobshchenii Dagestanskogo
        universiteta. Matematika i fizika (Collection of Scientific
        Communications of Dagestan University). Makhachkala, pp.
        67-83 (1965(66)).

1-10. A. Z. Efendiev and E. S. Iskenderova. Ibid., pp. 84-89.
1-11. A. Z. Efendiev et al. Ibid., No. 2, pp. 3-7 (1968).
1-12. A. Z. Efendiev et al. Izvestiya vuzov SSSR. Fizika, No. 11,
        pp. 155-157 (1968).
1-13. R. Mikhailova. Sbornik nauchnykh soobshchenii Dagestanskogo
        universiteta po estestvennym i tekhnicheskim naukam (Col-
        lection of Scientific Communications of Dagestan University
        in the Natural and Technical Sciences). Makhachkala,
        part 2, pp. 58-70 (1970).
1-14. E. S. Iskanderova. Ibid., part 1, pp. 146-157.
1-15. J. Dutton and W. Morris. Brit. Journal of Applied Physics,
        vol. 18, pp. 1115-1120 (1967).
1-16. T. N. Daniel et al. Proc. of the 9th Intern. Conf. on Phen-
        omena in Ionized Gases. Bucharest, pp. 260-265 (1969).
1-17. J. Pfaue. Z. für angewandte Physik, vol. 16, No. 1, pp. 15-
        23 (1963).
1-18. H. Tholl. Z. für Physik, vol. 172, pp. 536-555 (1963).
1-19. K. Richter. In Elektronnye laviny i proboi v gazah (Electron
        Avalanches and Breakdown in Gases). Moscow, Mir Publishers,
        pp. 293-321, by H.Raether (1968).
1-20. E. K. Müller. Z. für angewandte Physik, vol. 21, pp. 219-
        224, 475-479 and 554-558 (1966).
1-21. A. Z. Efendiev. Sbornik nauchnyk- soobshchenii Dagestanskogo
        universiteta. Matematika i fizika. Makhachkala, pp. 90-
        101 (1966(66)).
1-22. H. Tholl. In Elektronnye laviny i proboi v gazakh. Moscow,
        Mir Publishers, pp. 254-292 (1968).
1-23. H. Tholl.· Proc. of the 7th Intern. Conf. on Phenomena in
        Ionized Gases. Belgrade, vol. 1, pp. 620-624 (1966).
1-24. C. Driver. Z. für Naturforschung, vol. 19a, pp. 1327-1328
        (1964).
1-25. H. Raethjen. Z. für Physik, vol. 186, pp. 444-451 (1965).
1-26. T. H. Teich. Z. für Physik, vol. 199, pp. 378-394 (1967).
1-27. W. Hoffmann. Z. für Physik, vol. 200, pp. 287-303 (1967).
1-28. K. Wagner. In Elektronnye laviny i proboi v gazakh. Moscow,
        Mir Publishers, pp. 322-328 (1968).
1-29. K. H. Wagner. Z. für Physik, vol. 180, pp. 516-522 (1964).
1-30. K. H. Wagner. Z. für Physik, vol. 189, pp. 465-515 (1966).
1-31. K. H. Wagner. Proc. of the 7th Intern. Conf. on Phenomena in
        Ionized Gases, Belgrade, vol. 1, pp. 571-576 (1966).
1-32. K. H. Wagner. Z. für Physik, vol. 204, pp. 177-197 (1967).
1-33. W. Sroka. Physics Letters, vol. 25A, pp. 770-772 (1967).
1-34. P. Suleebka and R. S. N. Rau. Journal of Physics. D-Applied
        Physics. England, vol. 5, p. 2055 (1972).
1-35. K. Allen and K. Phillips. In Elektronnye laviny i proboi v
        gazakh. Moscow, Mir Publishers, pp. 221-253 (1968).
1-36. C. Driver. Z. für angewandte Physik, vol. 24, No. 1, pp.
        24-32 (1967).
1-37. A. A. Doran. Z. für Physik, vol. 208, No. 5, pp. 427-440 (1968).
1-38. K. Möstl and U. Timm. Z. für Physik, vol. 209, pp. 60-67 (1968).
1-39. M. C. Cavenor. Australian Journal of Physics, vol. 23, No. 6,
        pp. 953-965 (1970).
1-40. M. M. Kekez et al. Journal of Physics. D-Applied Physics.
        England, vol. 3, No. 12, pp. 1886-1898 (1970).

1-41. I. D. Chalmers and H. Duffy. Journal of Physics. D-Applied Physics. England, vol. 4, p. 1302 (1971).

1-42. W. Reinghaus. Journal of Physics. D-Applied Physics. England, vol. 5, p. 1448 (1972).

1-43. I. S. Marshak. Uspekhi fizicheskikh nauk, vol. 71, No. 4, pp. 631-675 (1960).

1-44. G. Raether. Elektronnye laviny i proboi v gazakh. Translation from English edited by V. S. Komel'kov, Moscow, Mir Publishers, 390 pages (1968).

1-45. J. D. Graggs. Surveys on Phenomena in Ionized Gases. Vienna, pp. 473-487 (1968).

1-46. Dzh. Dawson and W. Vinn. In Elektronnye laviny i proboi v gazakh. Moscow, Mir Publishers, pp. 357-371 (1968).

1-47. F. Llewellyn-Jones. Ionization, Avalanches, and Breakdown. London, Methuen, 103 pages (1967).

1-48. A. J. Davies and C. J. Evans. Computer Physics Communications. Holland, vol. 3, pp. 322-335 (1972).

1-49. A. Ward. In Elektronnye laviny i proboi v gazakh. Moscow, Mir Publishers, pp. 349-356 (1968).

1-50. L. E. Kline and J. G. Siambis. Proc. of the IEEE, vol. 59, No. 4, pp. 707-709 (1971).

1-51. E. P. Oppenheimer et al. Intern. Journal of Electronics. England, vol. 32, pp. 441-448 (1972).

1-52. E. D. Lozanskii. ZhTF, vol. 38, No. 9, pp. 1563-1567 (1968).

1-53. V. M. Kirilenko. ZhTF, vol. 40, No. 8, pp. 1756-1758 (1970).

1-54. A. T. Matyushin and V. T. Matyushin. O primenenii debaevskogo priblizheniya k perekhodu laviny v strimer (The Use of the Debye Approximation of the Streamer Transition of an Avalanche). Preprint No. P13-5504, Joint Institute for Nuclear Research (1970).

1-55. W. Legler. Z. für Physik, vol. 173, pp. 169-183 (1963).

1-56. T. Teich. Z. für Naturforschung, vol. 19a, pp. 1420-1421 (1964).

1-57. H. Krisch. Z. für Physik, vol. 178, pp. 354-364 (1964).

1-58. H. Krisch. Z. für Naturforschung, vol. 19a, pp. 1136-1137 (1964).

1-59. W. Köhrmann. Applied Sci. Research. Holland, vol. 5B, Nos. 1-4, pp. 288-290 (1955).

1-60. C. G. Sluijter. Actes 10 Congr. Intern. Cinematographie Ultra-Rapide, Paris, pp. 341-345 (1972).

1-61. N. Warmoltz. Philips Technische Rundschau, vol. 9, No. 4, pp. 105-113 (1947).

1-62. G. Hartmann. Elektrotechnische Z. Berlin West, vol. A78, pp. 694-699 (1957).

1-63. W. Rettner. Svetotekhnika, No. 6, pp. 1-4 (1966).

1-64. V. V. Nyubin. Svetotekhnika, No. 1, pp. 11-13 (1974).

1-65. I. S. Marshak. Sbornik materialov po vakuumnoi tekhnike (Collection of Materials on Vacuum Engineering). Moscow-Leningrad, Gosenergoizdat Publishing House, No. 13, pp. 12-27 (Electric Vacuum Devices Plant, Office of Technical Information) (1957).

1-66. I. S. Marshak and V. A. Subbotin. Ibid., pp. 28-41.

1-67. I. S. Marshak and L. J. Schukin. Journal of Society of Motion Picture and Television Engineers, vol. 70, pp. 169-176 (1961).

1-68.   A. Bielski and T. Krol. Acta physica polonica. A-General
        Physics. Polska, vol. 29, No. 4, pp. 571-572 (1966).

1-69.   Yu. G. Basov et al. Zhurnal prikladnoi spektroskopii, vol. 23,
        No. 6, pp. 1095-1097 (1975).

2-1.   I. S. Marshak. Uspekhi fizicheskikh nauk, vol. 77, No. 2,
        pp. 229-286 (1962).

2-2.   I. S. Marshak. Elektricheskii proboi gaza pri atmosfernom
        davlenii (Electric Breakdown of Gas at Atmospheric Pressure).
        Dissertation, Moscow Power Engineering Institute (1945).

2-3.   J. A.Fitzpatrick et al. Journal of Applied Physics, vol. 21,
        pp. 1269-1276 (1950).

2-4.   N. N. Ogurtsova and I. V. Podmoshenskii. Optika i spektros-
        kopiya, vol. 4, pp. 539-541 (1958).

2-5.   R. E. Rovinskii. Teplofizika vysokikh temperatur, vol. 10,
        No. 1, pp. 1-6 (1972).

2-6.   I. S. Marshak et al. Svetotekhnika, No. 4, pp. 8-17 (1961).

2-7.   I. S. Marshak. Sbornik materialov po vakuumnoi tekhnike.
        Moscow-Leningrad, Gosenergoizdat Publishing House, No. 22,
        pp. 27-37 (1960). (Electric Vacuum Devices Plant, Office
        of Technical Information).

2-8.   Ya. B. Zel'dovich and Yu. P. Raizer. Fizika udarnykh voln i
        vysokotemperaturnykh gidrodinamicheskikh yavlenii (The
        Physics of Shock Waves and High-Temperature Hydrodynamic
        Phenomena). Moscow, Nauka Publishers, 632 pages (1966).

2-9.   Yu. P. Raizer. Uspekhi fizichekikh nauk, vol. 108, p. 429
        (1972).

2-10.   S. I. Drabkina. JETP, vol. 21, No. 4, p. 473 (1951).

2-11.   S. I. Braginskii. JETP, vol. 34, No. 6, p. 1548 (1958).

2-12.   Yu. K. Bobrov. ZhTF, vol. 44, No. 11, p. 2340 (1974).

2-13.   S. I. Baranik et al. ZhTF, vol. 44, No. 11, p. 2352 (1974).

2-14.   B. L. Borovich et al. Trudy fizicheskogo instituta AN SSSR
        im. P. I. Lebedeva, vol. 76, pp. 3-35 (1974).

2-15.   A. F. Aleksandrov and A. A. Rukhadze. Uspekhi fizicheskikh
        nauk, vol. 112, No. 2, pp. 193-230 (1974).

2-16.   A. F. Alekdandrov et al. ZhTF, vol. 44, No. 11, p. 2414 (1974).

2-17.   A. F. Aleksandrov et al. JETP, vol. 61, No. 11, p. 1841 (1971).

2-18.   V. F..D'yachenko and V. S. Imshennik. In Voprosy teorii
        plazmy (Topics in Plasma Theory). Edited by M. A. Leon-
        tovich. Moscow, Atomizdat Publishing House, vol. 5, p.
        394 (1967).

2-19.   H. Fisher. Conf. on Extremely High Temperatures. New York,
        Wiley-London, Chapman and Hall, pp. 11-27 (1958).

2-20.   N. M. Gegechkori. JETP, vol. 21, No. 4, p. 493 (1951).

2-21.   G. Glaser. Z. fur Physik, vol. 143, p. 44 (1955).

2-22.   S. I. Andreev and V. E. Gavrilov. Teplofizika vysokikh tem-
        peratur, vol. 8, No. 1, p. 203 (1970).

2-23.   L. M. Biberman and G. E. Norman. Uspekhi fizicheskikh nauk,
        vol. 91, No. 2, p. 193 (1967).

2-24.   G. G. Dolgov and S. L. Mandel'shtam. JETP, vol. 24, p. 691
        (1953).

2-25. A. A. Vekhov et al. Zhurnal prikladnoi spektroskopii, vol.
       12, p. 978 (1970).
2-26. R. H. Dishington et al. Applied Optics, vol. 13, No. 10,
       pp. 2300-2312 (1974).
2-27. E. I. Asinovskii et al., Teplofizika vysokikh temperatur,
       vol. 11, No. 5, p. 939 (1973).
2-28. N. A. Kozlov and V. V. Fomin. Teplofizika vysokikh temperatur,
       vol. 14, No. 3, pp. 457-461 (1976).
2-29. L. V. Babin et al. Teplofizika vysokikh temperatur, vol. 7,
       No. 3, p. 570 (1969).
2-30. V. A. Gerasimov et al. Zhurnal prikladnoi spektoskopii,
       vol. 14, No. 6, p. 986 (1971).
2-31. V. G. Nikolaevskii et al. ZhTF, vol. 42, No. 2, p. 364 (1972).
2-32. A. N. Vasil'eva et al. Teplofizika vysokikh temperatur, vol.
       9, No. 5, p. 865 (1971).
2-33. A. N. Vasil'eva et al. Pis'ma v zhurnal eksperimental'noi
       i teoreticheskoi fiziki, vol. 15, No. 10, p. 613 (1972).
2-34. R. G. Vdovchenko et al. Sbornik materialov po vakuumnoi
       tekhnike, No. 21. Moscow-Leningrad, Gosenergoizdat
       Publishing House, pp. 17-36 (Electric Vacuum Devices Plant
       Office of Technical Information) (1959).
2-35. Yu. M. Vas'kovskii et al. Svetotekhnika, No. 10, p. 19 (1973).
2-36. I. V. Bykov et al. Kvantovaya elektronika, vol. 2, No. 1,
       pp. 181-184 (1975).
2-37. V. V. Ivanov and A. G. Rozanov. Teplofizika vysokikh temperatur,
       vol. 10, No. 5, pp. 1102 (1972).
2-38. M. P. Vanyukov et al. ZhTF, vol. 25, p. 1248 (1955).
2-39. M. P. Vanyukov and V. I. Isaenko. Svetotekhnika, No. 3, p. 7
       (1960).
2-40. K. S. Wul'fson and I. Sh. Libin. JETP, vol. 21, p. 510 (1951).
2-41. Yu. M. Vas'kovskii et al. Elektronnaya tekhnika, seriya IV.
       Elektrovakuumnye i gazorazryadnye pribory, No. 4, pp. 25-
       27 (1976).
2-42. M. P. Vanyukov and A. A. Mak. Uspekhi fizicheskikh nauk, vol.
       66, p. 301 (1958).
2-43. K. S. Wul'fson et al. Izvestiya AN SSSR, seriya fizicheskikaya,
       vol. 19, p. 61 (1955).
2-44. V. P. Kirsanov et al. Optika i spektroskopiya, vol. 13, p.
       276 (1962).
2-45. V. P. Kirsanov et al. In Espekhi nauchnoi fotografii, vol. 9,
       p. 109 (1964).
2-46. H. Fisher and W. Schwanzer. Applied Optics, vol. 8, No. 3,
       p. 697 (1969).
2-47. L. N. Bykhovskaya et al. Svetotekhnika, No. 10, p. 21 (1963).
2-48. F. A. Charnaya and Z. G. Yakob. Optika i spektroskopiya,
       vol. 18, No. 3, p. 530 (1965).
2-49. A. A. Bakeev and R. E. Rovinskii. Teplofizika vysokikh
       temperatur, vol. 8, No. 6, pp. 1121-1127 (1970).
2-50. I. Sh. Model'. JETP, vol. 32, p. 714 (1957).
2-51. M. P. Vanyukov et al. Optika i spektroskopiya, vol. 4, p. 90
       (1958).

2-52. V. P. Kirsanov. Teplofizika vysokikh temperatur, vol. 14,
        No. 5, p. 1130 (1976).

3-1.   G. Ecker and W. Weizel. Z. für Naturforschung, vol. 12a,
        No. 10, p. 859 (1957).

3-2.   G. Brunner. Z. für Physik, vol. 159, p. 288 (1960).

3-3.   V. Finkel'nburg† and R. Mekker.† Elektricheskie dugi i
        termicheskaya plazma (Electric Arcs and Thermal Plasma).
        Moscow, Foreign Literature Publishing House, 370 pages
        (1961).

3-4.   Diagnostika plazmy (Plasma Diagnostics). Edited by R. Huddle-
        ston and S. Leonard. Moscow, Mir Publishers, 515 pages
        (1967).

3-5.   G. Grim. Spektroskopiya plazmy (Plasma Spectroscopy). Moscow,
        Atomizdat Publishing House, 452 pages (1969).

3-6.   Metody issledovaniya plazmy (Methods of Plasma Research).
        Edited by V. Lokhte-Khol'tgreven.† Moscow, Mir Publishers,
        552 pages (1971).

3-7.   V.L. Granovskii. Elektricheskii tok v gaze (Electric Current
        in a Gas). Moscow, Nauka Publishers, 543 pages (1971).

3-8.   D. Sampson. Uravneniya perenosa energii i kolichestva dvizh-
        heniya v gazakh s uchetom izlucheniya (Equations of Energy
        Transport and Momentum Transfer in Gases with Allowance for
        Radiation). Moscow, Mir Publishers, 205 pages (1969).

3-9.   V. N. Egorov et al. Doklady AN SSSR, vol. 121, No. 3, p. 440
        (1958).

3-10. V. N. Kolesnikov. In Fizicheskaya optika (Physical Optics).
        Moscow, Nauka Publishers, vol. 30, p. 53 (1966).

3-11. L. Brober and R. S. Tankin. Journal of Quantum Spectroscopy
        and Radiative Transfer, vol. 10, No. 9, p. 991 (1970).

3-12. D. L. Evans et al. American Institute of Aeronautics and
        Astronautics-Journal, No. 42, p. 9 (1970).

3-13. W. Neuman. Beitr. aus der Plasmaphysik, vol. 11, No. 3, p.
        248 (1971).

3-14. G. Pichler and V. Vujnovic. Physical Letters, vol. A40, No.
        5, p. 397 (1972).

3-15. J. B. Schumaker and C. H. Popenoe. Journal of Research.
        National Bureau of Standards, vol. A76, No. 2, p. 71 (1972).

3-16. J. H. Goncz. Journal of Applied Physics, vol. 36, p. 742
        (1965).

3-17. K. Günther. Beitr. aus der Plasmaphysik, vol. 8, No. 5, p.
        383 (1968).

3-18. M. M. Popovich and V. V. Uroshevich. Teplofizika vysokikh
        temperatur, vol. 9, No. 3, p. 627 (1971).

3-19. S. I. Andreev and V. E. Gavrilov. Optika i spektroskopiya,
        vol. 26, No. 1, pp. 121-123 (1969).

3-20. K. Günther and G. Radtke. Beitr. aus der Plasmaphysik, vol.
        12, No. 1 (1972).

3-21. K. Günther. Beitr. aus der Plasmaphysik, vol. 10, No. 6,
        pp. 469-485 (1970).

3-22. A. A. Bakeev. Elektricheskie i opticheskie svoistva impul'
       snogo ksenovogo razryada vysokogo davleniya (The Electrical
       and Optical Properties of a High-Pressure Pulsed Xenon
       Discharge). Dissertation, Moscow Physicotechnical Institute,
       (1970).

3-23. V. L. Ginzburg. Rasprostranenie elektromagnitnykh voln v
       plazme (The Propagation of Electromagnetic Waves in Plasma).
       Moscow, Fizmatgiz Publishing House, 683 pages (1960).

3-24. Ergebnisse der Plasmaphysik und der Gaselektronik (Results of
       Plasma Physics and Gas Electronics). Berlin, Akademie-
       Verlag, vol. 1 (1967).

3-25. A. A. Bakeev et al. Pribory i tekhnika eksperimenta, No. 4,
       p. 166 (1969).

3-26. A. A. Bakeev et al. Radiotekhnika i elektronika, vol. 14,
       p. 1998 (1969).

3-27. A. A. Bakeev et al. Teplofizika vysokikh temperatur, vol. 9,
       No. 4, pp. 841-843 (1971).

3-28. J. Bowe. Physical Rev., vol. 117, No. 6, p. 1411 (1960).

3-29. N. L. Allen and B. A. Prew. Journal of Physics, Great Britain,
       vol. 3B, p. 1113 (1970).

3-30. A. A. Bakeev et al. Teplofizika vysokikh temperatur, vol. 7,
       No. 6, p. 1203 (1969).

3-31. V. V. Ivanov and A. G. Rozanov. Teplofizika vysokikh temperatur,
       vol. 10, No. 5, p. 1102 (1972).

3-32. A. A. Bakeev et al. Teplofizika vysokikh temperatur, vol. 11,
       No. 5, p. 1111 (1973).

3-33. A. N. Vorob'ev and E. V. Daniel'. Zhurnal prikladnoi spektro-
       skopii, vol. 12, No. 2, pp. 347-349 (1970).

3-33a. G. A. Volkova et al. Zhurnal prikladnoi spektroskopii, vol.
       24, No. 6, pp. 972-975 (1976).

3-34. C. Chapman and T. Cowling. Matematicheskaya teoriya neodnoro-
       dnykh gazov (The Mathematical Theory of Inhomogeneous Gases).
       Moscow, Foreign Literature Publishing House, 510 pages
       (1960).

3-35. I. Shkarovskii, T. Johnson, and M. Bachinskii. Kinetika chastits
       plazmy (The Kinetics of Plasma Particles). Moscow, Atomizdat
       Publishing House, 395 pages (1969).

3-36. B. A. Trubnikov. In Voprosy teorii plazmy, No. 1. Moscow,
       Atomizdat Publishing House, p. 98 (1963).

3-37. Kineticheskie protsessy v gazakh i plazme (Kinetic Processes
       in Gases and Plasma). Edited by Khokhshtim. Moscow,
       Atomizdat Publishing House, 368 pages (1972).

3-38. I. McDaniel, Protsessy stolknovenii v ionizovanntkh gazakh
       (Collisional Processes in Ionized Gases). Moscow, Mir
       Publishers, 832 pages (1967).

3-39. L. Spitzer. Fizika polnost'yu ionizovannogo gaza (The Physics
       of Fully Ionized Gas). Moscow, Mir Publishers, 212 pages
       (1965).

3-40. S. C. Lin et al. Journal of Applied Physics, vol. 26, No. 1,
       p. 95 (1955).

3-41. R. S. Devoto. Physics of Fluids, vol. 10, p. 354 (1967).

3-42. W. L. Nighan. Physics of Fluids, vol. 12, No. 1, p. 162 (1969).

3-43. V. S. Rogov. Transportnye koeffitsienty plazmy (Plasma Transport Coefficients). Preprint No. 47, Institute of Applied Mathematics, USSR AS, 19 pages (1969).

3-44. R. S. Devoto. Physics of Fluids, vol. 16, No. 5, p. 616 (1973).

3-45. R. S. Devoto and D. Mukherjee. Journal of Plasma Physics, vol. 9, No. 1, p. 65 (1973).

3-46. R. C. Devoto. American Institute of Aeronautics and Astronautics-Journal, vol. 7, No. 2, p. 199 (1969).

3-47. V. P. Kirsanov. Predel'nye kharakteristiki gazorazryadnykh impul'snykh istochnikov sveta (The Limiting Characteristics of Pulsed Gas-Discharge Light Sources). Dissertation, Physics Institute, USSR AS. Moscow (1970).

3-48. I. V. Demenik et al. ZhTF, vol. 33, No. 4, p. 489 (1963).

3-49. A. A. Bakeev and R. E. Rovinskii. Teplofizika vysokikh temperatur, vol. 8, No. 1, pp. 207-209 (1970).

3-50. S. I. Andreev and V. E. Gavrilov. ZhTF, vol. 40, No. 6, p. 1300 (1970).

3-51. C. H. Church et al. Journal of Quantitative Spectroscopy and Radiative Transfer, vol. 8, p, 403 (1968).

3-52. Yu. L. Klimontovich. Uspekhi fizicheskikh nauk, vol. 110, No. 4, p. 573 (1973).

3-53. S. M. Vukovich and M. M. Popovich. Teplofizika vysokikh temperatur, vol. 10, No. 2, p. 419 (1972).

3-54. J. Hackmann and J. Uhlenbusch. Proc. of the 10th Intern. Conf. on Phenomena in Ionized Gases. Oxford University Press, p. 260 (1971).

3-55. J. Buss et al. Z. für angewandte Physik, vol. 22, No. 4, p. 345 (1967).

3-56. J. C. Morris et al. Physics of Fluids, vol. 13, No. 3, p. 608 (1970).

3-57. V. S. Vorob'ev and A. L. Khomkin. Teplofizika vysokikh temperatur, vol. 10, No. 5, p. 939 (1972).

3-58. I. I. Litvinov. Teplofizika vysokikh temperatur, vol. 11, No. 4, p. 695 (1973).

3-59. Atomnye i molekulyarnye protsessy (Atomic and Molecular Processes). Edited by D. Bates. Moscow, Mir Publishers, 777 pages (1964).

3-60. S. E. Frish. Opticheskie spektry atomov (Optical Spectra of Atoms). Moscow-Leningrad, Fizmatgiz Publishing House, 640 pages (1963).

3-61. I. I. Sobel'man. Vvedenie v teoriyu atomnykh spektrov (Introduction to the Theory of Atomic Spectra). Moscow, Fizmatgiz Publishing House, 640 pages (1963).

3-62. I. B. Levinson and A. A. Nikitina. Rukovodstvo po vychisleniyu intensivnostei linii v atomnykh spektrakh (Guide to the Calculation of Line Intensities in Atomic Spectra). Leningrad State University Press, 57 pages (1962).

3-63. D. R. Bates and A. Damgaart. Philosophical Transactions of the
       Royal Society London, vol. 242A, No. 842, p. 101 (1949).

3-64. W. Finkelnburg and T. Peters. Handbuch der Physik. Berlin,
       Springer-Verlag, vol. 28 (1957).

3-65. T. Bethe and E. Solpiter. Kvantovaya mekhanika atomov s odnim
       i dvumya elektronami. Moscow, Fizmatgiz Publishing House,
       562 pages (1960).

3-66. M. J. Seaton. Monthly Notices of the Royal Astronomical Society,
       vol. 118, No. 5, pp. 504-518 (1958).

3-67. A. Burges and M. Seaton. Rev. of Modern Physics, vol. 30,
       No. 3, p. 992 (1958).

3-68. A. Burges and M. Seaton. Monthly Notices of the Royal Astro-
       nomical Society, vol. 120, No. 2, p. 121 (1960).

3-69. G. E. Norman. Optika i spektroskopiya, vol. 12, No. 3, p.
       333 (1962).

3-70. L. M. Biberman and G. E. Norman. Optika i spektroskopiya,
       vol. 8, p. 443 (1960).

3-71. L. M. Biberman and G. E. Norman. Journal of Quantitative
       Spectroscopy and Radiative Transfer, vol. 3, No. 3, p. 221
       (1963).

3-72. L. M. Biberman et al. Astronomicheskii zhurnal, vol. 39,
       No. 1, p. 107 (1962).

3-73. D. R. Ingliss and E. A. Teller. Astrophysical Journal, vol.
       90, No. 3, p. 439 (1939).

3-74. L. M. Biberman et al. Optika i spektroskopiya, vol. 15, No.
       3, p. 330 (1963).

3-75. I. T. Yakubov. Optika i spektroskopiya, vol. 19, No. 4, p.
       497 (1965).

3-76. Yu. V. Moskovin. Teplofizika vysokikh temperatur, vol. 6,
       No. 1, p. 1 (1968).

3-77. A. P. Sobolev. Optika i spektroskopiya, vol. 33, No. 6, p.
       1179 (1972).

3-78. S. E. Frish and O. P. Bochkova. Vestnik Leningradskogo
       gosudarstvennogo universiteta, vol. 16, No. 3, p. 40 (1961).

3-79. Spektroskopiya gazorazryadnoi plazmy (Spectroscopy of a Gas-
       Discharge Plasma). Edited by S. E. Frish. Leningrad,
       Nauka Publishers, 361 pages (1970).

3-80. A. A. Gershun. Izbrannye trudy po fotometrii i svetotedknike
       (Selected Works on Photometry and Illumination Engineering).
       Moscow, Fizmatgiz Publishing House, 548 pages (1968).

3-81. A. S. Doinikov et al. Svetotekhnika, No. 12, p. 13 (1971).

3-82. I. V. Podmoshenskii and L. D. Kondrasheva. Materialy X
       vsesoyuznoogo soveshchaniya po spektroskopii (Materials
       of the 10th All-Union Conference on Spectroscopy), vol. II.
       L'vov University Press, 204 pages (1958).

3-83. F. A. Charnaya and Z. G. Yakob. Optika i spektroskopiya,
       vol. 19, No. 2, pp. 181-185 (1965).

3-84. J. Emmett et al. Journal of Applied Physics, vol. 35, pp.
       2601-2604 (1964).

3-85. S. I. Andreev and O. G. Baikov. Optika i spektroskopiya, vol.
      25, No. 4, p. 481 (1968).
3-86. S. I. Andreev and V. E. Gavrilov. Optika i spektroskopiya,
      vol. 26, No. 4, pp. 665-667 (1969).
3-87. L. I. Gavrilova et al. In Impul'snaya fotometriya (Pulse
      Photometry), No. 1. Leningrad, Mashinostroenie Publishing
      House, pp. 136-144 (1969).
3-88. A. A. Bakeev et al. Optika i spektroskopiya, vol. 27, No. 2,
      pp. 215-220 (1969).
3-89. L. I. Gavrilova et al. In Impul'snaya fotometriya No. 4.
      Leningrad, Mahinostroenie Publishing House, pp. 122-125
      (1975).
3-90. L. I. Gavrilova et al. Zhurnal prikladnoi spektroskopii, vol.
      12, No. 3, p. 537 (1970).
3-91. V. V. Yankov. Optika i spektroskopiya, vol. 14, p. 29 (1963).
3-92. V. M. Batenin and P. V. Minaev. Teplofizika vysokikh temperatur,
      vol. 9, No. 4, p. 676 (1971).
3-93. G. V. Gimbarzhevskii et al. Optika i spektroskopiya, vol. 28,
      p. 1101 (1970).
3-94. A. A. Bakeev et al. Optika i spektroskopiya, vol. 28, p. 1101,
      (1970).
3-95. V. A. Golubev and V. F. Klimkin. Teplofizika vysokikh temp-
      eratur, vol. 9, No. 4, p. 683 (1971).
3-96. A. A. Bakeev et al. Optika i spektroskopiya, vol. 37, p.
      1165 (1974).
3-97. R. E. Rovinskii. ZhTF, vol. 45, No. 8, pp. 1782-1784 (1975).
3-98. G. I. Kozlov and D. I. Roitenburg. Optika i spektroskopiya,
      vol. 36, p. 850 (1974).
3-99. A. F. Aleksandrov et al. ZhTF, vol. 44, No. 3, pp. 491-501
      (1974).
4-1. V. E. Il'in and S. V. Lebedev. ZhTF, vol. 32, p. 986 (1962).
4-2. A. G. Goloveiko and S. P. Rzhevskaya. Inzhenerno-fizicheskii
      zhurnal, vol. 16, p. 1073 (1969).
4-3. J. M. Mitterauer. Proc. of the 8th Intern Conf. on Phenomena
      in Ionized Gases. Vienna, p. 90 (1967).
4-4. N. M. Zykova et al. ZhTF, vol. 40, p. 2361 (1970).
4-5. V. E. Grakov et al. ZhTF, vol. 40, p. 2134 (1970).
4-6. I. I. Beilis et al. Teplofizika vysokikh temperatur, vol 13,
      No. 4, pp. 701-705 (1975).
4-7. T. G. Kesaev. Katodnye protsessy elektricheskoi dugi (Cathode
      Processes of an Electric Arc). Moscow, Nauka Publishers
      (1968), 244 pages.
4-8. A. W. Hull. Journal of Applied Physics, vol. 35, p. 490 (1964).
4-9. G. Eckert and K. G. Müller. Journal of Applied Physics,
      vol. 30, p. 1466 (1959).
4-10. I. N. Ostretsov V. A. Petrosov, A. A. Paporotnikov, and Yu.
      A. Utkin. Tezisy dokladov na I Vsesoyuznoi konferentsii
      po plazmennym uskoritelyam (Abstracts of Papers at the First
      All-Union Conference on Plasma Accelerators). Moscow (1967).
4-11. I. Sh. Libin. Radiotekhnika i elektronika, vol. 4, p. 1026
      (1959).

4-12. K. H. Kington. Journal of Applied Physics, vol. 36, p. 1351 (1965).

4-13. V. I. Rakhovskii. ZhTF, vol. 35, p. 2228 (1965).

4-14. A. P. Nevskii. Teplofizika vysokikh temperatur, vol. 8, p. 898 (1970).

4-15. N. V. Afanas'ev. Zhurnal prikladnoi spektroskopii, vol. 5, No. 2, p. 138 (1966).

4-16. V. V. Kubyshkin et al. Elektronnaya tekhnika, ser. 3 Gazorazryadnye pribory, No. 2, p. 55 (1966).

4-17. K. Inoue. Japanese patent No. 49-16465, published April 22, (1974).

4-18. S. Oyama and T. Tanaka. Japanese patent No. 49-16994, published April 26 (1974).

4-19. R. Cosco and J. Pappas. US patent No. 3,849,690, published November 19 (1974).

4-20. S. I. Faifer et al. USSR patent No. 444,269, published September 25 (1974).

4-21. M. Oyama. Japanese patent No. 50-14069, published May 24 (1975).

4-22. E. S. Savranskaya et al. USSR patent No. 492,950, published December 25 (1975).

4-23. V. V. Kubyshkin. Elektronnaya tekhnika, ser. 3, Gazorazryadnye pribory, No. 1, p. 47 (1970).

4-24. J. M. Somerville and J. F. Williams. Proc. of the Physical Society of London, vol. 74, No. 477, p. 309 (1959).

5-1. V. G. Baryshnikov et al. In Impul'snaya fotometriya (Pulse Photometry). Leningrad, Mashinostroenie, No. 3, pp. 4-9 (1973).

5-2. Vocabulaire internationale de l'eclairage, 3rd ed. Bureau Central GIE. Paris (1970). (See Svetotekhnika, Nos. 1-12 (1973).)

5-3. V. V. Meshov, Osnovy svetotekhniki, part 1. Moscow-Leningrad, Gosenergoizdat Publishing House, 352 pages (1957).

5-4. R. A. Sapozhnikov. Teoreticheskaya fotometriya (Theoretical Photometry). Moscow, Energiya Publishers, p. 268 (1977).

5-5. M. M. Gurevich. Vvedenie v fotometriyu (Introduction to Photometry). Leningrad, Energiya Publishers, 244 pages (1968).

5-6. A. S. Doinikov and V. V. Ignat'ev. In Impul'snaya fotometriya (Pulse Photometry). Leningrad, Mashinostroenie, No. 2, pp. 121-125 (1972).

5-7. N. A. Tolstoi and M. V. Epifanov. Optika i spektroskopiya, vol. 16, No. 4, pp. 677-683 (1964).

5-7a. V. G. Baryshnikov. Svetotekhnika, No. 2, pp. 8-12 (1966).

5-7b. G. Glaser. Optik, vol. 7, No. 2, pp. 61-90 (1950).

5-8. E. V. Kuvaldin. In Impul'snaya fotometriya (Pulse Photometry). Leningrad, Mashinostroenie, No. 2, pp. 17-22 (1972).

5-9. B. Ya. Lutset et al. Svetotekhnika, No. 6, p. 19 (1972).

5-10. V. I. Vasil'ev et al. Optika i spektroskopiya, vol. 11, No. 1, pp. 118-122 (1961).

5-11. I. S. Marshak et al. In Impul'snaya fotometriya. Leningrad, Mashinostroenie Publishing House, pp. 122-129 (1969).

5-12. A. S. Doinikov and V. K. Pakhimov. In Impul'snaya fotometriya. Leningrad, Mashinostroenie Publishing House, pp. 129-132 (1969).

5-13. I. V. Deminik et al. Svetotekhnika, No. 8, pp. 10-12 (1969).

5-14. Yu. N. Vlasov et al. Svetotekhnika, No. 12, pp. 23-24 (1970).

5-15. L. I. Gavrilova et al. In Impul'snaya fotometriya. Leningrad, Mashinostroenie Publishing House, pp. 105-113 (1973).

5-16. V. P. Kirsanov and S. V. Troshkin. Svetotekhnika, No. 1, pp. 12-15 (1967).

5-17. Yu. A. Anan'ev et al. Optiko-mekhanicheskaya promyshlennost', No. 9, pp. 35-36 (1972).

5-18. A. S. Doinikov. Svetotekhnika, No. 6, pp. 12-14 (1970).

5-19. M. M. Gurevich. Svetotekhnika, No. 3, pp. 29-30 (1963).

5-20. A. D. Stokes. Journal of the Optical Soc. of America, vol. 57, No. 9, pp. 1100-1105 (1967).

5-21. A. S. Doinikov and V. G. Dorogov. Optika i spektroskopiya, vol. 31, No. 5, pp. 817-821 (1971).

5-22. A. A. Vekhov et al. Zhurnal prikladnoi spektroskopii, vol. 12, No. 6, pp. 979-983 (1970).

5-23. Yu. G. Anikiev et al. Optika i spektroskopiya, vol. 32, No. 2, pp. 392-395 (1972).

5-24. Yu. G. Basov et al. Zhurnal prikladnoi spektroskopii, vol. 21, No. 1, pp. 32-34 (1974).

5-25. V. P. Kirasnov et al. Svetotekhnika, No. 1, p. 12 (1963).

5-26. G. N. Rokhlin and L. I. Shchukin. Elektronnaya tekhnika, ser. IV, "Elektrovakuumnye i gazorazryadnye pribory", 1977, No. 6, pp. 23-27.

5-27. A. S. Doinikov and V. K. Pakhomov. In Impul'snaya fotometriya. Leningrad, Mashinostroenie Publishing House, No. 3, pp. 97-99 (1973).

5-28. A. S. Doinikov et al. In Impul'snaya fotometriya. Leningrad, Mashinostroenie Publishing House, No. 4, pp. 126-131 (1975).

5-29. F. A. Charnaya and L. N. Bykhovskaya. Optika i spektroskopiya, vol. 16, No. 2, pp. 365-367 (1964).

5-30. L. N. Bykhovskaya and T. V. Yalovega. In Impul'snaya fotometriya. Leningrad, Mashinostroenie Publishing House, pp. 153-159 (1972).

5-31. L. N. Bykhovskaya and V. P. Khaustova. In Impul'snaya fotometriya. Leningrad, Mashinostroenie Publishing House, No. 3, pp. 135-140 (1973).

5-32. S. I. Andreev and M. P. Vanyukov. Pribory i tekhnika eksperimenta, No. 4, pp. 76-79 (1961).

5-33. S. I. Andreev and M. P. Vanyukov. ZhTF, vol. 32, No. 6, pp. 738-745 (1962).

5-34. D. P. C. Thackeray. Journal of Sci. Instruments, vol. 35, pp. 206-211 (1958).

5-35. I. S. Marshak. Svetotekhnika, No. 6, pp. 17-19 (1959).

5-36. V. A. Gavanin. Svetotekhnika, No. 2, pp. 22-26 (1967).

5-37. A. S. Doinikov et al. In Impul'snaya fotometriya. Leningrad,
       Mashinostroenie Publishing House, No. 2, pp. 126-130 (1972).
5-38. L. I. Gavrilova and V. G. Ignat'ev. In Impul'snaya fotometriya.
       Leningrad, Mahinostroenie Publishing House, No. 2, pp. 139-
       144 (1972).
5-39. A. S. Doinikov. Issledovanie osnovnykh kharakteristik izkuch-
       eniya pryamykh trubchatykh ksenonovykh impul'snykh lamp
       (Investigation of the Main Radiation Characteristics of
       Straight Tubular Xenon Flashlamps). Author's abstract of
       dissertation. Moscow, Physics Institute, USSR AS, 24 pages
       (1972).
5-40. A. S. Doinikov. Obzory po elektronnoi tekhnike, ser. "Elek-
       trovakuumnye i gazorazryadnye pribory". Moscow, Elektronika
       Central Scientific Research Institute, No. 11 (154), 35
       pages (1973).
5-41. J. H. Goncz and P. B. Newell. Journal of the Optical Soc.
       of America, vol. 56, No. 1, pp. 87-92 (1966).
5-42. V. P. Kirsanov et al. Optika i spektroskopiya, vol. 13, No.
       3, pp. 442-446 (1962).
5-43. V. I. Bulykov et al. Svetotekhnika, No. 10, pp. 21-24 (1967).
5-44. L. I. Gavrilova et al. Svetotekhnika, No. 5, pp. 14-15 (1971).
5-45. Yu. G. Basov et al. Kvantovaya elektronika, vol. 2, No. 8,
       pp. 1840-1844 (1975).
5-45a. A. S. Kamrukov et al. Zhurnal prikladnoi spektroskopii, vol.
       23, No. 3, pp. 393-397 (1975).
5-46. F. A. Charnaya et al. Zhurnal prikladnoi spektroskopii, vol.
       11, No. 5, pp. 790-795 (1969).
5-46a. M. Gusinow. Journal of Applied Physics, vol. 46, No. 11,
       pp. 4847-4851 (1975).
5-47. V. G. Ignat'ev and N. V. Kamyshov. Elektronnaya tekhnika,
       ser. 3, No. 4, pp. 79-83 (1967).
5-48. S. N. Belov et al. Zhurnal prikladnoi spektroskopii, vol. 10,
       No. 3, pp. 408-412 (1969).
5-49. I. V. Podmoshenskii et al. Zhurnal prikladnoi spektroskopii,
       vol. 9, No. 1, pp. 96-98 (1968).
5-50. O. J. Edwards. Journal of the Optical Soc. of America, vol.
       56, No. 10, pp. 1314-1319 (1966).
5-51. A. N. Vorob'ev and E. V. Daniel'. Zhurnal prikladnoi spek-
       troskopii, vol. 12, No. 2, pp. 347-349 (1970).
5-52. S. I. Andreev and V. E. Gavrilov. Zhurnal prikladnoi spek-
       troskopii, vol. 20, No. 5, pp. 780-783 (1974).
5-53. A. N. Zaidel' and E. Ya. Shreider. Spektroskopiya vakuumnogo
       ul'trafioleta (Vacuum Ultraviolet Spectroscopy). Moscow,
       Nauka Publishers, 471 pages (1967).
5-54. V. G. Ignat'ev. In Impul'snaya fotometriya. Leningrad,
       Mashinostroenie Publishing House, No. 3, pp. 119-126 (1973).
5-55. L. I. Gavrilova et al. In Impul'snaya fotometriya. Leningrad,
       Mashinostroenie Publishing House, pp. 132-136 (1969).
5-56. V. G. Ignat'ev and V. M. Podgaetskii. Kvantovaya elektronika,
       No. 4, pp. 121-125 (1971).

5-57. V. G. Ignat'ev. In Impul'snaya fotometriya. Leningrad,
       Mashinostroenie Publishing House, No. 3, pp. 113-119 (1973).
5-58. V. G. Ignat'ev and O. M. Mikhailov. In Impul'snaya fotometriya.
       Leningrad, Mashinostroenie Publishing House, No. 3, pp.
       126-133 (1973).
5-59. O. M. Mikhailov. In Impul'snaya fotometriya. Leningrad,
       Mashinostroenie Publishing House, No. 3, pp. 126-133 (1973).
5-60. N. M. Galaktionova and A. A. Mak. Optika i spektroskopiya,
       vol. 16, No. 1, pp. 153-155 (1964).
5-61. O. M. Kutev. Optika i spektroskopiya, vol. 17, No. 2, pp.
       295-297 (1964).
5-62. V. F. Egorova et al. Zhurnal prikladnoi spektroskopii, vol.
       1, No. 4, pp. 294-298 (1964).
5-63. J. R. Oliver and F. S. Barnes. Proc. of the IEEE, vol. 59,
       No. 4, pp. 638-644 (1971).
5-64. J. H. Goncz and W. J. Mitchell, Jr. IEEE Journal of Quantum
       Electronics, vol. QE-3, No. 7, pp. 330-331 (1967).
5-65. V. G. Ignat'ev et al. In Impul'snaya fotometriya. Leningrad,
       Mashinostroenie Publishing House, No. 3, pp. 99-105 (1973).
5-66. J. R. Oliver and F. S. Barnes. IEEE Journal of Quantum
       Electronics, vol. QE-5, No. 5, pp. 232-237 (1969).
5-67. V. P. Kirsanov et al. Svetotekhnika, No. 7, pp. 5-7 (1966).
5-68. S. I. Andreev et al. Zhurnal prikladnoi spektroskopii,
       vol. 6, No. 1, pp. 27-32 (1967).
5-69. V. G. Nikiforov. Zhurnal prikladnoi spektroskopii. vol. 15,
       No. 1, pp. 151-153 (1971).
5-70. D. Roess and G. Zeidler. Electronics, vol. 39, No. 18, pp.
       115-118 (1966).
5-70a. Yu. G. Basov et al. Zhurnal prikladnoi spektroskopii, vol.
       23, No. 4, pp. 590-595 (1975).
5-71. G. A. Volkova and V. K. Prokof'ev. Optika i spektroskopiya,
       vol. 23, No. 4, pp. 640-641 (1967).
5-72. S. I. Levikov. Optiko-mekhanicheskaya promyshlennost',
       No. 8, pp. 54-63 (1969).
5-72a. M. Gusinow. Applied Optics, 1975, vol. 14, No. 11, pp. 2645-
       2649.
5-73. V. M. Gardash'yan et al. Voprosy radioelektroniki, seriya
       obshchetekhnicheskaya, No. 5, pp. 3-9 (1968).
5-74. V. M. Gardash'yan et al. Radiotekhnika i elektronika, vol.
       14, No. 6, pp. 1069-1071 (1969).
5-75. E. S. Kovalenko et al. Zhurnal prikladnoi spektroskopii, vol.
       15, No. 5, pp. 920-924 (1971).
5-76. V. N. Makarov et al. Radiotekhnika i elektronika, vol. 19,
       No. 1, pp. 119-122 (1974).
5-77. V. P. Kirsanov et al. Proc. of the 7th Intern. Conf. on
       Phenomena in Ionized Gases. Belgrade, pp. 790-791 (1965).
5-77a. Yu. A. Martsinkovskii. USSR patent No. 410,489, published
       May 12 (1974).
5-78. I. S. Marshak and L. I. Shchukin. Uspekhi nauchnoi fotografii,
       vol. 9, pp. 93-105 (1964).
5-79. V. V. Ivanov et al. Kvantovaya elektronika, vol. 1, No. 5,
       pp. 1283-1285 (1974).

5-80. A. Ya. Balagurov and V. V. Kubyshkin. Zhurnal prikladnoi spektroskopii, vol. 11, No. 3, pp. 503-507 (1973).

5-81. V. N. Budnik et al. Zhurnal prikladnoi spektroskopii, vol. 15, No. 4, pp. 617-621 (1971).

5-82. I. S. Marshak. Svetotekhnika, No. 1, pp. 17-20 (1957).

5-83. I. V. Kosinskaya and L. P. Polozova. Zhurnal prikladnoi spektroskopii, vol. 11, No. 6, pp. 1151-1152 (1969).

5-84. V. G. Ignat'ev and O. M. Mikhailov. Zhurnal prikladnoi spektroskopii, vol. 15, No. 1, pp. 46-50 (1971).

5-84a. L. I. Gavrilova et al. "Spectral distribution and UV energy of the radiation of unlimited pulsed discharges". Paper at the Sixth All-Union Seminar on Pulse Photometry. Moscow, March 29-April 2 (1976).

5-85. L. A. Vainshtein et al. JETP, vol. 24, No. 3, pp. 326-338 (1953).

5-86. A. A. Mak. Issledovanie izlucheniya intensivnogo iskrovogo razryada (Investigation of the Radiation from an Intense Spark Discharge). Dissertation, Leningrad State Optical Institute (1960).

5-87. S. L. Mandel'shtam and N. K. Sukhodrev. JETP, vol. 24, No. 6, pp. 701-707 (1953).

5-88. M. P. Vanyukov et al. Optika i spektroskopiya, vol. 6, No. 1, pp. 17-23 (1959).

5-89. M. P. Vanyukov et al. Optika i spektroskopiya, vol. 8, No. 4, pp. 439-445 (1960).

5-90. N. I. Falkowsky. Proc. of the 11th Intern. Conf. on Phenomena in Ionized Gases. Prague, s.a., pp. 235-240 (1973).

5-91. J. G. Edwards. Appl. Opt., vol. 6, No. 5, pp. 837-843 (1967).

5-92. E. V. Daniel' and I. V. Kolpakova. Zhurnal prikladnoi spektroskopii, vol. 10, No. 4, pp. 592-594 (1969).

5-93. V. G. Ignat'ev and A. N. Tokarev. Zhurnal prikladnoi spektroskopii, vol. 19, No. 4, pp. 632-635 (1973).

5-94. Yu. G. Anikiev and M. E. Zhabotinskii. Kvantovaya elektronika, vol. 1, No. 12, pp. 2557-2565 (1964).

5-95. I. I. Litvinov and V. V. Poduval'tsev. Kvantovaya elektronika, vol. 1, No. 1, pp. 211-215 (1974).

5-96. A. A. Mak and A. A. Shcherbakov. Kvantovaya elektronika, Moscow, Sovetskoe Radio Publishing House, No. 5 (17), pp. 68-76 (1973).

5-96a. Yu. G. Basov et al. Optika i spektroskopiya, vol. 38, No. 3, pp. 608-614 (1975).

5-97. V. P. Kirsanov et al. Kvantovaya elektronika, vol. 3, No. 2, p. 431 (1976).

5-97a. Yu. G. Basov et al. Zhurnal prikladnoi spektroskopii, vol. 24, No. 2, pp. 259-262 (1976).

5-97b. V. N. Makarov and Yu. G. Basov. Optika i spektroskopiya, vol. 40, No. 5, pp. 879-884 (1976).

5-98. H. E. Edgerton. Journal of the Optical Soc. of America, vol. 36, No. 7, pp. 390-399 (1946).

5-99.   I. S. Marshak. Sbornik materialov po vakuumnoi tekhnike
        (Collections of Materials on Vacuum Technology). Moscow-
        Leningrad, Gosenergoizdat Publishing House, No. 11, pp.
        3-23 (1957). (Moscow Electron-Tube Plant, Office of Tech-
        nical Information.)

5-100. V. A. Gavanin and M. S. Levchuk. Svetotekhnika, No. 4, pp.
        6-9 (1967).

5-101. V. G. Ignat'ev and L. A. Isaev. In Impul'snaya fotometriya.
        Leningrad, Mashinostroenie Publishing House, pp. 118-122
        (1969).

5-102. V. A. Gavanin. In Impul'snaya fotometriya. Leningrad, Mashino-
        stroenie Publishing House, pp. 111-118 (1969).

5-103. A. S. Doinikov et al. In Impul'snaya fotometriya. Leningrad,
        Mashinostroenie Publishing House, No. 4, pp. 69-72 (1975).

5-104. M. P. Vanyukov et al. ZhTF, vol. 32, No. 3, pp. 373-375 (1962).

5-105. M. P. Vanyukov et al. Uspekhi nauchnoi fotografii, vol. 9,
        pp. 116-120 (1964).

5-106. V. P. Zhil'tsov and I. V. Barchenko. Svetotekhnika, No. 2,
        pp. 14-16 (1967).

5-107. B. V. Kalachev et al. Svetotekhnika, No. 2, p. 14 (1974).

5-108. V. P. Kirsanov et al. Svetotekhnika, No. 10, pp. 18-20 (1963).

5-109. A. S. Doinikov et al. Svetotekhnika, No. 4, pp. 11-13 (1975).

6-1.   H. E. Edgerton et al. Proc. of the 6th Intern. Congr. on High-
        Speed Photography. The Hague, p. 143 (1963).

6-2.   L. Waszak. Microwaves. Laser Technology Section, vol. 8,
        No. 5, p. 130 (1969).

6-3.   V. P. Kirsanov et al. Kvantovaya elektronika, No. 6 (18), p.
        43 (1973).

6-4.   V. M. Podgaetskii and B. V. Skvortsov. Kvantovaya elektronika,
        No. 4, (10), p. 82 (1972).

6-5.   M. G. Byalko et al. Kvantovaya elektronika, vol. 1, No. 6,
        p. 1350 (1974).

6-6.   J. M. McMahon and J. L. Emmet. IEEE Journal of Quantum Elec-
        tronics, vol. QE-9, No. 10, p. 992 (1973).

6-7.   V. I. Vasil'ev and I. S. Marshak. Sbornik materialov po vakuum-
        noi tekhnikie. Moscow-Leningrad, Gosenergoizdat Publishing
        House, No. 14, pp. 19-52 (1958). (Electric-Vacuum Devices
        Plant, Office of Technical Information.)

6-8.   V. P. Kirsanov et al. Elektronnaya tekhnika, ser. 3, No. 2,
        p. 25 (1966).

6-9.   D. Rose. Optical Spectra, vol. 4, No. 2, pp. 43-47 (1970).

6-10.   L. E. Belousova et al. Inzhenerno-fizicheskii zhurnal, vol.
        9, No. 1, p. 105 (1965).

6-11.   J. H. Rosolowski and R. J. Charles. Journal of Applied Physics,
        vol. 36, No. 5, p. 1792 (1965).

6-12.   M. P. Vanyukov et al. Zhurnal prikladnoi spektroskopii, vol.
        11, No. 4, p. 726 (1969).

6-13.   K. R. Lang and F. S. Barnes. Journal of Applied Physics, vol.
        35, No. 1, p. 107 (1964).

6-14.   R. A. Dugdale et al. Brit. Journal of Applied Physics, vol.
        13, p. 508 (1962).

6-15. L. E. Belousova. Svetotekhnika, No. 2, p. 12 (1973).

6-16. Yu. V. Afanas'ev and O. N. Krokhin. JETP, vol. 52, No. 4, p. 966 (1967).

7-1. N. A. Kozlov et al. Elektronnaya tekhnika, ser. 10, Kvantovaya elektronika, No. 2, pp. 53-57 (1975).

7-2. P. N. Dashuk, S.L. Zaients, V.S. Komel'kov, et al. Tekhnika bol' shikh impul'snykh tokov i magnitnykh polei (The Technology of High Pulsed Currents and Magnetic Fields). Moscow, Atomizdat Publishing House, 472 pages (1970).

7-3. G. A. Vorob'ev and G. A. Mesyats. Tekhnika formirovaniya vysokovol'tnykh impul'sov (Techniques for Shaping High-Voltage Pulses). Moscow, Gosatomizdat Publishing House, 169 pages (1963).

7-4. P. I. Shkuropat. ZhTF, vol. 39, No. 7, pp. 1256-1263 (1969).

7-5. F. Fryungel'. Impul'snaya tekhnika (Pulse Technology). Moscow-Leningrad, Energiya Publishing House, 488 pages (1965).

7-6. P. I. Shkuropat. ZhTF, vol. 36, No. 6, pp. 1058-1064 (1966).

7-7. G. A. Mesyats et al. ZhTF, vol. 36, No. 6, pp. 1058-1064 (1966).

7-8. P. F. Plotnikov and R. M. Aronovich. Soviet patent No. 150,930, published in Byulleten' izobretenii i tovarnykh znakov, No. 20 (1962).

7-9. A. M. Berdichevskii et al. Obzory po elektronnoi teknike, ser. "Elektrovakuumny i gazorazryadnye pribory". Moscow, Elektronika Institute, No. 9 (62), 48 pages (1972).

7-10. Yu. V. Kiselev. Obzory po elektronnoi tekhnike, ser. "Gazorazryadnye pribory". Moscow, Elektronika Institute, No. 12 (81), pp. 13-18 (1969).

7-11. K. J. Germeshausen. British patent No. 892,022, published February 24 (1960).

7-12. K. J. Germeshausen. US patent No. 3,356,888, published December 5 (1967).

7-13. B. Ya. Lutset et al. Svetotekhnika, No. 8, pp. 14-16 (1974).

7-14. L. I. Shchukin. Vsesoyuznoy naucho-tekhnicheskaya konferentsiya "Sovremennoe sostoyanie i perspektivy razvitiya vysokoskorostnoi fotografii v kinematografii i metrologii bystroprotekayushchikh protsessov. Tezisy dokladov (All-Union Scientific and Technical Conference on the Current Status and Prospects for the Development of High-Speed Photography in the Filming and Metrology of High-Speed Processes. Abstracts of Papers). Exhibition of Achievements of the USSR National Economy — All-Union Scientific Research Institute of Optical and Physical Measurements. Moscow, p. 106 (1975).

7-15. B. Z. Gorbenko et al. Pribory i tekhnika eksperimenta, No. 2, pp. 169-173 (1968).

7-16. A. A. Golovanov et al. Elektronnaya tekhnika, ser. 3, No. 4, pp. 32-39 (1967).

7-17. V. I. Isaenko and G. N. Travleev. Pribory i tekhnika eksperimenta, No. 6 (1961).

7-18. M. P. Vanyukov et al. Uspekhi nauchnoi fotografii, vol. IX,
       pp. 121-125 (1964).
7-19. M. P. Vanyukov et al. ZhTF, vol. 32, No. 6, p. 747 (1962).
7-20. V. I. Isaenko. Issledovanie prostranstvenno-vremennykh i
       chastotnykh kharakteristik iskrovogo razryada (Investigation
       of the Spatiotemporal and Frequency Characteristics of a
       Spark Discharge). Dissertation. Leningrad State Optical
       Institute (1968).
7-21. V. P. Zhil'tsov. Svetotekhnika, No. 9, pp. 9-14 (1964).
7-22. V. P. Zhil'tsov et al. Svetotekhnika, No. 2, pp. 13-15 (1964).
7-23. V. P. Zhil'tsov and I. V. Barchenko. Svetotekhnika, No. 2,
       pp. 14-16 (1967).
7-24. V. P. Zhil'tsov. Puti povysheniya chastoty vspyshek impul'
       snykh istochnikov sveta (Ways of Increasing the Flash
       Frequency of Pulsed Light Sources). Dissertation. Moscow
       Power Engineering Institute (1965).
7-25. L. H. Barrett. Electronics, No. 32, p. 116 (1959).
7-26. I. S. Marshak et al. Svetotekhnika, No. 11, pp. 13-17 (1961).
7-27. V. P. Zhil'tsov. Soviet patent No. 152,040. Published in
       Byulleten' izobretenii i tovarnykh znakov, No. 23 (1962).
7-28. V. P. Zhil'tsov. Svetotekhnika, No. 7, pp. 17-22 (1963).
7-29. V. P. Zhilt'sov and L. F. Lobov. Pribory i tekhnika eksperi-
       menta, No. 1, pp. 101-104 (1963).
7-30. M. P. Vanyukov and V. I. Isaenko. Pribory i tekhnika eksperi-
       menta, No. 6, p. 50 (1958).
7-31. M. P. Vanyukov et al. Optiko-mekhanicheskaya promyshlennost',
       No. 1, pp. 78-82 (1963).
7-32. F. Früngel. Explosivstoffe, No. 4, pp. 1-12 (1957).
7-33. A. Stenzel and G. Thomer. Journal of the Society of Motion
       Pictures and Television Engineers, vol. 70, pp. 18-20
       (1961).
7-34. V. P. Zhil'tsov and E. Kh. Slutskii. Pribory i tekhnika
       eksperimenta, No. 4, pp. 132-135 (1963).
7-35. K. J. Germeshausen. Journal of the Society of Motion Pictures
       and Television Engineers, vol. 52, No. 5, pp. 24-32 (1949).
7-36. C. C. Rockwood and W. P. Harvey. Journal of the Society of
       Motion Pictures and Television Engineers, vol. 63, No. 8,
       pp. 64-66 (1954).
7-37. G. A. Golostenov. Zhurnal nauchnoi i prikladnoi fotografii
       i kinematografii, vol. 1, No. 4, pp. 286-294 (1956).
7-38. H. E. Edgerton. Proc. of the 3rd Intern. Congr. on High-Speed
       Photography. Butterworth, London, pp. 51-56 (1957).
7-39. M. P. Vanyukov and V. I. Isaenko. Svetotekhnika, No. 3,
       pp. 7-11 (1960).
7-40. A. A. Abramyan and M. M. Agababyan. Svetotekhnika, No. 4,
       p. 9 (1974).
7-41. V. P. Zhil'tsov. Svetotekhnika, No. 2, p. 5 (1965).
7-42. G. D. Salamadra, I. M. Naboko, and I. K. Sevast'yanova. Pribory
       i tekhnika eksperimenta, No. 2, pp. 124-127 (1959).

7-43. G. D. Salamandra and I. K. Sevast'yanova. Inzhenerno-fizicheskii
      zhurnal, vol. 3, No. 9, pp. 31-36 (1960).
7-44. M. P. Vanyukov et al. Svetotekhnika, No. 8, p. 20 (1963).
7-45. I. Sh. Libin. Nekotorye voprosy razrabotki stroboskopicheskikh
      priborov (Some Problems of the Development of Stroboscopic
      Devices). Dissertation, Moscow Power Engineering Institute
      (1962).
7-46. V. I. Vasil'ev et al. Sbornik materialov po vakuumnoi tekhnike.
      Moscow-Leningrad, Gosenergoizdat Publishing House, No. 24,
      pp. 43-58 (1960). (Electric-Vacuum Devices Plant, Office
      of Technical Information.)
7-47. G. A. Volkova. Svetotekhnika, No. 9, pp. 8-9 (1973).
7-48. Yu. P. Andreev et al. Vestnik Moskovskogo universiteta.
      Khimiya, No. 4, pp. 442-445 (1974).
7-49. O. K. Botvinkin and A. I. Zaporozhskii. Kvartsevoe steklo
      (Quartz Glass). Moscow, Stroiizdat Publishing House,
      260 pages (1965).
7-50. Deistvie ioniziruyushchikh izluchenii na neorganicheskie stekla
      (The Effect of Ionizing Radiations on Inorganic Glasses).
      Moscow, Atomizdat Publishing House, 242 pages (1968).
      Authors: G. V. Byurganovskaya, V. V. Vargin, N. A. Leko
      and N. F. Orlov.
7-51. B. Ya. Lutset, V. A. Samodergin, and L. I. Shchukin. Obzory
      po elektronnoi tekhnike, ser. "Elektrovakuumnye i gazoraz-
      ryadnye pribory". Moscow, Elektronika Institute, No. 4,
      (287), 56 pages (1975).
7-52. Edgerton, Germeshausen, and Greer Catalog of Spherical Flash-
      lamps, United States, Data Sheet No. F10050-1, 4/73.
7-53. Sims and Vasak. Elektronika (Translated from English), vol.
      41, No. 17, pp. 19-24 (1968).
7-54. Edgerton, Germeshausen, and Greer Catalog of Quartz-Envelope
      Xenon Flashlamps, United States, Data Sheet No. F1002-1,
      3/73.
7-55. ILC Technology Catalog of Quartz-Envelope Xenon Flashlamps,
      United States, bulletin 1524 (1974).
7-56. P. Hoekstra and C. Meyer. Philips Technical Review, vol. 21,
      No. 3, pp. 73-87 (1959).
7-57. R. L. Stephens and W. F. Hug. Laser Focus, vol. 8, No. 7,
      pp. 38-46 (1972).
7-58. V. G. Nikiforov. Elektronnaya tekhnika. Kvantovaya elektron-
      ika, series 10, No. 2, pp. 64-69 (1975).
7-59. B. R. Belostotskii, Yu. V. Lyubavskii, and V. M. Ovchinnikov.
      Osnovy lazernoi tekhniki (Principles of Laser Technology).
      Moscow, Sovetskoe Radio Publishing House, 408 pages (1972).
7-60. D. A. Goukhberg. Sbornik materialov po vakuumnoi tekhnike.
      Moscow-Leningrad, Gosenergoizdat Publishing House, No. 8,
      pp. 41-52 (1956). (Electric-Vacuum Devices Plant, Office
      of Technical Information.)
7-61. M. S. Morzeeva et al. Elektronnaya tekhnika, series 6,
      Materialy, No. 10, pp. 62-66 (1973).

7-62. B. Ya. Lutset et al. Svetotekhnika, No. 8, pp. 14-16 (1974).

8-1.   Istochniki vysokointensivnogo opticheskogo izlucheniya.
       Sistema uslovnykh oboznachenii (Sources of High-Intensity
       Optical Radiation. The System of Conventional Symbols).
       GOST 19,685-74.

8-2.   Istochniki vysokointensivnogo opticheskogo izlucheniya gazoraz-
       ryadnye impul'snye. Osnovnye parametry (Pulsed Gas-Discharge
       Sources of High-Intensity Optical Radiation). GOST 17,399-
       72.

8-3.   M. L. Lyubimov. Spai metalla so steklom (Metal-Glass Joints).
       Moscow, Energiya Publishing House, 280 pages (1968).

8-4.   Spai stekla s metallom i keramikoi (Joints between Glass and a
       Metal or a Ceramic). Part IV, Tematicheskii ukazatel'
       literatury, No. 4 (7). Moscow, Elektronika Central Scien-
       tific Research Institute, 25 pages (1972).

8-5.   Paika kvartsa s metallom. Bibliograficheskii ukazatel'
       literatury (Soldering of Quartz to Metal. A Bibliographic
       Guide to the Literature). Moscow, Elektronika Central
       Scientific Research Institute, 5 pages (1972).

8-6.   Yu. P. Andreev et al. Soviet patent No. 347,767. Published
       in Otkrytiya. Izobreteniya. Promyshlennye obraztsy.
       Tovarnye znaki, No. 23, p. 204 (1972).

8-7.   G. N. Rokhlin. Gazorazryadnye istochniki sveta (Gas-Discharge
       Light Sources). Moscow, Energiya Publishing House, 560
       pages (1966).

8-8.   A. M. Stommel. Signal (US), vol. 26, No. 6, pp. 32-33 (1972).

8-9.   L. Reed and R. MacRae. American Ceramic Society Bull.,
       pp. 611-613 (June 1969).

8-10. V. E. Mnuskin et al. Soviet patent No. 187,165. Published in
       Izobreteniya. Promyshlennye obraztsy. Tovarnye znaki, No.
       20, p. 83 (1966).

8-11. N. F. Kazakov et al. Elektronnaya tekhnika, series 14, No. 3,
       (11), p. 25 (1968).

8-12. V. G. Ignat'ev et al. Svetotekhnika, No. 11, pp. 17-18 (1969).

8-13. V. A. Malashenkov et al. Soviet patent No. 314,252. Published
       in Otkrytiya. Izobreteniya. Promyshlennye obraztsy.
       Tovarnye znaki, No. 27, p. 194 (1971).

8-14. V. A. Malashenkov et al. Soviet patent No. 378,997. Published
       in Otkrytiya. Izobreteniya. Promyshlennye obraztsy.
       Tovarnye znaki, No. 19, p. 152 (1973).

8-15. L. Verheiden. Journal of Physics, ser. E, vol. 1, p. 145
       (1968).

8-16. M. I. Rubtsov et al. Soviet patent No. 408,394. Published
       in Otkrytiya. Izobreteniya. Promyshlennye obraztsy.
       Tovarnye znaki, No. 47, p. 196 (1973).

9-1.   A. L. Vasserman and V. V. Skvortsov. In Uspekhi nauchnoi
       fotografii, vol. 9, pp. 20-25 (1964).

9-2.   "Economical electronic photoflashes", Elektronika (translated
       from English), No. 2, pp. 3-4 (1973).

9-3. N. N. Bogdanov et al. Moskovskii energeticheskii institut.
      Doklady nauchno-tekhnicheskaya konferentsiya po itogam
      nauchno-issledovatel'skikh rabot za 1966-1967 gg., sektsiya
      elektronnoi tekhniki, podsektsiya promyshlennoi elektroniki
      (Moscow Power Engineering Institute. Papers at the Scientific
      and Technical Conference on the Results of Scientific Research
      in the Period 1966-1967. Electronic Engineering Section.
      Industrial Electronics Subsection). Moscow, pp. 14-26 (1967).

9-4. N. N. Laptev et al. Soviet patent No. 151,382. Published in
      Byulleten' izobretenii i tovarnykh znakov, No. 21 (1962).

9-5. N. N. Laptev et al. Soviet patent No. 153,935. Published in
      Byulleten' izobretenii i tovarnykh znakov, No. 8 (1963).

9-6. I. V. Nezhdanov and V. S. Moin. Soviet patent No. 182,766.
      Published in Izobreteniya. Promyshlennye obraztsy. Tovarnye
      znaki, No. 12 (1968).

9-7. V. Yu. Roginskii. Sovremennye istochniki pitaniya (Modern Power
      Supplies). Moscow, Energiya Publishing House, p. 104 (1969).

9-8. A. I. Bertinov et al. Elektrichestvo, No. 8, pp. 54-61 (1967).

9-9. P. M. Mostov et al. Proc. of the Institute of Radio Engineering,
      No. 5, pp. 941-948 (1961).

9-10. S. M. Smirnov. Elektrichestvo, No. 10, pp. 60-64 (1961).

9-11. L. F. Lebedev. Issledovanie emkostnykh nakopitelei energii
      dlya impul'snykh istochnikov sveta (Investigation of Capaci-
      tive Energy Storage Devices for Pulsed Light Sources).
      Dissertation, Moscow Power Engineering Institute (1970).

9-12. L. F. Lebedev. Svetotekhnika, No. 3, pp. 10-14 (1969).

9-13. N. L. Ivanova and G. N. Senilov. Moskovskii energeticheskii
      institut. Doklady nauchno-tekhnicheskoi konferentsii po
      itogam nauchno-issledovatel'skikh razrabotok za 1966-1967
      gg. Subsection of the Problem-Oriented Permanent Magnets
      Laboratory, pp. 333-368 (1967).

9-14. G. N. Senilov et al. Soviet patent No. 288,596. Published
      in Otkrytiya. Izobreteniya. Promyshlennye obraztsy.
      Tovarnye znaki, No. 36 (1972).

9-15. L. F. Lebedev. Svetotekhnika, No. 7, pp. 18-20 (1969).

9-16. L. F. Lebedev. Svetotekhnika, No. 6, pp. 14-16 (1970).

9-17. A. E. Krasnopol'skii and L. F. Lebedev. Soviet patent No.
      307,470. Published in Otkrytiya. Izobreteniya. Promyshlen-
      nye obraztsy. Tovarnye znaki, No. 20 (1971).

9-18. B. E. Fridman. Trudy nauchno-issledovatel'skogo proektno-
      konstrukturskogo instituta tekhnologii mashinostroenie
      (Transactions of the Scientific Research, Planning and
      Design Institute of Mechanical Engineering). Leningrad,
      No. 4 (1967).

9-19. R. Freed and L. S. Klivans. Electronic Industries, vol. 20,
      No. 4, pp. 94-98 (1961).

9-20. L. T. Rees. Transactions of the IEEE on Industry and General
      Applications, vol. 5, No. 5, pp. 600-606 (1969).

9-21. L. T. Rees. "Capacitor charging from batteries with DC-DC
       convertors", Trans. IEEE on Industry and General Applications,
       Group 3rd Conference Rec., pp. 827-837 (Sept.-Oct. 3, 1968).
9-22. H. C. Early and R. C. Walker. Conference on Extremely High
       Temperatures. Boston, Mass. (March 1958).
9-23. W. Koch. Fourth Symposium on Engineering Problems in Thermo-
       nuclear Research. Frascati, Rome, p. 27 (May 23-27, 1966).
9-24. A. V. Reimers et al. Elektrichestvo, No. 4, pp. 84-85 (1970).
9-25. A. V. Reimers et al. Elektrichestvo, No. 6, pp. 81-82 (1969).
9-26. L. D. Eimin† and P. R. Viderkhol'd†. Pribory dlya nauchnykh
       issledovanii (translated from English), No. 6, pp. 78-82
       (1964).
9-27. L. N. Fedotov et al. Elektrichestvo, No. 10 (1971).
9-28. A. I. Bertinov et al. ZhTF, vol. 41, No. 7, pp. 1443-1451
       (1971).
9-29. Tezisy dokladov Vsesoyuznogo soveshchaniya po inzhenernym
       problemam upravlyaemogo termoyadernogo sinteza (Abstracts
       of Papers at the All-Union Conference on Engineering Problems
       of Controlled Thermonuclear Fusion). Leningrad, Publishing
       House of the D. V. Efremov Scientific Research Institute of
       Electrophysical Equipment, p. 312 (1974).
9-29a. Conger et al. Pribory dlya nauchnykh issledovanii (translated
       from English), No. 11, pp. 47-49 (1967).
9-30. A. E. Voitenko et al. Pribory i tekhnika eksperimenta, No. 3,
       p. 177 (1973).
9-31. E. A. Korolev and L. D. Khazov. Zhurnal prikladnoi spektro-
       skopii, vol. 6, No. 4, pp. 467-470 (1967).
9-32. W. J. Griffen. Journal of the Society of Motion Pictures and
       Television Engineers, vol. 66, pp. 127-129 (1957).
9-33. D. P. C. Thackery. Proc. of the 3rd Intern. Congr. on High-
       Speed Photography. Butterworth, London, pp. 21-29 (1957).
9-34. J. B. Gladis et al. Rev. Sci. Instruments, vol. 27, No. 2,
       p. 83 (1956).
9-35. S. I. Andreev and M. P. Vanyukov. Uspekhi nauchnoi fotografii,
       vol. 9, pp. 153-158 (1964).
9-36. S. I. Andreev et al. Uspekhi nauchnoi fotografii, vol. 9,
       pp. 147-150 (1964).
9-37. P. Bogen and H. Conrads. Proc. of the 7th Congr. on High-
       Speed Photography. Zurich, 1965. Darmstadt, Helwich,
       pp. 68-71 (1967).
9-38. S. I. Andreev et al. Zhurnal prikladnoi spektroskopii, vol.
       5, No. 6, pp. 712-717 (1966).
9-39. A. I. Salishchev. Uspekhi nauchnoi fotografii, vol. 6, p.
       155 (1959).
9-40. I. S. Marshak. Soviet patent No. 116,564. Published in
       Byulleten' izobretenii i tovarnykh znakov, No. 12, p. 165
       (1958).
9-41. K. Folrath. In Fizika bystroprotekayushchikh protsessov (The
       Physics of High-Speed Processes). (Translated from German).
       Edited by N. A. Zlatin. Moscow, Mir Publishers, vol. 1,
       pp. 96-199 (1971).

9-42. W. Thorwart et al. Proc. of the 7th congr. on High-Speed
      Photography. Zurich, 1965. Darmstadt, Helwich, pp. 51-55,
      (1967).
9-43. High-Voltage Capacitors. San Diego, California (1967).
      Maxwell Laboratories.
9-44. V. D. Bespalov and V. V. Konotop. Elektrotekhnicheskaya promy-
      shlennost'. Apparaty vysokogo napryazheniya, transformatory,
      silovye kondensatory (The Electronic Components Industry.
      High-Voltage Units, Transformers, Power Capacitors), No. 3
      (35), pp. 16-17 (1974).
9-45. V. D. Bespalov and V. V. Konotop. Elektrotekhnika, No. 1,
      pp. 51-54 (1974).
9-46. Yu. A. Anufriev and B. P. Kartyshev. Elektronnaya promyshlen-
      nost', No. 4, pp. 78-81 (1974).
9-47. L. N. Kosmarskii et al. In Elektrofizicheskaya apparatura i
      elektricheskaya izolyatsiya (Electrophysical Apparatus
      and Electrical Insulation). Moscow, Energiya Publishing
      House, pp. 170-176 (1970).
9-47a. Yu. V. Kiselev and V. P. Cherapanov. Iskrovye razryadniki
      (Spark Gaps). Moscow, Sovetskoe Radio Publishing House, 72
      pages (1976).
9-48. P. N. Dashuk and G. S. Kichaeva. Tezisy dokladov Vsesoyuznogo
      soveshchaniya po inzhenernym problemam upravlyaemogo termoya-
      dernogo sinteza. Nauchnoissledovatel'skii institut elektro-
      fizicheskoi apparatury im. D. V. Efremova. Leningrad, pp.
      141-142 (1974).
9-49. V. S. Potokin. Elektrichestvo, No. 6, pp. 74-77 (1973).
9-50. V. V. Konotop. Osnovy proektirovaniya vysokovol'tnykh impul'
      snykh ustroistv (Design Principles for High-Voltage Pulsed
      Devices). Part 1. Publishing House of the Khar'kov Poly-
      technical Institute, 142 pages (1973).
9-51. E. N. Prokhorov. Moscovskii energeticheskii institut. Doklady
      nauchno-tekhnicheskoi konferentsii po itogam NIR za 1966-
      1967 gg. (Electric Power Engineering Section, High-Voltage
      Equipment Subsection, pp. 127-131 (1967).
9-52. Tekhnika bol'shikh impul'snykh tokov i magnitnykh polei
      (The Technology of High Pulsed Currents and Magnetic Fields).
      Edited by V. S. Komel'kov. Moscow, Atomizdat Publishing
      House, 472 pages (1970).
9-53. A. M. Arsh, Yu. D. Khromoi, R. G. Antokhin and A. M. Serbinov.
      Tezisy dokladov Vsesoyuznogo soveshchaniya po inzhenernym
      problemam upravlyaemogo termoyadernogo sinteza. D. V.
      Efremova Scientific Research Institute of Electrophysical
      Equipment. Leningrad, pp. 194-195 (1974).
9-54. P. L. Kalantarov and L. A. Tseitlin. Raschet induktivnostei
      (The Calculation of Inductances). Leningrad, Energiya
      Publishing House, 416 pages (1970).
9-55. W. A. Ward. US Patent No. 3,355,625.
9-56. V. P. Zhil'tsov and Yu. P. Ivanov. Soviet patent No. 192,941.
      Published in Izobreteniya. Promyshlennye obraztsy. Tov-
      arnye znaki, No. 6 (1967).

9-57.   V. P. Zhil'tsov and D. P. Polosin. Svetotekhnika, No. 6,
        pp. 12-13 (1973).
9-58.   B. P. Aseev. Kolebatel'nye tsepi (Oscillatory Circuits),
        3rd ed. Moscow, Svyaz'izdat Publishing House, 462 pages
        (1955).
9-59.   N. N. Krylov. Teoreticheskie osnovy radiotekhniki (Theoretical
        Principles of Radio Engineering). Moscow-Leningrad, Mors-
        koi Transport Publishing House, 552 pages (1953).
9-60.   M. I. Kontorovich. Operatsionnoe ischislenie i nestatsionarnye
        yavleniya v elektricheskikh tsepyakh (Operational Calculus
        and Unsteady Phenomena in Electric Circuits). Moscow,
        Gostekhizdat Publishing House, 230 pages (1955).
9-61.   A. L. Vasserman. Issledovanie i metody rascheta puskoreguli-
        ruyushchikh ustroistv dlya moshchnykh trubchatykh kseno-
        novykh lamp (The Investigation and Methods of Design of
        Start-Control Devices for High-Power Tubular Xenon Lamps).
        Dissertation, Moscow Power Engineering Institute (1968).
9-62.   K. A. Zheltov. Elektrichestvo, No. 11, pp. 63-66 (1970).
9-62a. R. E. W. Pettifier et al. Journal of Physics, Scientific
        Instruments, vol. 8, No. 10, pp. 875-877 (1975).
9-63.   S. I. Andreev et al. Pribory i tekhnika eksperimenta, No. 3,
        pp. 89-92 (1962).
9-64.   L. P. Arkhina. "A pulsed transformer", Informatsionnyi listok.
        Moscow, No. 73, 1701, p. 2 (VIMI) (1973).
9-65.   Ya. A. Itskokhi. Impul'snye transformatory (Pulsed Transfor-
        mers). Moscow, Sovetskoe Radio Publishing House, 480 pages
        (1950).
9-66.   Ya. S. Itskhoki. Impul'snaya tekhnika (Pulse Technology).
        Moscow, Sovetskoe Radio Publishing House, 728 pages (1949).
9-67.   N. N. Evtikhiev et al. Soviet patent No. 415,761. Published
        in Otkrytiya. Izobreteniya. Promyshlennye obraztsy.
        Tovarnye znaki, No. 6 (1974).
9-68.   J. M. Girard. Lux, No. 69, pp. 322-332 (October 1972).
9-69.   A. Schmid and H. E. Scheddin. Konstr. Elem. Meth., vol. 8,
        No. 3, pp. 102-110 (1971).
9-70.   P. Schütze. Radio-Fernsehen-Elektronik, vol. 19, No. 74,
        pp. 814-816 (1970).
9-71.   A. A. Dul'zon et al. In Tekhnika vysokikh napryazjenii (High
        Voltage Technology). Tomsk University Press, pp. 101-104
        (1973).
9-72.   A. I. Mesenyashin. Izvestiya vuzov SSSR. Energetika, No. 7,
        pp. 26-32 (1973).
9-73.   A. I. Mesenyashin. Elektrotekhnika, No. 6, pp. 13-14 (1971).
9-74.   A. I. Mesenyashin. Izvestiya vuzov SSSR. Energetika, No. 9,
        pp. 36-41 (1970).
9-75.   G. M. Goncharenko. Trudy Moskovskogo energeticheskogo
        instituta. Elektroenergetika, No. 45, 248 pages (1963).
9-76.   G. S. Kuchinskii. Vysokovol'tnye impul'snye kondensatory
        (High-Voltage Pulsed Capacitors). Leningrad, Energiya
        Publishing House, 176 pages (1973).

9-77. P. V. Borisoglebskii and A. S. Kudratillaev. Izvestiya
       Akademii nauk UzSSR, seriya tekhnicheskikh nauk, No. 4,
       pp. 11-18 (1965).

9-78. A. S. Kudratillaev. Izvestiya Akademii nauk UzSSR, seriya
       tekhnicheskikh nauk, No. 1, pp. 16-21 (1967).

9-79. P. V. Borisoglebskii and A. S. Kudratillaev. Moskovskii
       energeticheskii institut. Doklady nauchno-tekhnicheskoi
       konferentsii po itogam NIR za 1966-1967 gg. Electrical
       Engineering Section, High-Voltage Equipment Subsection,
       pp. 56-62 (1967).

9-80. P. N. Dashuk. Issledovanie i razrabotka osnovnykh elementov
       emkostnykh generatorov bol'shikh impul'snykh tokov (The
       Investigation and Development of the Main Elements of
       Capacitative Generators of High Pulsed Currents). Dis-
       sertation. Leningrad Polytechnical Institute (1963).

9-81. A. M. Berdichevskii et al. Obzory po elektronnoi tekhnike, ser.
       Elektrovakuumnye i gazorazryadnye pribory, No. 9 (62).
       Moscow, Elektronika Institute, 50 pages (1972).

9-82. A. V. Korsuntsev. Tokoogranichivayushchie plavkie predokhran-
       iteli dlya zashchity kondensatornykh baterei vysokogo
       napryzheniya ot povrezhdenii pri proboe kondensatorov
       (Current-Limiting Fuses for Protection of High-Voltage
       Capacitor Banks against Damage upon Breakdown of Capacitors).
       Dissertation. Leningrad Polytechnical Institute (1952).

9-83. A. V. Podmazov et al. Elektrichestvo, No. 11, pp. 82-83
       (1972).

9-84. G. M. Goncharenko et al. Doklady nauchno-tekhnicheskoi
       konferentsii Moskovskogo energeticheskogo instituta po
       itogam NIR za 1966-1967. Electrical Engineering Section,
       High-Voltage Equipment Subsection. Moscow, Moscow Power
       Engineering Institute, pp. 143-151 (1967).

9-85. Pribory i tekhnika eksperimenta, No. 2, pp. 159-163 (1966).

9-86. A. A. Rudenko et al. Tezisy dokladov na Vsesoyuznom sovesh-
       chanii po inzhenernym problemam upravlyaemogo termoya-
       dernogo sinteza. Nauchno-issledovatel'skii institute
       elektrofizicheskoi apparatury. Leningrad, pp. 143-144
       (1974).

9-87. V. G. Geis,[†] Kh. K. Mor.[†] Vzryvayushchiesya provolochki
       (Exploding Wires). Translated from English. Moscow,
       Foreign Literature Publishing House, 341 pages (1963).

9-88. Yu. A. Anan'ev et al. Soviet patent No. 421,084. Published
       in Otkrytiya. Izobreteniya. Promyshlennye obraztsy.
       Tovarnye znaki, No. 11 (1974).

9-89. D. Braun.[†] Pribory dlya nauchnykh issledovanii (translated
       from English), No. 9, pp. 6-10 (1971).

9-90. D. G. Baratov et al. Pribory i tekhnika eksperimenta, No. 5,
       pp. 91-92 (1974).

10-1. A. A. Vol'kenshtein and E. V. Kuvaldin. Fotoelektricheskaya
       impul'snaya fotometriya (Photoelectric Pulse Photometry).
       Leningrad, Mashinostroenie Publishing House, 192 pages
       (1975).

10-2.   Fotoelektronnye pribory (Photoelectronic Devices). Moscow,
         Nauka Publishers, 592 pages (1965). Authors: N. A. Sobolev,
         A. G. Berkovskii, N. O. Chechek, and R. E. Eliseev.
10-3.   Impul'snaya fotometriya (Pulse Photometry), a collection of
         papers. Leningrad, Mashinostroenie Publishing House,
         152 pages (1969).
10-4.   Impul'snaya fotometriya, No. 2. Leningrad, Mashinostroenie
         Publishing House, 200 pages (1972).
10-5.   Impul'snaya fotometriya, No. 3. Leningrad, Mashinostroenie
         Publishing House, 183 pages (1973).
10-6.   Impul'snaya fotometriya, No. 4. Leningrad, Mashinostroenie
         Publishing House, 248 pages (1975).
10-7.   M. Moro-Ano. Fotometriya kratkovremennykh i peremennykh
         svetovykh yavlenii (The Photometry of Short-Duration and
         Variable Light Phenomena). Moscow-Leningrad, Oborongiz,
         119 pages (1939).
10-8.   R. Smith, F. Jones,[†] and R. Chesmer.[†] Obnaruzhenie i izmerenie
         infrakrasnogo izlucheniya (The Detection and Measurement
         of Infrared Radiation). Moscow, Foreign Literature
         Publishing House, 448 pages (1959).
10-9.   A. V. Luizov. Inertsiya zreniya (Persistence of Vision).
         Moscow, Oborongiz Publishing House, 248 pages (1961).
10-10.   A. A. Shishlovskii. Prikladnaya fizicheskaya optika (Applied
         Physical Optics). Moscow, Fizmatgiz Publishing House,
         822 pages (1961).
10-11.   I. M. Nagibina and V. K. Prokof'ev. Spektral'nye pribory i
         tekhnika spektroskopii (Spectral Instruments and Spectro-
         scopy). Moscow-Leningrad, Mashgiz Publishing House,
         271 pages (1963).
10-12.   A. N. Zaidel', G. V. Ostrovskaya, and Yu. I. Ostrovskii.
         Tekhnika i praktika spektroskopii (The Techniques and
         Practice of Spectroscopy). Moscow, Nauka Publishers,
         375 pages (1972).
10-13.   A. V. Pavlov and A. I. Chernikov. Priemniki izlucheniya
         avtomaticheskikh optiko-elektronnykh priborov (Radiation
         Detectors of Automatic Optoelectronic Devices). Moscow,
         Energiya Publishing House, 240 pages (1972).
10-14.   V. G. Baryshnikov. Voprosy prokhozhdeniya svetovogo i foto-
         elektricheskogo impul'sov cherez izmeritel'nyi trakt i
         razrabotka apparatury dlya fotometrirovaniya impul'snykh
         lamp (Problems of the Transmission of a Light Pulse and
         a Photoelectric Pulse through a Measuring Circuit, and the
         Development of Apparatus for Flashlamp Photometry).
         Dissertation. V. I. Lenin All-Union Electrical Engineering
         Institute. Moscow (1966).
10-15.   E. V. Kuvaldin. Fotoelektricheskoe fotometrirovanie istoch-
         nikov impul'snogo izlucheniya (Photoelectric Photometry
         of Pulsed Radiation Sources). Dissertation. Moscow Power
         Engineering Institute (1969).

10-16. H. E. Edgerton. Electronics, vol. 21, pp. 78-81 (June 1948).

10-17. N. A. Soboleva and A. E. Melamid. Fotoelektronnye pribory
        (Photoelectronic Devices). Moscow, Vysshaya Shkola Pub-
        lishing House, 376 pages (1974).

10-18. I. S. Marshak. Sbornik materialov po vakuumnoi tekhnike.
        Moscow-Leningrad, Gosenergoizdat Publishing House, No. 11,
        pp. 3-23 (1957). (Moscow Electric-Vacuum Devices Plant,
        Office of Technical Information.)

10-19. A. Blondel and J. Rey. Compte rendus d'Academie de Science,
        vol. 162, p. 861 (1916).

10-20. E. Baumgardt. Rev. d'optique, vol. 28, No. 12, pp. 661 (1949).

10-21. R. L. Fol'b. Osnovy vizual'noi probleskovoi signalizatsii
        (Principles of Visual Flash-Signaling). Moscow, Mashinos-
        troenie Publishing House, 200 pages (1964).

10-22. I. S. Marshak and M. G. Feigenbaum. Svetotekhnika, No. 4,
        pp. 16-19 (1958).

10-23. R. L. Fol'b. Svetotekhnika, No. 12, pp. 11-16 (1958).

10-24. I. S. Marshak and M. G. Feigenbaum. Svetotekhnika, No. 3,
        pp. 17-22 (1959).

10-25. I. S. Marshak and N. K. Tsypkin. Svetotekhnika, No. 6, pp.
        21-22 (1958).

10-26. R. L. Fol'b. Svetotekhnika, No. 8, pp. 14-19 (1960).

10-27. A. S. Doinikov et al. In Impul'snaya fotometriya, collection
        2. Leningrad, Mashinostroenie Publishing House, pp. 193-
        194 (1972).

10-28. P. M. Tikhodeev. Svetovye izmereniya v svetotekhnike (foto-
        metriya) (Measurements of Light in Illumination Engin-
        eering (Photometry)). Moscow-Leningrad, Gosenergizdat
        Publishing House, 464 pages (1962).

10-28a. V. Stas'kevich. Svetotekhnika, No. 4, pp. 10-12 (1973).

10-29. A. S. Doinikov et al. In the book Impul'snaya Fotometriya
        (Pulse Photometry), collection 4. Leningrad, Mashinos-
        troenie Publishing House, pp. 65-67 (1975).

10-30. M. I. Epshtein. Spektral'nye izmereniya v elektrovakuumnoi
        tekhnike (Spectrum Measurements in Electro-Vacuum Engin-
        eering). Moscow, Energiya (Energy), 144 pages (1970).

10-31. I. V. Demenik et al. Zhurnal prikladnoi spektroskopii, vol.
        13, No. 4, pp. 740-744 (1970).

10-32. M. A. Zakha. Electronics, No. 23, pp. 32-40 (1972)